STUDIES IN LOGIC

AND

THE FOUNDATIONS OF MATHEMATICS

VOLUME 70

NORTH-HOLLAND PUBLISHING COMPANY
AMSTERDAM • LONDON

THE THEORY OF SEMISETS

THE THEORY OF SEMISETS

PETR VOPĚNKA

Department of Mathematics,
Charles University, Prague

and

PETR HÁJEK

Mathematical Institute
of the Czechoslovak Academy of Sciences, Prague

1972

NORTH-HOLLAND PUBLISHING COMPANY
AMSTERDAM · LONDON

Library of Congress Catalog Card Number 77-157005
ISBN 0 7204 2267 1

PUBLISHERS:
NORTH-HOLLAND PUBLISHING COMPANY – AMSTERDAM • LONDON

ACADEMIA – PUBLISHING HOUSE
OF THE CZECHOSLOVAK ACADEMY OF SCIENCES • PRAGUE

PRINTED IN CZECHOSLOVAKIA

CONTENTS

INTRODUCTION

This book on the theory of semisets presents an attempt to create a theory whose universe of discourse extends that of set theory; thus the new theory admits the existence of certain objects which cannot exist from the point of view of set theory. Semisets are subclasses of sets; whereas in set theory the comprehension axiom ensures that every subclass of a set is a set, the theory of semisets admits the existence of semisets which are not sets (proper semisets).

Although the reader of this book must be able to deal with abstract generalizations, no particular mathematical knowledge is required for its study. In the present introduction we compare the theory of semisets with classical set-theoretical systems. This comparison should serve to explain the reasons for the existence of the theory of semisets.

The notion of set is of course a very familiar one. Cantor "defined" the notion of set as follows: "Unter einer Menge verstehen wir jede Zusammenfassung M von bestimmten wohlunterschiedenen Objekten m unserer Anschauung oder unseres Denkens (welche die Elemente von M genannt werden) zu einem Ganzen". ("A set is a collection into a whole of definite, distinct objects of our intuition or of our thought. The objects are called the elements (members) of the set.") However since intuitive Cantorian ideas on sets are not free from contradiction – in particular the idea that any property determines a set consisting of precisely the objects having that property – it was necessary to create an axiomatic theory of sets. Let us mention two such axiomatic systems which are widely considered today.

(a) Zermelo-Fraenkel set theory has as its basic notions "set" and "elementhood". Since it is provable in Zermelo-Fraenkel set theory that there is no set of all sets, it is not always possible to "collect into a whole" all objects having certain properties. In other words there are properties P (the property of being a set, for example) for which there is no set whose elements are precisely all objects having the property P.

(b) However the assumption of the existence of "large" collections of elements (e.g. the collection of all sets) is not contradictory, provided we suppose that these large collections are not sets and in particular that they cannot themselves be elements of any set. Bernays-Gödel set theory has as its basic notions "class" and "elementhood"; sets are defined as those classes which are elements of classes. Both systems of set theory are closely related; namely, any statement which concerns sets and elementhood alone (and which is therefore meaningful in both systems) is provable in one system if and only if it is provable in the other.

The known proofs of contradictions in intuitive set theory cannot be reproduced in Zermelo-Fraenkel or in Bernays-Gödel set theory. Nevertheless the question remains whether one can prove a contradiction in these systems in another way. It follows from the above that there are only two possibilities; either both systems are consistent or both are contradictory.

According to a famous result of Gödel (1931), there can be no absolute proof of consistency for any "sufficiently strong" theory; this applies in particular to any theory, like set theory, in which Peano arithmetic can be developed. Thus it is impossible to establish the consistency of any such theory without using stronger means that those available in the theory itself. In view of Gödel's result, the most that can be hoped for is a demonstration of relative consistency: if a certain theory **S** is consistent then so is another theory **T**. If we regard **S** as in some sense trustworthy and **T** as questionable, then a demonstration of relative consistency would tell us that **T** is in fact just as trustworthy as **S**; insofar as one believes in the consistency of **S** one must also believe in the consistency of **T**. Conversely if we suspect **S** to be inconsistent then the relative consistency result reduces the problem of finding a contradiction in **S** to the perhaps easier problem of finding a contradiction in **T**. Let us give three examples which have been important in the history of the subject.

(A) In 1922 A. Fraenkel showed that if set theory is consistent then it remains consistent upon the addition of certain statements contradicting the axiom of choice. Fraenkel's method was later placed on a firm logical foundation by Mostowski. A number of statements contradicting the axiom of choice have since been shown to be consistent using Fraenkel's method. However these results all have a certain unsatisfactory character in that the sets for which the axiom of choice fails always have "irregular" elements; thus for example the method cannot be used to prove the consistency of the hypothesis that the set of real numbers has no well-ordering. If we assume as an axiom of set theory that there are no "irregular" elements (one of the so-called regularity axioms), then Fraenkel's method gives no information as to the independence of the axiom of choice (cf. sub. (C)).

(B) In 1938 (and in a more detailed form in 1940), K. Gödel established the consistency of the axiom of choice and the generalized continuum hypothesis with the axioms of set theory (including the axiom of regularity) by means of his famous Δ-model. Other important consistency results have been obtained by a generalization of this construction (e.g. Hajnal).

(C) In 1963 P. J. Cohen established the independence of the continuum hypothesis and the axiom of choice in set theory with the axiom of regularity. Cohen's method was reworked independently by Scott-Solovay and by Vopěnka into the so-called method of Boolean-valued models.

We note that all the above-mentioned consistency results will be established in the course of this book.

In his consistency proof Gödel defines the notion of "constructible class" and shows that if the axioms of Bernays-Gödel set theory are interpreted by replacing the notion of "class" by the notion of "constructible class" (so that in particular "set" is replaced by "constructible set"), then the interpreted axioms are provable in Bernays-Gödel set theory and in addition the interpreted version of the axiom of choice is provable.

The constructible sets form a class **L** and the constructible classes are certain subclasses of **L**. It is important that in Gödel's interpretation classes are interpreted as constructible subclasses of **L** and not as arbitrary subclasses. If classes are interpreted as arbitrary subclasses of **L** then the sets in the sense of this new interpretation are still precisely the elements of **L** but we can no longer prove the comprehensivity axiom in this interpretation. Indeed, if a is an element of **L** and x is a subset of a then x is a subset of **L**, and hence a semiset from the point of view of our interpretation; however we cannot prove that x is an element of **L**, i.e. that x is a set in our interpretation. Although we cannot verify all theorems of set theory in the new interpretation, we can nevertheless verify many of them.

From the above remarks we see that one cannot verify the statement "every semiset is a set" in the interpretation, and so this interpretation allows us to "imagine" proper semisets.

Many other constructions invented for consistency proofs can be modified in a similar way, interpreting the notion of class more widely than in the original construction. In each case we obtain an interpretation in which proper semisets may (or do) exist. The question therefore arises whether it is possible to axiomatize this general situation in an appropriate way, i.e.

to choose appropriate statements verified in such interpretations and to take these as the axioms for a new theory. Such axioms will be given in this book. Here we shall outline the intuitive content of the axioms in order to explain why they were chosen as they were. The basic ideas on classes, sets and semisets are as follows:

1. We study *classes* and their *elements*; we consider the *membership relation* between elements and classes.

2. Classes are *identical* iff they have the same elements.

3. Certain classes may be elements of certain other classes. Those classes which are elements of other classes are called *sets*. We do not consider elements of classes other than sets. A class is called a *semiset* iff it is a subclass of a set.

4. For any two sets there is a *pair set*.

5. *Any "reasonable" property determines a class*, namely the class of all sets having that property. (It suffices to assume that each of seven particular "reasonable" properties determines a class.)

6. There exists an *infinite set*.

7. *"Small" cannot be mapped one-one onto "large"*. More precisely, given a one-one correspondence between certain sets and certain semisets, the number of sets is "small" iff the number of semisets is "small"; the precise definition of a one-one correspondence between sets and semisets and of the notion "small" will be found in Chapter I.

8. *"Reasonable" semisets are sets.*

On specifying the word "reasonable" in items 5 and 8 we obtain the theory of semisets; all of the assumptions $1-8$ will hold in the interpretations mentioned above. If in item 8 we specify the word "reasonable" in another way we obtain a theory of sets equivalent to the theory consisting of Gödel's axiom groups A, B, C. As far as items $1-7$ are concerned, the theory of sets and the theory of semisets are identical; the difference between the two theories is to be found in item 8. In the theory of sets all semisets are considered to be "reasonable" and so the notion of semiset coincides with that of set. On the other hand, in the theory of semisets the "reasonable" semisets are those determined by "reasonable" properties in the manner referred to in item 5. As in 5 it suffices to restrict ourselves to six "reasonable" properties and to postulate that the semisets determined by these are sets.

We shall investigate both the theory of semisets and the theory of sets. In addition we shall consider various extensions of these theories; this is because in some proofs one cannot dispense with axioms of regularity and/or choice (axioms which, as we shall show, are consistent with both theories). However in this introduction we shall not distinguish between the theory of semisets (sets) and extensions obtained by adding regularity and/or choice axioms. The reader will find precise formulation in the body of the text.

What is the relation between the theory of semisets and the theory of sets?

(1) As we have seen, the theory of sets is an extension of the theory of semisets obtained by adding the axiom "all semisets are sets". Moreover this extension is conservative w.r.t. statements concerning sets alone; i.e. such statements are provable in the theory of sets iff they are provable in the theory of semisets.

(2) The theory of semisets is thus a generalization of the theory of sets obtained by giving up the assumption "all semisets are sets". In this sense the universe of discourse of the theory of semisets extends the universe of discourse of the theory of sets.

(3) It follows from (1) that we have only two possibilities for the three theories – Zermelo-Fraenkel set theory, Bernays-Gödel set theory and the theory of semisets: either all three are consistent or all three are contradictory. Moreover, any statement about sets alone is consistent with one iff it is consistent with all.

(4) As an application, the theory of semisets yields an axiomatization of Cohen's method in the following sense. We formulate certain additional axioms on semisets, called the axioms of support; these axioms are consistent with the theory of semisets. With each statement φ which concerns sets alone, we associate its "semiset version" $\hat{\varphi}$ which concerns sets and semisets alone. This is done in such a way that:

(i) The statement φ holds in all Boolean-valued models if and only if the statement $\hat{\varphi}$ is provable in the theory of semisets with a general axiom of support.

(ii) The consistency of φ with set theory can be proved by means of a particular Boolean-valued model if and only if $\hat{\varphi}$ is provable in the theory of semisets with a particular axiom of support. Hence, having proved $\hat{\varphi}$ in the latter theory we have established the consistency of φ with set theory.

(5) Axioms of support extend the theory of semisets conservatively w.r.t. statements which concern sets alone. Hence, having proved a statement of this kind in the theory of semisets with an axiom of support, we know that

it is provable in the theory of semisets alone (and hence also in set theory). The reader will find applications of this method in the book.

(6) There are also other axioms on semisets which extend the theory of semisets conservatively w.r.t. statements which concern sets alone. For example, both the axiom of standardness and its negation extend the theory of semisets in this way. The axiom of standardness asserts that every non-empty semiset of ordinals has a first element; it follows from "reasonable" support axioms. The theory of semisets with the negation of the axiom of standardness will not be studied in the present work, but it can be used in the study of non-standard analysis for example.

In presenting the theory of semisets the authors hope to make some contribution to the task of breaking through the bars of the prison in which mathematicians find themselves. This prison is set theory and the authors believe that mathematicians will escape from it just as they escaped from the prison of three-dimensional space.

The book is devoted to mathematical and metamathematical investigations of the notion of semiset and related notions such as the notion of support. Particular consistency results will play the role of examples; for example the consistency and independence of the axiom of choice, of the generalized continuum hypothesis, of the axiom "every set is definable", the results of Jech and Sochor etc. To say that we shall be concerned with both mathematical and metamathematical results means that we shall not only prove theorems within the theory of semisets but we shall also concern ourselves with results about theories. Thus we shall consider what statements are or are not provable in a theory; in particular we shall be concerned with relative consistency results. The statement "the theory **T** remains consistent upon the addition of the axiom φ" is not a statement of **T** but a statement about **T**, a meta-statement. In this book, any statement about sets, semisets, classes, relations, etc. is a mathematical statement or, more strictly, a statement formulated within the theory of semisets or some similar theory: to prove such a statement we appeal to the axioms of the theory in question. On the other hand, statements about the statements of a theory, about provability and unprovability, are metamathematical statements; to demonstrate such statements (i.e. to convince ourselves of their truth) we shall use essentially weaker assumptions. In the metamathematical investigation of a theory we shall entirely ignore the usual mathematical meaning of the concepts of the theory and concern ourselves solely with the structure of statements and proofs; in other words, we shall be concerned only with syntactical aspects of theories. The possibility of abstracting from the meanings of concepts in this way we learn from mathematical logic. The aspects of mathematical theories which interest us here may be reduced to the consideration of symbols and their configurations; in a sense a theory is to be regarded as a purely formal game with symbols.

We shall exclude certain methods of proof, such as proof by contradiction, from metamathematical arguments. If we assert for example that a certain proof exists then we shall always give instructions for constructing that proof. If we speak of a mapping in a metamathematical context then we shall give instructions which enable us to construct the image of any given object. In metamathematics we deal only with finite objects, such as formulas and proofs for example, and so our method may be said to be finitistic. It does not seem to be necessary to specify in advance exactly what methods are to be regarded as finitistic; it will be clear from our arguments which methods are being used.

We shall establish the relative consistency of a theory **T** w.r.t. another theory **S** by giving instructions which would enable us to convert any proof of a contradiction in the theory **T** into a proof of a contradiction in the theory **S**. In this book we shall always understand the notion of model in this sense; that is, a model of one theory in another theory consists of directions for converting proofs in the first theory into proofs in the second theory. We usually model one theory in another by interpreting the basic notions of the first theory as certain notions of the second theory; if this interpretation is suitably defined then statements and proofs in the first theory may be converted in a natural way into statements and proofs in the second theory.

Since we are concerned only with syntactical aspects of proofs, the models which we consider may be called syntactical models, in contradistinction to semantical models based on Tarski's notions of satisfaction and truth. As particular examples of syntactical models we have the relative interpretations of Tarski.

Many results and ideas of other authors are used in this book. Above all should be mentioned the classical results of axiomatic set theory, due to Zermelo, Fraenkel, von Neumann, Bernays, Gödel, Mostowski, Specker, Lévy and many others, as well as the results and ideas obtained in the application of Cohen's method, due to Scott, Solovay, McAloon and many others, including the authors' colleagues, Balcar, Bukovský, Hrbáček, Jech, Polívka, Sochor and Štěpánek. The authors are very grateful to the latter for many fruitful discussions concerning the content of the book and for several improvements; in particular many proofs were simplified and corrected by B. Balcar and J. Polívka.

In developing the new theory the authors had to adapt the ideas of other authors for the purpose. In particular we mention McAloon's proof of the independence of the axiom of constructibility from the axiom of definability and the method of Jech and Sochor for using consistency proofs of the Fraenkel-Mostowski type to obtain consistency results for set theory with the axiom of regularity.

The authors were not able to assign every particular notion or result to its true author and for this reason references have been omitted altogether. The bibliography at the end contains only a selection of important works more or less related to the topics treated in the present book. Thus this book should be regarded as a result of the work of all the mathematicians mentioned above and of the many others whose results have become indispensable for anyone concerned with set theory.

The authors would also like to express their gratitude to T. Jech and G. Rousseau who prepared the translation of the book into English and to K. Čuda, Mich. Chytil, J. Mlček, J. Polívka and A. Sochor for their help in reading proofs.

P. V. and P. H.

CHAPTER I

SECTION 1

The theory of classes

In the present Section we introduce the fundamental fragment of the various (gödelian) theories of sets and consider a very general example of the notion of a model. The treatment in this Section will not be too formal. The reader should find the ideas natural, but on the other hand we will see the necessity for a systematic foundation. (Places where such a systematic foundation is necessary will be indicated by the sign [!].)

Systematic foundations will be laid in Section 2; the reader will then see that the considerations of Section 1 are consistent with the general metamathematical conception developed there; on the other hand, the considerations of Section 1 will serve to illustrate the general concepts introduced in Section 2. We shall come to theories dealing specifically with sets and semisets in Section 4.

a) Axioms of the theory of classes

1101. The fundamental concepts of the theory of classes are the concept of *class* and the concept of *membership*. We will denote arbitrary classes by the letters $X, Y, Z, \ldots$ (with additional indices if necessary); i.e., the letters $X, Y, Z, \ldots$ will be used as *variables for classes*. The statement "X belongs to Y" (or, "X is an element of Y") will be written as $X \in Y$; we call the symbol $\in$ the membership predicate. We shall use logical connectives $\&$, $\vee$, $\rightarrow$, $\equiv$, $\neg$ (and, or, implies, if and only if, not) and quantifiers $\forall$, $\exists$ (for every, for some). The formula $X = Y$ will stand for "X is equal to Y" ("X is the same class as Y"). For the time being, we shall treat these symbols as a standard vocabulary for expressing mathematical statements by formulas; e.g. the formula $(\forall Z)(Z \in X \equiv Z \in Y)$ will be considered to stand for the statement "for every Z, Z belongs to X if and only if Z belongs to Y".

1102. Axiom (F1) $(\exists X)(\exists Y)(X \in Y)$.

Axiom (F2) $(\forall Z)(Z \in X \equiv Z \in Y) \equiv X = Y$.

DEFINITION (d1) $M(X) \equiv (\exists Y)(X \in Y)$.
DEFINITION (d2) $[(\exists x)(x = X) \equiv M(X)] \& (\forall x)(\exists X)(x = X)$.

Axiom (F1) states that there exist classes X and Y such that X belongs to Y. Axiom (F2) (the *axiom of extensionality*) states that X is equal to Y if and only if X has the same elements as Y. One part of this axiom is obvious [!]; if X is the same as Y then for any Z, $Z \in X$ iff*) $Z \in Y$. This is a general property of the identity. The other part is nontrivial; if X and Y have the same elements then they are equal (i.e., different classes are distinguished by some element). The theory having F1 and F2 as its axioms is called the *fundamental* $\in$*-theory* **TE** (or, the fundamental theory of the epsilon predicate).

The predicate M defined in (d1) stands for "... is a set"; a *set* is defined as a class which is an element of some other class. Definition (d2) serves to introduce a new sort of variable; these are the *set-variables* (variables for sets), for which we use lower case letters; X is equal to some x iff it is a set. An immediate consequence of axiom (F1) is the statement $(\exists X)\, M(X)$, i.e. there exists at least one set.

1103. AXIOM (A1) $(\forall x, y)(\exists z)(\forall u)(u \in z \equiv . u = x \vee u = y)$.
This is the *pairing axiom*; for any two sets x, y there exists a set whose elements are just x and y. It follows from the axiom of extensionality that given two sets x and y, there can be at most one set having x and y as its only elements, i.e.

$$[(\forall u)(u \in z \equiv . u = x \vee u = y) \&$$
$$\& (\forall u)(u \in w \equiv . u = x \vee u = y)] \rightarrow z = w .$$

Thus we have proved from (A1) and (F2) that for each x and y there exists a unique z such that

$$(\forall u)(u \in z \equiv . u = x \vee u = y);$$

if the symbol $\exists!$ is used to stand for "there exists a unique", then we may write this as follows:

$$(\forall x, y)(\exists! z)(\forall u)(u \in z \equiv . u = x \vee u = y).$$

*) "iff" is an abbreviation for "if and only if".

We denote this unique z by $\{x, y\}$ [!] (the pair of x, y), i.e. we define the *pairing operation* as follows:

(d3) $u \in \{x, y\} \equiv (u = x \vee u = y)$.

We shall write $\{x\}$ instead of $\{x, x\}$.

1104. The *ordered pair* and the *ordered triple* are defined in the usual way as follows:

(d4) $\langle x, y \rangle = \{\{x\}, \{x, y\}\}$

(d5) $\langle x, y, z \rangle = \langle x, \langle y, z \rangle \rangle$.

1105. LEMMA. $\langle x, y \rangle = \langle u, v \rangle \equiv (x = u \,\&\, y = v)$.

Proof. Exercise.

1106. (d6) $\text{Rel}(X) \equiv (\forall z)(z \in X \rightarrow (\exists u, v)(z = \langle u, v \rangle)$.

X is a (*binary*) *relation* if all its elements are ordered pairs.

(d7) $\text{Rel}_3(X) \equiv (\forall z)(z \in X \rightarrow (\exists u, v, w)(z = \langle u, v, w \rangle))$.

X is a *ternary relation* if all its elements are ordered triples.

1107. We shall now present seven axioms which ensure the existence of certain classes. As we shall show later, these axioms enable us to prove for any "reasonable" property the existence of a class whose elements are just the sets with this property.

(B1) $(\exists Z)(\forall x)(x \in Z)$

(B2) $(\forall X)(\exists Z)(\forall x)(x \in Z \equiv (\exists u, v)(x = \langle u, v \rangle \,\&\, u \in v \,\&\, x \in X))$

(B3) $(\forall X, Y)(\exists Z)(\forall x)(x \in Z \equiv x \in X \,\&\, x \notin Y)$

(B4) $(\forall X)(\exists Z)(\forall x)(x \in Z \equiv (\exists y)(\langle y, x \rangle \in X))$

(B5) $(\forall X, Y)(\exists Z)(\forall x)(x \in Z \equiv (\exists u, v)(x = \langle u, v \rangle \,\&\, v \in Y \,\&\, x \in X))$

(B6) $(\forall X)(\exists Z)(\forall x)(x \in Z \equiv (\exists u, v)(x = \langle u, v \rangle \,\&\, \langle v, u \rangle \in X))$

(B7) $(\forall X)(\exists Z)(\forall x)(x \in Z \equiv (\exists u, v, w)(x = \langle u, v, w \rangle \,\&\, \langle v, w, u \rangle \in X))$.

Axiom (B1) ensures the existence of the class of all sets (the *universal class* or *universe*). Axiom (B2) ensures for any X the existence of the class whose elements are all ordered pairs $\langle u, v \rangle \in X$ such that u is an element of v; this class is called the *representation of* $\in$ *on* X. Axiom (B3) is the axiom of *complement*, and (B4) is the axiom of *domain* (or *projection*). (B5) is the axiom of *restriction*; the elements of Z are the ordered pairs $\langle u, v\rangle$ belonging to X such that $v \in Y$. Axioms (B6) and (B7) are the axioms of *converse*. It follows immediately that in each of these 7 cases, the class Z is determined uniquely by X in cases 2, 4, 6 and 7, and by X and Y in cases 3 and 5. Hence all of these axioms may be strengthened by writting $(\exists! Z)$ instead of $(\exists Z)$; these strengthened statements follows from axioms (B1) to (B7) and the axiom of extensionality. Hence the following definitions are justified:

1108. (d8) $(\forall x)(x \in \mathbf{V})$

V is a constant denoting the *universal class*.

(d9) $\mathrm{Rel}(\mathbf{E}(X)) \,\&\, (\forall u, v)(\langle u, v\rangle \in \mathbf{E}(X) \equiv \langle u, v\rangle \in X \,\&\, u \in v)$
(the *representation of membership on X*)

(d10) $x \in X - Y \equiv x \in X \,\&\, x \notin Y$
(the *complement of Y relative to X*)

(d11) $x \in \mathbf{D}(X) \equiv (\exists y)(\langle y, x\rangle \in X)$
(the *domain of X*)

(d12) $\mathrm{Rel}(X \restriction Y) \,\&\, (\forall u, v)(\langle u, v\rangle \in X \restriction Y \equiv \langle u, v\rangle \in X \,\&\, v \in Y)$
(the *restriction of X to Y*)

(d13) $\mathrm{Rel}(\mathbf{Cnv}(X)) \,\&\, (\forall u, v)(\langle u, v\rangle \in \mathbf{Cnv}(X) \equiv \langle v, u\rangle \in X)$
(the *converse of X*)

(d14) $\mathrm{Rel}_3(\mathbf{Cnv}_3(X)) \,\&\, (\forall u, v, w)(\langle u, v, w\rangle \in \mathbf{Cnv}_3(X) \equiv \langle v, w, u\rangle \in X)$
(the *ternary converse of X*).

1109. Axioms F1, F2, A1, B1 – B7 (together with definitions (d1) – (d14)) are the *axioms of the theory of classes* **TC**. We have defined one constant **V** and a number of operations, e.g. pairing, ∈-representation, complementation, domain, restriction and two converses. The operations just mentioned are called the (*fundamental*) *gödelian operations* and we shall sometimes denote them by $\mathbf{F}_1, \ldots, \mathbf{F}_7$.

1110. A number of useful definitions follows:

(15) $\mathrm{Cls}(X) \equiv X = X$

(X is *class*; every class satisfies Cls (X))

(d16a) $X \subseteq Y \equiv (\forall z)(z \in X \rightarrow z \in Y)$

(X is a *subclass* (a *part*) *of* Y, X is *included* in Y, Y *contains* X); $\subseteq$ is the *predicate of inclusion*

(d16b) $X \subset Y \equiv . X \subseteq Y \& X \neq Y$

(X is a *proper subclass of* Y); $X \neq Y$ means the same as $\neg (X = Y)$

(d17) $0 = \mathbf{V} - \mathbf{V}$

(the *empty class*)

(d18) $-X = \mathbf{V} - X$

(the *complement of* X)

(d19) $X \cap Y = X - (X - Y)$

(the *intersection of* X *and* Y)

(d20) $X \cup Y = -(-X \cap -Y)$

(the *union of* X *and* Y)

(d21) $X \times Y = (\mathbf{V} \upharpoonright Y) \cap \mathbf{Cnv}(\mathbf{V} \upharpoonright X)$

(the *product of* X *and* Y)

(d22) $\mathbf{E} = \mathbf{E}(\mathbf{V})$

(the *representation of membership*)

(d23) $\mathbf{W}(X) = \mathbf{D}(\mathbf{Cnv}(X))$

(the *range of* X)

(d24) $\mathbf{C}(X) = \mathbf{D}(X) \cup \mathbf{W}(X)$

(the *field of* X).

1111. LEMMA (**TC**). $\neg (\exists x)(x \in 0)$;

$x \in X \cap Y \equiv . x \in X \& x \in Y$;

$x \in -X \equiv x \notin X$;

$x \in X \cup Y \equiv . x \in X \vee x \in Y$;

$\mathrm{Rel}(X \times Y) \& (\forall u, v)(\langle u, v\rangle \in X \times Y \equiv u \in X \& v \in Y)$;

$\mathrm{Rel}(\mathbf{E}) \& (\forall u, v)(\langle u, v\rangle \in \mathbf{E} \equiv u \in v)$;

$x \in \mathbf{W}(X) \equiv (\exists y)(\langle x, y\rangle \in X)$.

Proof. Exercise.

1112. Finally we define

$$\langle x\rangle = x ,$$
$$\langle x_1, \ldots, x_{n+1}\rangle = \langle x_1, \langle x_2, \ldots, x_{n+1}\rangle\rangle$$

(this definition gives instructions for defining *quadruples*, *quintuples* etc.).

$$\mathbf{V}^1 = \mathbf{V}\,, \quad \mathbf{V}^{n+1} = \mathbf{V} \times (\mathbf{V}^n)$$

($\mathbf{V}^n$ is the class of all n-tuples).

b) *Representability of formulas*

In this book we shall be dealing not only with the theory **TC**, but also with certain other theories which have additional axioms (in certain cases we shall also consider theories with fewer axioms). Each of these theories will include the axioms F1, F2; in other words we will be concerned with extensions of the fundamental $\in$-theory **TE**. Each such theory has the predicates $\in$ and $=$ and the class-variables; however there may be additional predicates, operations and constants present, as well as partial variables (ranging over some of the classes). In fact we have already introduced for example the predicates M, Rel, Rel_3, $=$, $\subseteq$, $\subset$, the constants **V**, 0, **E**, the operations **E**, **D**, **Cnv**, $\mathbf{Cnv}_3$, $\{\ \}$, $\langle\ \rangle$, $-$, $\upharpoonright$ and the set-variables $x, y, z, \ldots$

1113. Given a theory **T**, a formula φ is said to be a *formula of* **T** or a **T**-formula if φ is formulated in the language of **T**. Each variable occurring in φ is either free or bound in φ. See Section 2 for the details; here we only stress the fact that, given two **T**-formulas φ and ψ, $\varphi \,\&\, \psi$ is a (meaningful) formula (and a **T**-formula) iff no variable free in one of the formulas φ, ψ is bound in the other. (Cf. the definition of normal formulas given below.) Similarly for $\varphi \vee \psi$, $\varphi \to \psi$, $\varphi \equiv \psi$. Expressions such as $\mathbf{D}(X)$, X, $\mathbf{W}(X \cap \cap\, \mathbf{Cnv}(Y))$ are called *terms* ("descriptions for classes").

By an *extension of a theory* **T** we mean a theory which has among its axioms all axioms of **T** (thus **T** is an extension of itself).

We promised above to furnish a proof that for any "reasonable" property the class of all sets having this property exists. To specify which properties we have in mind we introduce the metamathematical concept of normal **T**-formula. We simultaneously define what it is for a variable to be free or bound in a normal **T**-formula.

1114. METADEFINITION. Let **T** be an extension of **TE**.

(a) If x is a set variable and Y is an arbitrary variable or constant of **T** different from x (the last condition is automatically satisfied if Y is not a set variable) then $x \in Y$ is a *normal* **T**-*formula*; x is *free* in $x \in Y$; if Y is a variable then Y is *free* in $x \in Y$; no variable is *bound* in $x \in Y$.

(b) If φ and ψ are normal **T**-formulas and if no variable is free in one formula and bound in the other, then $\varphi \,\&\, \psi$ and $\neg \varphi$ are *normal* **T**-*formulas*; a variable is *free* (or *bound*) in $\varphi \,\&\, \psi$ if it is free (or bound) in either φ or ψ; a variable is *free* (*bound*) in $\neg \varphi$ if it is free (bound) in φ.

(c) If x is a variable not bound in a normal **T**-formula φ then $(\exists x)\, \varphi$ is a *normal* **T**-*formula*; x is *bound* in $(\exists x)\, \varphi$; other variables occurring in φ are *free* (*bound*) in $(\exists x)\, \varphi$ if they are free (bound) in φ.

(d) Every normal **T**-formula is obtained by a finite number of applications of rules b) and c), beginning with normal **T**-formulas described sub a).

Remark. If the theory **T** is fixed or is clear from the context we say "normal formula" of "NF" instead of "normal **T**-formula".

E.g. the formula

$$\neg(\exists z)\,(z \in X \,\&\, z \notin Y)\,,$$

which is equivalent to the formula $X \subseteq Y$, is a NF. Notice that no class-variables are bound in NFs. We shall now give the promised existence proof. For each normal formula we prove the existence of the corresponding class; moreover we describe this class.

1115. METADEFINITION. Let **T** be an extension of **TC** (e.g., **TC** itself) and let φ be a **T**-formula.

(1) A sequence $x_1, \ldots, x_n$ of distinct set variables is said to be *free for* φ iff no x_i $(1 \leq u \leq n)$ is bound in φ.

(2) Let $x_1, \ldots, x_n$ be free for φ. We say that φ is *representable in* **T** *w.r.t.* $x_1, \ldots, x_n$ if there is a term $\mathbf{T}$ of **T** such that the following is provable in **T**:

$$(*) \qquad (\forall x_1, \ldots, x_n)\,(\langle x_1, \ldots, x_n\rangle \in \mathbf{T} \equiv \varphi) \,\&\, \mathbf{T} \subseteq \mathbf{V}^n\,,$$

$\mathbf{T}$ is said to represent φ in **T** w.r.t. $x_1, \ldots, x_n$.

(3) φ is said to be *fully representable in* **T** if it is representable in **T** w.r.t. any sequence free for φ.

Remark. It is not required that the x_i actually occur in φ, but only that they should not be bound in φ. On the other hand it is possible that φ contains some free variables not occurring among the x_i's; let these be $Y_1, \ldots, Y_m$ (not necessarily class variables). Then we understand $(*)$ as saying the same as

$$(\forall Y_1, \ldots, Y_n)\,[(\forall x_1, \ldots, x_n)\,(\langle x_1, \ldots, x_n\rangle \in \mathbf{T} \equiv \varphi) \,\&\, \mathbf{T} \subseteq \mathbf{V}^n]\,.$$

(Cf. the axioms of group theory: one says "$x + (y + z) = (x + y) + z$ is an axiom of group theory" meaning "$(\forall x)(\forall y)(\forall z)(x + (y + z) = (x + y) + z)$ is an axiom of group theory".)

In order to demonstrate that every NF is fully representable in **TC** (and also for other purposes) we demonstrate some Metalemmas. In these Metalemmas **T** is an arbitrary extension of **TC**.

1116. METALEMMA. If φ is fully representable in **T** then $\neg\varphi$ is fully representable in **T**.

Demonstration. If $\mathbf{T}$ represents φ w.r.t. $x_1, \ldots, x_n$ then $\mathbf{V}^n - \mathbf{T}$ represents $\neg\varphi$ w.r.t. $x_1, \ldots, x_n$.

1117. METALEMMA. If φ, ψ are fully representable in **T** then $\varphi \& \psi$ is fully representable in **T**.

Demonstration. If $\mathbf{T}_1$ and $\mathbf{T}_2$ represent φ and ψ respectively w.r.t. $x_1, \ldots, x_n$ then $\mathbf{T}_1 \cap \mathbf{T}_2$ represents $\varphi \& \psi$ w.r.t. $x_1, \ldots, x_n$.

1118. METALEMMA. If φ is fully representable in **T** and x is a set variable not bound in φ then $(\exists x)\,\varphi$ is fully representable in **T**.

Demonstration. Let $x_1, \ldots, x_n$ be a sequence free for $(\exists x)\,\varphi$. Then $x, x_1, \ldots, x_n$ is free for φ and, by the full representability of φ, we have a term $\mathbf{T}$ such that $(\forall \ldots)(\langle x, x_1, \ldots, x_n\rangle \in \mathbf{T} \equiv \varphi) \& \mathbf{T} \subseteq \mathbf{V}^{n+1}$ is provable in **T**. Consequently, $(\forall \ldots)(\langle x_1, \ldots, x_n\rangle \in \mathbf{D}(\mathbf{T}) \equiv (\exists x)\,\varphi) \& \mathbf{D}(\mathbf{T}) \subseteq \mathbf{V}^n$ is provable in **T**.

Remark. $(\forall \ldots)\,\varphi$ denotes the closure of φ, i.e. $(\forall \ldots)\,\varphi$ is an abbreviation for $(\forall x) \ldots (\forall y)\,\varphi$ where $x, \ldots, y$ are all the free variables of φ.

1119. METALEMMA Let x be a set variable and let Y an arbitrary variable or constant of **T** distinct from x. Then $x \in Y$ is fully representable in **T**.

Demonstration. The demonstration is technically difficult but the reader is recommended to read it in order to become more familiar with operations we have introduced. Let $x_1, \ldots, x_n$ be a sequence of distinct set variables. First suppose that this sequence does not contain Y. (This is automatically the case if Y is not a set variable.)

(a) Consider the "pathological" case where x does not occur in $x_1, \ldots, x_n$. Then we are looking for a term $\mathbf{T}$ such that the following is provable: $\mathbf{T} = 0$ if $x \notin Y$, $\mathbf{T} = \mathbf{V}^n$ if $x \in Y$. Show that the term $\mathbf{D}((Y \cap \{x\}) \times \mathbf{V}^n)$ has this property.

(b) Now let x be x_i; we are looking for a term $\mathbf{T}$ such that $(\langle x_1, \ldots, x_n\rangle \in \mathbf{T} \equiv x_i \in Y) \& \mathbf{T} \subseteq \mathbf{V}^n$ is provable. If $n = 1$ take Y for $\mathbf{T}$. If $n > 1$ and

$i = 1$ take $Y \times \mathbf{V}^{n-1}$ for $\mathbf{T}$. If $n > 1$ and $i = n$ take $\underbrace{\mathbf{V} \times (\mathbf{V} \times \ldots \times (\mathbf{V}}_{n-1 \text{ times}} \times Y) \ldots)$ for $\mathbf{T}$. If $1 < i < n$ take $\underbrace{\mathbf{V} \times (\mathbf{V} \times \ldots \times (\mathbf{V}}_{i-1 \text{ times}} \times (Y \times \mathbf{V}^{n-i})) \ldots)$ for $\mathbf{T}$.

Now let Y be x_j.

(c) If x does not occur in $x_1, \ldots, x_n$ then we are looking for a term such that the following is provable: $\mathbf{T}$ is the class of all n-tuples such that the j-th element contains x. For $n = 1$ we must find a term $\mathbf{S}$ defining the class of all sets containing x (as an element). Show that $\mathbf{W}\,((\mathbf{Cnv}\,(\mathbf{E})) \upharpoonright \{x\})$ has this property. For $n > 1$ continue as in (b) above writing $\mathbf{W}\,((\mathbf{Cnv}\,(\mathbf{E})) \upharpoonright \{x\})$ instead of Y.

(d) Finally, let x be x_i and let Y be x_j. We have assumed that x and Y are distinct, so that $i \neq j$. If $i < j$ let $\mathbf{T}_1$ be the constant $\mathbf{E}$; if $i > j$ let $\mathbf{T}_1$ be the term $\mathbf{Cnv}\,(\mathbf{E})$. Put $\bar{\imath} = \min(i, j)$, $\bar{\jmath} = \max(i, j)$. It is provable in **TC** that $\langle x_{\bar{\imath}}, x_{\bar{\jmath}} \rangle \in \mathbf{T}_1 \equiv x_i \in x_j$. If $\bar{\jmath} = n$ let $\mathbf{T}_2$ be $\mathbf{T}_1$; otherwise let $\mathbf{T}_2$ be the term $\mathbf{Cnv}_3\,(\mathbf{Cnv}_3\,(\underbrace{(\mathbf{V} \times \ldots \times \mathbf{V})}_{(n-\bar{\jmath}) \text{ times}} \times \mathbf{T}_1))$; it is provable that $\langle x_{\bar{\imath}}, x_{\bar{\jmath}}, x_{\bar{\jmath}+1}, \ldots, x_n \rangle \in \mathbf{T}_2 \equiv x_i \in x_j$. If $\bar{\jmath} = \bar{\imath} + 1$, let $\mathbf{T}_3$ be $\mathbf{T}_2$ if $\bar{\jmath} \neq \bar{\imath} + 1$ notice first that the following is provable:

$$\langle x_{\bar{\imath}} \langle x_{\bar{\jmath}}, \ldots, x_n \rangle \rangle \in \mathbf{T}_2 \equiv \langle \langle x_{\bar{\jmath}}, \ldots, x_n \rangle, x_{\bar{\imath}} \rangle \in \mathbf{Cnv}\,(\mathbf{T}_2) \equiv$$
$$\equiv \langle x_{\bar{\jmath}-1}, \langle x_{\bar{\jmath}}, \ldots, x_n \rangle, x_{\bar{\imath}} \rangle \in \mathbf{V} \times \mathbf{Cnv}\,(\mathbf{T}_2) \equiv$$
$$\equiv \langle x_{\bar{\imath}}, x_{\bar{\jmath}-1}, x_{\bar{\jmath}}, \ldots, x_n \rangle \in \mathbf{Cnv}_3\,(\mathbf{V} \times \mathbf{Cnv}\,(\mathbf{T}_2)) .$$

We have succeeded in putting the variable $x_{\bar{\jmath}-1}$ inside. If we repeat this device $(\bar{\jmath} - \bar{\imath} - 1)$ times, we obtain a term $\mathbf{T}_3$ such that the following is provable:

$$\langle x_{\bar{\imath}}, x_{\bar{\imath}+1}, \ldots, x_n \rangle \in \mathbf{T}_3 \equiv x_i \in x_j .$$

Hence we let $\mathbf{T}$ be the term $\underbrace{\mathbf{V} \times (\mathbf{V} \times \ldots (\mathbf{V}}_{(\bar{\imath}-1) \text{ times}} \times \mathbf{T}_3))$. This completes the demonstration.

1120. METATHEOREM. If **T** is any extension of **TC** then every normal **T**-formula is fully representable in **T**.

This is an immediate consequence of the preceding lemmas. In fact, we have shown even more: if each variable free in φ is one of the x_i then the operation $\{\ \}$ is not used in the construction of $\mathbf{T}$. Thus, define:

1121. METADEFINITION. Let **T** be an extension of **TC**. (a) Every variable and constant of **T** is a *gödelian term* of **T**. (b) If $\mathbf{T}_1$, $\mathbf{T}_2$ are gödelian terms of **T** then $\mathbf{E}(\mathbf{T}_1)$, $\mathbf{T}_1 - \mathbf{T}_2$, $\mathbf{D}(\mathbf{T}_1)$, $\mathbf{T}_1 \upharpoonright \mathbf{T}_2$, $\mathbf{Cnv}(\mathbf{T}_1)$, $\mathbf{Cnv}_3(\mathbf{T}_1)$ are *gödelian terms* of **T**. (Briefly, gödelian terms.)

In the following corollaries **T** means an arbitrary extension of **TC**.

1122. COROLLARY. If φ is a normal **T**-formula and $x_1, \ldots, x_n$ are all the free variables of φ then there is a gödelian term representing φ in **T** w.r.t. $x_1, \ldots, x_n$.

Remark. If $\varphi \equiv \psi$ is provable in **T** and φ is fully representable in **T** then obviously ψ is fully representable in **T**.

1123. METADEFINITION. Let **T** be an extension of **TC** and let ψ be a **T**-formula. ψ is said to be *normal in* **T** if there is a normal **T**-formula φ such that $\varphi \equiv \psi$ is provable in **T**.

1124. COROLLARY. Every formula normal in **T** is fully representable in **T**.

Example. Show that formulas $x = y$, $x = Y$ are normal in **TC**.

1125. The term which represents a NF $\varphi(x_1, \ldots, x_n, X_1, \ldots, X_m)$ w.r.t. the sequence $x_1, \ldots, x_n$ of its free set variables will often be denoted by the expression

$$\{\langle x_1, \ldots, x_n \rangle; \quad \varphi(x_1, \ldots, x_n, X_1, \ldots, X_m)\} .$$

In what follows we shall work in the theory **TC**.

1126. METALEMMA. Let **T** be an extension of **TC**; let φ be a normal **T**-formula, and let Y be a variable free in φ. If **G** is a gödelian term of **T** such that no variable occurring in **G** is bound in φ, then the result of replacing all occurrences of Y in φ by **G** is a formula normal in **T**.

Demonstration. Y can occur either in the context $x \in Y$ or, if Y is a set variable, in the context $Y \in Z$. The last formula is equivalent to $(\exists u)(u = Y \,\&\, u \in Z)$ where u is a set variable not occurring in φ, and $u = Y$ is equivalent to $(\forall v)(v \in u \equiv v \in Y)$. Hence it suffices to show that $u \in \mathbf{G}$ is a formula normal in **T**. This is done by induction. (a) If **G** is a variable or a constant then $u \in \mathbf{G}$ is normal.

(b) Let us show that the formula $\langle z_1, z_2 \rangle \in R$ is normal in **TC**. This formula is equivalent (in **TC**) to $(\exists x)(x = \langle z_1, z_2 \rangle \,\&\, x \in R)$, hence it suffices to show that $x = \langle z_1, z_2 \rangle$ is normal. But (in **TC**) $x = \langle z_1, z_2 \rangle \equiv$

$\equiv x = \{\{z_1\}, \{z_1, z_2\}\} \equiv (\exists u, v)(x = \{u, v\} \,\&\, u = \{z_1, z_1\} \,\&\, v = \{z_1, z_2\}$, hence it suffices to show that $x = \{u, v\}$ is normal. Now (in **TC**), $x = \{u, v\} \equiv$ $\equiv (\forall y)(y \in x \equiv . \, y = u \vee y = v)$ and the right hand side is normal because $y = u$ is normal.

Analogously one can show that $\langle z_1, z_2, z_3\rangle \in R$ is normal. Moreover, if $\mathbf{t}$ is a term and $\mathbf{t} \in R$ is normal then $\mathbf{t} = R$ is also normal (for it is equivalent to $(\forall v)(v \in \mathbf{t} \equiv v \in R)$). Now, let $\mathbf{G}_1, \mathbf{G}_2$ be gödelian terms of **T** and suppose that $u \in \mathbf{G}_1$, $u \in \mathbf{G}_2$ are normal. Then $u \in \mathbf{E}(\mathbf{G}_1)$, $u \in (\mathbf{G}_1 - \mathbf{G}_2)$ etc. are also normal as may be seen from the following equivalences provable in **T**:

$$u \in \mathbf{E}(\mathbf{G}_1) \equiv (\exists v, w)(u = \langle v, w\rangle \,\&\, v \in w \,\&\, u \in \mathbf{G}_1),$$

$$u \in \mathbf{G}_1 - \mathbf{G}_2 \equiv u \in \mathbf{G}_1 \,\&\, u \notin \mathbf{G}_2, \quad \text{etc.}$$

1127. A number of useful definitions follows:

(d25) $\mathsf{U}(X) = \{z; (\exists y)(z \in y \,\&\, y \in X)\}$
(the *sum of* X (the *union of* X), sometimes denoted by $\bigcup X$)

(d26) $\mathbf{P}(X) = \{z; z \subseteq X\}$
(the *power-class of* X)

(d27) $X``Y = \mathbf{W}(X \restriction Y) = \{u; (\exists v \in Y)(\langle u, v\rangle \in X)\}$

(d28) $\mathrm{Un}(X) \equiv \mathrm{Rel}(X) \,\&\, (\forall u, v, w)(\langle v, u\rangle \in X \,\&\, \langle w, u\rangle \in X . \rightarrow v = w)$
(X is a *function* (or *mapping*))

(d29) $\mathrm{Un}(X, Y, Z) \equiv . \, \mathrm{Un}(X) \,\&\, \mathbf{D}(X) = Y \,\&\, \mathbf{W}(X) \subseteq Z$
(X is a *mapping of* Y *into* Z)

(d30) $\mathrm{Un}_2(X) \equiv . \, \mathrm{Un}(X) \,\&\, \mathrm{Un}(\mathbf{Cnv}(X))$
(X is a *one-to-one mapping*)

(d31) $\mathrm{Un}_2(X, Y, Z) \equiv . \, \mathrm{Un}(X, Y, Z) \,\&\, \mathrm{Un}(\mathbf{Cnv}(X), Z, Y)$
(X is a *one-to-one mappings of* Y *onto* Z)

(d32) $F`y = z$ if $\mathrm{Un}(F)$ and $\langle z, y\rangle \in F$
$F`y = 0$ otherwise
(z is the *image of* y *in* F)

c) *A model*

1128. We shall now be concerned with the study of relations.

(d33) If R is a relation and $x \in \mathbf{C}(R)$, then

$$\mathbf{Ext}_R(x) = \{y; \langle y, x \rangle \in R\}$$

(the extension of x in R)

(d34) $\mathrm{Sat}_R(X) \equiv X \subseteq \mathbf{C}(R) \,\&\, (\forall x, y \in \mathbf{C}(R))(x \in X \,\&\, \mathbf{Ext}_R(x) = \mathbf{Ext}_R(y) \mathbin{.\!\!\to} y \in X)$

(X is a *saturated part of* $\mathbf{C}(R)$).

(d35) For $X \subseteq \mathbf{C}(R)$

$$\mathbf{SAT}_R(X) = \{z \in \mathbf{C}(R); (\exists x \in X)(\mathbf{Ext}_R(x) = \mathbf{Ext}_R(z))\}$$

(the *saturation of* X).

1129. LEMMA (**TC**). If R is a relation and X and Y are saturated parts of its field then $X \cap Y$, $X \cup Y$ and $X - Y$ are also saturated. The saturation of a subclass of $\mathbf{C}(R)$ is saturated.

1130. DEFINITION (**TC**).

(d36) A relation R is *extensional* ($\mathrm{Extl}(R)$) if

$$\langle x, z \rangle \in R \,\&\, \mathbf{Ext}_R(x) = \mathbf{Ext}_R(y) \to \langle y, z \rangle \in R$$

for all x, y, z in $\mathbf{C}(R)$.

1131. LEMMA (**TC**). R is extensional if and only if for each $x \in \mathbf{C}(R)$ the extension of x in R is saturated.

1132. DEFINITION (**TC**).

(d37) $\mathrm{Prg}(R) \equiv (\forall x, y \in \mathbf{C}(R))(\exists z \in \mathbf{C}(R))(\mathbf{Ext}_R(z) = \mathbf{SAT}_R(\{x, y\}))$
(R is *pairing*)

(d38) $\mathrm{Elk}(R) \equiv \mathrm{Rel}(R) \,\&\, R \neq 0 \,\&\, \mathrm{Extl}(R) \,\&\, \mathrm{Prg}(R)$.

Thus a non-empty relation R is **E**-*like* if it is extensional and pairing.

(d39) $\mathrm{sExtl}(R) \equiv (\forall x, y \in \mathbf{C}(R)(x \neq y \to \mathbf{Ext}_R(x) \neq \mathbf{Ext}_R(y))$
(R is *strongly extensional*).

1133. LEMMA (**TC**). (a) Any strongly extensional relation is extensional.

(b) A relation R is strongly extensional iff every part of $\mathbf{C}(R)$ is saturated.

Proof. (a) If a relation R is strongly extensional, then $\mathbf{Ext}_R(x) = \mathbf{Ext}_R(y)$ implies $x = y$, and hence $\langle x, z \rangle \in R$ implies $\langle y, z \rangle \in R$. (b) If $x \neq y$ and

$\mathbf{Ext}_R(x) = \mathbf{Ext}_R(y)$ then $\{x\}$ is not saturated. Hence if every part of $\mathbf{C}(R)$ is saturated then R is strongly extensional; the converse is obvious.

1134. LEMMA (**TC**). $\mathbf{E}$ is an $\mathbf{E}$-like relation.

Proof. Since there exists at least one set x and we have $\langle x, \{x\}\rangle \in \mathbf{E}$ by A1, the relation $\mathbf{E}$ is non-empty. $\mathbf{E}$ is strongly extensional, since $\mathbf{Ext}_{\mathbf{E}}(x) = x$ for any x. Finally, $\mathbf{E}$ is pairing, since for $z = \{x, y\}$ we have $\mathbf{Ext}_{\mathbf{E}}(z) = = \{x, y\} = \mathbf{SAT}_{\mathbf{E}}(\{x, y\})$.

1135. In what follows we denote by the constant $\mathbf{R}$ some fixed (arbitrarily chosen) $\mathbf{E}$-like relation. By that we enlarge **TC** in an inessential manner [!]; in the remainder of this Section we shall denote by $\mathbf{TC}_1$ the theory **TC** with the constant $\mathbf{R}$ and with the assumption Elk $(\mathbf{R})$ (and possibly with further definitions in which the constant $\mathbf{R}$ is involved).

1136. DEFINITION ($\mathbf{TC}_1$).

$$(\exists X^*)(X^* = X) \equiv \mathrm{Sat}_{\mathbf{R}}(X)\,;$$

the variables $X^*, Y^*, \ldots$ range over saturated parts of $\mathbf{C}(\mathbf{R})$;

$$X^* \in^* Y^* \equiv (\exists z \in Y^*)(X^* = \mathbf{Ext}_{\mathbf{R}}(z))\,.$$

1137. Remark. 1) $X^* = \mathbf{Ext}_{\mathbf{R}}(z)$ means that X^* is the extension of z (in $\mathbf{R}$). We shall also say that z is a code of X^* (or, precisely, and $\mathbf{R}$-code of X^*; this phrase will be often used). Then the preceding definition reads: X^* is an $*$-element of Y^* iff there is a code of X^* which is an element of Y^*. Thus $\mathbf{R}$ "codes" the new membership.

2) If $\mathbf{R} = \mathbf{E}$ then the variables X^* range over all classes and $X \in^* Y$ is equivalent to $X \in Y$. (Prove.)

We have introduced new variables X^* and a new predicate $\in^*$; we shall regard the X^* as new classes and $\in^*$ as a new membership predicate. We may ask whether we are justified in doing so, whether the axioms of **TC** hold in the new sense. We shall show that this is indeed the case.

Moreover, in the course of the verification we shall make clear the precise meaning of the phrase "the axioms hold in the new sense".

1138. LEMMA ($\mathbf{TC}_1$). $(\exists X^*, Y^*)(X^* \in^* Y^*)$.
(Axiom (F1) in the new sense.)

Proof. Since $\mathbf{R}$ is nonempty there exists some $\langle x, y\rangle \in \mathbf{R}$ and hence $x \in \mathbf{Ext_R}(y)$. The extensions of both x and y are saturated, since $\mathbf{R}$ is extensional; hence $\mathbf{Ext_R}(x) \in^* \mathbf{Ext_R}(y)$.

1139. LEMMA ($\mathbf{TC}_1$). $(\forall X^*, Y^*)(X^* = Y^* \equiv (\forall Z^*)(Z^* \in^* X^* \equiv Z^* \in^* Y^*))$. (Axiom (F2) in the new sense.)

Proof. The implication from left to right is trivial; hence suppose that $X^* \neq Y^*$, e.g. that $X^* - Y^* \neq 0$. Let $z \in X^*$ and $z \notin Y^*$; we have $\mathbf{Ext_R}(z) \in^* X^*$; if $\mathbf{Ext_R}(z) \in^* Y^*$, then $\mathbf{Ext_R}(z) = \mathbf{Ext_R}(u)$ for some $u \in Y^*$, so that $z \in Y^*$ since Y^* is saturated; this is a contradiction, so we have $Z^* \in^* X^*$ and $Z^* \notin^* Y^*$, where $Z^* = \mathbf{Ext_R}(z)$. Hence $\neg(\forall Z^*)(Z^* \in^* X^* \equiv Z^* \in^* Y^*)$.

1140. DEFINITION ($\mathbf{TC}_1$).

$$M^*(X^*) \equiv (\exists Y^*)(X^* \in^* Y^*) .$$

(Definition (d1) in the new sense.)

1141. LEMMA ($\mathbf{TC}_1$). $M^*(X^*) \equiv (\exists z \in \mathbf{C}(\mathbf{R}))(X^* = \mathbf{Ext_R}(z))$.

Proof. The implication from left to right follows from the definition of $\in^*$; suppose that $X^* = \mathbf{Ext_R}(z)$ for some $z \in \mathbf{C}(\mathbf{R})$. Since $\mathbf{C}(\mathbf{R})$ is saturated we have $X^* \in^* \mathbf{C}(\mathbf{R})$ and $(\exists Y^*)(X^* \in^* Y^*)$; hence $M^*(X^*)$.

1142. DEFINITION ($\mathbf{TC}_1$).

$$(\forall X^*)[(\exists x^*)(x^* = X^*) \equiv M^*(X^*)] \,\&\, (\forall x^*)(\exists X^*)(x^* = X^*)$$

1143. LEMMA ($\mathbf{TC}_1$).

$$(\forall x^*, y^*)(\exists z^*)(\forall u^*)(u^* \in^* z^* \equiv . \, u^* = x^* \vee u^* = y^*) .$$

(Axiom A1 in the new sense.)

Proof. Suppose that $x^* = \mathbf{Ext_R}(p)$, $y^* = \mathbf{Ext_R}(q)$ and let z be such that $\mathbf{Ext_R}(z) = \mathbf{SAT_R}(\{p, q\})$ (such a z exists since $\mathbf{R}$ is pairing). For $z^* = \mathbf{Ext_R}(z)$ we have $u^* \in^* z^* \equiv . \, u^* = x^* \vee u^* = y^*$.

1144. DEFINITION ($\mathbf{TC}_1$).

$$z^* \in^* \{x^*, y^*\}^* \equiv . \, z^* = x^* \vee z^* = y^* .$$

1145. METADEFINITION. If φ is a formula of $\mathbf{TC}$ then let φ^* be the formula which is obtained by adding asterisks to all predicates other than $=$ and to all operations and variables; i.e. we replace $\in$ by $\in^*$, X by X^*, x by x^*, M by M* etc.

To prove (B1)*, …, (B7)* in $\mathbf{TC}_1$ we need the following

1146. METALEMMA. If $\varphi(x_1, \bullet, X_1, \bullet)$ is a NF then

$$(*) \qquad (z_1 \in \mathbf{C}(\mathbf{R}) \,\&\, \bullet) \to \varphi^*(\mathbf{Ext}_{\mathbf{R}}(z_1), \bullet, X_1^*, \bullet)$$

is normal in $\mathbf{TC}_1$.

(The formula $\varphi^*(\mathbf{Ext}_{\mathbf{R}}(z_1), \bullet, X_1^*, \bullet)$ is the result of substituting of $\mathbf{Ext}_{\mathbf{R}}(z_1), \bullet$ for $x_1^*, \bullet$ in $\varphi^*(x_1^*, \bullet, X_1^*, \bullet)$. $x_1, \bullet$ is an abbreviation for a finite sequence of variables, say, $x_1, \ldots, x_n$. Analogously in other cases.)

Demonstration. For each NF φ we shall find a formula ψ normal in $\mathbf{TC}_1$ and such that

$$(z_1 \in \mathbf{C}(\mathbf{R}) \,\&\, \bullet) \to [\varphi^*(\mathbf{Ext}_{\mathbf{R}}(z_1), \bullet, X_1^*, \bullet) \equiv \psi(z_1, \bullet, X_1^*, \bullet, \mathbf{R})]$$

is provable in $\mathbf{TC}_1$. Then the formula $(*)$ is equivalent to

$$(z_1 \in \mathbf{C}(\mathbf{R}) \,\&\, \bullet) \to \psi(z_1, \bullet, X_1^*, \bullet, \mathbf{R})$$

which is normal in $\mathbf{TC}_1$. If φ is $x_1 \in x_2$ take $\langle z_1, z_2\rangle \in \mathbf{R}$ for ψ; if φ is $x_1 \in X_1$ take $z_1 \in X_1^*$ for ψ. The induction step for & and $\neg$ presents no difficulties. Let $\varphi(y, x_1, \bullet, X_1, \bullet)$ be a NF and suppose

$$(w \in \mathbf{C}(\mathbf{R}) \,\&\, z_1 \in \mathbf{C}(\mathbf{R}) \,\&\, \bullet) \to$$
$$\to [\varphi^*(\mathbf{Ext}_{\mathbf{R}}(w), \mathbf{Ext}_{\mathbf{R}}(z_1), \bullet, X_1^*, \bullet) \equiv \psi(w, z_1, \bullet, X_1^*, \bullet, \mathbf{R})]$$

is provable in $\mathbf{TC}_1$ where ψ is normal in $\mathbf{TC}_1$. Then

$$(z_1 \in \mathbf{C}(\mathbf{R}) \,\&\, \bullet) \to [(\exists y^*)\, \varphi^*(y^*, \mathbf{Ext}_{\mathbf{R}}(z_1), \bullet, X_1^*, \bullet) \equiv$$
$$\equiv (\exists w \in \mathbf{C}(\mathbf{R}))\, \psi(w, z_1, \bullet, X_1^*, \bullet, \mathbf{R})]$$

is provable in $\mathbf{TC}_1$ and the formula $(\exists w \in \mathbf{C}(\mathbf{R}))\, \psi(w, z_1, \bullet, X_1^*, \bullet, \mathbf{R})$ is normal in $\mathbf{TC}_1$.

1147. Remark. The normality of the formula $\langle z_1, z_2\rangle \in \mathbf{R}$ in $\mathbf{TC}_1$ can be established as in the demonstration of 1126.

1148. METATHEOREM. (B1)* to (B7)* are provable in $\mathbf{TC}_1$.

Demonstration. a) (B1)*. We can prove in $\mathbf{TC}_1$ that $\mathbf{C}(\mathbf{R})$ is saturated and that $(\forall x^*)\,(x^* \in^* \mathbf{C}(\mathbf{R}))$; hence $(\exists Z^*)\,(\forall x^*)\,(x^* \in^* Z^*)$.

b) (B2)* – (B7)*. Axioms (B2) to (B7) have the form $(\forall X_1, \ldots)(\exists Z)(\forall x)(x \in Z \equiv \varphi_i(x, X_1, \bullet))$ where the φ_i are normal in **TC**. Hence we may argue as follows in $\mathbf{TC}_1$: For any X_1^*, $\bullet$ let $z \in Z \equiv . z \in \mathbf{C}(\mathbf{R})\,\&$ $\&\ \varphi_i^*(\mathbf{Ext_R}(z), X_1^*, \bullet)$; such a class Z exists in view of 1124 and 1146.

From the defining property of Z we see that if $\mathbf{Ext_R}(z) = \mathbf{Ext_R}(w)$, then $z \in Z$ implies $w \in Z$; hence Z is saturated.

Thus we may set $Z^* = Z$, so that $x^* \in^* Z^* \equiv \varphi_i^*(x^*, X_1^*, \ldots)$. We have the following

1149. METATHEOREM. For any axiom φ of **TC**, φ^* is provable in $\mathbf{TC}_1$.

1150. We see therefore that the sort X^* and the predicate $\in^*$ furnish a model of **TC** in $\mathbf{TC}_1$. We assigned to each formula φ of **TC** its counterpart φ^* (the same formula in a new sense). Evidently [!] the counterparts of provable formulas are provable; for we have shown that the axioms hold in the new sense and hence each proof may be performed in the new sense. In this way we obtain a number of provable statements. Notice that if φ is a formula of **TC** then φ^* tells us something about a fixed **E**-like relation **R**.

If we carry out the development in some extension **T** of the theory $\mathbf{TC}_1$, everything proceeds in the same way: we introduce the asterisked concepts and prove in **T** all asterisked axioms of **TC**. We may of course impose on **R** stronger conditions than those mentioned above; everything proceeds in the same way as before, provided that we guarantee (in **T**) the existence of some **R** which satisfies these requirements. It may turn out that, in addition to the asterisked axioms of **T**, we can also prove in **T** some φ^* for which it is not known whether φ is provable in **T**. It may then be asserted that if **T** is consistent [!], then it remains consistent on the addition of φ as a new axiom. For if a contradiction would be proved in this extended theory, then an analogous proof could be carried out for the asterisked concepts, and so a contradiction could be proved in the theory **T** (in which φ^* is provable).

This example gives us an indication of what models are and of the significance of constructins models; the construction of a model serves at least as a means for generating new statements (and notions) and as a method for providing relative consistency proofs. The next section will provide a formal foundation for the developments carried out so far; however the results of the present section serve not only as an illustration but play an important role throughout the book. After reading Section 2 the reader may return to Section 1; he will then see that the present section is consistent with the formal conception to be presented in Section 2.

SECTION 2

Logical foundations

1201. In this section we shall provide a formal foundation for certain informal considerations, including those of the preceding section. We shall analyse precisely various concepts such as "theorem", "proof", "definition", "model"; thus we shall not take these notions for granted but shall derive them from simpler concepts and so reduce to a minimum the number of notions which are taken to be intuitively known. In doing this, we shall try to observe two requirements: the requirement of exactness – we shall for example define precisely and in an entirely abstract way the notion of a model – and, on the other hand, the requirement of adequacy – we shall require that the defined notions adequately represent the intuitive notions. It turns out that essentially, it suffices to take as a basis for our considerations the notion of symbols and their finite configurations. The developments of Section 1 serve as an illustration; on the other hand, these developments satisfy the formal requirements of the present section.

In what follows, a large number of important concepts will be introduced. To avoid confusion we shall first indicate the most important of these. Firstly, the notion of an *axiomatic theory* is defined; each theory has its *language* (basic concepts) and *axioms* (concerning these concepts). Further, we define what it is for a statement of the language (a formula) to be *provable* in the theory. A *model* of one theory in the other assigns to each formula of the first theory a formula of the second theory in such a way that provability is preserved (in a certain well-defined sense). The concept of a *faithful model* is of special importance; a few criteria for models to be faithful will be given. *Direct models* are a very simple type of model which so to speak only change the notation of basic concepts. *Definitions* are axioms of a certain type which introduce a new basic concept; it follows that definitions do not enrich the theory, i.e. a definitional extension of a theory does not make any new formulas of the original theory provable. The theory of classes and other theories investigated in this book are built up from two concepts – membership relation and class – by adding axioms which either do not enrich the language or are definitions (or "quasi-definitions"). Such theories are called *∈-theories.* It turns out that for any ∈-theory **T**, it suffices to "model" (in some theory **S**) only the basic concepts mentioned; this already induces a direct model of **T** in a certain definitional extension of **S**.

We shall now carry out in detail the programme outlined above. If the notions informally explained in the outline are already sufficiently clear to the reader, he may omit the rest of the section, returning as occasion demands.

a) *Formulas, Axioms, Theories*

1202. The *symbols* which will appear in our work are the following:

a) *Variables* of various *sorts* (e.g. *class-variables* $X_1, X_2, \ldots$, *set-variables* $x_1, x_2, \ldots$); a sort is assumed to be an integer; for each pair of integers p and q there is a unique variable which is called the q-th variable of sort p.

b) *Predicates* of various *arity* (e.g. the unary (1-ary) predicate **M**, "... is a set", the binary (2-ary) predicate ∈ "... is an element of ..."); for each pair of integers $p > 0$ and q, there is a unique predicate called the q-th p-ary predicate.

c) *Operations* of various *arity* (e.g. the 0-ary operation **V**, "the universal class", the unary operation **D**, "the domain of ...", the binary operation ∩, "the intersection of ... and ..."); for each pair of integers p and q there is a unique operation, called the q-th p-ary operation; 0-ary operations are called *constants*.

d) *Connectives:* binary &, ∨, →, ≡ ("and", "or", "implies", "is equivalent to") and unary ¬ ("not").

e) *Quantifiers:* ∀, ∃ (universal, existential).

f) *Equality predicate* = : a particular binary predicate (different from all predicates sub b)).

g) *Parentheses* (,).

It is assumed further that all these symbols are distinct, that none of them is an integer and that for each symbol it can be decided to which of the groups a) to g) it belongs; if a symbol belongs to one of the groups a), b), c) then the corresponding integers p and q are uniquely determined.

1203. A *language* is any finite sequence of distinct predicates, operations and sorts; we assume that this sequence does not contain the equality predicate and that it contains at least one sort. In referring to the language we shall often represent a sort by some variable of this sort, e.g. by the first one. For simplicity it is sometimes assumed that each predicate precedes each operation and each operation precedes each sort.

A language $\mathbf{L}_1$ is a *sublanguage* of a language $\mathbf{L}_2$ if each predicate (operation, sort) of $\mathbf{L}_1$ is a predicate (operation, sort) of $\mathbf{L}_2$.

Two languages $\mathbf{L}_1$, $\mathbf{L}_2$ are *similar* if they have the same length and if, for each integer p, the p-th symbol in $\mathbf{L}_1$ is a k-ary predicate (k-ary operation, sort) if and only if the p-th symbol in $\mathbf{L}_2$ is a k-ary predicate (k-ary operation, sort). For example, the language $(\in, X_1)$ is a sublanguage of $(\in, \subseteq, X_1, x_1)$; the languages $(\in, X_1)$ and $(\subseteq, x_1)$ are similar.

Now we define recursively the concepts of term and formula of a language.

1204. (a) Each variable of each sort in $\mathbf{L}$ is an *atomic term* of the language $\mathbf{L}$; each constant in $\mathbf{L}$ is an *atomic term* of $\mathbf{L}$.

(b) If $\mathbf{F}$ is a p-ary operation in the language $\mathbf{L}$ and $\mathbf{t}_1, \ldots, \mathbf{t}_p$ are terms of $\mathbf{L}$, then the configuration $\mathbf{F}(\mathbf{t}_1, \ldots, \mathbf{t}_p)$ is a *term* of $\mathbf{L}$.

(c) There are no other terms than those obtained from atomic terms by applying the rule (b) finitely many times.

1205. (a) If $\mathbf{t}_1$ and $\mathbf{t}_2$ are terms of $\mathbf{L}$ then the configuration $\mathbf{t}_1 = \mathbf{t}_2$ is an *atomic formula* of the language $\mathbf{L}$; each variable occurring in this formula is *free* and no variable is *bound* in this formula. If P is a p-ary predicate in $\mathbf{L}$ and $\mathbf{t}_1, \ldots, \mathbf{t}_p$ are terms (of $\mathbf{L}$) then the configuration $\mathrm{P}(\mathbf{t}_1, \ldots, \mathbf{t}_p)$ is an *atomic formula* of $\mathbf{L}$; each variable occurring in this formula is *free* in $\mathrm{P}(\mathbf{t}_1, \ldots, \mathbf{t}_p)$ and no variable is *bound* in this formula.

(b) If φ and ψ are formulas of $\mathbf{L}$ and if no variable is simultaneously free in φ and bound in ψ or conversely, then each of the following configurations is a formula of $\mathbf{L}$:

$$(\varphi \,\&\, \psi), \ (\varphi \vee \psi), \ (\varphi \to \psi), \ (\varphi \equiv \psi);$$

a variable is *free* (*bound*) in such a formula if it is free (bound) either in φ or in ψ.

(c) If φ is a formula of **L** then so is $\neg\varphi$; a variable is *free* (*bound*) in $\neg\varphi$ if it is free (bound) in φ.

(d) If φ is a formula of **L** and x is a variable of a sort in **L** and if x is not bound in φ, then $(\forall x)\,\varphi$ and $(\exists x)\,\varphi$ are formulas of **L**; x is *bound* in this formula and each variable other than x is *free* (*bound*) in this formula, it if is free (bound) in φ.

1206. Remark. If P is a binary predicate and if $\mathbf{t}_1$ and $\mathbf{t}_2$ are terms, then we often write $\mathbf{t}_1\ \mathrm{P}\ \mathbf{t}_2$ instead of $\mathrm{P}(\mathbf{t}_1, \mathbf{t}_2)$ (e.g. $\mathbf{t}_1 \in \mathbf{t}_2$ instead of $\in(\mathbf{t}_1, \mathbf{t}_2)$); similarly for binary operations (we write $X \cap Y$ instead of $\cap(X, Y)$).

1207. A formula is *closed* if no variable is free in it. We shall use lower-case Greek letters $\varphi, \psi, \ldots$ to denote formulas; if φ is a formula such that $x, \ldots, y$ include all variables free in φ then φ may be written as $\varphi(x, \ldots, y)$.

1208. Remark. The notion of a formula (in particular, a closed formula) adequately represents the intuitive notion of "statement" or "proposition". We shall now define the notions of proof and provability. We follow here the natural intuitive idea of a proof; i.e. a succession (sequence) of statements which are either evident or evidently follow from their predecessors.

1209. LOGICAL AXIOMS.

I. If φ, ψ, χ are formulas then the following configurations are axioms, provided they are formulas

$$\begin{aligned}
&\varphi \to (\psi \to \varphi)\\
&(\varphi \to (\psi \to \chi)) \to ((\varphi \to \psi) \to (\varphi \to \chi))\\
&(\neg\varphi \to \neg\psi) \to (\psi \to \varphi)\\
&(\varphi \equiv \psi) \to (\varphi \to \psi)\\
&(\varphi \equiv \psi) \to (\psi \to \varphi)\\
&(\varphi \to \psi) \to ((\psi \to \varphi) \to (\varphi \equiv \psi))\\
&(\varphi \vee \psi) \equiv (\neg\varphi \to \psi)\\
&(\varphi \,\&\, \psi) \equiv \neg(\neg\varphi \vee \neg\psi)\,.
\end{aligned}$$

II. a) If φ is a formula and x, y are variables of the same sort which are not bound in φ, then

$$(\forall x)\,\varphi(x) \to \varphi(y)$$

is an axiom ($\varphi(y)$ is the result of substituting y for all occurrences of x in φ).

b) If φ, ψ are formulas then the configuration

$$(\forall x)(\varphi \to \psi) \to (\varphi \to (\forall x)\psi)$$

is an axiom (provided it is a formula; i.e. provided x does not occur in φ).

c) If $(\exists x)\varphi$ is a formula then

$$(\exists x)\varphi \equiv \neg(\forall x)\neg\varphi$$

is an axiom.

III. If $\mathbf{t}, \mathbf{s}, \mathbf{r}, \mathbf{t}_1, \mathbf{s}_1, \ldots$ are terms, while P is a predicate and **F** is an operation, then the following formulas are axioms:

$$\begin{aligned}
&\mathbf{t} = \mathbf{t}\\
&\mathbf{t} = \mathbf{s} \to \mathbf{s} = \mathbf{t}\\
&(\mathbf{t} = \mathbf{s} \,\&\, \mathbf{s} = \mathbf{r}) \to \mathbf{t} = \mathbf{r}\\
&(\mathbf{t}_1 = \mathbf{s}_1 \,\&\, \bullet) \to (\mathrm{P}(\mathbf{t}_1, \bullet) \equiv \mathrm{P}(\mathbf{s}_1, \bullet))\\
&(\mathbf{t}_1 = \mathbf{s}_1 \,\&\, \bullet) \to \mathbf{F}(\mathbf{t}_1, \bullet) = \mathbf{F}(\mathbf{s}_1, \bullet)
\end{aligned}$$

(if we suppose that P and **F** are p-ary then $\mathbf{t}_1 = \mathbf{s}_1 \,\&\, \bullet$ is an abbreviation for $\mathbf{t}_1 = \mathbf{s}_1 \,\&\, \ldots \,\&\, \mathbf{t}_p = \mathbf{s}_p$, and similarly, $(\mathbf{t}_1, \bullet)$ is an abbreviation for $(\mathbf{t}_1, \ldots, \mathbf{t}_p)$).

1210. RULES OF INFERENCE

I. *Detachment Rule.* Let $\varphi \to \psi$ be a formula. Infer ψ from φ and $(\varphi \to \psi)$.

II. *Generalization Rule.* Let $(\forall x)\varphi$ be a formula. Infer $(\forall x)\varphi$ from φ.

1211. AXIOM SYSTEM; AXIOMATIC THEORY

If **L** is a language then any finite (possibly empty) sequence of formulas of **L** is called an *axiom system* in **L**. An (axiomatic) theory **T** consists of a language **L** and of an axiom system in **L**. The language **L** is called the *language of the theory* **T** and the members of the axiom system are called the (*nonlogical*) *axioms of* **T**. Instead of "a formula of the language of the theory **T**" we often say briefly "a formula of the theory **T**" or "a **T**-formula".

1212. If **T** is a theory then a sequence $\varphi_1, \ldots, \varphi_p$ of **T**-formulas is a *proof* in the theory **T** if for each $i = 1, \ldots, p$, φ_i is either a logical axiom or an axiom of **T** or is inferred from some of the formulas $\varphi_1, \ldots, \varphi_{i-1}$ by a rule of inference. A **T**-formula φ is *provable* in **T**

$$(\mathbf{T} \vdash \varphi)$$

if there exists a proof in $\mathbf{T}$ which has φ as its last element.

Example. Let $\mathbf{t}$ be a term, let x be a variable of a sort s and let $\varphi(x)$ be a formula such that no variable occurring in $\mathbf{t}$ is bound in φ. Denote by $\varphi(\mathbf{t})$ the result of substituting $\mathbf{t}$ for all occurrences of x in φ. Let y be a variable of the sort s and distinct from all variables occurring in $\mathbf{t}$. The formula $(\exists y)(\mathbf{t} = y)$ can be read "$\mathbf{t}$ is of the sort s". It is a matter of routine to show that the formula

$$(*) \qquad ((\forall x)\, \varphi(x) \,\&\, (\exists y)(\mathbf{t} = y)) \rightarrow \varphi(\mathbf{t})$$

is provable in each theory $\mathbf{T}$ such that $(*)$ is a $\mathbf{T}$-formula.

The metamathematical definition stated above constitutes one possible formalization of the notion of provability. Since the number of accepted "evident statements" and "evident inferences" is minimized, the formal proofs are very lengthy; however, the methods used in informal proofs are mirrored by a number of metatheorems on formal provability; this convinces us of the following

1213. Adequacy Hypothesis:

The formal notion of provability adequately represents the intuitive notion, i.e. a mathematical proof is correct if and only if each of its parts can be formalized.

We shall use this hypothesis in two ways. Firstly in proving some mathematical statement (e.g. in the set theory) we shall proceed informally (in words); having proved the statement informally we deduce that a formal proof exists and that it would be only a matter of time to find it. On the other hand we shall demonstrate some metatheorems on formal provability; we deduce from them the rules for informal mathematical proof.

1214. Variables of a sort i are denoted (if necessary) by x^i, y^i, x_1^i etc. If $\mathbf{T}$ is a theory and i, j are sorts in its language, then i is *subordinate* to j in $\mathbf{T}$ if $\mathbf{T} \vdash (\forall x^i)(\exists y^j)(x^i = y^j)$. The sort i is *universal* in $\mathbf{T}$ if for each term $\mathbf{t}$ of $\mathbf{T}$ and each variable x^i not occurring in $\mathbf{t}$ we have

$$\mathbf{T} \vdash (\exists x^i)(x^i = \mathbf{t})\,;$$

it follows that all sorts in the language of $\mathbf{T}$ are subordinate to i.

1215. LEMMA. Every sort is subordinate to itself; if a sort i is subordinate to a sort j and j is subordinate to a sort k, then i is subordinate to k. (Obvious.)

1216. LEMMA. Let **T** be a theory and let the sort i be subordinate to the sort j in **T**; let $\varphi(y^j)$ be a **T**-formula and suppose that x^i is not bound in φ. Then $\mathbf{T} \vdash (\forall y^j)\, \varphi(y^j) \to (\forall x^i)\, \varphi(x^i)$. (Exercise.)

1217. If **T** and **S** are theories, then **T** is *stronger* than **S** if every **S**-formula provable in **S** is provable in **T**.

(Corollary: the language of **S** is a sublanguage of the language of **T**, i.e. each symbol of the language of **S** is in the language of **T**.) In particular, **T** is an *extension* of **S** if the sequence of axioms of **S** is an initial segment of the sequence of axioms of **T**. The extension of **S** by an axiom φ is denoted by $(\mathbf{S}, \varphi)$ or $(\mathbf{S} + \varphi)$ or simply $\mathbf{S}, \varphi$. **T** is *equivalent* to **S** if **T** is stronger than **S** and **S** is stronger than **T**.

1218. METATHEOREM (on deduction). Let **T** be a theory. If $\varphi \to \psi$ is a formula and φ is closed, then

$$\mathbf{T} \vdash \varphi \to \psi \quad \text{iff} \quad \mathbf{T}, \varphi \vdash \psi \,.$$

Demonstration. If $\mathbf{T} \vdash \varphi \to \psi$ then $\mathbf{T}, \varphi \vdash \varphi \to \psi$ and $\mathbf{T}, \varphi \vdash \varphi$; hence $\mathbf{T}, \varphi \vdash \psi$ by Detachment Rule. Conversely, if $\mathbf{T}, \varphi \vdash \psi$ then one can verify $\mathbf{T} \vdash \varphi \to \psi$ by induction on the length of the proof of φ. (Technicalities omitted.)

b) *Models*

1219. Let **T** and **S** be theories. A mapping*) $\mathfrak{M}$ which assigns to each **T**-formula φ an **S**-formula $\varphi^{\mathfrak{M}}$ is called a (*syntactic*) *model* of **T** in **S** if the following conditions hold:

1) $\mathfrak{M}$ *preserves axioms*; i.e.

$$\mathbf{S} \vdash \varphi^{\mathfrak{M}}$$

whenener φ is a logical axiom or an axiom of **T**,

2) $\mathfrak{M}$ *preserves implication, quantifiers and negation*, i.e., for any **T**-formulas φ and ψ and variable x,

*) When we speak of a mapping in metamathematics we assume always that the value to be assigned to each argument is prescribed explicitly.

$$\mathbf{S}, (\varphi \to \psi)^{\mathfrak{M}}, \varphi^{\mathfrak{M}} \vdash \psi^{\mathfrak{M}},$$
$$\mathbf{S}, \varphi^{\mathfrak{M}} \vdash ((\forall x)\, \varphi)^{\mathfrak{M}},$$
$$\mathbf{S}, (\neg \varphi)^{\mathfrak{M}} \vdash \neg (\forall \ldots)(\varphi^{\mathfrak{M}}).$$

(We suppose that $(\varphi \to \psi)$ and $(\forall x)\, \varphi$ are formulas.)

1220. **T** is *contradictory* if there exists a **T**-formula φ such that both φ and $\neg \varphi$ are provable in **T**. **T** is *consistent* if it is not contradictory.

1221. LEMMA. **T** is contradictory if and only if every **T**-formula is provable in **T**. (Exercise.)

1222. The Provability and Consistency Principles. Let $\mathfrak{M}$ be a model of **T** in **S**.

1) For any **T**-formula φ, $\mathbf{T} \vdash \varphi$ implies $\mathbf{S} \vdash \varphi^{\mathfrak{M}}$; moreover, for any **T**-formula ψ, $\mathbf{T}, \psi \vdash \varphi$ implies $\mathbf{S}, \psi^{\mathfrak{M}} \vdash \varphi^{\mathfrak{M}}$. (The Provability Principle.)

2) If **T** is contradictory then so is **S**; i.e. if **S** is consistent then so is **T**. (The Consistency Principle.) These are simple but fundamental principles.

Demonstration. 1) Let $\varphi_1, \ldots, \varphi_n$ be a proof of φ in $(\mathbf{T}, \psi)$; we show by induction that $\mathbf{S}, \psi^{\mathfrak{M}} \vdash \varphi^{\mathfrak{M}}$. If φ_i is a logical axiom, an axiom of **T** or the formula ψ, then $\mathbf{S}, \psi^{\mathfrak{M}} \vdash \varphi_i^{\mathfrak{M}}$ by the definition of a model. If $\varphi_1^{\mathfrak{M}}, \ldots, \varphi_{i-1}^{\mathfrak{M}}$ have been proved in $(\mathbf{S}, \psi^{\mathfrak{M}})$ and φ_i immediately follows from φ_j, φ_k by the Detachment Rule φ_k being $\varphi_j \to \varphi_i$, $(j, k < i)$ then $\mathbf{S}, \psi^{\mathfrak{M}} \vdash \varphi_i^{\mathfrak{M}}$, since $\mathfrak{M}$ preserves implication; if φ_i immediately follows from φ_j by the Generalization Rule (φ_i being $(\forall x)\, \varphi_j$, $j < i$) then $\mathbf{S}, \psi^{\mathfrak{M}} \vdash \varphi_i^{\mathfrak{M}}$ since $\mathfrak{M}$ preserves the quantifier.

2) Suppose $\mathbf{T} \vdash \varphi$, $\mathbf{T} \vdash \neg\varphi$. Then, by 1), $\mathbf{S} \vdash \varphi^{\mathfrak{M}}$ and $\mathbf{S} \vdash (\neg\varphi)^{\mathfrak{M}}$; hence $\mathbf{S} \vdash (\forall \ldots)(\varphi^{\mathfrak{M}})$ and $\mathbf{S} \vdash \neg(\forall \ldots)(\varphi^{\mathfrak{M}})$, since $\mathfrak{M}$ preserves the negation.

Unprovability Principle. Let $\mathfrak{M}$ be a model of **T** in **S**, let φ be a **T**-formula such that $\mathbf{S} \vdash \neg(\varphi^{\mathfrak{M}})$. If **S** is consistent then φ is not provable in **T**.

Indeed, if $\mathbf{T} \vdash \varphi$ then we would have $\mathbf{S} \vdash \varphi^{\mathfrak{M}}$ and so **S** would be contradictory.

Remark. If $(\neg\varphi)^{\mathfrak{M}}$ is **S**-provable then $\mathfrak{M}$ is a model of $(\mathbf{T}, \neg\varphi)$ in **S** and, if **S** is consistent, φ cannot be **T**-provable by the Consistency Principle. But it is possible that $\neg(\varphi^{\mathfrak{M}})$ is **S**-provable and $(\neg\varphi)^{\mathfrak{M}}$ is not; moreover, it is possible that both $\neg(\varphi^{\mathfrak{M}})$ and $\neg((\neg\varphi)^{\mathfrak{M}})$ are **S**-provable. In these cases we obtain the unprovability of φ in **T** by the unprovability principle. (Note that in most cases we are then able to construct another model $\mathfrak{N}$ of $(\mathbf{T}, \neg\varphi)$ in **S**.)

1223. LEMMA. If $\mathfrak{M}$ is a model of $\mathbf{T}$ in $\mathbf{S}$ and φ is a $\mathbf{T}$-formula, then $\mathfrak{M}$ is a model of $(\mathbf{T}, \varphi)$ in $(\mathbf{S}, \varphi^{\mathfrak{M}})$.

1224. Let $\mathfrak{M}$ be a model of $\mathbf{T}$ in $\mathbf{S}$ and let φ be a $\mathbf{T}$-formula; we say that φ *holds* (*is valid*) *in* $\mathfrak{M}$ if $\mathbf{S} \vdash \varphi^{\mathfrak{M}}$.

1225. Let $\mathbf{T}$ be a theory. The mapping which assigns to each $\mathbf{T}$-formula φ the formula φ itself is called the *identical model of* $\mathbf{T}$ and is denoted by $\mathfrak{Id}_{\mathbf{T}}$.

1226. Let $\mathbf{T}$, $\mathbf{S}$, $\mathbf{U}$ be theories; suppose that $\mathfrak{M}_1$ is a model of $\mathbf{T}$ in $\mathbf{S}$ and that $\mathfrak{M}_2$ is a model of $\mathbf{S}$ in $\mathbf{U}$. The mapping which assigns to each $\mathbf{T}$-formula φ the $\mathbf{U}$-formula $(\varphi^{\mathfrak{M}_1})^{\mathfrak{M}_2}$ is a model of $\mathbf{T}$ in $\mathbf{U}$; it is called the *composition of the models* $\mathfrak{M}_1$, $\mathfrak{M}_2$ or the model $\mathfrak{M}_1$ *constructed within the model* $\mathfrak{M}_2$; it is denoted by $\mathfrak{M}_1 * \mathfrak{M}_2$.

1227. We see that theories and models behave as objects and morphisms in the sense of category theory. We shall not study categorical properties of syntactic models here; but sometimes we shall use diagrams to describe relations among various models. E.g. Lemma 1223 means that in the diagram

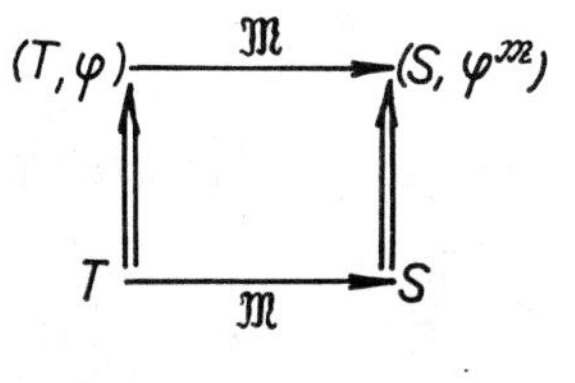

the upper arrow is a model (and obviously the diagram is commutative). Double arrows denote identity mappings.

1228. LEMMA. Let $\mathfrak{M}$ be a model of $\mathbf{T}$ in $\mathbf{S}$ and let $\mathfrak{N}$ be a model of $\mathbf{S}$ in $\mathbf{U}$. If a $\mathbf{T}$-formula φ is valid in $\mathfrak{M}$ then it is also valid in $\mathfrak{M} * \mathfrak{N}$.

Demonstration. Let φ hold in $\mathfrak{M}$, i.e. $\mathbf{S} \vdash \varphi^{\mathfrak{M}}$. Then $\mathbf{U} \vdash (\varphi^{\mathfrak{M}})^{\mathfrak{N}}$, i.e. φ holds in $\mathfrak{M} * \mathfrak{N}$.

1229. Let $\mathbf{T}$ and $\mathbf{S}$ be theories and let $\mathfrak{M}_1$ and $\mathfrak{M}_2$ be models of $\mathbf{T}$ in $\mathbf{S}$. We say that $\mathfrak{M}_1$ is *weaker than* $\mathfrak{M}_2$ ($\mathfrak{M}_1 \prec \mathfrak{M}_2$)*) if for any $\mathbf{T}$-formula φ, $\varphi^{\mathfrak{M}_2}$ is provable in $(\mathbf{S}, \varphi^{\mathfrak{M}_1})$. $\mathfrak{M}_1$ is *equivalent* to $\mathfrak{M}_2$, if $\mathfrak{M}_1 \prec \mathfrak{M}_2$ and $\mathfrak{M}_2 \prec \mathfrak{M}_1$.

*) To be exact, we should say that $\mathfrak{M}_1$ as a model of $\mathbf{T}$ in $\mathbf{S}$ is weaker than $\mathfrak{M}_2$ since $\mathfrak{M}_1$ and $\mathfrak{M}_2$ can be models of other theories as well; see the following lemma.

1230. LEMMA. a) $\mathbf{S}$ is stronger than $\mathbf{T}$ iff the identical mapping $\mathfrak{Id}_{\mathbf{T}}$ of $\mathbf{T}$-formulas is a model of $\mathbf{T}$ in $\mathbf{S}$.

b) Let $\mathfrak{M}$ be a model of $\mathbf{T}$ in $\mathbf{S}$ and let $\hat{\mathbf{S}}$ be stronger than $\mathbf{S}$; then $\mathfrak{M}$ is a model of $\mathbf{T}$ in $\mathbf{S}$.

c) Let $\mathfrak{M}$ be a model of $\mathbf{T}$ in $\mathbf{S}$ and let $\mathbf{T}$ be stronger than $\mathbf{T}_0$; then the restriction of $\mathfrak{M}$ to $\mathbf{T}_0$-formulas is a model of $\mathbf{T}_0$ in $\mathbf{S}$.

d) Let $\mathfrak{M}_1$, $\mathfrak{M}_2$ be models of $\mathbf{T}$ in $\mathbf{S}$ and let $\mathfrak{M}_1 \prec \mathfrak{M}_2$; then each $\mathbf{T}$-formula valid in $\mathfrak{M}_1$ is valid in $\mathfrak{M}_2$.

e) Let $\mathfrak{M}_1$ and $\mathfrak{M}_2$ be models of $\mathbf{T}$ in $\mathbf{S}$ such that $\mathfrak{M}_1 \prec \mathfrak{M}_2$; if φ is a $\mathbf{T}$-formula then $\mathfrak{M}_1$ as a model of $(\mathbf{T}, \varphi)$ in $(\mathbf{S}, \varphi^{\mathfrak{M}_1})$ is weaker than $\mathfrak{M}_2$ as a model of $(\mathbf{T}, \varphi)$ in $(\mathbf{S}, \varphi^{\mathfrak{M}_1})$.

The demonstrations follow from the definitions; e.g. $\mathfrak{M}$ as mapping of $\mathbf{T}$-formulas into $\hat{\mathbf{S}}$-formulas is the composition of two arrows in the diagram

$$T \xrightarrow[\mathfrak{M}]{} S \Longrightarrow \hat{S}$$

where both arrows are models; hence $\mathfrak{M}$ is a model of $\mathbf{T}$ in $\hat{\mathbf{S}}$.

1231. Let $\mathfrak{M}$ be a model of $\mathbf{T}$ in $\mathbf{S}$ and $\mathfrak{N}$ a model of $\mathbf{S}$ in $\mathbf{T}$. We say that $\mathfrak{N}$ is an *inverse* of $\mathfrak{M}$ if $\mathfrak{M} * \mathfrak{N}$ is equivalent to the identical model $\mathfrak{Id}_{\mathbf{T}}$. Theories $\mathbf{T}$, $\mathbf{S}$ are *isomorphic* if there are models $\mathfrak{M}$ (of $\mathbf{T}$ in $\mathbf{S}$) and $\mathfrak{N}$ (of $\mathbf{S}$ in $\mathbf{T}$) such that each is inverse to the other.

1232. Let $\mathfrak{M}$ be a model of $\mathbf{T}$ in $\mathbf{S}$. We say that $\mathfrak{M}$ is *faithful* if for any $\mathbf{T}$-formula φ, $\mathbf{S} \vdash \varphi^{\mathfrak{M}}$ implies $\mathbf{T} \vdash \varphi$. If, for every $\mathbf{T}$-formula φ, the model $\mathfrak{M}$ considered as a model of $(\mathbf{T}, \varphi)$ in $(\mathbf{S}, \varphi^{\mathfrak{M}})$ is faithful, then we say that $\mathfrak{M}$ is *essentially faithful.*

If a model $\mathfrak{M}$ of $\mathbf{T}$ in $\mathbf{S}$ is faithful, then only provable $\mathbf{T}$-formulas are valid in $\mathfrak{M}$ and hence instead of proving a $\mathbf{T}$-formula φ in $\mathbf{T}$ we may prove $\varphi^{\mathfrak{M}}$ in $\mathbf{S}$; in many cases this is easier. This notion is also of importance in proving theories to be consistent; for if $\mathfrak{M}$ is a model of $\mathbf{T}$ in $\mathbf{S}$ and $\mathbf{S}$ is consistent then so is $\mathbf{T}$. We may ask whether we obtain more by the construction of $\mathfrak{M}$, whether $\mathfrak{M}$ is a model of some extension $(\mathbf{T}, \varphi)$ of $\mathbf{T}$ where φ is unprovable in $\mathbf{T}$. The answer is negative if the model $\mathfrak{M}$ of $\mathbf{T}$ in $\mathbf{S}$ is faithful; the model $\mathfrak{M}$ yields us no other consistency results relative to $\mathbf{S}$; in other words, the model has been exploited to the full.

1233. LEMMA. Let $\mathfrak{M}$ be a model of $\mathbf{T}$ in $\mathbf{S}$, $\mathfrak{N}$ a model of $\mathbf{S}$ in $\mathbf{U}$.

(1) If $\mathfrak{M}$, $\mathfrak{N}$ are (essentially) faithful then $\mathfrak{M} * \mathfrak{N}$ is (essentially) faithful.

(2) If $\mathfrak{M} * \mathfrak{N}$ is (essentially) faithful then $\mathfrak{M}$ is (essentially) faithful.

Demonstration. (1) If $\mathbf{U} \vdash (\varphi^{\mathfrak{M}})^{\mathfrak{N}}$ and $\mathfrak{N}$ is faithful then $\mathbf{S} \vdash \varphi^{\mathfrak{M}}$; if $\mathfrak{M}$ is also faithful, we obtain $\mathbf{T} \vdash \varphi$. (2) If $\mathbf{S} \vdash \varphi^{\mathfrak{M}}$ then $\mathbf{U} \vdash (\varphi^{\mathfrak{M}})^{\mathfrak{N}}$ and therefore $\mathbf{T} \vdash \varphi$. Similarly for essentially faithful models.

1234. LEMMA. Let $\mathfrak{M}$, $\mathfrak{N}$ be models of $\mathbf{T}$ in $\mathbf{S}$. (1) If $\mathfrak{M} \prec \mathfrak{N}$ and $\mathfrak{N}$ is faithful then $\mathfrak{M}$ is also faithful. (2) If $\mathfrak{M}$ is equivalent to $\mathfrak{N}$ and $\mathfrak{N}$ is essentially faithful then $\mathfrak{M}$ is also essentially faithful.

Demonstration. (1) If φ is a $\mathbf{T}$-formula and $\mathbf{S} \vdash \varphi^{\mathfrak{M}}$ then $\mathbf{S} \vdash \varphi^{\mathfrak{N}}$ and so $\mathbf{T} \vdash \varphi$. (2) If φ, ψ are $\mathbf{T}$-formulas such that $\mathbf{S}, \varphi^{\mathfrak{M}} \vdash \psi^{\mathfrak{M}}$ then $\mathbf{S}, \varphi^{\mathfrak{N}} \vdash \varphi^{\mathfrak{M}}$, $\mathbf{S}, \varphi^{\mathfrak{M}} \vdash \psi^{\mathfrak{M}}$, $\mathbf{S}, \psi^{\mathfrak{M}} \vdash \psi^{\mathfrak{N}}$; hence $\mathbf{S}, \varphi^{\mathfrak{N}} \vdash \psi^{\mathfrak{N}}$ and $\mathbf{T}, \varphi \vdash \psi$ since $\mathfrak{N}$ is essentially faithful.

1235. LEMMA. Let $\mathfrak{M}$ be a model of $\mathbf{T}$ in $\mathbf{S}$ and let $\mathfrak{N}$ be an inverse of $\mathfrak{M}$; if φ is a $\mathbf{T}$-formula then $\mathfrak{N}$ is a model of $(\mathbf{S}, \varphi^{\mathfrak{M}})$ in $(\mathbf{T}, \varphi)$ and $\mathfrak{N}$ is an inverse of $\mathfrak{M}$ considered as a model of $(\mathbf{T}, \varphi)$ in $(\mathbf{S}, \varphi^{\mathfrak{M}})$.

Demonstration. To verify that $\mathfrak{N}$ is a model of $(\mathbf{S}, \varphi^{\mathfrak{M}})$ in $(\mathbf{T}, \varphi)$ it suffices to prove $(\mathbf{T}, \varphi) \vdash (\varphi^{\mathfrak{M}})^{\mathfrak{N}}$. Since $\mathfrak{M} * \mathfrak{N}$ is equivalent to $\mathfrak{Id}_{\mathbf{T}}$ we have $\mathbf{T}, \varphi \vdash (\varphi^{\mathfrak{M}})^{\mathfrak{N}}$. It remains to verify that for any $\mathbf{T}$-formula ψ we have $\mathbf{T}, \varphi, \psi \vdash \psi^{\mathfrak{M}*\mathfrak{N}}$ and $\mathbf{T}, \varphi, \psi^{\mathfrak{M}*\mathfrak{N}} \vdash \psi$; this follows from the fact that $\mathbf{T}, \psi^{\mathfrak{M}*\mathfrak{N}} \vdash \psi$ and $\mathbf{T}, \psi \vdash \psi^{\mathfrak{M}*\mathfrak{N}}$.

1236. LEMMA. Suppose that $\mathfrak{M}$ is a model of $\mathbf{T}$ in $\mathbf{S}$.

(1) If $\mathfrak{M}$ has an inverse then it is essentially faithful.

(2) If $\mathbf{T}$ is stronger than $\mathbf{S}$ and if the model $\mathfrak{M}$ considered as a model of $\mathbf{T}$ in $\mathbf{T}$ is (essentially) faithful then $\mathfrak{M}$ is (essentially) faithful as a model of $\mathbf{T}$ in $\mathbf{S}$.

Demonstration. (1) The identical model is faithful; if $\mathfrak{N}$ is an inverse of $\mathfrak{M}$ then $\mathfrak{M} * \mathfrak{N}$ is faithful and $\mathfrak{M}$ is faithful by Lemma 1233. Moreover, if φ is a $\mathbf{T}$-formula and if $\mathfrak{M}$ has an inverse, then $\mathfrak{M}$ considered as a model of $(\mathbf{T}, \varphi)$ in $(\mathbf{S}, \varphi^{\mathfrak{M}})$ has an inverse by Lemma 1235 and is therefore faithful.

(2) The identical model of $\mathbf{S}$ is a model of $\mathbf{S}$ in $\mathbf{T}$; since $\mathfrak{M} * \mathfrak{Id}_{\mathbf{S}}$ considered as a model of $\mathbf{T}$ in $\mathbf{T}$ is faithful, the model $\mathfrak{M}$ considered as a model of $\mathbf{T}$ in $\mathbf{S}$ is also faithful.

1237. METATHEOREM. Suppose that the languages $\mathbf{L}$ and $\mathbf{L}^*$ are similar, where $\mathbf{L}$ is $(\mathrm{P}_1, \bullet, \mathbf{F}_1, \bullet, s_1, \bullet)$ and $\mathbf{L}^*$ is $(\mathrm{P}_1^*, \bullet, \mathbf{F}_1^*, \bullet, s_1^*, \bullet)$. If $\Lambda_{\mathbf{L}}$ $(\Lambda_{\mathbf{L}^*})$ is the theory with the empty axiom system and the language $\mathbf{L}$ $(\mathbf{L}^*)$, then $\Lambda_{\mathbf{L}}$ is isomorphic to $\Lambda_{\mathbf{L}^*}$.

Demonstration. Assign to each **L**-formula φ an **L***-formula φ^* simply by adding asterisks to all predicates, operations and sorts with the exception of the equality predicate. This mapping is one-to-one from **L**-formulas onto **L***-formulas, assigns logical axioms to logical axioms and preserves $\rightarrow, \forall, \neg$. The inverse mapping has the same properties and hence each of these models is an inverse of the other.

1238. If **L** and **L*** are similar languages then the mapping defined in the course of the preceding demonstration is called the *direct translation* of **L**-formulas onto **L***-formulas; we denote it by $\mathfrak{Dir}_{\mathbf{L},\mathbf{L}^*}$.

1239. LEMMA. Suppose that **L** is the language of a theory **T**, that **L*** is a sublanguage of the language of a theory **S** and that **L** and **L*** are similar. Then the direct translation of **L**-formulas onto **L***-formulas is a model of **T** in **S** if and only if, for any axiom φ of **T**, φ^* is provable in **S**. (This follows from Metatheorem 1237).

1240. If the direct translation of **T**-formulas onto formulas of some sublanguage of the language of **S** is a model of **T** in **S**, then we call it a *direct model of* **T** *in* **S**.

1241. LEMMA. (1) The composition of direct models is a direct model. (2) If a direct model is faithful then it is essentially faithful.

Demonstration. (1) is obvious. (2) Let $\mathfrak{M}$ be a direct model of **T** in **S** and suppose $\mathfrak{M}$ to be faithful. Let φ be a closed **T**-formula. If **S**, $\varphi^{\mathfrak{M}} \vdash \psi^{\mathfrak{M}}$ then **S** $\vdash \varphi^{\mathfrak{M}} \rightarrow \psi^{\mathfrak{M}}$ by the Deduction theorem; hence **S** $\vdash (\varphi \rightarrow \psi)^{\mathfrak{M}}$ because $\varphi^{\mathfrak{M}} \rightarrow \psi^{\mathfrak{M}}$ is the same as $(\varphi \rightarrow \psi)^{\mathfrak{M}}$ for direct models. Since $\mathfrak{M}$ is a faithful model of **T** in **S** it follows that **T** $\vdash \varphi \rightarrow \psi$ and so **T**, $\varphi \vdash \psi$; thus $\mathfrak{M}$ is a faithful model of **T**, φ in **S**, $\varphi^{\mathfrak{M}}$. Note that it really suffices to consider closed φ; for $((\forall \ldots)\, \varphi)^{\mathfrak{M}}$ is the same as $(\forall \ldots)\, \varphi^{\mathfrak{M}}$, so that if **S**, $\varphi^{\mathfrak{M}} \vdash \psi^{\mathfrak{M}}$ then **S**, $(\forall \ldots)\, \varphi^{\mathfrak{M}} \vdash \psi^{\mathfrak{M}}$, hence **T**, $(\forall \ldots)\, \varphi \vdash \psi$ and **T**, $\varphi \vdash \psi$.

c) *Conservative extensions*

1242. Let **S** be an extension of a theory **T**. **S** is said to be a *conservative extension* of **T** if the identical model of **T** in **S** is faithful.

(Commentary). Note that $\mathfrak{Id}_{\mathbf{T}}$ considered as a model of **T** in **S** is faithful iff each **T**-formula φ provable in **S** is provable in **T**. We shall be interested in certain extensions which enrich the language; from the point of view of the original theory, new symbols can be treated as a means for simplifying formulas and proofs.

1243 CONVENTION. In the rest of the book *we restrict ourselves to theories having a universal sort.* In each theory **T** we fix a sort which is universal in **T** and call it *the universal sort of* **T**.

1244. We shall consider now four particular methods for obtaining conservative extensions of a given theory **T**:

a) Let $\varphi(x_1, \ldots, x_n)$ be a **T**-formula having free variables $x_1, \ldots, x_n$ of arbitrary sorts (where $n > 0$) and let P be an n-ary predicate not in the language of **T**. The formula

$$(\forall x_1, \bullet)\,(\mathrm{P}(x_1, \bullet) \equiv \varphi(x_1, \bullet))\,\&$$
$$\&\,(\forall X_1, \bullet)\,[(\neg\,(\exists x_1)\,(X_1 = x_1) \vee \bullet) \rightarrow \neg\,\mathrm{P}(X_1, \bullet)]$$

(where $X_1, \bullet$ are variables of the universal sort and not occurring in $\varphi(x_1, \bullet)$) is called the *definition of* P *in* **T** *by* φ.

b) Let $\varphi(y, x_1, \ldots, x_n)$ be a **T**-formula having free variables $y, x_1, \ldots, x_n$ of arbitrary sorts (where $n \geqq 0$) and suppose that

$$\mathbf{T} \vdash (\exists\,!\,y)\,\varphi(y, x_1, \bullet)\,;$$

let **F** be an n-ary operation not in the language of **T**. The formula

$$(\forall y, x_1, \bullet)\,(y = \mathbf{F}(x_1, \bullet) \equiv \varphi(y, x_1, \bullet))\,\&$$
$$\&\,(\forall X_1, \bullet)\,[(\neg\,(\exists x_1)\,(x_1 = X_1) \vee \bullet) \rightarrow \mathbf{F}(X_1, \bullet) = X_1]$$

(where $X_1, \bullet$ are variables of the universal sort and not occurring in $\varphi(y, x_1, \bullet)$) is called the *definition of* **F** *in* **T** *by* φ.

c) Let $\varphi(x^i)$ be a **T**-formula having one free variable x^i and let j be a sort not in the language of **T**; suppose that

$$\mathbf{T} \vdash (\exists x^i)\,\varphi(x^i)\,.$$

The formula

$$(\forall x^i)\,[(\exists y^j)\,(x^i = y^j) \equiv \varphi(x^i)]\,\&\,(\forall y^j)\,(\exists x^i)\,(x^i = y^j)$$

is called the *definition of the sort* j *in* **T** *by* φ.

d) Let $\varphi(x_1, \ldots, x_n)$ be a **T**-formula having free variables $x_1, \ldots, x_n$ of arbitrary sorts (where $n > 0$) and suppose that

$$\mathbf{T} \vdash (\exists x_1, \bullet)\, \varphi(x_1, \bullet)\,;$$

let $\mathbf{A}_1, \ldots, \mathbf{A}_n$ be constants not in the language of **T**. The formula

$$\varphi(\mathbf{A}_1, \bullet)\,\&\,(\exists x_1, \bullet)\,(\mathbf{A}_1 = x_1\,\&\,\bullet)$$

is called the *fixing of constants* $\mathbf{A}_1, \bullet$ *in* **T** *by* φ.

Remark. Each definition is a conjunction of two parts. The first part is the interesting part of the definition and tells us what the new symbol means for variables of sorts we are interested in. The second part is the securing part; in the case of a predicate and of a function it defines the meaning of the new symbol for each tuple of objects containing an element of a sort uninteresting w.r.t. the definition, and in the case of a sort the securing part says that the sort j is subordinated to the sort i. Similarly, each fixing consists of two parts; the first (interesting) part tells us the property that the new constants have, and the second (securing) part says that the new constants are objects of appropriate sorts. The securing part will be abbreviated by *sec*. Hence we write e.g. a definition of a predicate in the form

$$(\forall x_1, \bullet)\,(\mathrm{P}(x_1\,\bullet,) \equiv \varphi(x_1, \bullet))\,\&\,sec\,.$$

Note that *sec* is empty if we define a 0-ary function, i.e. a constant. The part *sec* will be often omitted at all, in particular if we will formulate a definition informally. The string of quantifiers in the first part of a definition will be usually also omitted. (Hence we write e.g. only $\mathrm{P}(x_1, \bullet) \equiv \varphi(x_1, \bullet)$.)

1245. METATHEOREM. (1) Let **T** be a theory and let **S** be an extension of **T** obtained by adding the definition of a precidate, operation or sort. Then **T** and **S** are isomorphic; there is a model $\mathfrak{R}$ of **S** in **T** such that $\mathfrak{R}$ and $\mathfrak{Id}_{\mathbf{T}}$ (as a model of **T** in **S**) are inverse each to the other. (2) Let **T** be a theory and let **S** be an extension of **T** obtained by adding the fixing of some constants. Then $\mathfrak{Id}_{\mathbf{T}}$ as a model of **T** in **S** has an inverse. (3) Consequently, the extension of a theory **T** obtained by adding a definition or fixing is conservative.

The technical details of the demonstration will be omitted; we give here an outline for the case of a predicate and of constants.

(a) To each **S**-formula ψ we assign a **T**-formula $\psi^{\mathfrak{R}}$ in such a way that $\mathrm{P}(\mathbf{t}_1, \bullet)$ is replaced by $(\exists x_1, \bullet)(\mathbf{t}_1 = x_1 \,\&\, \bullet \,\&\, \varphi(x_1, \bullet))$; note that it is necessary to rename some of the variables. It can be seen that $\psi^{\mathfrak{R}}$ is a **T**-formula and that $\mathbf{S} \vdash \psi \equiv \psi^{\mathfrak{R}}$. Moreover, $\mathfrak{R}$ is a model of **S** in **T**, and $\mathfrak{Id}_{\mathbf{T}}$ and $\mathfrak{R}$ are inverse each to the other. (It follows that both models are essentially faithful.)

(b) Let **S** be the extension of **T** by the fixing of constants $\mathbf{A}_1, \ldots, \mathbf{A}_n$ by $\varphi(x_1, \ldots, x_n)$. If $\psi(\mathbf{A}_1, \ldots, \mathbf{A}_n, y_1, \ldots, y_m)$ is an **S**-formula then we define $\psi^{\mathfrak{R}}$ as the **T**-formula $\varphi(x_1, \ldots, x_n) \rightarrow \psi(x_1, \ldots, x_n, y_1, \ldots, y_m)$ (some variables must be renamed). Then $\mathfrak{R}$ is a model of **S** in **T** and is inverse to $\mathfrak{Id}_{\mathbf{T}}$. It is easy to see that $\mathfrak{R}$ is a faithful model of **T** in **S**, i.e.

$$\mathbf{S} \vdash \psi(\mathbf{A}_1, \ldots, y_1, \ldots) \quad \text{iff} \quad \mathbf{T} \vdash \varphi(x_1, \ldots) \rightarrow \psi(x_1, \ldots, y_1, \ldots)\,.$$

Thus we observe that the fixing of constants corresponds to the usual mathematical reasoning: "there exist $x_1, \ldots, x_n$ such that $\varphi(x_1, \ldots, x_n)$; *choose some such* $x_1, \ldots, x_n$ *and denote them by* $\mathbf{A}_1, \ldots, \mathbf{A}_n$". If we prove something for fixed but arbitrarily chosen $\mathbf{A}_1, \ldots, \mathbf{A}_n$ then we conclude that we have proved it for all $x_1, \ldots, x_n$ such that $\varphi(x_1, \ldots, x_n)$.

Note that $\mathfrak{R}$ is not necessarily essentially faithful.

1246. LEMMA. a) Let **S** be the extension of **T** by the definition of a sort j subordinate to the sort i by $\varphi(x^i)$. If $\psi(x^i)$ is an **S**-formula and y^j is not bound in $\psi(x^i)$, then

$$\mathbf{S} \vdash (\forall y^j)\, \psi(y^j) \equiv (\forall x^i)\,(\varphi(x^i) \rightarrow \psi(x^i))\,,$$
$$\mathbf{S} \vdash (\exists y^j)\, \psi(y^j) \equiv (\exists x^i)\,(\varphi(x^i) \,\&\, \psi(x^i))\,.$$

1247. LEMMA. Let $\mathfrak{M}$ be a direct model of **T** in **S** and let $\hat{\mathbf{T}}$ be the extension of **T** (a) by the definition of a predicate P by $\varphi(x_1, \ldots, x_n)$, (b) by the definition of an operation **F** by $\varphi(y, x_1, \ldots, x_n)$, (c) by the definition of a sort j by $\varphi(x^i)$, or (d) by the fixing of constants by $\varphi(x_1, \ldots, x_n)$. Then in **S**, we can use the formula $\varphi^{\mathfrak{M}}$ (a) to define an n-ary predicate by $\varphi^{\mathfrak{M}}$, (b) to define an n-ary operation by $\varphi^{\mathfrak{M}}$, (c) to define a sort subordinate to i^* by $\varphi^{\mathfrak{M}}$, or (d) to fix constants by $\varphi^{\mathfrak{M}}$.

Demonstration. For (a) it is sufficient to note that the number of free variables in $\varphi^{\mathfrak{M}}$ is the same as in φ and that their sorts correspond. For (b) it is sufficient to observe that if $\mathbf{T} \vdash (\exists! y)\, \varphi(y, x_1, \bullet)$ then $\mathbf{S} \vdash (\exists! y^*)\, \varphi^*(y^*, x_1^*, \bullet)$. Cases (c) and (d) are treated similarly.

We are led to introduce the following concepts:

1248. Let $\mathfrak{Dir}_{\mathbf{L},\mathbf{L}^*}$ be a direct model of $\mathbf{T}$ in $\mathbf{S}$ and let $\hat{\mathbf{T}}$ be the extension of $\mathbf{T}$ (a) by the definition of an n-ary predicate P by φ, (b) by the definition of an n-ary operation $\mathbf{F}$ by φ, (c) by the definition of a sort j by φ, or (d) by the fixing of constants $\mathbf{A}_1, \bullet$ by φ. Let φ^* be the image of φ in $\mathfrak{Dir}_{\mathbf{L},\mathbf{L}^*}$. Suppose (a) that P* is an n-ary predicate not in the language of $\mathbf{S}$, (b) that $\mathbf{F}^*$ is an n-ary operation not in the language of $\mathbf{S}$, (c) that j^* is a sort not in the language of $\mathbf{S}$ or (d) that $\mathbf{A}_1^*, \bullet$ are constants not in the language of $\mathbf{S}$. The extension $\hat{\mathbf{S}}$ of $\mathbf{S}$ (a) by the definition of P* by φ^*, (b) by the definition of $\mathbf{F}^*$ by φ^*, (c) by the definition of j^* by φ^* or (d) by the fixing of $\mathbf{A}_1^*, \bullet$ by φ^* is called the *extension of* $\mathbf{S}$ *induced by the model* $\mathfrak{Dir}_{\mathbf{L},\mathbf{L}^*}$ *and the extension* $\hat{\mathbf{T}}$. The direct translation (a) of ($\mathbf{L}$, P) onto ($\mathbf{L}^*$, P*), (b) of ($\mathbf{L}$, $\mathbf{F}$) onto ($\mathbf{L}^*$, $\mathbf{F}^*$), (c) of ($\mathbf{L}$, j) onto ($\mathbf{L}^*$, j^*) or (d) of ($\mathbf{L}$, $\mathbf{A}_1$, $\bullet$) onto ($\mathbf{L}^*$, $\mathbf{A}_1^*$, $\bullet$) is called the *extension of the model* $\mathfrak{Dir}_{\mathbf{L},\mathbf{L}^*}$ *induced by the extension* $\hat{\mathbf{T}}$ *and the theory* $\mathbf{S}$.

1249. LEMMA. The extension of $\mathfrak{Dir}_{\mathbf{L},\mathbf{L}^*}$ induced by $\hat{\mathbf{T}}$ and $\mathbf{S}$ is a model of $\hat{\mathbf{T}}$ in the extension $\hat{\mathbf{S}}$ of $\mathbf{S}$ induced by $\mathfrak{Dir}_{\mathbf{L},\mathbf{L}^*}$ and $\hat{\mathbf{T}}$.

Demonstration. We denote the extension of $\mathfrak{Dir}_{\mathbf{L},\mathbf{L}^*}$ by $\hat{\mathfrak{D}}$. Clearly $\hat{\mathfrak{D}}$ is a model of $\mathbf{T}$ in $\hat{\mathbf{S}}$; to prove that it is a model of $\hat{\mathbf{T}}$, it suffices to verify the validity in $\hat{\mathfrak{D}}$ of the added definition (fixing of constants). This follows from the construction of $\hat{\mathbf{S}}$; indeed, the translation of the definition (fixing of constants) is implied by the definition (fixing of constants) which is added to $\mathbf{S}$.

d) Epsilon-theories

1250. We now choose a binary predicate (different from $=$) and denote it $\in$ (the *membership predicate*). Further, we choose a sort with variables $X, Y, \dots$ which we call class-variables. The language $(\in, X)$ is called the *fundamental language*, and each language similar to the fundamental language is called F-*like*. Formulas of the fundamental language will be called fundamental formulas (FF). The *fundamental* $\in$*-theory* **TE** is the theory having the fundamental language and the following axioms:

(F1) $(\exists X, Y)(X \in Y)$,

(F2) $(\forall Z)(Z \in X \equiv Z \in Y) \equiv X = Y$.

Let $\mathbf{T}$ be an extension of the fundamental $\in$-theory and let φ be a $\mathbf{T}$-formula. The formula φ is called *fundamental in* $\mathbf{T}$ if there exists some fundamental φ_0 such that $\mathbf{T} \vdash \varphi \equiv \varphi_0$.

The concept of an $\in$-*theory* is defined recursively:

1) The fundamental $\in$-theory is an $\in$-theory.

2) If $\mathbf{T}$ is an $\in$-theory and φ a definition (of predicate, operation or sort) or a fixing of constants in $\mathbf{T}$, then $(\mathbf{T}, \varphi)$ is an $\in$-theory.

3) If $\mathbf{T}$ is an $\in$-theory and φ is a $\mathbf{T}$-formula fundamental in $\mathbf{T}$, then $(\mathbf{T}, \varphi)$ is an $\in$-theory.

1251. Let φ be an axiom of an $\in$-theory $\mathbf{T}$. If φ is one of the axioms (F1), (F2) or if φ is formulated in the $\in$-theory whose axioms coincide with the sequence of axioms of $\mathbf{T}$ preceding φ then φ is called a *proper axiom of* $\mathbf{T}$. In the opposite case φ is a definition or fixing of constants and is called an *improper axiom of* $\mathbf{T}$.

When we speak of the language $\mathbf{L}_\mathbf{T}$ of an $\in$-theory $\mathbf{T}$ we shall assume that the symbols in $\mathbf{L}_\mathbf{T}$ are ordered according to their first occurrences in the axioms of $\mathbf{T}$, i.e. $\mathbf{L}_\mathbf{T}$ begins with X, $\in$, and continues with notions introduced by successive improper axioms of $\mathbf{T}$.

It can be said that in $\in$-theories we are interested mainly in the fundamental formulas, i.e. in the properties of the membership-predicate. The defined symbols play an auxiliary role, although they greatly simplify the development. Theoretically, we could in fact restrict ourselves to theories having the fundamental language:

1252. LEMMA. If $\mathbf{T}$ is an $\in$-theory then there exists an $\in$-theory $\mathbf{T}_0$ having the fundamental language and such that $\mathbf{T}$ is equivalent to some conservative extension of $\mathbf{T}_0$.

Demonstration. By induction. 1) The lemma is obvious if $\mathbf{T}$ is the fundamental $\in$-theory. 2) Let $\mathbf{T}$ be an $\in$-theory. Let $\mathbf{T}_0$ be an $\in$-theory having the fundamental language and suppose that $\mathbf{T}_0'$ is a conservative extension of $\mathbf{T}_0$ such that $\mathbf{T}_0'$ is equivalent to $\mathbf{T}$. If φ is a definition or a fixing of constants then $(\mathbf{T}_0', \varphi)$ is a conservative extension of $\mathbf{T}_0$ and is equivalent to $(\mathbf{T}, \varphi)$. 3) Suppose that $\mathbf{T}$, $\mathbf{T}_0$ and $\mathbf{T}_0'$ are as before. Let φ be a $\mathbf{T}$-formula fundamental in $\mathbf{T}$; i.e. there exists a fundamental φ_0 such that $\mathbf{T} \vdash \varphi \equiv \varphi_0$ and so $\mathbf{T}_0' \vdash \varphi \equiv \varphi_0$. Denote by $\mathbf{T}_1$ the theory $(\mathbf{T}_0, \varphi_0)$ and by $\mathbf{T}_1'$ the theory $(\mathbf{T}_0', \varphi)$. Then $\mathbf{T}_1'$ is a conservative extension of $\mathbf{T}_1$ and is equivalent to $(\mathbf{T}, \varphi)$. To prove this it suffices to show that $\mathbf{T}_1' \vdash \varphi$ and that $(\mathbf{T}, \varphi) \vdash \varphi_0$; these facts however are immediate.

1253. Let $\mathbf{S}$ be some fixed $\in$-theory and suppose that $\mathbf{L}$ is an F-like language $(X^*, \in^*)$ which is a sublanguage of the language of $\mathbf{S}$. Given an $\in$-theory $\mathbf{T}$ we ask if there is a conservative extension $\hat{\mathbf{S}}$ of $\mathbf{S}$ such that there is a direct model of $\mathbf{T}$ in $\hat{\mathbf{S}}$ translating X as X^* and $\in$ as $\in^*$. If $\mathbf{L}_1$ is the language of $\mathbf{T}$ and $\mathbf{L}_2$ is a language similar to $\mathbf{L}_1$ and beginning with $X^*, \in^*$, then we can add to $\mathbf{S}$ translations of all improper axioms of $\mathbf{T}$ by $\mathfrak{Dir}_{\mathbf{L}_1, \mathbf{L}_2}$ and we obtain an extension $\hat{\mathbf{S}}$; we also have the direct translation $\mathfrak{Dir}_{\mathbf{L}_1, \mathbf{L}_2}$ of $\mathbf{T}$-formulas into $\hat{\mathbf{S}}$-formulas. But two questions arise: (1) whether $\hat{\mathbf{S}}$ is a conservative extension of $\mathbf{S}$ and, moreover, whether $\hat{\mathbf{S}}$ is an $\in$-theory; (2) whether $\mathfrak{Dir}_{\mathbf{L}_1, \mathbf{L}_2}$ is a model of $\mathbf{T}$ in $\hat{\mathbf{S}}$. This leads to the following definitions.

1254. Let $\mathbf{L}$ be an F-like language and let $\mathbf{T}$ be an $\in$-theory with language $\mathbf{L}_1$. A *derivation* of $\mathbf{T}$ w.r.t. $\mathbf{L}$ is the sequence of translations of all improper axioms of $\mathbf{T}$ by $\mathfrak{Dir}_{\mathbf{L}_1, \mathbf{L}_2}$ where $\mathbf{L}_2$ is any language similar to $\mathbf{L}_1$ and beginning with $\mathbf{L}$. An arbitrary derivation of $\mathbf{T}$ w.r.t. $\mathbf{L}$ is denoted by $\partial\mathbf{T}/\partial\mathbf{L}$. The language $\mathbf{L}_2$ used for the construction of $\partial\mathbf{T}/\partial\mathbf{L}$ is uniquely determined by $\mathbf{L}$ and $\partial\mathbf{T}/\partial\mathbf{L}$; hence we denote $\mathfrak{Dir}_{\mathbf{L}_1, \mathbf{L}_2}$ by $\mathfrak{Dir}\,(\partial\mathbf{T}/\partial\mathbf{L})$ and call it a *direct translation given by* $\mathbf{L}$ *and* $\mathbf{T}$.

1255. Let $\partial\mathbf{T}/\partial\mathbf{L}$ be an arbitrary derivation of $\mathbf{T}$ w.r.t. $\mathbf{L}$. We say that $\partial\mathbf{T}/\partial\mathbf{L}$ is *compatible with* $\mathbf{S}$ if the language $\mathbf{L}_2$ used for the construction of $\partial\mathbf{T}/\partial\mathbf{L}$ has no symbols in common with the language of $\mathbf{S}$ except those of $\mathbf{L}$. Suppose that $\mathbf{T}$ has n axioms; $\mathbf{T}_m$ denotes the sequence consisting of the first m axioms of $\mathbf{T}$.

Given a derivation $\partial\mathbf{T}/\partial\mathbf{L}$ of $\mathbf{T}$ w.r.t. $\mathbf{L}$ we denote by $\partial\mathbf{T}_m/\partial\mathbf{L}$ the segment of $\partial\mathbf{T}/\partial\mathbf{L}$ which is a derivation of $\mathbf{T}_m$ w.r.t. $\mathbf{L}$.

1256. We say that $\mathbf{L}$ *determines a model of* $\mathbf{T}$ *in* $\mathbf{S}$ if there is a derivation of $\mathbf{T}$ w.r.t. $\mathbf{L}$ compatible with $\mathbf{S}$ such that, for every $m < n$ such that φ_{m+1} is a proper axiom of $\mathbf{T}$, φ_{m+1} holds in $\mathfrak{Dir}\,(\partial\mathbf{T}_m/\partial\mathbf{L})$ as a model in $(\mathbf{S}, \partial\mathbf{T}_m/\partial\mathbf{L})$.

To justify the definition we prove the following

1257. Lemma. If $\mathbf{L}$ determines a model of $\mathbf{T}$ in $\mathbf{S}$ then, for any derivation $\partial\mathbf{T}/\partial\mathbf{L}$ compatible with $\mathbf{S}$, $(\mathbf{S}, \partial\mathbf{T}/\partial\mathbf{L})$ is a conservative extension of $\mathbf{S}$ (in addition, an $\in$-theory) and $\mathfrak{Dir}\,(\partial\mathbf{T}/\partial\mathbf{L})$ is a model of $\mathbf{T}$ in $(\mathbf{S}, \partial\mathbf{T}/\partial\mathbf{L})$. Moreover, if $(\partial\mathbf{T}/\partial\mathbf{L})_1, (\partial\mathbf{T}/\partial\mathbf{L})_2$ are two derivations of $\mathbf{T}$ w.r.t. $\mathbf{L}$ compatible with $\mathbf{S}$ and arrows 1, 2 in the diagram 1258 are the corresponding direct translations then there is a direct model $\mathfrak{I}$ of $\mathbf{S}, (\partial\mathbf{T}/\partial\mathbf{L})_1$ in $\mathbf{S}, (\partial\mathbf{T}/\partial\mathbf{L})_2$ which is an isomorphism and which makes the diagram commute.

1258. *1259.*

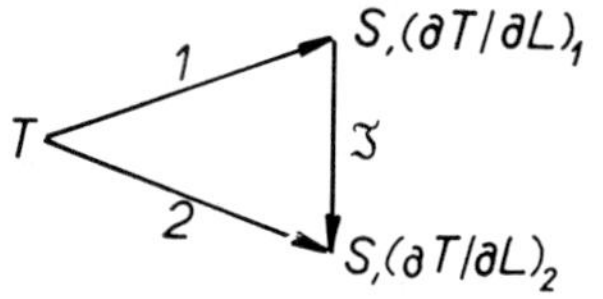

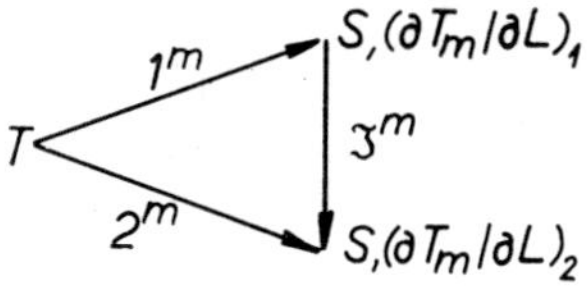

Demonstration. Let $(\partial\mathbf{T}/\partial\mathbf{L})_1$ be a derivation of $\mathbf{T}$ w.r.t. $\mathbf{L}$ compatible with $\mathbf{S}$. Further let $(\partial\mathbf{T}/\partial\mathbf{L})_2$ be another derivation of $\mathbf{T}$ w.r.t. $\mathbf{L}$ compatible with $\mathbf{S}$. Then for each $m \leqq n$ the languages of $\mathbf{S}$, $(\partial\mathbf{T}_m/\partial\mathbf{L})_1$ and of $\mathbf{S}$, $(\partial\mathbf{T}_m/\partial\mathbf{L})_2$ are similar and we have direct translations $\mathfrak{J}^m = \mathfrak{Dir}_{\mathbf{L}_m^{(1)},\, \mathbf{L}_m^{(2)}}$ and $\mathfrak{J}^m = \mathfrak{Dir}_{\mathbf{L}_m^{(2)},\, \mathbf{L}_m^{(1)}}$ ($\mathbf{L}_m^{(i)}$ is the language of $\mathbf{S}$, $(\partial\mathbf{T}_m/\partial\mathbf{L})_i$). Evidently, these translations are models of $\mathbf{S}, (\partial\mathbf{T}_m/\partial\mathbf{L})_1$ in $\mathbf{S}, (\partial\mathbf{T}_m/\partial\mathbf{L})_2$ and of $\mathbf{S}, (\partial\mathbf{T}_m/\partial\mathbf{L})_2$ in $\mathbf{S}, (\partial\mathbf{T}_m/\partial\mathbf{L})_1$ respectively and their compositions are the corresponding identities. Hence $\mathfrak{J}$ is the inverse of $\mathfrak{J}$ and $\mathfrak{J}$ is the inverse of $\mathfrak{J}$; the diagram 1258 commutes.

Hence it remains to demonstrate that $\mathbf{S}, (\partial\mathbf{T}/\partial\mathbf{L})_1$ is a conservative extension of $\mathbf{S}$ and that the arrow 1 in Diagram 1258 is a model of $\mathbf{T}$. Write $\partial\mathbf{T}/\partial\mathbf{L}$ instead of $(\partial\mathbf{T}/\partial\mathbf{L})_1$. We use induction on m. Set $m = 2$. Then $\mathbf{T}_m$ is the fundamental $\in$-theory and so all of its axioms hold in $\mathfrak{Dir}\,(\partial\mathbf{T}_2/\partial\mathbf{L})$ as a model in $\mathbf{S}$ because they are proper axioms. (Note that in this case $\partial\mathbf{T}_2/\partial\mathbf{L}$ is empty). Suppose now that $\mathbf{S}, (\partial\mathbf{T}_m/\partial\mathbf{L})$ is a conservative extension of $\mathbf{S}$ (and an $\in$-theory) and that $\mathfrak{Dir}\,(\partial\mathbf{T}_m/\partial\mathbf{L})$ is a model of $\mathbf{T}_m$ in $\mathbf{S}, (\partial\mathbf{T}_m/\partial\mathbf{L})$. Take $\mathbf{T}_{m+1} = (\mathbf{T}, \varphi)$. If φ is a proper axiom then $\partial\mathbf{T}_{m+1}/\partial\mathbf{L}$ is the same as $\partial\mathbf{T}_m/\partial\mathbf{L}$; by assumption φ holds in $\mathfrak{Dir}\,(\partial\mathbf{T}_m/\partial\mathbf{L})$ as a model in $\mathbf{S}, (\partial\mathbf{T}_m/\partial\mathbf{L})$; hence $\mathbf{S}, (\partial\mathbf{T}_{m+1}/\partial\mathbf{L})$ is a conservative extension of $\mathbf{S}$, and $\mathfrak{Dir}\,(\partial\mathbf{T}_{m+1}/\partial\mathbf{L})$ is a model of $\mathbf{T}_{m+1}$ in $\mathbf{S}, (\partial\mathbf{T}_{m+1}/\partial\mathbf{L})$. Finally if φ is an improper axiom then $\mathbf{S}, (\partial\mathbf{T}_{m+1}/\partial\mathbf{L})$ is the extension of $\mathbf{S}, (\partial\mathbf{T}_m/\partial\mathbf{L})$ induced by $\mathbf{T}_{m+1}$ and $\mathfrak{Dir}\,(\partial\mathbf{T}_m/\partial\mathbf{L})$, while $\mathfrak{Dir}\,(\partial\mathbf{T}_{m+1}/\partial\mathbf{L})$ is the extension of $\mathfrak{Dir}\,(\partial\mathbf{T}_m/\partial\mathbf{L})$ induced by $\mathbf{T}_{m+1}$ and $\mathbf{S}, (\partial\mathbf{T}_m/\partial\mathbf{L})$. Hence, by Lemma 1247, $\mathbf{S}, (\partial\mathbf{T}_{m+1}/\partial\mathbf{L})$ is a conservative extension of $\mathbf{S}$ (and an $\in$-theory) and $\mathfrak{Dir}\,(\partial\mathbf{T}_{m+1}/\partial\mathbf{L})$ is a model of $\mathbf{T}_{m+1}$ in $\mathbf{S}, (\partial\mathbf{T}_{m+1}/\partial\mathbf{L})$. Thus the lemma is proved.

1260. We illustrate the step-by-step construction of $\mathbf{S}, (\partial\mathbf{T}/\partial\mathbf{L})$ and the decision whether $\mathbf{L}$ determines a model of $\mathbf{T}$ in $\mathbf{S}$ by the following example (cf. Diagram 1261):

1261.

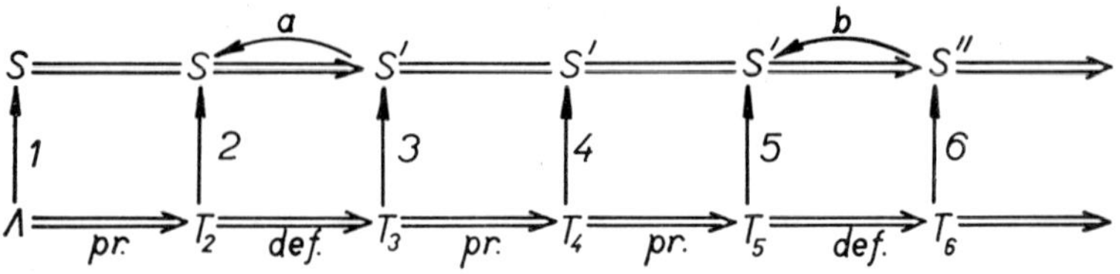

Suppose that the axioms of **T** are the axioms of **TE** followed by a definition, two proper axioms, a definition, etc. Given the language **L** (a sublanguage of the language of **S**) we have $\mathfrak{Dir} = \mathfrak{Dir}_{\mathbf{F},\mathbf{L}}$ (**F** being the fundamental language); this is certainly a model of $\Lambda_{\mathbf{F}}$ in **S** (arrow 1). If we succeed in proving the translations of the axioms of **TE** $(= \mathbf{T}_2)$ in **S** we know that **L** determines a model of $\mathbf{T}_2$ in **S** (arrow 2). $\mathbf{T}_3$ is a conservative extension of $\mathbf{T}_2$ by a definition, hence it induces together with $\mathfrak{Dir}$ a conservative extension $\mathbf{S}'$ of **S** and a direct model $\mathfrak{Dir}'$ of $\mathbf{T}_3$ in $\mathbf{S}'$ (arrow 3); moreover, the identity $\mathfrak{Id}_{\mathbf{S}}$ as a model in $\mathbf{S}'$ has an inverse (arrow *a*); the composition of arrows 3 and *a* is a model of $\mathbf{T}_3$ in **S**. $\mathbf{T}_4$ is an extension of $\mathbf{T}_3$ by a proper axiom; if we show that this axiom holds in $\mathfrak{Dir}'$ we know that $\mathfrak{Dir}'$ is a model of $\mathbf{T}_4$ in $\mathbf{S}'$ (arrow 4). The same can be said for arrow 5. The explanation of arrows 6 and *b* is the same as that for arrows 3 and *a*. Finally we see that $\mathfrak{Dir}''$ (i.e. $\mathfrak{Dir}\,(\partial\mathbf{T}_6/\partial\mathbf{L})$, arrow 6) is a model of $\mathbf{T}_6$ in $\mathbf{S}''$ (i.e. in $\mathbf{S}, \partial\mathbf{T}_6/\partial\mathbf{L}$); if we want we can compose the model $\mathfrak{Dir}''$ with arrows *b* and *a* and obtain a model of $\mathbf{T}_6$ in **S**. This justifies our term "determines a model in **S**"; but it is more convenient to consider the model $\mathfrak{Dir}\,(\partial\mathbf{T}_6/\partial\mathbf{L})$ as a model in **S**, $(\partial\mathbf{T}_6/\partial\mathbf{L})$ itself, keeping in mind the theoretical possibility of obtaining *mutatis mutandis* a model in **S**.

More generally, suppose that we have **T** and **S** and we want first to define a sort and a binary predicate in **S** and then ask if the F-like language so defined determines a model of **T** in **S** (extended by our definitions). Consider the general form of a definition of an F-like language.

1262. A pair of formulas

(1) $(\forall X)\,[(\exists X^*)\,(X^* = X) \equiv \chi(X)]\;\&\; sec,$

(2) $(\forall X^*, Y^*)(\,X^* \in^* Y^* \equiv \varepsilon(X^*, Y^*))\;\&\; sec$

is a *definition of an* F-*like language* $((X^*, \in^*)$ by means of $\chi, \varepsilon)$ if (a) χ is a formula with exactly one free variable X (class variable) and not containing any variable of the sort of X^* nor containing the binary predicate $\in^*$, (b) ε is a formula with exactly two free variables X^*, Y^* and not containing the predicate $\in^*$.

Every definition of an F-like language (briefly an F-*definition*) is uniquely determined by the formulas χ, ε and by the language $(X^*, \in^*)$. Conversely, given an F-definition Δ, the formulas χ, ε and the language $\mathbf{L}_\Delta = (X^*, \in^*)$ are uniquely determined.

1263. Given an F-definition Δ we say that Δ is an F-*definition in* $\mathbf{S}$ if (a) χ is an $\mathbf{S}$-formula such that $\mathbf{S} \vdash (\exists X)\ \chi(X)$, (b) ε is an $(\mathbf{S}, (1))$-formula and (c) neither X^* nor $\in^*$ belong to the language of $\mathbf{S}$. Then obviously (1) is a definition of a sort in $\mathbf{S}$ and (2) is a definition of a binary predicate in $(\mathbf{S},(1))$. In this case $(\mathbf{S}, \Delta)$ is an $\in$-theory provided that $\mathbf{S}$ is; furthermore, in $(\mathbf{S}, \Delta)$ we have the language $\mathbf{L}_\Delta = (X^*, \in^*)$ and may ask if it determines a model of $\mathbf{T}$ in $(\mathbf{S}, \Delta)$. We write $\partial\mathbf{T}/\partial\Delta$ instead of $(\Delta, \partial\mathbf{T}/\partial\mathbf{L}_\Delta)$ and say that $\partial\mathbf{T}/\partial\Delta$ is *compatible with* $\mathbf{S}$ if Δ is an F-definition in $\mathbf{S}$ and if $\partial\mathbf{T}/\partial\mathbf{L}_\Delta$ is compatible with $(\mathbf{S}, \Delta)$. We also say that Δ *determines a model of* $\mathbf{T}$ *in* $\mathbf{S}$ instead of saying that Δ is an F-definition in $\mathbf{S}$ and that $\mathbf{L}_\Delta$ determines a model of $\mathbf{T}$ in $(\mathbf{S}, \Delta)$. (Draw a diagram similar to 1261!) Finally we write $\mathfrak{Dir}\,(\partial\mathbf{T}/\partial\Delta)$ instead of $\mathfrak{Dir}\,(\partial\mathbf{T}/\partial\mathbf{L}_\Delta)$. The last translation is often denoted by Δ if $\mathbf{T}$ and $\mathbf{S}$ are evident from the context.

1264. The notion "Δ is an F-definition in $\mathbf{S}$" can be generalized in the following way: We say that π is a *specification for* Δ *in* $\mathbf{S}$ if there are constants $\mathbf{A}_1, \bullet$ occurring in χ and/or in ε such that π is a fixing of $\mathbf{A}_1, \bullet$ in $\mathbf{S}$ and if Δ is an F-definition in $(\mathbf{S}, \pi)$; in this case we also say that Δ with π is a *parametric* F-*definition* in $\mathbf{S}$. We say that Δ with the specification π *determines a model of* $\mathbf{T}$ *in* $\mathbf{S}$ if π is a specification for Δ in $\mathbf{S}$ and if Δ determines a model of $\mathbf{T}$ in $(\mathbf{S}, \pi)$ (i.e. π is a fixing of $\mathbf{A}_1, \bullet$ in $\mathbf{S}$, Δ is an F-definition in $(\mathbf{S}, \Delta)$ and $\mathbf{L}_\Delta$ determines a model of $\mathbf{T}$ in $(\mathbf{S}, \pi, \Delta)$). If it is the case then $\mathfrak{Dir}\,(\partial\mathbf{T}/\partial\Delta)$ is called the *direct model of* $\mathbf{T}$ *determined by* Δ.

1265. Given $\mathbf{S}$ and Δ we can have various specifications for Δ in $\mathbf{S}$. If π_1, π_2 are two such specifications then we say that π_1 is *stronger than* π_2 in case $\mathbf{S}, \pi_1 \vdash \pi_2$. We have the following useful facts:

1266. Lemma. (a) If Δ with the specification π_2 determines a model of $\mathbf{T}$ in $\mathbf{S}$ and if π_1 is a specification for Δ in $\mathbf{S}$ stronger than π_2, then Δ with the specification π_1 determines a model of $\mathbf{T}$ in $\mathbf{S}$; every formula which holds in $\mathfrak{Dir}\,(\partial\mathbf{T}/\partial\Delta)$ as a model in $\mathbf{S}, \pi_2, \partial\mathbf{T}/\partial\Delta$ holds also in $\mathfrak{Dir}(\partial\mathbf{T}/\partial\Delta)$ as a model in $\mathbf{S}, \pi_1, \partial\mathbf{T}/\partial\Delta$.

(b) Let π_1, π_2 be specifications for Δ in $\mathbf{S}$ and suppose that π_1 is stronger than π_2. If $\mathfrak{Dir}\,(\partial\mathbf{T}/\partial\Delta)$ is a model of $\mathbf{T}$ in $\mathbf{S}, \pi_2, \partial\mathbf{T}/\partial\Delta$ and if it is a faithful model of $\mathbf{T}$ in $\mathbf{S}, \pi_1, \partial\mathbf{T}/\partial\Delta$ then it is also faithful as a model in $\mathbf{S}, \pi_2, \partial\mathbf{T}/\partial\Delta$.

Demonstration. (a) follows from the fact that **S**, π_1, $\partial\mathbf{T}/\partial\Delta$ is stronger than **S**, π_2, $\partial\mathbf{T}/\partial\Delta$ so that all arrows in the following commutative diagram are models:

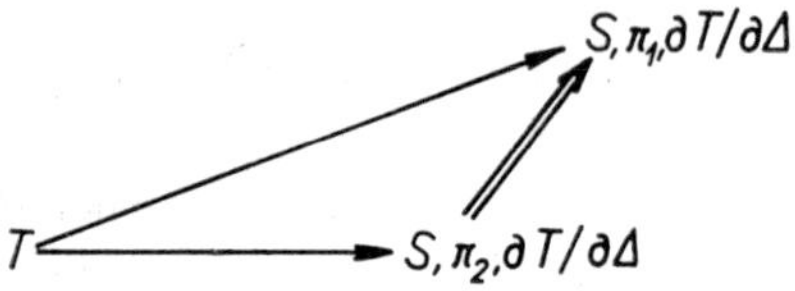

(b) follows from this diagram by Lemma 1233.

1267. (Absoluteness.) Let **T**, **S** be two $\in$-theories such that the language of **T** is a sublanguage of the language of **S**. Let $\mathfrak{D}$ be a direct model of **T** in **S** and let $\varphi(U, \bullet)$ be a **T**-formula (with free variables U, $\bullet$ of arbitrary sorts). Denote $[\varphi(U, \bullet)]^{\mathfrak{D}}$ by $\varphi^*(U^*, \bullet)$ and suppose that the sort of U^* is subordinate (in **S**) to the sort of U and similarly for other free variables of φ. (The last condition is trivially satisfied for every variable U which is a class variable.) Then we say that

(a) φ is *absolute in* **S** *w.r.t.* $\mathfrak{D}$ if

$$\mathbf{S} \vdash (U = U^* \,\&\, \bullet) \rightarrow [\varphi^*(U^*, \bullet) \equiv \varphi(U, \bullet)] \,;$$

(b) φ is *absolute from above in* **S** *w.r.t.* $\mathfrak{D}$ if

$$\mathbf{S} \vdash (U = U^* \,\&\, \bullet) \rightarrow [\varphi(U, \bullet) \rightarrow \varphi^*(U^*, \bullet)] \,;$$

(c) φ *is absolute from below* in **S** w.r.t. $\mathfrak{D}$ if

$$\mathbf{S} \vdash (U = U^* \,\&\, \bullet) \rightarrow [\varphi^*(U^*, \bullet) \rightarrow \varphi(U, \bullet)] \,.$$

1268. Lemma. (a) If $\varphi(U, \bullet, V)$ is absolute from below then so is $(\exists V)\,(\varphi(U, \bullet, V))$; if $\varphi(U, \bullet, V)$ is absolute from above then so is $(\forall V)\,(\varphi(U, \bullet, V)$.

(b) $\varphi(U, \bullet)$ is absolute iff it is absolute both from below and from above. (Obvious.)

We close this Section with a useful test for the faithfulness of certain direct models.

1269. LEMMA. Let **T**, **S** be $\in$-theories such that **S** is a conservative extension of **T**. Let **L** be an F-like sublanguage $(X^*, \in^*)$ of the language of **S** determining a model of **T** in **S**. Suppose that there is an **S**-formula $\mu(X, X^*)$ with exactly two free variables such that

$$\mathbf{S} \vdash (\forall X)(\exists!\, X^*)\, \mu(X, X^*)\,; \tag{3}$$

$$\mathbf{S} \vdash (\forall X^*)(\exists!\, X)\, \mu(X, X^*)\,; \tag{4}$$

$$\mathbf{S} \vdash \mu(X, X^*) \,\&\, \mu(Y, Y^*)\,. \rightarrow [X \in Y \equiv X^* \in^* Y^*]\,. \tag{5}$$

If all improper axioms of **T** are definitions (no fixing of constants) then $\mathfrak{Dir}\,(\partial\mathbf{T}/\partial\mathbf{L})$ and $\mathfrak{Id}_{\mathbf{T}}$ are equivalent as models in **S**, $\partial\mathbf{T}/\partial\mathbf{L}$ (consequently, $\mathfrak{Dir}\,(\partial\mathbf{T}/\partial\mathbf{L})$ is a faithful model of **T** in **S**, $\partial\mathbf{T}/\partial\mathbf{L}$).

Demonstration. We proceed by induction on the "segment theories" $\mathbf{T}_n$. Set $n = 2$; then all $\mathbf{T}_2$-formulas are fundamental formulas and it follows easily by induction on the structure of an arbitrary $\mathbf{T}_2$-formula that the following is provable in **S**:

$$(\mu(X_1, X_1^*) \,\&\, \bullet) \rightarrow [\varphi(X_1, \bullet) \equiv \varphi^*(X_1^*, \bullet)]\,. \tag{6}$$

Hence **S**, $\varphi \vdash \varphi^*$ and **S**, $\varphi^* \vdash \varphi$ so that $\mathfrak{Dir}\,(\partial\mathbf{T}_2/\partial\mathbf{L})$ is equivalent to $\mathfrak{Id}_{\mathbf{T}_2}$ as a model in **S**. Now suppose that for some n we already know that (6) is provable in **S**, $\partial\mathbf{T}_n/\partial\mathbf{L}$ for every $\mathbf{T}_n$-formula φ. If φ_{n+1} is a proper axiom of **T** then $\mathbf{T}_{n+1}$-formulas are the same as $\mathbf{T}_n$-formulas so that the induction step is trivial. If φ_{n+1} is a definition in $\mathbf{T}_n$ by means of ψ then we have

$$\mathbf{S}, \partial\mathbf{T}_n/\partial\mathbf{L} \vdash (\mu(X_1, X_1^*) \,\&\, \bullet) \rightarrow [\psi(X_1, \bullet) \equiv \psi^*(X_1^*, \bullet)]$$

which has the following consequences:

(a) If φ_{n+1} is a definition of a predicate P then

$$\mathbf{S}, \partial\mathbf{T}_{n+1}/\partial\mathbf{L} \vdash (\mu(X_1, X_1^*) \,\&\, \bullet) \rightarrow [\mathrm{P}(X_1, \bullet) \equiv \mathrm{P}^*(X_1^*, \bullet)]\,.$$

(b) If φ_{n+1} is a definition of a function **F** then

$$\mathbf{S}, \partial\mathbf{T}_{n+1}/\partial\mathbf{L} \vdash (\mu(X_1, X_1^*) \,\&\, \bullet) \rightarrow [X_1 \equiv \mathbf{F}(X_2, \bullet) \equiv X_1^* = \mathbf{F}^*(X_2^*, \bullet)]\,.$$

(c) If φ_{n+1} is a definition of a new variable u then

$$\mathbf{S}, \partial\mathbf{T}_{n+1}/\partial\mathbf{L} \vdash (\forall u)(\exists!\, u^*)\, \mu(u, u^*) \,\&\, (\forall u^*)(\exists!\, u)\, \mu(u, u^*)\,.$$

Now it follows by induction on the structure of an arbitrary formula φ that (6) is provable in $\mathbf{S}, \partial\mathbf{T}_{n+1}/\partial\mathbf{L}$ for each $\mathbf{T}_{n+1}$-formula. Hence the lemma is demonstrated.

1270. Remarks. (1) We also see the difficulty with fixings: if the induction assumption holds for $\mathbf{T}_n$ and if φ_{n+1} is a fixing of constant **A**, say, by means of $\chi(X)$ then in $\mathbf{S}, \partial\mathbf{T}_{n+1}/\partial\mathbf{L}$ we cannot deduce the formula $\mu(X_1, X_1^*) \to \to [X_1 = \mathbf{A} \equiv X_1^* = \mathbf{A}^*]$ (or, equivalently, $\mu(\mathbf{A}, \mathbf{A}^*)$) from the formula $\mu(X_1, X_1^*) \to [\chi(X_1) \equiv \chi^*(X_1^*)]$ (which is certainly provable in $\mathbf{S}, \partial\mathbf{T}_n/\partial\mathbf{L}$) because we do not know if $(\exists!\, X)\, \chi(X)$ is provable in $\mathbf{S}, \partial\mathbf{T}_n/\partial\mathbf{L}$. However $\mu(\mathbf{A}, \mathbf{A}^*)$ is consistent with $\mathbf{S}, \partial\mathbf{T}_n/\partial\mathbf{L}$ since it is a fixing of the constant $\mathbf{A}^*$, and hence $\mathfrak{Dir}\,(\partial\mathbf{T}/\partial\mathbf{L})$ is a faithful model of $\mathbf{T}_{n+1}$ in $\mathbf{S}, \partial\mathbf{T}_n/\partial\mathbf{L}, \mu(\mathbf{A}, \mathbf{A}^*)$, the latter theory being stronger than $\mathbf{S}, \partial\mathbf{T}_{n+1}/\partial\mathbf{L}$.

(2) If $\mu(X, X^*)$ is $X = X^*$ then (3) and (4) reduce to $\mathbf{S} \vdash (\forall X)(\exists X^*)(X = X^*)$.

SECTION 3

The theory of classes continued

The theory of classes was described in Section 1; in the present Section we shall investigate this theory from the point of view of Section 2. We shall show among other things that **TC** really is an $\in$-theory and that the model described in Section 1 is given by a certain F-definition, called the normal F-definition. We shall investigate some of the properties of this F-definition and we shall introduce another related F-definition, called the standard F-definition. Finally, we shall consider absoluteness in models determined by these F-definitions with certain specifications; of particular interest here are the restricted formulas. The general results obtained in this Section will be used throughout the book.

a) **TC** *as an* $\in$*-theory*

1301. It follows immediately from the definition 1250 of an $\in$-theory that **TE** (i.e. the theory with axioms (F1, F2)) is an $\in$-theory. Moreover, (d1) is a definition in **TE** and (d2) is a definition in (**TE**, d1), so that (**TE**, d1, d2) is an $\in$-theory. If we already know that a theory **T** stronger than (**TE**, d1, d2) is an $\in$-theory and want to add an axiom φ as a proper axiom we must verify that it is fundamental in **T**. The following Metalemma is sometimes useful.

1302. METALEMMA. Let **T** be an $\in$-theory stronger than (**TE**, d1, d2). If φ is a formula normal in **T** whose only free variables are class variables and which does not contain any constants then φ is fundamental in **T**.

Demonstration. We show by induction that for every normal **T**-formula $\varphi(x, \bullet, X, \bullet)$ whose only free variables are class and/or set variables and which does not contain any constants there is a fundamental formula $\psi(Z, \bullet, X, \bullet)$ such that

$$(\mathbf{TE}, \mathrm{d1}, \mathrm{d2}) \vdash (x = Z \,\&\, \bullet) \rightarrow [\varphi(x, \bullet, X, \bullet) \equiv \psi(Z, \bullet, X, \bullet)] .$$

If φ is $x_1 \in x_2$ or $x_1 \in X_1$ then for ψ take $Z_1 \in Z_2$ or $Z_1 \in X_1$ respectively. The induction step for & and $\neg$ is obvious. Finally, if

$$(\mathbf{TE}, \mathrm{d1}, \mathrm{d2}) \vdash (y = U \,\&\, x = Z \,\&\, \bullet) \rightarrow$$
$$\rightarrow [\varphi(y, x, \bullet, X, \bullet) \equiv \psi(U, Z, \bullet, X, \bullet)] ,$$

then

$$(\mathbf{TE}, \mathrm{d1}, \mathrm{d2}) \vdash (x = Z \,\&\, \bullet) \rightarrow [(\exists y)\, \varphi(y, x, \bullet, X, \bullet) \equiv$$
$$\equiv (\exists U, W)(U \in W \,\&\, \psi(U, Z, \bullet, X, \bullet)] .$$

1303. METATHEOREM. **TC** is an $\in$-theory.

Demonstration. **TC** is an extension of (**TE**, d1, d2). Axiom (A1) is a closed formula normal in (**TE**, d1, d2) and is therefore fundamental in this theory. (d3)–(d6) are again definitions. The axioms (B1) to (B7) are closed formulas of the form $(\forall X, \bullet)(\exists Z)\, \varphi(X, \bullet, Z)$ where φ is normal in (**TE**, d1, d2, A1, d3); for, by 1126 the formula $\langle z_1, z_2 \rangle \in Y$ is normal in this theory. All free variables of φ are class-variables and no constants occur in φ; therefore axioms (B1) to (B7) are fundamental formulas in the latter theory. Thus **TC** is an $\in$-theory since all of the remaining axioms are definitions.

Remark. All extensions of **TC** considered in the sequel are $\in$-theories; this fact will be evident in each particular case and will not require explicit verification.

Gödelian terms will play an important role in our considerations. For the moment we shall only prove a lemma concerning certain "commutativities" between gödelian operations and the operation of intersection.

1304. LEMMA (**TC**). The following holds for arbitrary classes X, Y, Z:

$$\mathbf{E}(X) \cap Z = \mathbf{E}(X \cap Z) ; \quad (X - Y) \cap Z = (X \cap Z) - (Y \cap Z) ;$$
$$(X \upharpoonright Y) \cap Z = (X \cap Z) \upharpoonright (Y \cap \mathbf{D}(Z)) ; \quad \mathbf{Cnv}(X) \cap Z = \mathbf{Cnv}(X \cap \mathbf{Cnv}(Z)) ;$$
$$\mathbf{Cnv}_3(X) \cap Z = \mathbf{Cnv}_3(X \cap \mathbf{Cnv}_3(\mathbf{Cnv}_3(Z))) .$$

The proofs of these equations are elementary and can be left to the reader. Note that we do not state anything about the operation **D**.

b) *The normal* F-*definition*

1305. METADEFINITION. The pair of formulas

(1) $$(\forall X)[(\exists X^*)(X^* = X) \equiv \mathrm{Sat}_{\mathbf{R}}(X)] \,\&\, sec\,,$$

(2) $$(\forall X^*, Y^*)(X^* \in^* Y^* \equiv (\exists z \in Y^*)(X^* = \mathbf{Ext}_{\mathbf{R}}(z))) \,\&\, sec$$

is called the normal F-definition (**R** being a constant) and is denoted by $\mathfrak{N}(\mathbf{R})$.

Thus classes in the sense of $\mathfrak{N}(\mathbf{R})$ are saturated parts of $\mathbf{C}(\mathbf{R})$.

1306. METATHEOREM. $\mathfrak{N}(\mathbf{R})$ with the specification Elk $(\mathbf{R})$ determines a model of **TC** in **TC**. Moreover, $\mathfrak{Dir}(\partial\mathbf{TC}/\partial\mathfrak{N}(\mathbf{R}))$ is a faithful model of **TC** in (**TC**, Elk $(\mathbf{R})$, $\partial\mathbf{TC}/\partial\mathfrak{N}(\mathbf{R})$).

Demonstration. We demonstrated already in Section 1 that $\mathfrak{Dir}(\partial\mathbf{TC}/\partial\mathfrak{N}(\mathbf{R}))$ is a model of **TC** in (**TC**, Elk $(\mathbf{R})$, $\partial\mathbf{TC}/\partial\mathfrak{N}(\mathbf{R})$);*) it remains to demonstrate that it is faithful. Observe that $\mathbf{R} = \mathbf{E}$ is a specification for $\mathfrak{N}(\mathbf{R})$ in **TC** stronger than Elk $(\mathbf{R})$ and such that $\mathfrak{Dir}(\partial\mathbf{TC}/\partial\mathfrak{N}(\mathbf{R}))$ as a model in (**TC**, $\mathbf{R} = \mathbf{E}$, $\partial\mathbf{TC}/\partial\mathfrak{N}(\mathbf{R})$) is equivalent to the identical model. Indeed, it is easy to see that the following formulas are provable in (**TC**, $\mathbf{R} = \mathbf{E}$, $\partial\mathbf{TC}/\partial\mathfrak{N}(\mathbf{R})$):

$$(\forall X)(\exists X^*)(X^* = X)\,, \quad X^* \in^* Y^* \equiv X^* \in Y^*\,.$$

Hence, by 1269 and 1270 the direct model of **TC** determined by $\mathfrak{N}(\mathbf{R})$ with the specification $\mathbf{R} = \mathbf{E}$ is faithful. By 1266(b) the direct model of **TC** determined by $\mathfrak{N}(\mathbf{R})$ with the specification Elk $(\mathbf{R})$ is also faithful.

We shall often denote $\mathfrak{Dir}(\partial\mathbf{T}/\partial\mathfrak{N}(\mathbf{R}))$ simply by $\mathfrak{N}(\mathbf{R})$ if the theory **T** is clear from the context.

1307. METALEMMA. The formulas $X \subseteq Y$ and $U = X - Y$ are absolute in $\mathfrak{N}(\mathbf{R})$ as a model in **TC**, Elk $(\mathbf{R})$, $\partial\mathbf{TC}/\partial\mathfrak{N}(\mathbf{R})$.

Demonstration. We proceed in the theory indicated. For any X^*, Y^*, $X^* \subseteq^* Y^* \equiv (\forall x^*)(x^* \in^* X^* \rightarrow x^* \in^* Y^*) \equiv (\forall z \in \mathbf{C}(\mathbf{R}))(z \in X^* \rightarrow z \in Y^*) \equiv$ $\equiv X^* \subseteq Y^*$ (recall that X^*, Y^* are saturated subclasses of $\mathbf{C}(\mathbf{R})$!). Further, we want to prove $U^* = X^* -^* Y^* \equiv U^* = X^* - Y^*$. This formula is equivalent to $(\forall u \in \mathbf{C}(\mathbf{R}))(u \in X^* -^* Y^* \equiv u \in X^* - Y^*)$. But $u \in X^* -^*$

*) The specification should be written in the form Elk(**R**) & *sec*, i.e. Elk (**R**) & $(\exists X)(\mathbf{R} = X)$; but we shall always omit the part *sec* in specifications.

$-^* Y^* \equiv \mathbf{Ext}_{\mathbf{R}}(u) \in^* X^* -^* Y^* \equiv \mathbf{Ext}_{\mathbf{R}}(u) \in^* X^* \,\&\, \mathbf{Ext}_{\mathbf{R}}(u) \notin^* Y^* \equiv u \in X^*$ $\&\, u \notin Y^* \equiv u \in X^* - Y^*$.

Unfortunately the formula Rel (X) is not absolute. However the following device enables us to handle relations in the sense of $\mathfrak{N}(\mathbf{R})$.

1308. DEFINITION. (**TC**, Elk $(\mathbf{R})$, ∂**TC**$/\partial\mathfrak{N}(\mathbf{R})$). For any X^* such that Rel* (X^*) we denote by **Dec** (X^*) the unique relation Z defined as follows:

$$\langle u, v\rangle \in Z \equiv .\ u, v \in \mathbf{C}(\mathbf{R}) \,\&\, \langle \mathbf{Ext}_{\mathbf{R}}(u), \mathbf{Ext}_{\mathbf{R}}(v)\rangle^* \in^* X^* .$$

Dec (X^*) is called the *decoded relation.*

1309. LEMMA (**TC**, Elk $(\mathbf{R})$, ∂**TC**$/\partial\mathfrak{N}(\mathbf{R})$). If Rel* (X^*) and $Z = \mathbf{Dec}(X^*)$ then

(a) $\mathbf{D}^*(X^*) = \mathbf{D}(Z), \quad \mathbf{W}^*(X^*) = \mathbf{W}(Z), \quad \mathbf{C}^*(X^*) = \mathbf{C}(Z);$

(b) $\mathbf{Ext}^*_{X^*}(\mathbf{Ext}_{\mathbf{R}}(u)) = \mathbf{Ext}_Z(u)$ for any $u \in \mathbf{C}(\mathbf{R})$.

Proof. (a) If $u \in \mathbf{D}^*(X^*)$ then $\mathbf{Ext}_{\mathbf{R}}(u) \in^* \mathbf{D}^*(X^*)$ and there exists v such that $\langle \mathbf{Ext}_{\mathbf{R}}(v), \mathbf{Ext}_{\mathbf{R}}(u)\rangle^* \in^* X^*$; hence $\langle v, u\rangle \in Z$, i.e. $u \in \mathbf{D}(Z)$. The proof of the converse is analogous, and similarly for **W** and **C**.

(b) $v \in \mathbf{Ext}_Z(u)$ iff $\langle v, u\rangle \in Z$ iff $\langle \mathbf{Ext}_{\mathbf{R}}(v), \mathbf{Ext}_{\mathbf{R}}(u)\rangle^* \in^* X^*$ iff $\mathbf{Ext}_{\mathbf{R}}(v) \in^*$ $\in^* \mathbf{Ext}^*_{X^*}(\mathbf{Ext}_{\mathbf{R}}(u))$ iff $v \in \mathbf{Ext}^*_{X^*}(\mathbf{Ext}_{\mathbf{R}}(u))$.

1310. DEFINITION (**TC**). A relation H is a *morphism* between two relations R_1 and R_2 if $\mathbf{D}(H) = \mathbf{C}(R_1)$, $\mathbf{W}(H) = \mathbf{C}(R_2)$, and for all $x, y \in \mathbf{C}(R_1)$ and all $u, v \in \mathbf{C}(R_2)$ $uHx \,\&\, vHy$ implies $xR_1y \equiv uR_2v$. If H is one-to-one we call it an *isomorphism between R_1 and R_2*. In this case, if $P_1 \subseteq \mathbf{C}(R_1)$, $P_2 \subseteq \mathbf{C}(R_2)$ and $H''P_1 = P_2$, then H is also called an *isomorphism between P_1 and P_2 with respect to R_1 and R_2*.

1311. LEMMA (**TC**). If H is a morphism between relations R_1 and R_2 then

(1) **Cnv** (H) is a morphism between R_2 and R_1,

(2) uHx implies $\mathbf{Ext}_{R_2}(u) = H''\,\mathbf{Ext}_{R_1}(x)$,

(3) $\mathrm{Sat}_{R_1}(X)$ implies $\mathrm{Sat}_{R_2}(H''X)$,

(4) R_1 is extensional iff R_2 is extensional.

Proof. (1) Obvious. (2) Suppose first that $v \in \mathbf{Ext}_{R_2}(u)$. We have $\langle v, u\rangle \in$

$\in R_2$; if vHy then $\langle y, x\rangle \in R_1$, i.e. $y \in \mathbf{Ext}_{R_1}(x)$. Hence $v \in H\text{''}\,\mathbf{Ext}_{R_1}(x)$. Now suppose that $v \in H\text{''}\,\mathbf{Ext}_{R_1}(x)$. We have vHy for some y such that $\langle y, x\rangle \in R_1$; hence $\langle v, u\rangle \in R_2$, i.e. $v \in \mathbf{Ext}_{R_2}(u)$. (3) Let X be a saturated part of $\mathbf{C}(R_1)$, let $x \in X$ and let uHx; suppose that $\mathbf{Ext}_{R_2}(u) = \mathbf{Ext}_{R_2}(v)$ and that vHy. By (1) and (2) we have $\mathbf{Ext}_{R_1}(x) = (\mathbf{Cnv}(H))\text{''}(\mathbf{Ext}_{R_2}(u)) = (\mathbf{Cnv}(H))\text{''}(\mathbf{Ext}_{R_2}(v)) = \mathbf{Ext}_{R_1}(y)$; hence $y \in X$ and $v \in H\text{''}X$. Thus $H\text{''}X$ is a saturated part of $\mathbf{C}(R_2)$. (4) Suppose that R_1 is extensional, i.e. that $(\forall x \in \mathbf{C}(R_1))(\mathrm{Sat}_{R_1}(\mathbf{Ext}_{R_1}(x)))$. If $u \in \mathbf{C}(R_2)$ and uHx then $\mathbf{Ext}_{R_2}(u) = H\text{''}(\mathbf{Ext}_{R_1}(x))$ and $\mathrm{Sat}_{R_2}(H\text{''}(\mathbf{Ext}_{R_1}(x)))$. Hence R_2 is extensional.

We sometimes encounter the following situation: we have a specification $\pi_1(\mathbf{R})$ for $\mathfrak{N}(\mathbf{R})$ in a theory $\mathbf{S}$ stronger than $\mathbf{TC}$ and another specification $\pi_2(\mathbf{R})$ for $\mathfrak{N}(\mathbf{R})$ and we can prove in $\mathbf{S}$ that for every relation X such that $\pi_2(X)$ there is a morphic relation Y such that $\pi_1(Y)$. We ask what can be said about the corresponding direct models.

1312. METATHEOREM. Let $\mathbf{S}$ be an $\in$-theory stronger than $\mathbf{TC}$ and let $\pi_1(\mathbf{R})$, $\pi_2(\mathbf{R})$ be specifications for $\mathfrak{N}(\mathbf{R})$ in $\mathbf{S}$. Suppose that the following is provable in $\mathbf{S}$: For every X such that $\pi_2(X)$ there is a Y such that $\pi_1(Y)$ and a morphism H between X and Y. Then any formula which holds in $\mathfrak{N}(\mathbf{R})$ as a model (of the theory with the fundamental language and no special axioms) in $\mathbf{S}$, $\pi_1(\mathbf{R})$, $\mathfrak{N}(\mathbf{R})$ also holds in $\mathfrak{N}(\mathbf{R})$ as a model in $\mathbf{S}$, $\pi_2(\mathbf{R})$, $\mathfrak{N}(\mathbf{R})$.

Demonstration. Consider the conservative extension $\hat{\mathbf{S}} = (\mathbf{S}, \pi_2(\mathbf{R}), \mathfrak{N}(\mathbf{R}))$ of $\mathbf{S}$. In $\hat{\mathbf{S}}$ we have variables X^* and the predicate $\in^*$. Let $\delta(\mathbf{R}, X, Z)$ be a formula saying "(Z is a morphism of $\mathbf{R}$ and X) and $\pi_1(X)$". By assumption, $\hat{\mathbf{S}} \vdash (\exists X, Z)\,\delta(\mathbf{R}, X, Z)$; fix $\mathbf{S}, \mathbf{H}$ in $\hat{\mathbf{S}}$ by $\delta(\mathbf{R}, \mathbf{S}, \mathbf{H})$. Let $\mathfrak{N}(\mathrm{S})$ be the F-definition obtained from $\mathfrak{N}(\mathbf{R})$ by replacing $\mathbf{R}$ by S, X^* and Y^* by $X^{\square}$ and $Y^{\square}$ respectively, and $\in^*$ by $\in^{\square}$; thus $\mathfrak{N}(\mathrm{S})$ is a "copy" of $\mathfrak{N}(\mathbf{R})$. We have $\hat{\mathbf{S}}, \delta(\mathbf{R}, \mathrm{S}, \mathbf{H}) \vdash \pi_1(\mathrm{S})$ and obviously $\mathfrak{N}(\mathrm{S})$ is an F-definition in $(\hat{\mathbf{S}}, \delta(\mathbf{R}, \mathrm{S}, \mathbf{H}))$; furthermore, $(\hat{\mathbf{S}}, \delta(\mathbf{R}, \mathrm{S}, \mathbf{H}), \mathfrak{N}(\mathrm{S}))$ is stronger than $(\mathbf{S}, \pi_1(\mathrm{S}), \mathfrak{N}(\mathrm{S}))$. (See diagram 1313.)

1313.

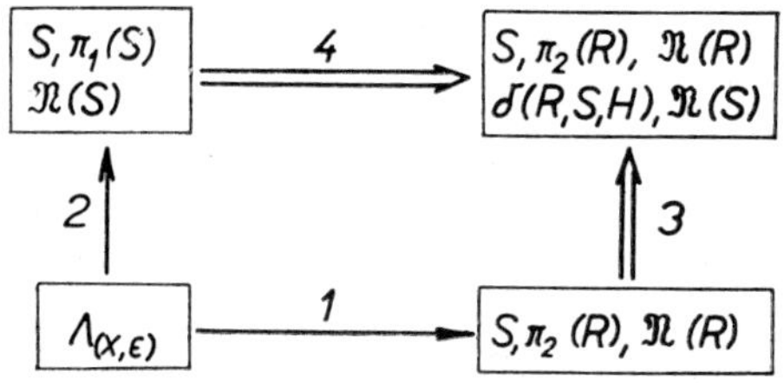

From Lemma 1311 we easily deduce that the following is provable in $(\hat{\mathbf{S}}, \delta(\mathbf{R}, \mathbf{S}, \mathbf{H}), \mathfrak{N}(\mathbf{S}))$:

(a) $(\forall X^*)(\exists! X^\square)(X^\square = \mathbf{H}\text{''}X^*), (\forall X^\square)(\exists! X^*)(X^\square = \mathbf{H}\text{''}X^*)$,

(b) $(X^\square = \mathbf{H}\text{''}X^* \,\&\, Y^\square = \mathbf{H}\text{''}Y^*) \rightarrow [X^* \in^* Y^* \equiv X^\square \in^\square Y^\square]$.

Consequently, if $\varphi(X, \ldots, Y)$ is a fundamental formula then

(c) $\hat{\mathbf{S}}, \delta(\mathbf{R}, \mathbf{S}, \mathbf{H}), \mathfrak{N}(\mathbf{S}) \vdash (X^\square = \mathbf{H}\text{''}X^* \,\&\, \bullet) \rightarrow [\varphi^\square(X^\square, \bullet) \equiv \varphi^*(X^*, \bullet)]$ is provable. If φ is a FF holding in $\mathfrak{N}(\mathbf{S})$ as a model in $(\mathbf{S}, \pi_1(\mathbf{S}), \mathfrak{N}(\mathbf{S}))$ (arrow 2) then φ holds in $\mathfrak{N}(\mathbf{S})$ as a model in $(\hat{\mathbf{S}}, \delta(\mathbf{R}, \mathbf{S}, \mathbf{H}), \mathfrak{N}(\mathbf{S}))$ (arrows $2 * 4$), i.e. $\varphi^\square$ is provable in the latter theory. By (a) and (c), φ^* is also provable in this theory. But φ^* is an $\hat{\mathbf{S}}$-formula and $(\hat{\mathbf{S}}, \delta(\mathbf{R}, \mathbf{S}, \mathbf{H}), \mathfrak{N}(\mathbf{S}))$ is a conservative extension of $\hat{\mathbf{S}}$ (arrow 3), hence $\hat{\mathbf{S}} \vdash \varphi^*$ i.e. φ holds in 1.

c) *The standard* F-*definition*

We shall often be concerned with relations which are sub-relations of the "membership" relation **E**. Thus we now state a theorem which gives sufficient conditions for such a relation to be **E**-like.

1314. DEFINITION (**TC**). X is a **V**-like class (Vlk (X)) if it has the following properties:

(i) X is not empty,

(ii) $u, v \in X \equiv \{u, v\} \in X$,

(iii) $(u, v \in X \,\&\, u \neq v) \rightarrow u \cap X \neq v \cap X$.

1315. LEMMA (**TC**). (a) If X is **V**-like then $\mathbf{E} \cap X$ is **E**-like (and, moreover, strongly extensional) and $X^2 = X \cap \mathbf{V}^2$. (b) **V** is **V**-like.

Proof. Suppose X is **V**-like. From (ii) we have $X^2 = X \cap \mathbf{V}^2$. Consequently, $\mathbf{E} \cap X = \mathbf{E} \cap X^2$ and $\mathbf{Ext}_{\mathbf{E} \cap X}(u) = u \cap X$ for all $u \in X$. Thus $\mathbf{E} \cap X$ is strongly extensional by (iii). By (i) X is non-empty; if $u \in X$ then by (ii) $\{u\} \in X$ and so $\langle u, \{u\} \rangle \in \mathbf{E} \cap X$. Hence $\mathbf{E} \cap X$ is non-empty. If $u, v \in X$ then $\{u, v\} \in X$ and $\mathbf{Ext}_{\mathbf{E} \cap X}(\{u, v\}) = \{u, v\}$ so that $\mathbf{E} \cap X$ is a pairing relation. Hence $\mathbf{E} \cap X$ is **E**-like. (b) is obvious.

1316. METADEFINITION. The pair of formulas

(3) $$(\forall X)[(\exists X^*)(X^* = X) \equiv X \subseteq \mathbf{M}] \,\&\, sec\,,$$

(4) $$(\forall X^*, Y^*)(X^* \in^* Y^* \equiv (\exists z \in Y^*)(X^* = z \cap \mathbf{M}))\ \&\ sec$$

is called the *standard* F-*definition* and denoted by $\mathfrak{St}(\mathbf{M})$.

Thus classes in the sense of $\mathfrak{St}(\mathbf{M})$ are subclasses of **M**.

1317. METATHEOREM. The F-definition $\mathfrak{St}(\mathbf{M})$ with the specification Vlk $(\mathbf{M})$ determines a model of **TC** in **TC**. Moreover, the model $\mathfrak{Dir}(\partial\mathbf{TC}/\partial\mathfrak{St}(\mathfrak{M}))$ is a faithful model of **TC** in **TC**, Vlk $(\mathbf{M})$, $\mathfrak{St}(\mathbf{M})$.

Demonstration. (1) We observe that the standard F-definition is closely related to the normal F-definition; it is in a sense a particular case. Consider Diagram 1317a:

1317a

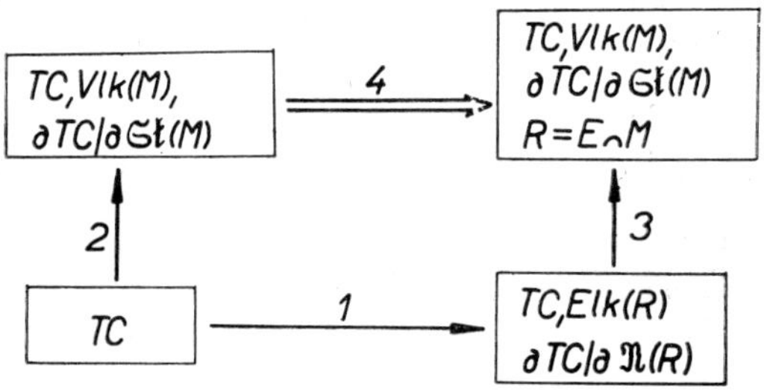

Arrow 1 is $\mathfrak{Dir}(\partial\mathbf{TC}/\partial\mathfrak{N}(\mathbf{R}))$; which we know to be a model. Arrow 2 is $\mathfrak{Dir}(\partial\mathbf{TC}/\partial\mathfrak{St}(\mathbf{M}))$; we want to demonstrate that it is a model. Arrow 4 is evidently a faithful identity model. Arrow 3 is the direct translation which translates the symbols of (**TC**, Elk $(\mathbf{R})$) identically and the symbols defined by $\partial\mathbf{TC}/\partial\mathfrak{N}(\mathbf{R})$ onto the corresponding symbols defined by $\partial\mathbf{TC}/\partial\mathfrak{St}(\mathbf{M})$. The diagram obviously commutes because the image of a **TC**-formula by 2 (or, equivalently, by $2 * 4$) coincides with the image by $1 * 3$. We demonstrate that arrow 3 is a model. All axioms of **TC** hold in 3 since they are translated identically. By Lemma 1315 Elk $(\mathbf{E} \cap \mathbf{M})$ is provable in the terminal theory of 3. Now consider the F-definition $\mathfrak{N}(\mathbf{R})$ (cf. the formulas (1), (2) above). To prove the images of these formulas in 3 it suffices to show that $X \subseteq \mathbf{M} \equiv . X \subseteq \mathbf{C}(\mathbf{R})\ \&\ \mathrm{Sat}_{\mathbf{R}}(X)$, $(\exists z \in Y^*)(X^* = \mathbf{Ext}_{\mathbf{R}}(z)) \equiv (\exists z \in Y^*)(X^* = z \cap \mathbf{M})$. But this follows (in the terminal theory of 3) from the fact that $\mathbf{R} = \mathbf{M} \cap \mathbf{E}$, sExtl $(\mathbf{R})$ and $\mathbf{Ext}_{\mathbf{R}}(z) = z \cap \mathbf{M}$. Hence both (1) and (2) hold in 3. Then all members of $\partial\mathbf{TC}/\partial\mathfrak{N}(\mathbf{R})$ hold in 3 because the definitions following $\mathfrak{N}(\mathbf{R})$ in $\partial\mathbf{TC}/\partial\mathfrak{N}(\mathbf{R})$ are translated into precisely the definitions following $\mathfrak{St}(\mathbf{M})$ in $\partial\mathbf{TC}/\partial\mathfrak{St}(\mathbf{M})$. Thus 3 is a model. Therefore all axioms of **TC** hold in $2 * 4$ (which is equivalent to $1 * 3$); but the image of an arbitrary axiom φ of **TC** in $2 * 4$ does not contain the constant **R** and, since 4 is faithful, φ holds in 2. This demonstrates that 2 is a model of **TC**.

(2) There is a specification for $\mathfrak{St}(\mathbf{M})$ in **TC** stronger than $\mathrm{Vlk}(\mathbf{M})$, namely $\mathbf{M} = \mathbf{V}$. By 1269, 1270 $\mathfrak{Dir}(\partial\mathbf{TC}/\partial\mathfrak{St}(\mathbf{M}))$ is equivalent to the identity model as a model of **TC** in **TC**, $\mathbf{M} = \mathbf{V}$, $\mathfrak{St}(\mathbf{M})$. Hence by 1266(b) $\mathfrak{Dir}(\partial\mathbf{TC}/\partial\mathfrak{St}(\mathbf{M}))$ is a faithful model of **TC** in **TC**, $\mathrm{Vlk}(\mathbf{M}), \partial\mathbf{TC}/\partial\mathfrak{St}(\mathbf{M})$.

The next Metatheorem tells more on the relation of the standard F-definition to the normal F-definition.

1318. METATHEOREM. Let **T** be a theory stronger than **TC**, let $\pi(\mathbf{R})$ be a specification for $\mathfrak{N}(\mathbf{R})$ in **T** stronger than $\mathrm{Elk}(\mathbf{R})$ and let $\psi(\mathbf{M})$ be a specification for $\mathfrak{St}(\mathbf{M})$ in **T** stronger than $\mathrm{Vlk}(\mathbf{M})$. Suppose **T**, $\psi(\mathbf{M}) \vdash \pi(\mathbf{E} \cap \mathbf{M})$. Then for any **TC**-formula φ we have:

(a) If φ holds in $\mathfrak{Dir}(\partial\mathbf{TC}/\partial\mathfrak{N}(\mathbf{R}))$ as a model in **T**, $\pi(\mathbf{R}), \partial\mathbf{TC}/\partial\mathfrak{N}(\mathbf{R})$ then φ holds in $\mathfrak{Dir}(\partial\mathbf{TC}/\partial\mathfrak{St}(\mathbf{M}))$ as a model in **T**, $\psi(\mathbf{M}), \partial\mathbf{TC}/\partial\mathfrak{St}(\mathbf{M})$.

(b) If φ is absolute from above (from below) in $\mathfrak{Dir}(\partial\mathbf{TC}/\partial\mathfrak{N}(\mathbf{R}))$ as a model in **T**, $\pi(\mathbf{R}), \partial\mathbf{TC}/\partial\mathfrak{N}(\mathbf{R})$) then φ is absolute from above (from below) in $\mathfrak{Dir}(\partial\mathbf{TC}/\partial\mathfrak{St}(\mathbf{M}))$ as a model in **T**, $\psi(\mathbf{M}), \partial\mathbf{TC}/\partial\mathfrak{St}(\mathbf{M})$.

Demonstration. Consider the following diagram:

1318a

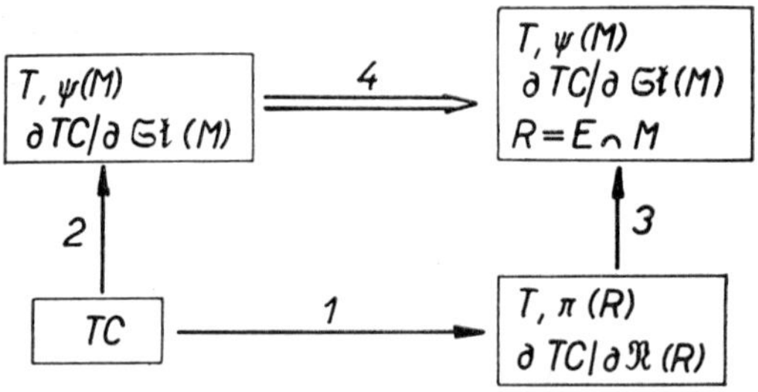

Here the arrows have the same meaning as in Diagram 1317a. We may suppose w.l.o.g. that the language of the derivation $\partial\mathbf{TC}/\partial\mathfrak{N}(\mathbf{R})$ is the same as the language of $\partial\mathbf{TC}/\partial\mathfrak{St}(\mathbf{M})$ and we denote images of **TC**-formulas in the corresponding direct model by asterisk. All arrows in Diagram 1318a are models (the assumption **T**, $\psi(\mathbf{M}) \vdash \pi(\mathbf{E} \cap \mathbf{M})$ ensures that 3 is a model). Neither the image φ^* of φ nor the formula $\varphi(X^*, \bullet) \to \varphi^*(X^*, \bullet)$ expressing the absoluteness from above contains the constant **R**. Hence if either formula is provable in the terminal theory of 1 then it is provable in the terminal theory of 3 (because it is translated identically by 3). Since 4 is faithful, this formula is provable in the terminal theory of 2 which completes the demonstration.

According to Lemma 1304 the following equalities may be proved in **TC**: $\mathbf{E}(X) \cap Z = \mathbf{E}(X \cap Z)$, $(X - Y) \cap Z = (X \cap Z) - (Y \cap Z)$.
Assuming that Z is **V**-like we can prove more:

1319. LEMMA (**TC**). If Z is **V**-like then

$$(X \upharpoonright Y) \cap Z = (X \cap Z) \upharpoonright (Y \cap Z), \quad \mathbf{Cnv}(X) \cap Z = \mathbf{Cnv}(X \cap Z),$$
$$\mathbf{Cnv}_3(X) \cap Z = \mathbf{Cnv}_3(X \cap Z).$$

(We may summarize these equalities in the form $\mathbf{F}_i(X, Y) \cap Z = \mathbf{F}_i(X \cap Z, Y \cap Z)$ for $i = 2, 3, 5, 6, 7$; where $\mathbf{F}_i$ denotes the i-th gödelian operation.)

Proof. Suppose Z is **V**-like. Since $\langle y, x \rangle \in Z \rightarrow x \in Z$ we have $\mathbf{D}(Z) \subseteq Z$; hence, by 1304,

$$(X \upharpoonright Y) \cap Z = (X \cap Z) \upharpoonright (Y \cap \mathbf{D}(Z)) \subseteq (X \cap Z) \upharpoonright (Y \cap Z) \subseteq$$
$$\subseteq (X \upharpoonright Y) \cap Z.$$

We have $\mathbf{Cnv}(Z) \subseteq Z$, $U \subseteq Z \rightarrow \mathbf{Cnv}(U) \subseteq Z$; hence

$$\mathbf{Cnv}(X) \cap Z = \mathbf{Cnv}(X \cap \mathbf{Cnv}(Z)) \subseteq \mathbf{Cnv}(X \cap Z) \subseteq \mathbf{Cnv}(X) \cap Z.$$

$\mathbf{Cnv}_3$ is treated similarly.

1320. THEOREM (**TC**). Let Z be **V**-like. If $X, Y \subseteq Z$ then $\mathbf{F}_i(X, Y) \subseteq Z$ for $i = 2, \ldots, 7$.

Proof. For $i \neq 4$ this follows from the preceding lemma; indeed, if $X, Y \subseteq Z$ then $X \cap Z = X$, and $Y \cap Z = Y$ so that $\mathbf{F}_i(X, Y) = \mathbf{F}_i(X, Y) \cap Z$, i.e. $\mathbf{F}_i(X, Y) \subseteq Z$. Furthermore, $X \subseteq Z \rightarrow \mathbf{D}(X) \subseteq Z$ follows from the fact that if $\langle u, v \rangle \in Z$ then $v \in Z$.

1321. LEMMA (**TC**). If Z_1, Z_2 are **V**-like classes and if F is an isomorphism of Z_1, Z_2 w.r.t. **E**, **E** then for arbitrary $X, Y \subseteq Z_1$ we have

$$F''(\mathbf{F}_i(X, Y)) = \mathbf{F}_i(F''X, F''Y) \quad \text{for} \quad i = 2, \ldots, 7.$$

Proof. This is trivial for $i = 3$. For other cases note that if $\{u, v\} \in Z_1$ then $\{F'u, F'v\} \in Z_2$ and $F'\{u, v\} = \{F'u, F'v\}$. Hence if $\langle u, v\rangle \in Z_1$ then $F'\langle u, v\rangle = \langle F'u, F'v\rangle$. The lemma follows by an easy examination of cases $(i = 2, 4, \ldots, 7)$. E.g.

$$x \in F''\mathbf{D}(X) \equiv (\exists y \in \mathbf{D}(X))(x = F'y) \equiv (\exists y, u)(\langle u, y\rangle \in X \,\&\, x = F'y) \equiv$$
$$\equiv (\exists y, u)(\langle F'u, F'y\rangle \in F''X \,\&\, x = F'y) \equiv$$
$$\equiv (\exists v)(\langle v, x\rangle \in F''X) \equiv x \in \mathbf{D}(F''X).$$

1322. METATHEOREM. In the model $\mathfrak{Dir}(\partial\mathbf{TC}/\partial\mathfrak{St}(\mathbf{M}))$ as a model in **TC**, Vlk $(\mathbf{M})$, $\partial\mathbf{TC}/\partial\mathfrak{St}(\mathbf{M})$ (a) the formulas $X \subseteq Y$, Rel (X) are absolute, (b) the gödelian operations $\mathbf{F}_i(X, Y)$ $(i = 2, \ldots, 7)$ are absolute (i.e., the formulas $U = \mathbf{F}_i(X, Y)$ are absolute).

Demonstration. (a) The absoluteness of $X \subseteq Y$ is trivial. Let us proceed in the theory **TC**, Vlk $(\mathbf{M})$, $\partial\mathbf{TC}/\partial\mathfrak{St}(\mathbf{M})$. Recall that sets in the sense of $\mathfrak{St}(\mathbf{M})$ are of the form $x \cap \mathbf{M}$ where $x \in \mathbf{M}$. Further, if $\{u, v\} \in \mathbf{M}$ then $u, v \in \mathbf{M}$, $\{u, v\} \cap \mathbf{M} = \{u, v\}$ and $\{u, v\} = \{u \cap \mathbf{M}, v \cap \mathbf{M}\}^*$. Indeed, $u \cap \mathbf{M} \in^* \{u, v\}$, $v \cap \mathbf{M} \in^* \{u, v\}$ and if $w \cap \mathbf{M} \in^* \{u, v\}$ then either $w \cap \mathbf{M} = u \cap \mathbf{M}$ or $w \cap \mathbf{M} = v \cap \mathbf{M}$, hence, by (iii) in Definition 1314 $w = u$ or $w = v$. Consequently $\langle u, v\rangle = \langle u \cap \mathbf{M}, v \cap \mathbf{M}\rangle^*$ and $\mathrm{Rel}(X^*) \equiv \mathrm{Rel}^*(X^*)$ follows.

(b) The absoluteness of $U = \mathbf{F}_i(X, \bullet)$ means that $\mathbf{F}_i(X^*, \bullet) = \mathbf{F}_i^*(X^*, \bullet)$ is provable. This is obvious for $X - Y$. For the other operations recall

(1) $$u, v \in \mathbf{M} \rightarrow [u \in v \equiv u \cap \mathbf{M} \in^* v \cap \mathbf{M}],$$

(2) $$u, v \in \mathbf{M} \equiv \langle u, v\rangle \in \mathbf{M},$$

(3) $$u, v \in \mathbf{M} \rightarrow \langle u, v\rangle = \langle u \cap \mathbf{M}, v \cap \mathbf{M}\rangle^*,$$

(4) $$\mathbf{F}_i(X^*, \bullet) \subseteq \mathbf{M}.$$

Let us consider $\mathbf{E}(X)$ and $\mathbf{D}(X)$ for example. We suppose $x \in \mathbf{M}$.

$$x \in \mathbf{E}(X^*) \,.\equiv.\, (\exists u, v)(x = \langle u, v\rangle \,\&\, u \in v \,\&\, x \in X^*) \equiv$$
$$\equiv.\, (\exists u, v \in \mathbf{M})(x = \langle u, v\rangle \,\&\, u \in v \,\&\, x \in X^*) \equiv$$
$$\equiv.\, (\exists u^*, v^*)(x = \langle u^*, v^*\rangle^* \,\&\, u^* \in^* v^* \,\&\, x \in X^*) \equiv$$
$$\equiv.\, x \in^* \mathbf{E}^*(X^*) \,.\equiv.\, x \in \mathbf{E}^*(X^*);$$

$$\begin{aligned} x \in \mathbf{D}(X^*) &\equiv .\, (\exists y)(\langle y, x\rangle \in X^*) \equiv \\ &\equiv .\, (\exists y \in \mathbf{M})(\langle y, x\rangle \in X^*) \equiv \\ &\equiv .\, (\exists y^*)(\langle y^*, x \cap \mathbf{M}\rangle^* \in^* X^*) \equiv \\ &\equiv .\, (x \cap \mathbf{M}) \in^* \mathbf{D}^*(X^*) \,.\equiv .\, x \in \mathbf{D}^*(X^*)\,. \end{aligned}$$

d) A stronger specification for $\mathfrak{St}(\mathbf{M})$; *restricted formulas*

1323. Definition (**TC**). X is a *complete class* (Comp (X)) if X contains with every element u all elements of u, i.e. $(\forall u, v)\,(u \in v \,\&\, v \in X \,.\rightarrow u \in X)$.

Note that **TC** $\vdash$ Comp $(X) \equiv \bigcup X \subseteq X \equiv X \subseteq \mathbf{P}(X)$. Evidently, **TC** $\vdash$ $\vdash$ Comp $(\mathbf{V})$. Consequently, the formula Vlk $(\mathbf{M})$ & Comp $(\mathbf{M})$ is a specification for $\mathfrak{St}(\mathbf{M})$ in **TC** which is stronger than the specification Vlk $(\mathbf{M})$. First observe that in the theory **TC**, (Vlk $(\mathbf{M})$ & Comp $(\mathbf{M})$), $\mathfrak{St}(\mathbf{M})$) (which we denote by $\mathbf{TC}_{\mathrm{Comp}}$ in the present discussion) we can replace the definition of $\in^*$ by the following equivalent but simpler definition: $X^* \in^* Y^* \equiv X^* \in Y^*$. Indeed, **TC**, Comp $(\mathbf{M}) \vdash z \in \mathbf{M} \rightarrow z \cap \mathbf{M} = z$. Consequently the sort of variables for sets in the sense of $\mathfrak{St}(\mathbf{M})$ is subordinate to the sort of set variables, i.e. $\mathbf{TC}_{\mathrm{Comp}} \vdash \mathrm{M}^*(X^*) \rightarrow \mathrm{M}(X^*)$; thus sets in the sense of $\mathfrak{St}(\mathbf{M})$ are simply elements of $\mathbf{M}$. The formulas $x \in y$, $x \in Y$ are absolute. Further recall that Gödelian operations $\mathbf{F}_2$ to $\mathbf{F}_7$ are absolute (by 1322). We shall find a whole system of formulas (having a certain syntactic structure) which are absolute in $\mathfrak{St}(\mathbf{M})$ with the specification Vlk $(\mathbf{M})$ & Comp $(\mathbf{M})$.

1324. Metadefinition (restricted formulas). Let **T** be an $\in$-theory stronger than (**TE**, d1, d2).

I. (a) Every atomic normal **T**-formula is a *restricted* **T**-*formula.* (b) If φ, ψ are restricted **T**-formulas and x, y are set variables then $\varphi \,\&\, \psi$, $\neg\, \varphi$, $(\exists x)\,(x \in y \,\&\, \varphi)$ are *restricted* **T**-*formulas* (provided they are formulas). (c) Every restricted **T**-formula is obtained by a finite number of applications of the rule (b) beginning with formulas described sub (a). We say "RF" instead of "restricted **T**-formula".

II. A **T**-formula φ is said to be *restricted in* **T** if there is a restricted **T**-formula ψ such that **T** $\vdash \varphi \equiv \psi$.

1325. Metadefinition (set formulas). Set formulas are formulas of the language $(x, \in)$. "SF" means "set formula".

1326. Remarks. 1) We compare the notions FF, SF, NF and RF. The FF's are the formulas of the language $(X, \in)$, while the SF's are the formulas of the language $(x, \in)$. A formula φ is NF iff the two following conditions hold: (1) each atomic subformula of φ has the form $x \in Y$ where x is a set variable and Y is a constant or a variable different from x (thus no equality sign occurs); (2) only set variables are quantified. If, in addition, all quantifiers are restricted to set variabes then φ is a RF.

2) By a normal set formula (NSF) we shall mean a formula which is both NF and SF. Similarly we shall speak of RSF's.

1327. METATHEOREM. Every restricted **TC**-formula without constants is absolute in $\mathfrak{Dir}(\partial\mathbf{TC}/\partial\mathfrak{St}(\mathbf{M}))$ as a model of **TC** in **TC**, Vlk $(\mathbf{M})$ & & Comp $(\mathbf{M})$, $\partial\mathbf{TC}/\partial\mathfrak{St}(\mathbf{M})$.

Demonstration. The metatheorem is true for atomic RF's and the induction step is obvious for &, $\neg$. Suppose that $\varphi(y, x, \bullet)$ is absolute and consider $(\exists y \in x)\, \varphi(y, x, \bullet)$. This formula is absolute from below (cf. 1268); hence it remains to demonstrate that it is absolute from above. We proceed in $\mathbf{TC}_{\mathrm{Comp}}$. Suppose $x, \bullet \in \mathbf{M}$, $X, \bullet \subseteq \mathbf{M}$, $x^* = x, \bullet$, $X^* = X, \bullet$ and $(\exists y \in x)\, \varphi(y, x, \bullet)$. Then we have $(\exists y \in \mathbf{M})\,(y \in x \,\&\, \varphi(y, x, \bullet))$ and hence $(\exists y^*)\,(y^* \in^* x^* \,\&\, \varphi^*(y^*, x^*, \bullet))$.

RF's are very important. In the remainder of this Section we shall show that quite a number of formulas are restricted in **TC** (or in stronger theories). Evidently, $x = y$ is restricted in **TC** because it is equivalent to $(\forall u \in x)$ $(u \in y) \,\&\, (\forall u \in y)\,(u \in x)$. Further we have the following

1328. METALEMMA. Let us say – in the context of this metalemma only – that a term **t** is a *pairing term* if it is built up from set variables and the operation $\{\ \}$. Then the following holds:

(a) For any pairing term **t**, the formulas $y \in \mathbf{t}$ and $y = \mathbf{t}$ are RSF's in **TC**.

(b) Let **t** be a pairing term which is not a variable. If $u_1, \ldots, u_n$ are variables in **t** and if φ is a RSF in **TC** then $(\exists u_1, \bullet)\,(x = t \,\&\, \varphi)$ and $(\forall u_1, \bullet)\,(x = \mathbf{t} \rightarrow \varphi)$ are RSF's in **TC**.

Demonstration. (a) If **t** is a set-variable then $y \in \mathbf{t}$ and $y = \mathbf{t}$ are RSF in **TC**. Suppose that **t** is $\{\mathbf{t}_1, \mathbf{t}_2\}$ and that the assertion holds for $\mathbf{t}_1$ and $\mathbf{t}_2$. The formula $y \in \mathbf{t}$ is equivalent to $y = \mathbf{t}_1 \vee y = \mathbf{t}_2$ which is a RSF in **TC**; the formula $y = \mathbf{t}$ is equivalent to the formula $(\forall u \in y)\,(u \in \mathbf{t}) \,\&\, \mathbf{t}_1 \in y \,\&\, \mathbf{t}_2 \in y$ which is a RSF in **TC**, since $\mathbf{t}_i \in y$ is equivalent to a RSF in **TC**, namely $(\exists z \in y)\,(z = \mathbf{t}_i)$.

(b) If $\mathbf{t}$ is a pairing term which is not a variable then it has the form $\{\mathbf{t}_1, \mathbf{t}_2\}$ for some pairing terms $\mathbf{t}_1$ and $\mathbf{t}_2$.

We distinguish three cases as follows:

(i) if $\mathbf{t}_1$ is u_1 and $\mathbf{t}_2$ is u_2 then the formula $(\exists u_1, u_2)(x = \{u_1, u_2\} \,\&\, \varphi)$ is equivalent to the formula $(\exists u_1 \in x)(\exists u_2 \in x)(x = \{u_1, u_2\} \,\&\, \varphi)$ which is a RSF in **TC**; analogously for the formula $(\exists u_1)(x = \{u_1, u_2\} \,\&\, \varphi)$;

(ii) if $\mathbf{t}_1$ is u_1 and $\mathbf{t}_2$ is not a variable then the formula $(\exists u_1, \bullet)(x = \{u_1, \mathbf{t}_2\} \,\&\, \varphi)$ is equivalent to the formula $(\exists u_1 \in x)(\exists v_1 \in x)(x = \{u_1, v_1\} \,\&\, (\exists u_2, \bullet)(v_1 = \mathbf{t}_2 \,\&\, \varphi))$ which is a RSF in **TC** by induction hypothesis; the demonstration is analogous if $\mathbf{t}_2$ is a variable and $\mathbf{t}_1$ is not;

(iii) if neither $\mathbf{t}_1$ nor $\mathbf{t}_2$ is a variable then the formula $(\exists u_1, \bullet)(x = \mathbf{t} \,\&\, \varphi)$ is equivalent to the formula $(\exists v_1 \in x)(\exists v_2 \in x)(x = \{v_1, v_2\} \,\&\, (\exists u_1, \ldots, u_i)$ $(v_1 = \mathbf{t}_1 \,\&\, (\exists u_{i+1}, \bullet)(v_2 = \mathbf{t}_2 \,\&\, \varphi)))$ which is a RSF in **TC** by induction hypothesis. The assertion for $\forall$ can be derived by considering the formula $\neg \varphi$.

1329. Corollary. If the formula $\varphi(x, u_1, u_2)$ is a RSF in **TC** then the formula $(\exists u_1, u_2)(x = \langle u_1, u_2\rangle \,\&\, \varphi(x, u_1, u_2))$ is a RSF in **TC**; similarly for $\langle u_1, u_2, u_3\rangle$.

1330. Metatheorem. The following formulas are RSF's in **TC**:

$$y \in \mathbf{E}(x), \quad y \in x_1 - x_2, \quad y \in \mathbf{D}(x), \quad y \in x_1 \upharpoonright x_2,$$
$$y \in \mathbf{Cnv}(x), \quad y \in \mathbf{Cnv}_3(x), \quad y \in \bigcup(x), \quad y \in \mathbf{P}(x).$$

Demonstration. All of the equivalences below are provable in **TC** and their right-hand sides are RSF's in **TC** by the preceding metatheorem.

$$y \in \mathbf{E}(x) \equiv .\, (\exists u, v)(y = \langle u, v\rangle \,\&\, u \in v) \,\&\, y \in x,$$
$$y \in x_1 - x_2 \equiv .\, y \in x_1 \,\&\, y \notin x_2,$$
$$y \in \mathbf{D}(x) \equiv (\exists z \in x)(\exists u)(z = \langle u, y\rangle),$$
$$y \in x_1 \upharpoonright x_2 \equiv (\exists u, v)(y = \langle u, v\rangle \,\&\, y \in x_1 \,\&\, v \in x_2),$$
$$y \in \mathbf{Cnv}(x) \equiv (\exists u, v)(y = \langle u, v\rangle \,\&\, (\exists z \in x)(z = \langle v, u\rangle)),$$

$$y \in \mathbf{Cnv}_3(x) \equiv (\exists u, v, w)(y = \langle u, v, w\rangle \,\&\, (\exists z \in x)(z = \langle w, v, u\rangle)),$$
$$y \in \bigcup(x) \equiv (\exists u \in x)(y \in u),$$
$$y \in \mathbf{P}(x) \equiv (\forall u \in y)(u \in x).$$

1331. METATHEOREM. The following formulas are RSF's in **TC**:

$$y = \mathbf{E}(x)\,,\ y = x_1 - x_2\,,\ y = \mathbf{D}(x)\,,\ y = x_1 \upharpoonright x_2\,,$$
$$y = \mathbf{Cnv}(x)\,,\ y = \mathbf{Cnv}_3(x)\,,\ y = \bigcup(x)\,.$$

Demonstration. All of the equivalences below are provable in **TC**:

$$y = \mathbf{E}(x) \equiv .\,(\forall u \in y)(u \in \mathbf{E}(x))\,\&\,(\forall u \in x)(u \in \mathbf{E}(x) \rightarrow u \in y)\,,$$

similarly for $x_1 - x_2$ and $x_1 \upharpoonright x_2$,

$$y = \mathbf{D}(x) \equiv .\,(\forall u \in y)(u \in \mathbf{D}(x))\,\&$$
$$\&\,(\forall u \in x)(\forall s, t)(u = \langle s, t\rangle \rightarrow t \in x)\,,$$

$$y = \mathbf{Cnv}(x) \equiv .\,(\forall u \in y)(u \in \mathbf{Cnv}(x))\,\&$$
$$\&\,(\forall z \in x)(\forall s, t)(z = \langle s, t\rangle \rightarrow (\exists u \in y)(u = \langle t, s\rangle))\,,$$

similarly for $\mathbf{Cnv}_3(x)$,

$$y = \bigcup(x) \equiv .\,(\forall u \in y)(u \in \bigcup(x))\,\&\,(\forall v \in x)(\forall u \in v)(u \in y)\,.$$

(Note that nothing is asserted about $y = \mathbf{P}(x)$.)

1332. METALEMMA. (a) The formula $\mathrm{Un}(f)$ is a RSF in **TC**.

(b) If $\varphi(x, y_1, \bullet)$ is a RSF in **TC** then $\mathrm{Un}(f)\,\&\,\varphi(f\text{'}z, y_1, \bullet)$ is a RSF in **TC**.

Demonstration. The following equivalences are provable in **TC**:

$$\mathrm{Un}(f) \equiv (\forall u \in f)(\exists x, y)(u = \langle y, x\rangle)\,\&$$
$$\&\,(\forall u, v \in f)(\forall x, y, z)(u = \langle y, x\rangle\,\&\,v = \langle z, x\rangle\,.\rightarrow y = z))\,,$$

$$\mathrm{Un}(f)\,\&\,\varphi(f\text{'}z, y_1, \bullet)\,. \equiv$$
$$\equiv .\,\mathrm{Un}(f)\,\&\,(\exists u \in f)(\exists x)(u = \langle x, z\rangle\,\&\,\varphi(x, y_1, \bullet))\,.$$

SECTION 4

The theory of semisets

Semisets will now be defined as those classes which are included in sets (as subclasses). We shall present some axioms for semisets and collect them together into the axiom system of the theory of semisets (**TSS**). We shall also present a theory stronger than **TSS** and called the theory of sets (**TS**). The theory of semisets is unusual in several aspects; however it plays a fundamental role throughout the book. The theory of sets is equivalent to the usual (Bernays-Gödel) set theory but it will only be used exceptionally. From the point of view of the theory of semisets **TS** seems to be too restrictive, assuming simply that all semisets are sets. In this way some very interesting developments are excluded, especially the construction of certain models. On the other hand, the connection between **TSS** and **TS** is quite close. (1) These theories are equiconsistent, i.e. **TSS** is consistent iff **TS** is. (2) Moreover, upon the addition to **TSS** and **TS** of a very weak "regularity" axiom (D1) whose consistency with **TSS** (and with **TS**) will be proved, then we obtain theories **TSS**′ and **TS**′ in which the same statements concerning sets (set formulas) are provable. Thus nothing concerning sets is lost in **TSS**′. Indeed, one can construct a model of **TS**′ in **TSS**′ which, so to speak, only excludes certain proper classes. This construction which might be called "downward", will be described in the present Section. (The demonstration of the consistency of (D1) with **TSS** will be postponed to Chapt. III Sect. 1.) (3) There is, however, another important method for obtaining models of **TS** in certain extensions of **TSS** (consistent relative to **TSS**) which will be presented in Chapt. V Sect. 1; this is the "upward" construction. Here one constructs a (faithful) model of a certain extension of **TS**; this model has, roughly speaking, more sets than the original theory of semisets. (In particular, the independence of the continuum hypothesis can be proved in this way.) It is this possibility for constructing models which gives the theory of semisets its importance.

a) Semisets

1401. Definition (**TC**). A class X is a *semiset* ($\mathrm{Sm}(X)$) if it is a subclass of a set, i.e. $(\exists y)(X \subseteq y)$. We use variables σ, ϱ (and others if necessary) for semisets.

1402. Definition (**TC**). A class X is a *real class* ($\mathrm{Real}(X)$) if its intersection with every set is a set, i.e. $(\forall x)\, \mathrm{M}(X \cap x)$.

1403. Lemma (**TC**). (a) $\mathbf{V}$ is a real class. (b) 0 is a semiset. (c) Every set is a semiset. (d) Every subclass of a semiset is a semiset. (e) The intersection of a class with a set is a semiset. (f) If a class is a real class and a semiset then it is a set.

Proof. (a) to (e) are obvious. (f) Suppose a semiset σ is a real class. There is a set y such that $\sigma \subseteq y$ and so $\sigma \cap y = \sigma$. Since σ is real, $\sigma \cap y$ is a set.

We now introduce some notions concerning relations in the context of semisets. A function was defined in the usual way as a many-one relation ($\mathrm{Un}(R)$). This definition can be generalized so that every relation can be considered as a "mapping which assigns to each $y \in \mathbf{D}(R)$ the class $\mathbf{Ext}_R(x)$". Such a generalized mapping might be called a *functor*. The following lemma shows that this really is a generalization of the definition of a function:

1404. Lemma (**TC**). If F is a function then there is a unique relation R such that $\mathbf{D}(R) \subseteq \mathbf{D}(F)$ and

$$(\forall x \in \mathbf{D}(F))(F\text{'}x = \mathbf{Ext}_R(x)) .$$

Proof. Let $\langle z, x\rangle \in R \equiv z \in F\text{'}x$; clearly $\mathbf{Ext}_R(x) = F\text{'}x$. If $F\text{'}x = 0$ then $x \notin \mathbf{D}(R)$ and $\mathbf{Ext}_R(x) = 0 = F\text{'}x$.

Notice that the values of a functor may be proper classes. Functors whose values are semisets play an important role in our further investigations. We call such functors *regular relations* or *semiset-valued functors*.

1405. Definition (**TC**).

$$\mathrm{Reg}(R) \equiv .\ \mathrm{Rel}(R)\,\&\,(\forall x \in \mathbf{D}(R))(\mathrm{Sm}(\mathbf{Ext}_R(x))) .$$

Remark. The predicate Reg (R) is normal since it is equivalent in **TC** to the formula

$$\mathrm{Rel}(R) \,\&\, (\forall x \in \mathbf{D}(R))\,(\exists z)\,(\forall y)\,(\langle y, x\rangle \in R \rightarrow y \in z)\,.$$

Functors which are "one-to-one" will be called *nowhere constant.*

1406. DEFINITION (**TC**).

$\mathrm{Ncon}(R) \equiv .\ \mathrm{Rel}(R) \,\&\, (\forall x, y \in \mathbf{D}(R))\,(x \neq y \rightarrow \mathbf{Ext}_R(x) \neq \mathbf{Ext}_R(y))\,.$

1407. LEMMA (**TC**). (a) $\mathrm{sExtl}(R) \rightarrow \mathrm{Ncon}(R)$,

(b) $\mathrm{Un}(R) \rightarrow (\mathrm{Un}_2(R) \equiv \mathrm{Ncon}(R))$.

Proof. (a) $\mathrm{sExtl}(R)$ means that any two different elements of $\mathbf{C}(R)$ have different extensions; $\mathrm{Ncon}(R)$ follows immediately. (b) Suppose that R is many-one; then $\mathbf{Ext}_R(x) = \{R'x\}$. If $\mathrm{Un}_2(R)$ then for $x, y \in \mathbf{D}(R)$, $x \neq y$ we have $R'x \neq R'y$ and hence $\{R'x\} \neq \{R'y\}$; if $\mathrm{Ncon}(R)$ then for $x \neq y$ we have $\{R'x\} \neq \{R'y\}$ and so $R'x \neq R'y$.

1408. DEFINITION (**TC**). A relation R is an *exact functor* ($\mathrm{Exct}(R)$) if it is regular and nowhere constant.

1409. LEMMA (**TC**). $\mathrm{Un}_2(R) \rightarrow \mathrm{Exct}(R)$.

b) *Exact functors*

1410. We now present two axioms which will be denoted by (C1) and (C2).

(C1) Axiom of Infinity

$$(\exists x)\,(0 \in x \,\&\, (\forall y)\,(y \in x \rightarrow \{y\} \in x))\,.$$

(C2) Axiom of Exact Functors

$$\mathrm{Exct}(X) \rightarrow (\mathrm{Sm}(\mathbf{D}(X)) \equiv \mathrm{Sm}(\mathbf{W}(X)))\,.$$

According to (C1) there exists a nonempty set which contains the empty class and for each element y the one-element set $\{y\}$. We shall see later that this set is – in a precisely defined sense – infinite. However this axiom will not be used for the present; when it is used we shall mention the fact explicitly. Axiom (C2) is of basic importance. According to this axiom an exact functor has a small domain iff it has a small range, where by a small class we mean a semiset.

1411. THEOREM (**TC**, C2). For any semisets σ, ϱ the classes $\bigcup \sigma$, $\sigma \cup \varrho$, $\mathbf{P}(\sigma)$ and $\sigma \times \varrho$ are semisets.

Proof. (a) Let σ be a semiset; we prove that its sum is also a semiset. Since $\mathbf{E}$ is an exact functor, $\mathbf{E} \upharpoonright \sigma$ is also an exact functor and since $\mathbf{D}(\mathbf{E} \upharpoonright \sigma) \subseteq \sigma$ we have by (C2) that $\mathbf{W}(\mathbf{E} \upharpoonright \sigma)$ is a semiset. This proves the assertion, since $\bigcup \sigma = \mathbf{W}(\mathbf{E} \upharpoonright \sigma)$.

(b) Let $\sigma \subseteq x$, $\varrho \subseteq y$. Then $\sigma \cup \varrho \subseteq \bigcup(\{x, y\})$, the latter class is a semiset.

(c) Let σ be a semiset; we prove that $\mathbf{P}(\sigma)$ is a semiset. Set $\mathbf{E}_1 = \{\langle u, v\rangle; u \in v \,\&\, v \subseteq \sigma\}$. The relation $\mathbf{E}_1$ is an exact functor and $\mathbf{W}(\mathbf{E}_1) \subseteq \sigma$, hence $\mathbf{W}(\mathbf{E}_1)$ is a semiset. If the empty class 0 is not a set then $\mathbf{D}(\mathbf{E}_1) = \mathbf{P}(\sigma)$ and we are done. In the oppossite case $\mathbf{D}(\mathbf{E}_1) = \mathbf{P}(\sigma) - \{0\}$ and $\mathbf{P}(\sigma) = \mathbf{D}(\mathbf{E}_1) \cup \{0\}$, hence $\mathbf{P}(\sigma)$ is the union of two semisets and the assertion follows by (b).

(d)
$$\sigma \times \varrho \subseteq \mathbf{P}(\mathbf{P}(\sigma \cup \varrho)) .$$

1412. THEOREM (**TC**, C2). For any semisets σ, ϱ the following classes are semisets:

$$\mathbf{E}(\sigma), \ \sigma - \varrho, \ \mathbf{D}(\sigma), \ \sigma \upharpoonright \varrho, \ \mathbf{Cnv}(\sigma), \ \mathbf{Cnv}_3(\sigma) .$$

Proof. Let σ and ϱ be arbitrary semisets. Since $\mathbf{E}(\sigma)$, $\sigma - \varrho$ and $\sigma \upharpoonright \varrho$ are included in σ it follows (without using (C2)) that these classes are semisets. To prove that $\mathbf{D}(\sigma)$, $\mathbf{Cnv}(\sigma)$ and $\mathbf{Cnv}_3(\sigma)$ are semisets we use the preceding theorem; indeed, $\mathbf{D}(\sigma) \subseteq \bigcup \bigcup \sigma$, $\mathbf{Cnv}(\sigma) \subseteq \mathbf{PP}\bigcup\bigcup \sigma$ and $\mathbf{Cnv}_3(\sigma) \subseteq \mathbf{PP}(\sigma_0 \cup \mathbf{PP}\sigma_0)$ where $\sigma_0 = \bigcup\bigcup\sigma \cup \bigcup\bigcup\bigcup\bigcup\sigma$.

1413. COROLLARY (**TC**, C2). For any semisets σ, ϱ the following classes are semisets: $\sigma \cap \varrho$, $\mathbf{W}(\sigma)$, $\mathbf{C}(\sigma)$, $\sigma''\varrho$. (Recall that $\sigma \cap \varrho = \sigma - (\sigma - \varrho)$, $\mathbf{W}(\sigma) = \mathbf{D}(\mathbf{Cnv}(\sigma))$, $\mathbf{C}(\sigma) = \mathbf{W}(\sigma) \cup \mathbf{D}(\sigma)$, $\sigma''\varrho = \mathbf{W}(\sigma \upharpoonright \varrho)$.)

Axiom (C2) also has the following consequences:

1414. LEMMA (**TC**, C2).

(a)
$$\operatorname{Ncon}(X) \,\&\, \operatorname{Sm}(\mathbf{W}(X)) \,.\!\rightarrow \operatorname{Sm}(\mathbf{D}(X)) ,$$

(b) $$\operatorname{Reg}(X) \,\&\, \operatorname{Sm}(\mathbf{D}(X)) \,.\to \operatorname{Sm}(\mathbf{W}(X)),$$

(c) $$\operatorname{Reg}(X) \,\&\, \operatorname{Sm}(\mathbf{D}(X)) \,.\to \operatorname{Sm}(X),$$

(d) $$\operatorname{Ncon}(X) \,\&\, \operatorname{Sm}(\mathbf{W}(X)) \,.\to \operatorname{Sm}(X).$$

Proof. (a) If X is a relation and $\mathbf{W}(X)$ is a semiset then X is trivially regular; hence if X is nowhere constant then it is exact and the assertion follows by (C2). (b) Let X be a regular relation with a semiset domain. Define $\bar{R} = \{\langle\langle y, x\rangle, x\rangle; \langle y, x\rangle \in R\}$. Then $\mathbf{D}(R) = \mathbf{D}(\bar{R})$ and so $\mathbf{D}(\bar{R})$ is a semiset. For $x \in \mathbf{D}(R)$, $\mathbf{Ext}_{\bar{R}}(x) = \mathbf{Ext}_R(x) \times \{x\}$, hence $\mathbf{Ext}_{\bar{R}}(x)$ is a semiset by 1411. If $x \neq y$ then $\mathbf{Ext}_{\bar{R}}(x) \cap \mathbf{Ext}_{\bar{R}}(y) = 0$ and so $\bar{R}$ is nowhere constant. By (C2), $\mathbf{W}(\bar{R})$ is a semiset and $\mathbf{W}(R) = \mathbf{W}(\mathbf{W}(\bar{R}))$ is also a semiset by 1413. (c) If X is regular and $\mathbf{D}(X)$ is a semiset then, by (b), $\mathbf{W}(X)$ is a semiset; further $X \subseteq \mathbf{W}(X) \times \mathbf{D}(\mathrm{R})$ and so X is a semiset by 1411. (d) is proved similarly.

There is a specification for the F-definition $\mathfrak{N}(\mathbf{R})$ such that $\mathfrak{N}(\mathbf{R})$ with this specification determines a model of **TC**, C2 in **TC**, C2.
Define in **TC**:

1415. DEFINITION (**TC**). A relation X is *almost universal* ($\operatorname{AUn}(X)$) if for every semiset $\sigma \subseteq \mathbf{C}(X)$ there exists $z \in \mathbf{C}(X)$ such that $\sigma \subseteq \mathbf{Ext}_X(z)$.

Remarks (**TC**). 1) Note that $\mathbf{E}$ is regular and almost universal. 2) The formula $\operatorname{AUn}(X)$ is normal in **TC** because it is equivalent to $(\forall x)(\exists y \in \mathbf{C}(X))(x \cap \mathbf{C}(X) \subseteq \mathbf{Ext}_X(y))$. 3) If $\mathbf{C}(X)$ is real then X is almost universal iff $(\forall x \subseteq \mathbf{C}(X))(\exists y \in \mathbf{C}(X))(x \subseteq \mathbf{Ext}_X(y))$.

1416. METATHEOREM. Consider the formula

$$\operatorname{sExtl}(\mathbf{R}) \,\&\, \operatorname{Prg}(\mathbf{R}) \,\&\, \operatorname{Reg}(\mathbf{R}) \,\&\, \operatorname{AUn}(\mathbf{R})$$

which in the context of the present Metatheorem will be denoted by $\operatorname{Mrel}(\mathbf{R})$ ("$\mathbf{R}$ is a model relation"). $\operatorname{Mrel}(\mathbf{R})$ is a specification for $\mathfrak{N}(\mathbf{R})$ in **TC**, C2 such that $\mathfrak{N}(\mathbf{R})$ with this specification determines a model of **TC**, C2 in **TC**, C2. Moreover, $\mathfrak{Dir}(\partial\mathbf{TC}/\partial\mathfrak{N}(\mathbf{R}))$ is a faithful model of **TC**, C2 in **TC**, C2, $\operatorname{Mrel}(\mathbf{R})$, $\partial\mathbf{TC}/\partial\mathfrak{N}(\mathbf{R})$ and the formula $\operatorname{Sm}(X)$ is absolute in this model.

Demonstration. $\operatorname{Mrel}(\mathbf{R})$ is a specification because **TC**, C2 $\vdash$ $(\exists X)\operatorname{Mrel}(X)$; indeed, **TC** $\vdash \operatorname{Mrel}(\mathbf{E})$. Moreover, this specification is stronger than Elk $(\mathbf{R})$ since $\mathbf{TC} \vdash \mathrm{AUn}(X) \rightarrow X \neq 0$; indeed, 0 is a semiset, $0 \subseteq \subseteq \mathbf{C}(X)$ and there is a $y \in \mathbf{C}(X)$ such that $0 \subseteq \mathbf{Ext}_X(y)$. Hence $\mathfrak{N}(\mathbf{R})$ with Mrel $(\mathbf{R})$ certainly determines a model of **TC** in **TC**, C2 by Metatheorem 1266. It remains to prove (C2)* in the theory **TC**, C2, Mrel $(\mathbf{R})$, $\partial\mathbf{TC}/\partial\mathfrak{N}(\mathbf{R})$.

First we prove $\mathrm{Sm}(X^*) \equiv \mathrm{Sm}^*(X^*)$. If X^* is a semiset then by almost-universality there exists $z \in \mathbf{C}(\mathbf{R})$ such that $X^* \subseteq \mathbf{Ext}_{\mathbf{R}}(z)$; we have $\mathrm{M}^*(\mathbf{Ext}_{\mathbf{R}}(z))$ and $X^* \subseteq^* \mathbf{Ext}_{\mathbf{R}}(z)$, since $\subseteq$ is absolute; hence $\mathrm{Sm}^*(X^*)$. Conversely, if $\mathrm{Sm}^*(X^*)$ then $X^* \subseteq^* z^*$ and $z^* = \mathbf{Ext}_{\mathbf{R}}(z)$ for some z; by regularity, z^* is a semiset and hence X^* is also a semiset. Suppose now that Z^* is some relation in the model and consider the decoded relation $W = = \mathbf{Dec}(Z^*)$. First we show that $\mathrm{Exct}^*(Z^*)$ implies $\mathrm{Exct}(W)$. If $\mathrm{Reg}^*(Z^*)$ then (all extensions are semisets)*; for $u \in \mathbf{D}(W)$ we have $\mathbf{Ext}_{\mathbf{R}}(u) \in^* \in^* \mathbf{D}^*(Z^*)$, $\mathbf{Ext}_W(u) = \mathbf{Ext}^*_{Z^*}(\mathbf{Ext}_{\mathbf{R}}(u))$ and $\mathrm{Sm}^*(\mathbf{Ext}_W(u))$; hence $\mathrm{Sm}(\mathbf{Ext}_W(u))$ and so W is regular. If $\mathrm{Ncon}^*(Z^*)$, we let u and v be distinct elements of $\mathbf{D}(W)$; since $\mathbf{R}$ is strongly extensional we have $u^* = \mathbf{Ext}_{\mathbf{R}}(u) \neq \neq \mathbf{Ext}_{\mathbf{R}}(v) = v^*$ and $\mathbf{Ext}^*_{Z^*}(u^*) \neq \mathbf{Ext}^*_{Z^*}(v^*)$; hence $\mathbf{Ext}_W(u) \neq \mathbf{Ext}_W(v)$ and W is nowhere constant. Thus $\mathrm{Exct}^*(Z^*)$ implies $\mathrm{Exct}(W)$. Further we have $\mathbf{D}^*(Z^*) = \mathbf{D}(W)$ and $\mathbf{W}^*(Z) = \mathbf{W}(W)$; hence $\mathrm{Sm}^*(\mathbf{D}^*(Z^*)) \equiv \equiv \mathrm{Sm}(\mathbf{D}(W)) \equiv \mathrm{Sm}(\mathbf{W}(W)) \equiv \mathrm{Sm}^*(\mathbf{W}^*(Z^*))$. We have proved that $\mathfrak{N}(\mathbf{R})$ with the specification Mrel $(\mathbf{R})$ determines a model of **TC**, C2 in **TC**, C2. The faithfulness of the corresponding direct model is proved in exactly the samy way as in the demonstration of Metatheorem 1306.

1417. Definition (**TC**). A class X is called an *almost universal class* (AUncl (X)) if every subsemiset of X is a subsemiset of an element of X,i.e.

$$(\forall \sigma \subseteq X)(\exists y \in X)(\sigma \subseteq y)\,.$$

Evidently, X is an almost universal class iff $\mathbf{E} \cap X^2$ is an almost universal relation; $\mathbf{V}$ is an almost universal class.

1418. Metatheorem. The F-definition $\mathfrak{St}(\mathbf{M})$ with the specification Vlk $(\mathbf{M})$ & AUncl $(\mathbf{M})$ determines a model of **TC**, C2 in **TC**, C2. The model $\mathfrak{Dir}(\partial\mathbf{TC}/\partial\mathfrak{St}(\mathbf{M}))$ is a faithful model of **TC**, C2 in **TC**, C2, Vlk $(\mathbf{M})$ & & AUncl $(\mathbf{M})$, $\partial\mathbf{TC}/\partial\mathfrak{St}(\mathbf{M})$ and the formula $\mathrm{Sm}(X)$ is absolute in this model.

The demonstration is analogous to the demonstration of the preceding Metatheorem but is much simpler because Rel (X) is absolute. Thus it can be left to the reader.

c) *Comprehensiveness; the theory of semisets*

1419. METADEFINITION. Let **T** be a theory stronger than **TC** and let φ be a normal **T**-formula. Let $X, \bullet$ be all free variables of φ which are not set variables and let $x_1, \bullet$ be a sequence of distinct set variables which is free for φ (cf. 1115). If **T** is any term representing φ in **T** w.r.t. $x_1, \bullet$ then φ is said to be *comprehensive in* **T** w.r.t. $x_1, \bullet$ in case

$$(*) \qquad \mathbf{T} \vdash (\mathrm{Real}\,(X) \,\&\, \bullet) \rightarrow \mathrm{Real}\,(\mathbf{T})\,.$$

(If φ has no free non-set variables replace $(*)$ by $\mathbf{T} \vdash \mathrm{Real}\,(\mathbf{T})$.)

The formula φ is *fully comprehensive in* **T** if it is comprehensive w.r.t. any sequence of distinct variables free for φ.

1420. Note that if $(*)$ holds then

$$(**) \qquad \mathbf{T} \vdash (\mathrm{Real}\,(X) \,\&\, \bullet) \rightarrow (\forall a)\,(\exists b)\,(\forall u)\,(u \in b \equiv \\ \equiv .\, u \in a \,\&\, (\exists x_1, \bullet)\,(u = \langle x_1, \bullet \rangle \,\&\, \varphi))\,;$$

for, given a we may take $b = \mathbf{T} \cap a$. Thus, given real classes $X, \bullet$ and a set a, there is a subset b of a which consists of all tuples $\langle u_1, \bullet \rangle \in a$ satisfying φ. On the other hand, if $(**)$ holds and **T** is a term representing φ in **T** w.r.t. $x_1, \bullet$ then we can proceed as follows in **T**: If $X, \bullet$ are real classes, a is an arbitrary set and b is the set satisfying the condition of $(**)$ then necessarily $b = \mathbf{T} \cap a$; hence **T** is a real class. Thus we see that the conditions $(*)$ and $(**)$ are equivalent and, in particular, the notion of comprehensiveness w.r.t. $x_1, \bullet$ does not depend on the particular choice of the term **T**.

1421. We now add to **TC**, C2 a group of axioms which will ensure that in the extended theory every RSF is fully comprehensive. The situation here is analogous to the introduction of the axioms B1 – B7 to ensure the "class comprehension scheme"

$$(\exists A)\,(\forall x)\,(x \in A \equiv (\exists u_1, \bullet)\,(x = \langle u_1, \bullet \rangle \,\&\, \varphi))$$

(Metatheorem 1120 and Metadefinition 1115).

There it was sufficient to take as axioms seven particular cases of the scheme. For the restricted comprehension scheme it suffices to take six particular cases; since they are analogous to the axioms B2 to B7 we shall denote them by A2 to A7. We already know that the formulas $y \in \mathbf{F}_i$ ($i = 2, \ldots, 7$), where $\mathbf{F}_i$ are the gödelian terms $\mathbf{E}(x)$, $x_1 - x_2$, $\mathbf{D}(x)$, $x_1 \upharpoonright x_2$, $\mathbf{Cnv}(x)$ and $\mathbf{Cnv}_3(x)$, are restricted in **TC**; further we know that in **TC** + C2, the classes corresponding to these terms are semisets. Thus if the formulas $y \in \mathbf{F}_i$ are comprehensive in some extension **T** of **TC** + C2 then one can prove in **T** that the $\mathbf{F}_i$ are sets (see 1330, 1412, 1403(f)). This is precisely the content of the following axioms:

1422. Axioms.

(A2) $(\forall x)(\exists z)(\forall u)(u \in z \equiv (\exists v, w)(u = \langle v, w\rangle \mathbin{\&} u \in x \mathbin{\&} v \in w))$,

(A3) $(\forall x, y)(\exists z)(\forall u)(u \in z \equiv u \in x \mathbin{\&} u \notin y)$,

(A4) $(\forall x)(\exists z)(\forall u)(u \in z \equiv (\exists v)(\langle v, u\rangle \in x))$,

(A5) $(\forall x, y)(\exists z)(\forall u)(u \in z \equiv (\exists v, w)(u = \langle v, w\rangle \mathbin{\&} u \in x \mathbin{\&} w \in y))$,

(A6) $(\forall x)(\exists z)(\forall u)(u \in z \equiv (\exists v, w)(u = \langle v, w\rangle \mathbin{\&} \langle w, v\rangle \in x))$,

(A7) $(\forall x)(\exists z)(\forall u)(u \in z \equiv (\exists v, w, t)(u = \langle v, w, t\rangle \mathbin{\&} \langle w, t, v\rangle \in x))$.

1423. The *theory of semisets* **TSS** is the extension of **TC** by (A2) to (A7), (C1) and (C2).

Axioms (A2) to (A7) differ from (B2) to (B7) only by having $(\forall x)(\exists z)\ldots$ and $(\forall x, y)(\exists z)\ldots$ instead of $(\forall X)(\exists Z)\ldots$ and $(\forall X, Y)(\exists Z)\ldots$; hence they assert nothing more nor less than the fact that for any set x (or any sets x, y) the class given by the axiom (B_i) is a set.

1424. We shall now show that in **TSS** every RF is comprehensive. We shall not need the axiom (C1) of infinity; to stress this fact we denote by $\mathbf{TSS}_{-\infty}$ the theory **TSS** with the axiom (C1) removed. First some lemmas.

1425. Lemma ($\mathbf{TSS}_{-\infty}$). For any sets x, y: $\mathbf{E}(x)$, $x - y$, $\mathbf{D}(x)$, $x \upharpoonright y$, $\mathbf{Cnv}(x)$, $\mathbf{Cnv}_3(x)$ are sets. Immediate from the axioms A2 to A7.

1426. Lemma ($\mathbf{TSS}_{-\infty}$). (a) Every set is a real class. (b) If classes X, Y are real then the following classes are also real:

$$\mathbf{E}(X), \; X - Y, \; X \upharpoonright Y, \; \mathbf{Cnv}(X), \; \mathbf{Cnv}_3(X).$$

(Note that nothing is asserted about $\mathbf{D}(X)$.)

Proof. (a) If x and y are sets then $x \cap y = x - (x - y)$ is a set. (b) The assertion follows from the equalities 1304 writing z instead of Z).

1427. LEMMA ($\mathbf{TSS}_{-\infty}$) (a) 0 is a set. (b) For any sets x, y the following classes are sets: $x \cap y$, $\mathbf{W}(x)$, $x''y$, $\bigcup x$, $x \cup y$, $\mathbf{C}(x)$, $x \times y$.

Proof. (a) $0 = x - x$. (b) The assertion concerning $x \cap y$, $\mathbf{W}(x)$, $x''y$ follows immediately from Lemma 1425. For $\bigcup x$ recall that in the proof of Lemma 1411 we showed that $\bigcup x = \mathbf{W}(\mathbf{E} \upharpoonright x)$. Now, $\mathbf{E} \upharpoonright x$ is a real class and, in addition, a semiset (by Lemma 1414). It follows that $\mathbf{E} \upharpoonright x$ is a set and hence $\mathbf{W}(\mathbf{E} \upharpoonright x) \equiv \bigcup x$ is a set. Consequently, $x \cup y = \bigcup(\{x, y\})$ is also a set. Finally, $\mathbf{C}(x) = \mathbf{D}(x) \cup \mathbf{W}(x)$ and $x \times y$ is a real class and a semiset.

We are now going to prove that every RF without constants is comprehensive in $\mathbf{TSS}_{-\infty}$. However it is possible to prove the full comprehensiveness of all formulas of a family of NF's substantially larger than the family of RSF's. (Cf. Metatheorem 1438.)

1428. METALEMMA. Every atomic NF without constants is fully comprehensive.

Demonstration. We follow the demonstration of Metalemma 1119. (a) For the term $\mathbf{T} = \mathbf{D}((Y \cap \{x\}) \times \mathbf{V}^n)$ we have $\mathbf{TSS}_{-\infty} \vdash \mathbf{T} = 0 \vee \mathbf{T} = \mathbf{V}^n$ and so $\mathbf{TSS}_{-\infty} \vdash \text{Real}(\mathbf{T})$ trivially. (b) The operation $\mathbf{D}$ is not used in the construction of the term $\mathbf{T}$ and so $\mathbf{T}$ is comprehensive by Lemma 1426. (c) It suffices to prove in $\mathbf{TSS}_{-\infty}$ that $\{y; x \in y\}$ is real. We proceed in $\mathbf{TSS}_{-\infty}$. Denote $\{y; x \in y\}$ by Y; let a be a set. Evidently $a \cap Y = \{y \in a; x \in y\} = \mathbf{D}(\mathbf{E} \cap (\{x\} \times a))$, while $\mathbf{E} \cap (\{x\} \times a)$ is a real semiset and therefore a set. It follows that $a \cap Y$ is a set because it is the domain of a set. Hence Y is real. (d) The operation $\mathbf{D}$ is not used in the construction of the term $\mathbf{T}$.

1429. METALEMMA. If φ, ψ are fully comprehensive then $\varphi \mathbin{\&} \psi$, $\neg \varphi$ are fully comprehensive.

This follows trivially from the demonstration of Metalemmas 1116, 1117.

1430. METALEMMA. If φ is fully comprehensive and x, y are set variables not bound in φ then $(\exists y \in x)\, \varphi$ is fully comprehensive.

Demonstration. Let $u_1, \bullet$ be free for $(\exists y \in x)\, \varphi$. We want to prove in $\mathbf{TSS}_{-\infty}$ that $a \cap \{\langle u_1, \bullet\rangle; (\exists y \in x)\, \varphi\}$ is a set for any a. Obviously $y, u_1, \bullet$

is free for φ and also for $y \in x \,\&\, \varphi$. Let us proceed in $\mathbf{TSS}_{-\infty}$. Suppose $(\mathrm{Real}(X) \,\&\, \bullet)$. Let a be any set. Then

$$a \cap \{\langle u_1, \bullet\rangle ;\ (\exists y \in x)\, \varphi\} = a \cap \mathbf{D}(\{\langle y, u_1, \bullet\rangle ;\ y \in x \,\&\, \varphi\}) .$$

By the full comprehensiveness of $y \in x \,\&\, \varphi$ the class $A = \{\langle y, u_1, \bullet\rangle; y \in \in x \,\&\, \varphi\}$ is real; we want to prove that $\mathbf{D}(A)$ is real, i.e. that $a \cap \mathbf{D}(A)$ is a set. (a) If x does not occur in the sequence $u_1, \bullet$ then we reason as follows: If $\langle u_1, \bullet\rangle \in a \cap \mathbf{D}(A)$ then there exists $y \in x$ such that

$$\langle y, u_1, \bullet\rangle \in A \cap (x \times a), \quad \text{i.e.} \quad a \cap \mathbf{D}(A) = a \cap \mathbf{D}(A \cap (x \times a)) .$$

But $x \times a$ is a set by 1427, hence $A \cap (x \times a)$ is a set and $a \cap \mathbf{D}(A)$ is a set. (b) Suppose now that x occurs among $u_1, \bullet$ so that x is u_1 and $\langle u_1, \bullet\rangle$ is $\langle u_1, u_2\rangle$, say. If $\langle u_1, u_2\rangle \in a \cap \mathbf{D}(A)$ then $u_1 \in \bigcup\bigcup(a)$ and if $y \in u_1$ then $\langle y, u_1, u_2\rangle \in \bigcup\bigcup\bigcup(a) \times a$. But $\bigcup\bigcup\bigcup(a) \times a$ is a set (cf. 1427), so that $\bigcup\bigcup\bigcup(a) \times a = c$, say. Hence $a \cap \mathbf{D}(A) = a \cap \mathbf{D}(A \cap c)$ and $a \cap \mathbf{D}(A)$ is a set.

Consequently, we have the following

1431. METATHEOREM. Every RF without constants is fully comprehensive in $\mathbf{TSS}_{-\infty}$.

1432. THEOREM ($\mathbf{TSS}_{-\infty}$). For every set x, the class $\mathbf{P}(x)$ is a set.

Proof. By 1411 $\mathbf{P}(x)$ is a semiset. If a is a set such that $\mathbf{P}(x) \subseteq a$ then $\mathbf{P}(x) = a \cap \{y; y \subseteq x\}$. The formula $y \subseteq x$ is a RF without constants and so it follows by the preceding Metatheorem that $\mathbf{P}(x)$ is a set.

We shall now establish a generalization of Metatheorem 1431; the rather technical demonstration may be omitted on first reading.

1433. METALEMMA. If φ is fully comprehensive and x, y, z are not bound in φ then $(\exists y \in \{x, z\})\, \varphi$, $(\exists y \in \bigcup(x))\, \varphi$ and $(\exists y \in \mathbf{P}(x))\, \varphi$ are fully comprehensive.

Demonstration. (1) If the sequence $u_1, \bullet$ is free for $(\exists y \in \{x, z\})\, \varphi(y)$ then it is free for $\varphi(x) \vee \varphi(z)$, the latter formula is fully comprehensive and $\mathbf{TSS}_{-\infty} \vdash (\exists y \in \{x, z\})\, \varphi \equiv .\ \varphi(x) \vee \varphi(z)$.

(2) If $u_1, \bullet$ is free for $(\exists y \in \bigcup(x))\, \varphi$ and v is a new set variable not bound in φ then $u_1, \bullet$ is free for $(\exists v \in x)\,(\exists y \in v)\, \varphi$, the latter formula is fully comprehensive and $\mathbf{TSS}_{-\infty} \vdash (\exists y \in \bigcup(x))\, \varphi \equiv (\exists v \in x)\,(\exists y \in v)\, \varphi$.

(3) Let $u_1, \bullet$ be free for $(\exists y \in \mathbf{P}(x))\,\varphi$. We prove in $\mathbf{TSS}_{-\infty}$ that $(\mathrm{Real}(X) \,\&\, \bullet) \rightarrow \mathrm{Real}(\{\langle u_1, \bullet\rangle; (\exists y \in \mathbf{P}(x))\,\varphi\})$. Suppose $\mathrm{Real}(X) \,\&\, \bullet$, let a be an arbitrary set. We have: $a \cap \{\langle u_1, \bullet\rangle; (\exists y \in \mathbf{P}(x))\,\varphi\} = a \cap \mathbf{D}(\{\langle y, u_1, \bullet\rangle;\; y \in \mathbf{P}(x) \,\&\, \varphi\})$. Set $A = \{\langle y, u_1, \bullet\rangle; y \in \mathbf{P}(x) \,\&\, \varphi\}$; A is real, since the formula $y \in \mathbf{P}(x) \,\&\, \varphi$ is fully comprehensive. We show that $\mathbf{D}(A)$ is real, i.e. that $a \cap \mathbf{D}(A)$ is a set.

(a) If x does not occur in $u_1, \bullet$ then as in the proof of Metalemma 1430 we have $a \cap \mathbf{D}(A) = a \cap \mathbf{D}(A \cap (\mathbf{P}(x) \times a))$, $a \cap \mathbf{D}(A)$ is a set.

(b) If x occurs in $u_1, \bullet$, $x = u_i$, say, then we reason as follows: If $\langle u_1, \bullet\rangle \in a \cap \mathbf{D}(A)$ then there exists $y \in \mathbf{P}(u_i)$ such that $\langle y, u_1, \bullet\rangle \in A$; but then $y \in \mathbf{P}(\bigcup \ldots \bigcup(a))$, hence $\langle y, u_1, \bullet\rangle \in \mathbf{P}(\bigcup \ldots \bigcup(a)) \times a$ and $a \cap \mathbf{D}(A) = a \cap \mathbf{D}(A \cap (\mathbf{P}(\bigcup \ldots \bigcup(a)) \times a))$; this last is a set.

1434. METALEMMA. $y = \mathbf{P}(x)$ is fully comprehensive.

Demonstration. $\mathbf{TSS}_{-\infty} \vdash y = \mathbf{P}(x) \equiv (\forall u \in y)\,(u \in \mathbf{P}(x)) \,\&\, (\forall u \in \mathbf{P}(x))\,(u \in y)$. The right hand side is fully comprehensive by the preceding Metalemma.

1435. METALEMMA. Let φ be fully comprehensive and suppose that x, y, z are not bound in φ. Then $(\exists y = \{x, z\})\,\varphi$, $(\exists y = \bigcup(x))\,\varphi$, $(\exists y = \mathbf{P}(x))\,\varphi$ are fully comprehensive.

Demonstration. Let $\mathbf{T}$ be $\{x, z\}$, $\bigcup(x)$, $\mathbf{P}(x)$ respectively, and let $u_1, \bullet$ be free for $(\exists y = \mathbf{T})\,\varphi$. Put $A = \{\langle y, u_1, \bullet\rangle; y = \mathbf{T} \,\&\, \varphi\}$. We want to prove $(\mathrm{Real}(X) \,\&\, \bullet) \rightarrow \mathrm{Real}(\mathbf{D}(A))$ in $\mathbf{TSS}_{-\infty}$. (Recall that $X, \bullet$ are the class variables free in φ.) If x (and z) do not occur in $u_1, \bullet$ then $\mathbf{TSS}_{-\infty} \vdash a \cap \mathbf{D}(A) = a \cap \mathbf{D}(A \cap (\{\mathbf{T}\} \times a))$ hence $\mathbf{TSS}_{-\infty} \vdash (\mathrm{Real}(X) \,\&\, \bullet) \rightarrow \mathrm{Real}(\mathbf{D}(A))$. If x (and z) occur in $u_1, \bullet$ then $\mathbf{TSS}_{-\infty} \vdash a \cap \mathbf{D}(A) = a \cap \mathbf{D}(A \cap (\mathbf{S}(a) \times a)$ for an appropriate term $\mathbf{S}(a)$ built up from $a, \bigcup, \mathbf{P}$; hence $\mathbf{TSS}_{-\infty} \vdash (\mathrm{Real}(X) \,\&\, \bullet) \rightarrow \mathrm{Real}(\mathbf{D}(A))$. (In case where $\mathbf{T}$ is $\{x, z\}$ and x occurs in $u_1, \bullet$ but z does not we find a term $\mathbf{S}(a, z)$ such that $\mathbf{TSS}_{-\infty} \vdash a \cap \mathbf{D}(A) = a \cap \mathbf{D}(A \cap (\mathbf{S}(a, z) \times a))$.)

1436. METALEMMA. Let φ be fully comprehensive, and let $\mathbf{T}$ be a term built up from set variables not bound in φ and from the operations $\{\ \}$, $\bigcup$, $\mathbf{P}$. Then (a) $(\exists y \in \mathbf{T})\,\varphi$ is fully comprehensive, (b) $(\exists y = \mathbf{T})\,\varphi$ is fully comprehensive.

Demonstration by simultaneous induction. Let $\mathbf{T}_1$, $\mathbf{T}_2$ be terms for which (a), (b) hold. Then

$$\mathbf{TSS}_{-\infty} \vdash (\exists y \in \{\mathbf{T}_1, \mathbf{T}_2\})\,\varphi \equiv (\exists u = \mathbf{T}_1)\,(\exists v = \mathbf{T}_2)\,(\exists y \in \{u, v\})\,\varphi,$$

the right hand side is fully comprehensive. Similarly for $=$. If $\mathbf{F}$ is either $\bigcup$ or $\mathbf{P}$ then

$$\mathbf{TSS}_{-\infty} \vdash (\exists y \in \mathbf{F}(\mathbf{T}_1))\, \varphi \equiv (\exists u = \mathbf{T}_1)(\exists y \in \mathbf{F}(u))\, \varphi\,,$$

$$\mathbf{TSS}_{-\infty} \vdash (\exists y = \mathbf{F}(\mathbf{T}_1))\, \varphi \equiv (\exists u = \mathbf{T}_1)(\exists y = \mathbf{F}(u))\, \varphi\,.$$

We are now able to formulate the generalization of Metatheorem 1431.

1437. METADEFINITION. Define PUP-*formulas* inductively: (a) Every atomic NF is a PUP-formula. (b) If φ, ψ are PUP-formulas then $\varphi \,\&\, \psi$, $\neg\varphi$ are. (c) If φ is a PUP-formula, $\mathbf{T}$ is a term built up from set variables $u_1, \bullet$ free for φ and from the operations $\{\ \}$, $\bigcup$, $\mathbf{P}$ and if y is a variable not bound in φ and different from $u_1, \bullet$ then $(\exists y \in \mathbf{T})\, \varphi$ is a PUP-formula.

1438. METATHEOREM. Every PUP-formula without constants is fully comprehensive in $\mathbf{TSS}_{-\infty}$.

1439. METALEMMA. Let $\mathbf{T}$ be a theory stronger than $\mathbf{TSS}_{-\infty}$, let φ be a PUP-formula, and let x be a variable free in φ; let $x_1, \bullet$ be a sequence free for φ and let $\mathbf{G}$ be a term constructed from the variables $x_1, \bullet$ and the operations $\{\ \}$, $\mathbf{E}$, $-$, $\mathbf{D}$, $\restriction$, $\mathbf{Cnv}$, $\mathbf{Cnv}_3$, $\bigcup$, $\mathbf{P}$. Then the result of replacing all occurrences of x in φ by $\mathbf{G}$ is equivalent to a PUP-formula in $\mathbf{T}$.

Demonstration. Write $\varphi(x)$ instead of φ and $\varphi(\mathbf{G})$ instead of the formula described above. $\varphi(\mathbf{G})$ is equivalent in $\mathbf{T}$ to $(\exists x)\,(x = \mathbf{G} \,\&\, \varphi(x))$. Using the technique of the demonstration of Metatheorem 1438 we see that it suffices to show that $(\exists x)\,(x = \mathbf{F}_i \,\&\, \varphi(x))$ is a PUP-formula in $\mathbf{T}$ where $\mathbf{F}_i$ is one of the terms $\{u, v\}$, $\mathbf{E}(u)$, $\ldots$, $\bigcup(u)$, $\mathbf{P}(u)$. This can be left to the reader as an exercise.

The following Lemma provides a useful test for the realness of classes:

1440. LEMMA ($\mathbf{TSS}_{-\infty}$). Let X be a class which is not a semiset and suppose that for all $x, y \in X$ either $x \subseteq y$ or $y \subseteq x$. Then $\bigcup(X)$ is a real class.

Proof. Let a be a set; we prove that $a \cap \bigcup(X)$ is a set. Define $\langle x, u\rangle \in R \equiv$ $\equiv (u \in a \cap \bigcup(X) \,\&\, x \in X \,\&\, u \notin x)$. R is regular: if $u \in a \cap \bigcup(X)$ then there exists $x_0 \in X$ such that $u \in x_0$; hence for $\langle x, u\rangle \in R$ we have $x \subseteq x_0$ and therefore $\mathbf{Ext}_R(u) \subseteq \mathbf{P}(x_0)$. Further, $\mathbf{D}(R) \subseteq a \cap \bigcup(X)$, so that $\mathbf{D}(R)$ is a semiset. It follows from (C2) that $\mathbf{W}(R)$ is a semiset. But $\mathbf{W}(R) \subseteq X$ and X is not a semiset; consequently, there exists $x_1 \in X - \mathbf{W}(R)$. We have $(a \cap \bigcup(X)) \subseteq x_1$; hence $a \cap \bigcup(X) = a \cap x_1$ and $a \cap x_1$ is a set.

d) *Model-classes*

1441. Definition ($\mathbf{TSS}_{-\infty}$). A class X is *closed* (Clos (X)) if, for all $x, y \in X$, $\mathbf{F}_i(x, y) \in X$ ($i = 1, \ldots, 7$), i.e. $\mathrm{Clos}(X) \equiv (\forall x, y \in X)(\{x, y\}, \mathbf{E}(x), x - y, \mathbf{D}(x), x \upharpoonright y, \mathbf{Cnv}(x), \mathbf{Cnv}_3(x) \in X)$.

Note that $\mathbf{TSS}_{-\infty} \vdash \mathrm{Clos}(\mathbf{V})$.

1442. Definition ($\mathbf{TSS}_{-\infty}$). A class X is a *model-class* (Mcl (X)) if it is a complete, closed and almost universal class, i.e. $\mathrm{Mcl}(X) \equiv . \mathrm{Comp}(X) \& \& \mathrm{Clos}(X) \& \mathrm{AUncl}(X)$.

1443. Lemma $(\mathbf{TSS})_{-\infty}$. (a) $\mathbf{V}$ is a model-class. (b) Every model-class is $\mathbf{V}$-like.

Proof. (a) is obvious. (b) Suppose X is a model-class; then X is non-empty because it is an almost universal class. Further, $x, y \in X \equiv \{x, y\} \in X$ because X is closed ($\rightarrow$) and complete ($\leftarrow$). It follows by completeness that $x \cap X = x$ for $x \in X$; hence

$$x, y \in X \& x \neq y . \rightarrow x \cap X \neq y \cap X .$$

1444. Theorem ($\mathbf{TSS}_{-\infty}$). Every model-class is a real class.

Proof. Let P be a model-class and let x be a set. Then $x \cap P$ is a semiset, $(x \cap P) \subseteq P$; since P is almost universal class there is an $y \in P$ such that $x \cap P \subseteq y$. It follows, using the completeness of P, that $x \cap P = x \cap P \cap \cap y = x \cap y$ and so $x \cap P$ is a set.

1445. Metatheorem. $\mathfrak{St}(\mathbf{M})$ with the specification Mcl $(\mathbf{M})$ determines a model of $\mathbf{TSS}_{-\infty}$ in $\mathbf{TSS}_{-\infty}$. $\mathfrak{Dir}(\partial\mathbf{TSS}/\partial\mathfrak{St}(\mathbf{M}))$ is a faithful model of $\mathbf{TSS}_{-\infty}$ in $\mathbf{TSS}_{-\infty}$, Mcl $(\mathbf{M})$, $\partial\mathbf{TSS}/\partial\mathfrak{St}(\mathbf{M})$; the formulas $\mathrm{M}(X)$, Real (X) are absolute from below and the formula Sm (X) is absolute.

By Metatheorems 1317, 1318, 1416 $\mathfrak{St}(\mathbf{M})$ with Mcl $(\mathbf{M})$ determines a model of $\mathbf{TC}$, C2 in $\mathbf{TSS}_{-\infty}$, and Sm (X) is absolute. Hence we must prove (A2)* to (A7)* in $\mathbf{TSS}_{-\infty}$, Mcl $(\mathbf{M})$, $\partial\mathbf{TSS}/\partial\mathfrak{St}(\mathbf{M})$. Denote the last theory by $\mathbf{TSS}_{\mathrm{Mcl}}$ (in the context of the present metatheorem only). We know (Metatheorem 1322) that gödelian operations are absolute, i.e. $\mathbf{TSS}_{\mathrm{Mcl}} \vdash \mathbf{F}_i^*(X^*, Y^*) = \mathbf{F}_i(X^*, Y^*)$. In particular, $\mathbf{TSS}_{\mathrm{Mcl}} \vdash \mathbf{F}_i^*(x^*, y^*) = = \mathbf{F}_i(x^*, y^*)$ hence it remains to show that $\mathbf{TSS}_{\mathrm{Mcl}} \vdash \mathrm{M}^*(\mathbf{F}_i(x^*, y^*))$. But sets in the sense of $\mathfrak{St}(\mathbf{M})$ are elements of $\mathbf{M}$; since $\mathbf{M}$ is closed it follows that $\mathbf{TSS}_{\mathrm{Mcl}} \vdash \mathbf{F}_i(x^*, y^*) \in \mathbf{M}$ and so $\mathbf{TSS}_{\mathrm{Mcl}} \vdash \mathrm{M}^*(\mathbf{F}_i^*(x^*, y^*))$. This proves (A2)* to (A7)*.

Specifying by $\mathbf{M} = \mathbf{V}$ we obtain the result concerning faithfulness. The fact that $\mathrm{M}(X)$ is absolute from below is trivial. Finally we prove $\mathbf{TSS}_{\mathbf{Mcl}} \vdash \vdash \mathrm{Real}^* (X^*) \rightarrow \mathrm{Real}(X^*)$. We proceed in $\mathbf{TSS}_{\mathbf{Mcl}}$. Suppose $\mathrm{Real}^* (X^*)$ and let x be a set; since $x \cap X^* \subseteq \mathbf{M}$ there exists $y \in \mathbf{M}$ such that $x \cap X^* \subseteq \subseteq y$. Thus $x \cap X^* = x \cap X^* \cap y$; X^* is a real class in the sense of $\mathfrak{St}(\mathbf{M})$ and y is a set in the sense of $\mathfrak{St}(\mathbf{M})$; hence there is a $u \in \mathbf{M}$ such that $u = = X^* \cap y$ and we obtain $x \cap X^* = x \cap u$ which is a set. Consequently, X^* is a real class.

e) *The theory of sets*

1446. The *theory of sets* **TS** is the extension of **TC**, C1, C2 by the axiom

$$\text{(C3)} \qquad (\forall \sigma)(\exists x)(\sigma = x)\,.$$

Thus (C3) says that every semiset is a set. We denote by $\mathbf{TS}_{-\infty}$ the theory **TS** with the axiom (C1) removed.

1447. METATHEOREM. ($\mathbf{TS}_{-\infty}$) is stronger than $\mathbf{TSS}_{-\infty}$.

Demonstration. It suffices to verify that (A2) to (A7) are provable in $\mathbf{TS}_{-\infty}$. But in **TC**, C2 we proved that $\mathbf{E}(x)$, $x - y$, ..., $\mathbf{Cnv}_3(x)$ are semisets; hence by (C3) these classes are sets.

Thus we have the following result in $\mathbf{TS}_{-\infty}$:

1448. THEOREM ($\mathbf{TS}_{-\infty}$). For any x, $\bigcup x$ and $\mathbf{P}(x)$ are sets; if F is a function then $F''x$ is a set for any x.

1449. Remark. Our theory **TC** is equivalent to the theory with Gödel's axiom system A, B. The formulas $\mathrm{M}(\bigcup(x))$, $\mathrm{M}(\mathbf{P}(x))$, $\mathrm{Un}(F) \rightarrow \mathrm{M}(F''x)$ are in fact Gödel's axioms C2, C3 and C4; we denote them by GC2 to GC4. Gödel's C1 has the same meaning as ours. We have seen that GC2 to GC4 are provable in $\mathbf{TS}_{-\infty}$. We now show that C2 and C3 are provable in **TC**, GC2, GC3, GC4.

For any X we define a relation I by $\langle x, y\rangle \in I \equiv y \in X \,\&\, x = y$; I is a function and if X is a semiset then $X \subseteq z$ for some z; since $X = I''z$, it follows that X is a set. Thus we have proved C3. Let R be an exact functor; we prove that $\mathbf{D}(R)$ is a set iff $\mathbf{W}(R)$ is a set. This will prove (C2), since by (C3), $\mathrm{Sm}(X) \equiv \mathrm{M}(X)$. R is nowhere constant and $(\forall x)(\mathrm{M}(\mathbf{Ext}_R(x))$. We define a function F by $\langle x, y\rangle \in F \equiv x \in \mathbf{D}(R) \,\&\, y = \mathbf{Ext}_R(x)$. F is one-to-one, $\mathbf{D}(F) = \mathbf{D}(R)$ and $\bigcup(\mathbf{W}(F)) = \mathbf{W}(R)$. If $\mathbf{D}(R)$ is a set then $\mathbf{D}(F)$ is a set, $\mathbf{W}(F) = F''(\mathbf{D}(F))$ is a set and hence $\mathbf{W}(R) = \bigcup(\mathbf{W}(F))$ is a set. Conversely, if $\mathbf{W}(R)$ is a set then $\mathbf{P}(\mathbf{W}(R))$, $\mathbf{W}(F) \subseteq \mathbf{P}(\mathbf{W}(R))$ and $\mathbf{D}(F) = (\mathbf{Cnv}(F))''(\mathbf{W}(F))$ are sets; hence $\mathbf{D}(R)$ is a set.

Thus we have shown that **TS** is equivalent to Gödel's axiom system A, B, C.

f) The first axiom of regularity; the real F-*definition*

1450. We shall now extend **TSS** (or **TS**) by the addition of the following axiom (the *first axiom of regularity*):

$$(\mathrm{D1}) \qquad (\forall X)(\forall x)(\exists y)(\mathbf{D}(X) \cap x = \mathbf{D}(X \cap y)) .$$

The theory **TSS**, D1 is denoted by **TSS′**; similarly, **TS**, D1 is denoted by **TS′**. $\mathbf{TSS}'_{-\infty}$ and $\mathbf{TS}'_{-\infty}$ denote **TSS′** and **TS′** with the axiom of infinity removed.

1451. THEOREM ($\mathbf{TSS}'_{-\infty}$). The domain of a real class is also a real class.

Proof. If X is a real class and if x is a set then by D1 there is a set y such that $\mathbf{D}(X) \cap x = \mathbf{D}(X \cap y)$, hence $\mathbf{D}(X) \cap x$ is a set by (A4).

1452. THEOREM ($\mathbf{TSS}_{-\infty}$). The axiom (D1) is equivalent to the following assertion: For every relation X whose domain is a semiset there is a subrelation $Y \subseteq X$ which is a semiset and has the same domain.

Proof. Suppose (D1). If X is a relation and $\mathbf{D}(X) \subseteq x$ then $\mathbf{D}(X) \cap x = \mathbf{D}(X)$ and by (D1) there is a y such that $\mathbf{D}(X) = \mathbf{D}(X \cap y)$. Hence $X \cap y$ is the required semiset. Conversely, assume the assertion of the theorem. Given X and x, consider $X' = X \upharpoonright x$. The domain of X' is a semiset and so there is a semiset Y such that $Y \subseteq X'$ and $\mathbf{D}(Y) = \mathbf{D}(X')$. If $Y \subseteq y$ then $\mathbf{D}(X) \cap x = \mathbf{D}(X') = \mathbf{D}(X \cap (y \upharpoonright x))$. We have proved (D1).

1453. METATHEOREM (Comprehension scheme). Every normal formula without constants is fully comprehensive in $\mathbf{TSS}'_{-\infty}$.

This is equivalent to the following assertion which may be proved by induction using Theorems 1426, 1451.

1454. METATHEOREM: For any gödelian term $\mathbf{T}(X_1, \bullet)$ without constants we have $\mathbf{TSS}'_{-\infty} \vdash (\mathrm{Real}(X_1) \,\&\, \bullet) \rightarrow \mathrm{Real}(\mathbf{T}(X_1, \bullet))$.

1455. METADEFINITION. The pair of formulas

$$(5) \qquad (\forall X)\,[(\exists X^*)\,(X^* = X) \equiv \mathrm{Real}\,(X)]\,\&\, sec\,,$$

$$(6) \qquad (\forall X^*, Y^*)\,(X^* \in^* Y^* \equiv X^* \in Y^*)\,\&\, sec$$

is called the *real* F-*definition* and denoted by $\mathfrak{Real}$.

Thus classes in the sense of $\mathfrak{Real}$ are real classes.

1456. METATHEOREM. The F-definition $\mathfrak{Real}$ determines a model of $\mathbf{TS}'_{-\infty}$ in $\mathbf{TSS}'_{-\infty}$ and of $\mathbf{TS}'$ in $\mathbf{TSS}'$. The model $\mathfrak{Dir}\,(\partial\mathbf{TS}/\partial\mathfrak{Real})$ is a faithful model of $\mathbf{TS}'_{-\infty}$ in $\mathbf{TSS}'_{-\infty}$, $\partial\mathbf{TS}/\partial\mathfrak{Real}$ and of $\mathbf{TS}'$ in $\mathbf{TSS}'$, $\partial\mathbf{TS}/\partial\mathfrak{Real}$.*) All normal $\mathbf{TS}$-formulas without constants are absolute; in particular, if φ is a closed set formula then

$$\mathbf{TSS}'_{-\infty}, \partial\mathbf{TS}/\partial\mathfrak{Real} \vdash \varphi \equiv \varphi^*\,.$$

Demonstration. By (6), the formula $X \in Y$ is absolute. Since $\mathbf{TSS}'_{-\infty} \vdash$ $\vdash (\forall x)\,\mathrm{Real}\,(x)$, (F1) holds in the model. Moreover, denoting the theory $\mathbf{TSS}'_{-\infty}, \partial\mathbf{TS}/\partial\mathfrak{Real}$ by $\mathbf{TSS}_{\mathrm{real}}$ (in the context of the present Metatheorem) we have $\mathbf{TSS}_{\mathrm{real}} \vdash \mathrm{M}(X) \equiv \mathrm{M}^*(X)$, i.e. sets of the model are exactly all sets. This implies immediately that all NF's without constants are absolute. Axiom F2 holds in $\mathfrak{Real}$, since $\mathbf{TSS}_{\mathrm{real}} \vdash (\forall x^*)\,(x^* \in^* X^* \equiv x^* \in^* Y^*) \equiv$ $\equiv (\forall x)(x \in X^* \equiv x \in Y^*)$. By absoluteness of NF's, (A1) holds in the model and if C1 is assumed in the theory we are dealing with then it holds in the model. The axioms of group B have the form $(\forall X, \bullet)\ (\exists Z)\ \varphi(X, \bullet, Z)$ where φ is normal in $\mathbf{TE}$. In $\mathbf{TSS}_{\mathrm{real}}$, we have for arbitrary $X^*, \bullet$:
$\mathrm{Real}\,(\mathbf{F}_i\,(X^*, \bullet))$, i.e. $(\exists Z^*)\ (Z^* = \mathbf{F}_i\,(X^*, \bullet))$; by the absoluteness of NF's we obtain $(\exists Z^*)\ \varphi^*\ (X^*, \bullet, Z^*)$. This proves (B1)* to (B7)*. Axiom C2 has the form $(\forall X)\ \varphi(X)$ where φ is normal in $\mathbf{TC}$, hence (C2) is absolute from above and we have $\mathbf{TSS}_{\mathrm{real}} \vdash (\mathrm{C2})^*$. The same can be said of (D1). Finally, since semisets of the model are real subclasses of sets, they are sets and therefore sets in the sense of the model. We have $\mathbf{TSS}_{\mathrm{real}} \vdash (\mathrm{C3})^*$. $\mathfrak{Real}$ is a model of $\mathbf{TS}'_{-\infty}$ in $\mathbf{TSS}_{\mathrm{real}}$.

*) This can be also formulated as follows: The model $\mathfrak{Real}$ makes $\mathbf{TSS}'$ to a conservative extension of $\mathbf{TS}'$ by identifying class variables in $\mathbf{TS}'$ with variables for real classes in $\mathbf{TSS}'$.

To demonstrate the faithfulness consider the diagram:

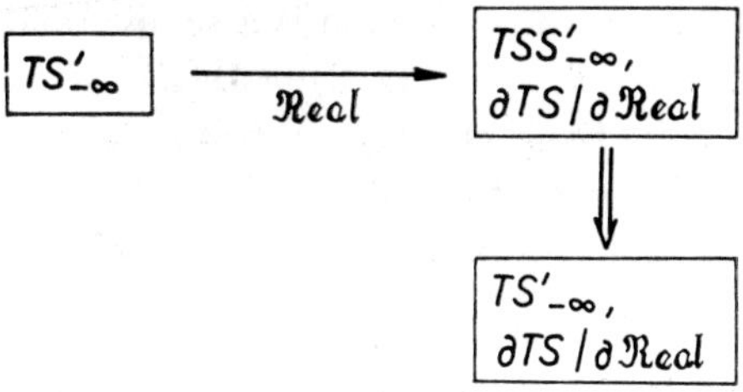

The composition of these two arrows is evidently equivalent to the identity model of the first theory in the third one and therefore, by 1233 (2), $\mathfrak{Real}$ is faithful.

1457. METACOROLLARY. (*The equiprovability principle.*) If φ is a SF then $\mathbf{TSS}'_{-\infty} \vdash \varphi$ iff $\mathbf{TS}'_{-\infty} \vdash \varphi$; more generally, if Γ is a sequence of closed SF's, then $\mathbf{TSS}'_{-\infty}, \Gamma \vdash \varphi$ iff $\mathbf{TS}'_{-\infty}, \Gamma \vdash \varphi$.

Demonstration. In Diagram 1458 the arrow 1 is a faithful model; since it is direct it is essentially faithful; thus 2 is a faithful model. Since NF's and therefore also SF's are absolute, we have $\mathbf{TSS}_{\text{real}} \vdash \psi \equiv \psi^{\mathfrak{Real}}$ for every SF ψ. This means that $\Gamma^{\mathfrak{Real}}$ in the terminal theory of 2 can be replaced by Γ. By faithfulness, $\mathbf{TS}'_{-\infty}\ \Gamma \vdash \varphi$ iff $\mathbf{TSS}_{\text{real}}, \Gamma \vdash \varphi^{\mathfrak{Real}}$; however $\mathbf{TSS}_{\text{real}}, \Gamma \vdash \varphi^{\mathfrak{Real}}$ means exactly the same as $\mathbf{TSS}_{\text{real}}, \Gamma \vdash \varphi$ which is equivalent to $\mathbf{TSS}'_{-\infty}, \Gamma \vdash \varphi$ because the former theory is a conservative extension of the latter.

1458.

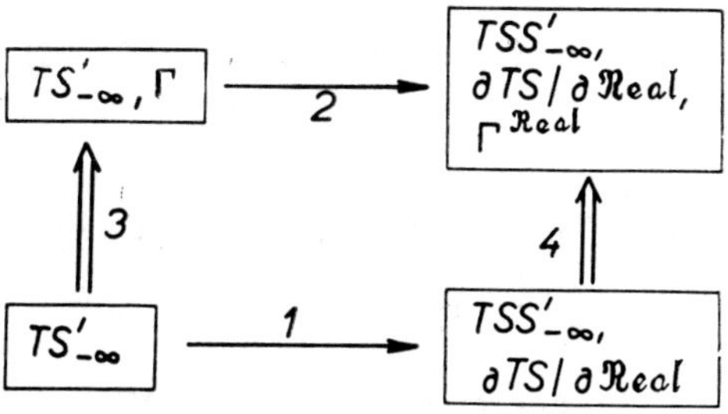

1459. METATHEOREM. $\mathfrak{N}(\mathbf{R})$ with the specification sExtl $(\mathbf{R})$ & Prg $(\mathbf{R})$ & & Reg $(\mathbf{R})$ & AUn $(\mathbf{R})$ (cf. Metatheorem 1416) determines a model of $\mathbf{TC}$, C2, D1 in $\mathbf{TC}$, C2, D1.

Demonstration. We proceed in $\mathbf{TC}$, C2, D1, Mrel $(\mathbf{R})$, $\partial\mathbf{TC}/\partial\mathfrak{N}(\mathbf{R})$ (where

Mrel $(\mathbf{R})$ is an abbreviation for the specification above). Let X^* be such that $\mathrm{Rel}^*(X^*)$ and $\mathrm{Sm}^*(\mathbf{D}^*(X^*))$. We denote by Z the decoded relation $\mathbf{Dec}(X^*)$; Z is a relation whose domain is a semiset and hence there exists $S \subseteq Z$ such that $\mathrm{Sm}(S)$ and $\mathbf{D}(S) = \mathbf{D}(Z)$. If we define $u \in Y^* \equiv$ $\equiv (\exists x, y)(\langle x, y\rangle \in S \,\&\, \langle \mathbf{Ext_R}(x), \mathbf{Ext_R}(y)\rangle^* = \mathbf{Ext_R}(u))$, then $S = \mathbf{Dec}(Y^*)$, $Y^* \subseteq^* X^*$ and $\mathbf{D}^*(Y^*) = \mathbf{D}^*(X^*)$. It remains only to verify that $\mathrm{Sm}^*(Y^*)$; to do this it suffices to show that Y^* is a semiset. We define a relation H as follows:

$$\langle u, v\rangle \in H \equiv (\exists x, y)(v = \langle x, y\rangle \in S \,\&\, \mathbf{Ext_R}(u) = $$
$$= \langle \mathbf{Ext_R}(x), \mathbf{Ext_R}(y)\rangle^*)\,;$$

H is one-to-one and hence an exact functor; since $\mathbf{D}(H) = S$ and $\mathbf{W}(H) =$ $= Y^*$, Y^* is a semiset.

g) *The notion of support; the support* F-*definition*

To end this Section we generalize the construction of the real model. The notion of a support which we shall define here will play an extremely important role in Chapter IV et seq.

1460. DEFINITION. A class X is *dependent on* a class Z ($\mathrm{Dep}(X, Z)$) if there is a set r such that $X = r''Z$.

1461. LEMMA ($\mathbf{TSS}_{-\infty}$).

(1) $$Z \neq 0 \to (\forall x)\,\mathrm{Dep}(x, Z)\,,$$

(2) $$\mathrm{Dep}(X, Y) \,\&\, \mathrm{Dep}(Y, Z)\,. \to \mathrm{Dep}(X, Z)\,,$$

(3) $$\mathrm{Dep}(X, Z) \to \mathrm{Sm}(X)\,,$$

(4) $$\mathrm{Dep}(X, Z) \,\&\, \mathrm{Dep}(Y, Z)\,. \to \mathrm{Dep}(X \cup Y, Z)\,,$$

(5) $$\mathrm{Dep}(\sigma, \sigma)\,,$$

(6) $$\mathrm{M}(Z) \,\&\, \mathrm{Dep}(X, Z)\,. \to \mathrm{M}(X)\,.$$

Proof. (1) We choose some $y \in Z$ and let $r = x \times \{y\}$; then $x = r''Z$. (2) If $X = r''Y$ and $Y = s''Z$ then we define $\langle x, y\rangle \in t \equiv (\exists u)(\langle x, u\rangle \in$ $\in r \,\&\, \langle u, y\rangle \in s)$ so that $X = t''Z$. (3) If $X = r''Z$ then $X \subseteq \mathbf{W}(r)$. (4) If $X = r''Z$ and $Y = s''Z$ then $X \cup Y = (r \cup s)''\,Z$. (5) If $\sigma \subseteq a$ and $r =$

$= \mathbf{I} \upharpoonright a$ (where $\mathbf{I} = \{\langle x, x\rangle; x \in \mathbf{V}\}$) then $\sigma = r''\sigma$. (6) If $Z = z$ then $r''Z = r''z = \mathbf{W}(r \upharpoonright z)$ is a set.

1462. DEFINITION (**TSS**$_{-\infty}$).

$$\mathrm{Supp}(Z) \equiv . Z \neq 0 \,\&\, (\forall \sigma, \varrho)(\mathrm{Dep}(\sigma, Z) \,\&\, \mathrm{Dep}(\varrho, Z) . \rightarrow$$
$$\rightarrow \mathrm{Dep}(\sigma - \varrho, Z)),$$

i.e. Z is a *support* if Z is a non-empty class such that $\sigma - \varrho$ is dependent on Z whenever both σ and ϱ are dependent on Z.

Remark. There exists at least one support; e.g. $\{0\}$ is a support and X is dependent on $\{0\}$ iff X is a set.

1463. METADEFINITION. The pair of formulas

$$(7) \qquad (\forall X)[(\exists X^*)(X^* = X) \equiv (\forall x)\,\mathrm{Dep}(X \cap x, \mathbf{Z})] \,\&\, sec\,,$$

$$(8) \qquad (\forall X^*, Y^*)(X^* \in^* Y^* \equiv X^* \in Y^*) \,\&\, sec$$

is called the *support* F-*definition* and denoted by $\mathfrak{Supp}(\mathbf{Z})$.

Thus classes in the sense of $\mathfrak{Supp}(\mathbf{Z})$ are classes which intersect each set in some semiset dependent on $\mathbf{Z}$.

1464. METATHEOREM. The F-definition $\mathfrak{Supp}(\mathbf{Z})$ with the specification $\mathrm{Supp}(\mathbf{Z})$ determines a model of **TSS**$'_{-\infty}$ in **TSS**$'_{-\infty}$ (and of **TSS**$'$ in **TSS**$'$).

Demonstration. We proceed in (**TSS**$'_{-\infty}$, $\mathrm{Supp}(\mathbf{Z})$, ∂**TSS**$/\partial\mathfrak{Supp}(\mathbf{Z})$). Consider the support $\mathbf{Z}$; the classes of the model are all classes which satisfy the condition $(\forall x)\,\mathrm{Dep}(X \cap x, \mathbf{Z})$. By Lemma 1461, every set is a class of the model. It suffices to show that $\mathbf{F}_i(X, Y)$, $i = 2, \ldots, 7$, is a class of the model whenever X and Y are classes of the model. In this case we may follow a procedure analogous to that used in proving that the real model is a model. It suffices to prove

$$\mathrm{Dep}(X, \mathbf{Z}) \,\&\, \mathrm{Dep}(Y, \mathbf{Z}) . \rightarrow \mathrm{Dep}(\mathbf{F}_i(X, Y), \mathbf{Z}) \quad (i = 2, \ldots, 7)\,;$$

(cf. 1304 and Axiom (D1)). The case $i = 3$ follows from the definition of support. In case $i = 2, 4, 6, 7$ we show that $\mathbf{F}_i(X)$ is dependent on X. Since X is dependent on $\mathbf{Z}$, X is a semiset and hence $X \subseteq u$ for some u; we have $\mathbf{E}(X) = r_2''X$ where $r_2 = \mathbf{I} \upharpoonright \mathbf{E}(u)$, $\mathbf{D}(X) = r_4''X$ where $\langle x, \langle y, x\rangle\rangle \in r_4 \equiv \langle y, x\rangle \in u$ and $\mathbf{Cnv}(X) = r_6''X$ where $\langle\langle y, x\rangle, \langle x, y\rangle\rangle \in$ $\in r_6 \equiv \langle x, y\rangle \in u$; similarly for $i = 7$. Finally in case $i = 5$, let X and Y be two classes dependent on **Z**; we prove that $X \upharpoonright Y$ is dependent. Since X and Y are semisets, $X \subseteq u$ and $Y \subseteq v$ for some u and v. If we let $z = \mathbf{W}(u)$ then $X \upharpoonright Y = X \cap (z \times Y)$ and if we define $\langle\langle y, x\rangle, x\rangle \in a \equiv$ $\equiv . y \in z \,\&\, x \in v$ then we have $z \times Y = a''Y$. Hence $z \times Y = a''Y$ is dependent, and $X \upharpoonright Y = X \cap (z \times Y) = X - (X - (z \times Y))$ is also dependent. This completes the proof.

1465. Remark. $\mathfrak{Dir}\,(\partial\mathbf{TSS}/\partial\mathfrak{Supp}\,(\mathbf{Z}))$ as a model in $\mathbf{TSS}'_{-\infty}$, $\mathbf{Z} = \{0\}$, $\partial\mathbf{TSS}/\partial\mathfrak{Supp}\,(\mathbf{Z})$ is equivalent to $\mathfrak{Dir}\,(\partial\mathbf{TSS}/\partial\mathfrak{Real})$ as a model in this theory. (Exercise.)

1466. Lemma ($\mathbf{TSS}'_{-\infty}$). A class $Z \neq 0$ is a support iff it has the following property:

$$(\forall\sigma)\,(\mathrm{Dep}\,(\sigma, Z) \rightarrow (\exists a)\,(\sigma \subseteq a \,\&\, \mathrm{Dep}\,(a - \sigma, Z)))\,.$$

Proof. Clearly every support has this property. Conversely, suppose that a class Z has the property. If σ_1, σ_2 are two semisets both dependent on Z then there are a_1, a_2 such that $\sigma_1 \subseteq a_1, \sigma_2 \subseteq a_2$ and $\mathrm{Dep}\,(a_1 - \sigma_1, Z)$ and $\mathrm{Dep}\,(a_2 - \sigma_2, Z)$. Set $a_3 = a_1 \cup a_2$; we have $\mathrm{Dep}\,(a_3 - \sigma_1, Z)$ and by 1461 we obtain $\mathrm{Dep}\,((a_3 - \sigma_1) \cup \sigma_2, Z)$. Put $\sigma_3 = (a_3 - \sigma_1) \cup \sigma_2$ and observe that $a_3 \supseteq \sigma_3$. We show $\mathrm{Dep}\,(a_3 - \sigma_3, Z)$. Indeed, there is an $a_4 \supseteq$ $\supseteq \sigma_3$ such that $\mathrm{Dep}\,(a_4 - \sigma_3, Z)$. Furthermore, $a_3 - \sigma_3 = ((a_4 - \sigma_3) \cup$ $\cup (a_3 - a_4)) - (a_4 - a_3)$ because $a_3 \supseteq \sigma_3$. But $a_3 - \sigma_3 = a_3 - ((a_3 -$ $- \sigma_1) \cup \sigma_2) = \sigma_1 \cap (a_3 - \sigma_2) = \sigma_1 - \sigma_2$. This proves that Z is a support.

1467. For the sake of those readers who prefer axiom-schemes, we shall now collect the axioms of **TSS** and **TS** into the following system:

TSS {
- (i) Axiom of Existence (F1) } **TE** (with (ii))
- (ii) Axiom of Extensionality (F2)
- (iii) Axiom of Pairing (A1)
- (iv) Class Comprehension Scheme (B1 – B7) } **TC** ((i)–(iv))
- (v) Axiom of Infinity (C1)
- (vi) Axiom of Exact Functor (C2)
- (vii) Restricted Comprehension Scheme (A2 – A7) ‖ (viii) Axiom "Semisets are sets" (C3)

} **TS** ((i)–(vi), (viii))

All models to be considered in this book are determined by one of the F-definitions $\mathfrak{R}\,(\mathbf{R})$, $\mathfrak{St}(\mathbf{M})$, $\mathfrak{Supp}\,(\mathbf{Z})$ with suitable specification or by the F-definition $\mathfrak{Real}$ or are compositions of such models.

CHAPTER II

In Chapter I we introduced the fundamental Gödelian theory of semisets **TSS** and the fundamental Gödelian theory of sets **TS**. We also considered some general models of these theories. Since we have not proved very many theorems in these theories, we have only been able to establish general properties of the models. In this Chapter we shall therefore define a number of set-theoretical concepts and prove a number of theorems in **TSS**; the models will be dealt with in later chapters.

We investigate ordinal numbers and well-orderings in Sect. 1, and equivalence and subvalence of sets in Sect. 2. Sect. 3 is devoted to a weak form of the axiom of choice and its consequences; in particular we consider the consequences for powers of sets. Sect. 4 is devoted to Boolean algebras and Sect. 5 to ordered sets.

SECTION 1

Ordinal numbers

Throughout this and following sections we shall work within the theory of semisets **TSS** or one of its extensions. We first consider the theory of ordinal numbers due to von Neumann. Let us recall here the following remark of Gödel:

2101. "The ordinal α will be the class of all ordinals less than α. For instance, $0 =$ the null (empty) set, $1 = \{0\}$, $2 = \{0, 1\}$, $\omega =$ the set of all integers, etc. In this way, the class of ordinals will be well ordered by the $\in$-relation, so that $\alpha < \beta$ corresponds to $\alpha \in \beta$. Any ordinal will itself be well ordered by the $\in$-relation since an ordinal is a class of ordinals. Moreover, any element of an ordinal must be identical with the segment generated by itself, since this segment is the class of all smaller ordinals."

2102. Remark. If R is a relation then xRy means the same as $\langle x, y \rangle \in R$.

2103. DEFINITION (**TSS**). a) A relation R is an *ordering* (Ordg (R)) if it has the following properties:

(1) $(\forall x, y, z)(xRy \,\&\, yRz \,.\!\rightarrow xRz)$ (*transitivity*),

(2) $(\forall x)(\neg xRx)$ (*irreflexivity*).

b) A relation R is a *linear ordering* (LOrdg (R)), if it has the following properties:

(1) R is an ordering,

(2) $(\forall x, y \in \mathbf{C}(R))(xRy \vee yRx \vee x = y)$ (*trichotomy*).

2104. DEFINITION (**TSS**). a) X is *ordered* (*linearly ordered*) by a relation R if the relation $R \cap X^2$ is an ordering (a linear ordering) and $X \subseteq \mathbf{C}(R)$.

b) A one-to-one mapping F of X onto Y is called an isomorphism between X and Y w.r.t. the relations R and S if

$$(\forall x, y \in X)(xRy \equiv (F\text{'}x)S(F\text{'}y)) .$$

2105. DEFINITION (**TSS**). X is an *ordinal* (Ord (X)) if it has the following properties:

a) X is complete,

b) $(\forall x, y \in X)(x \in y \vee y \in x \vee x = y)$,

c) $(\forall u \subseteq X)(u \neq 0 \rightarrow (\exists z \in u)(z \cap u = 0))$.

2106. Remark. The predicate Ord (X) is obviously normal: it is easy to find a PUP-formula equivalent to Ord (X).

2107. DEFINITION (**TSS**). $x \in \mathbf{On} \equiv$ Ord (x). **On** is the class of all *ordinal numbers*; ordinal numbers are the ordinals which are sets; we use lower-case letters from the beginning of the Greek alphabet to denote ordinal numbers.

2108. LEMMA (**TSS**). (a) Ord (0).

(b) If Ord (X), Ord (Y), Ord (Z) then

$$X \notin X\,; \quad \neg(X \in Y \,\&\, Y \in X)\,; \quad \neg(X \in Y \,\&\, Y \in Z \,\&\, Z \in X)\,.$$

(c) If Ord (X) and $y \in X$ then Ord (y).

(d) If X, Y are ordinals, then $X \cap Y$ is an ordinal.

Proof. (a) is obvious. (b) If X is an ordinal and $X \in X$ then $\{X\} \subseteq X$; putting $u = \{X\}$ we obtain a contradiction with (c) in the definition of an ordinal. Similarly put $u = \{X, Y\}$, $u = \{X, Y, Z\}$ in the remaining cases. (c) Suppose Ord (X) and $y \in X$. Then y is complete. For suppose $u \in v \in y$, then, by (b) in the definition of an ordinal, $u \in y \vee u = y \vee y \in u$; the last two cases contradict (b) of the present lemma and so $u \in y$. If $u, v \in y$ then $u, v \in X$ since X is complete; hence $u \in v \vee u = v \vee v \in u$. Finally, if $0 \neq u \subseteq y$ then $u \subseteq X$ and therefore by (c) in the definition there is a $z \in u$ such that $z \cap u = 0$. (d) is obvious.

2109. LEMMA (**TSS**). Let Y be a non-empty real subclass of **On**. Then there exists $z \in Y$ such that $z \cap Y = 0$.

Proof. Choose $y \in Y$ and set $u = y \cap Y$. If $u = 0$ we are finished If $u \neq 0$ then $u \subseteq y$ and thus there exists $z \in u$ such that $z \cap u = 0$, i.e. $z \cap y \cap Y = 0$. Since $z \in y$ we have $z \subseteq y$, so $z \cap Y = 0$.

2110. LEMMA (**TSS**). If X and Y are real ordinals and if Y is a proper subclass of X then $Y \in X$.

Proof. Since $X - Y$ is non-empty there exists $z \in X - Y$ such that $z \cap (X - Y) = 0$; hence $z \subseteq Y$. It suffices to prove $z = Y$. Suppose $z \subset Y$. Then there exists $z_1 \in Y - z$ such that $z_1 \cap (Y - z) = 0$; hence $z_1 \subseteq z$. If $z_1 = z$ then $z \in Y$, a contradiction; thus $z_1 \subset z$. Since $z, z_1 \in X$ and X is an ordinal we have either $z_1 \in z$ or $z \in z_1$. But $z \in z_1$ implies $z \subseteq z_1$, which contradicts $z_1 \subset z$, and so $z_1 \in z$. Since $z_1 \in Y$ we have $z_1 \notin Y - z$, a contradiction.

2111. LEMMA (**TSS**). If X and Y are different real ordinals then either $X \in Y$ or $Y \in X$.

Proof. Set $Z = X \cap Y$. Z is a real ordinal and $Z \subseteq X$, $Z \subseteq Y$. If $Z = X$ then $X \subseteq Y$; hence $X \subset Y$ and so $X \in Y$ by Lemma 2110. Similarly if $Z = Y$ then $Y \in X$. The remaining possibility $Z \subset X \,\&\, Z \subset Y$ can be excluded; for by Lemma 2110 it implies $Z \in X \,\&\, Z \in Y$, i.e. $Z \in X \cap Y = Z$ which contradicts Lemma 2108.

2112. LEMMA (**TSS**). Let Y be a non-empty real subclass of **On**. Then there is exactly one element z of Y such that $z \cap Y = 0$.

Proof. By 2109 there is at least one such z. Now we prove the uniqueness. Let $z_1, z_2 \in Y$, $z_1 \neq z_2$, $z_1 \cap Y = 0$, $z_2 \cap Y = 0$. Then by the preceding Lemma e.g. $z_1 \in z_2$ so that $z_1 \in z_2 \cap Y$, a contradiction. Thus the lemma is proved.

2113. Definition (**TSS**). Let Y be a non-empty real subclass of **On**. An element z of Y is called *minimal* ($z = \mathbf{Min}(Y)$) if $z \cap Y = 0$.

2114. Lemma (**TSS**). The class **On** is linearly ordered by the $\in$-relation **E**.

Proof. The irreflexivity of **E** on **On** ($\alpha \notin \alpha$) follows from Lemma 2108. For the transitivity suppose $\alpha \in \beta$ and $\beta \in \gamma$; then $\beta \subseteq \gamma$ and hence $\alpha \in \gamma$. The trichotomy follows from Lemma 2111.

2115. Lemma (**TSS**). (a) **On** is an ordinal; it is a real class and is not a set (consequently **On** is not a semiset).

(b) If X is an ordinal and X is not a semiset then $X = \mathbf{On}$.

Proof. (a) **On** is an ordinal by Lemma 2108(c), Lemma 2111 and Lemma 2109. **On** is a real class by Metatheorem 1438. (For Ord (X) is a PUP-formula.) **On** is not a set by Lemma 2108(b).

(b) Let X be an ordinal and suppose X is not a semiset. Then $X \subseteq \mathbf{On}$ and, for every ordinal number α, $X \not\subseteq \alpha$, i.e., for any α there is a β such that $\alpha \in \beta \in X$. By the completeness of X every α is an element of X and so $X = \mathbf{On}$.

2116. *Remark* We have seen that there are ordinals which are sets (the ordinal numbers) and that there is just one ordinal (the class **On**) which is not a semiset. Nothing is asserted about ordinals which are semisets but not sets; it can be shown by the ultraproduct model that the existence of such ordinals is consistent but we shall not be interested in them in the present book.

2117. Lemma (**TSS**). Let α be an ordinal number. Then

(a) $\alpha \cup \{\alpha\}$ is an ordinal number,

(b) there is no β such that

$$\alpha \in \beta \,\&\, \beta \in (\alpha \cup \{\alpha\}) .$$

Proof. (a) Obvious.

(b) Let $\alpha \in \beta$, $\beta \in \alpha \cup \{\alpha\}$. Then either $\beta \in \alpha$ or $\beta \in \{\alpha\}$, i.e. $\beta = \alpha$. Since $\alpha \in \beta$ it follows in either case that $\alpha \in \alpha$ – a contradiction.

2118. DEFINITION (**TSS**). $\alpha + 1 = \alpha \cup \{\alpha\}$. (the *successor* of α).

2119. METADEFINITION. We define in **TSS**:

$$1 = 0 + 1\,, \quad 2 = 1 + 1\,, \quad 3 = 2 + 1 \quad \text{etc.}$$

2120. LEMMA (**TSS**). If X is a class of ordinal numbers then its sum $\bigcup(X)$ is an ordinal. (Obvious.)

2121. DEFINITION (**TSS**). For $X \subseteq \mathbf{On}$ define

$$\mathbf{Sup}\,(X) = \bigcup(X) \quad (\text{the } \textit{supremum} \text{ of } X)\,.$$

If $\mathbf{Sup}\,(X) \in X$, we write $\mathbf{Max}\,(X)$ instead of $\mathbf{Sup}\,(X)$ (the *maximum* of X).

2122. LEMMA (**TSS**). The supremum of a set of ordinal numbers is an ordinal number.

2123. THEOREM (**TSS**). (The principle of the Transfinite Induction.) Let X be a real class satisfying the following conditions:

(a) $0 \in X$,

(b) $(\forall\alpha)\,(\alpha \in X \rightarrow (\alpha + 1) \in X)$,

(c) $(\forall x)\,(x \subseteq X \rightarrow \mathbf{Sup}\,(x) \in X)$.

Then X contains all ordinal numbers.

Proof. Suppose that not all ordinal numbers are in X. Then $\mathbf{On} - X$ is a non-void real class of ordinal numbers; let $z = \mathbf{Min}\,(\mathbf{On} - X)$. We will prove $z \in X$, a contradiction. Obviously $z \subseteq \mathbf{On} \cap X$. Set $z_1 = \mathbf{Sup}\,(z)$. By (c) $z_1 \in X$. If $z = z_1$ we are finished. If $z \in z_1$ then $z \subseteq z_1$ and $z_1 \subseteq z$; hence $z \in z$, a contradiction. If $z_1 \in z$ set $z_2 = z_1 + 1$. By (b) $z_2 \in X$. We claim that $z_2 = z$. Otherwise we would have $z_2 \in z$. Then $z_1 \in z_2 \in z$ and hence $z_1 \in \mathbf{Sup}\,(z) = z_1$, a contradiction.

2124. THEOREM (**TSS**). (Construction by Transfinite Recursion.) If G is a real function (i.e., a function and a real class) then there is exactly one real function F on the class $\mathbf{On}$ with the following property:

$$(\forall\alpha)\,(F\text{'}\alpha = G\text{'}(F\text{''}\alpha))\,.$$

Proof. Let K be the class of all set functions f such that $\mathbf{D}\,(f)$ is an ordinal number and $f\text{'}\alpha = G\text{'}(f\text{''}\alpha)$ for each $\alpha \in \mathbf{D}\,(f)$. It is easy to check that the condition defining K can be written as a PUP-formula, hence K is a real class. Now we prove that, for any $f, g \in K$, $\alpha \in \mathbf{D}(f) \cap \mathbf{D}(g)$ implies $f'\alpha = g'\alpha$. Assume $f, g \in K$ and $f'\alpha \neq g'\alpha$ for some $\alpha \in \mathbf{D}(f) \cap \mathbf{D}(g)$. Let $Y = \{\alpha \in \mathbf{D}(f) \cap \mathbf{D}(g); f'\alpha \neq g'\alpha\}$. Y is real by Metatheorem 1438. Hence, there is a least $\alpha \in \mathbf{D}(f) \cap \mathbf{D}(g)$ with $f'\alpha \neq g'\alpha$. Then $f''\alpha = g''\alpha$ and so $f'\alpha = G'(f''\alpha) = G'(g''\alpha) = g'\alpha$, a contradiction. Set $F = \bigcup(K)$; obviously, F is a function and $F'\alpha = G'F''\alpha$ whenever $\alpha \in \mathbf{D}(F)$. We prove that F is not a set. If F is a set then $\mathbf{D}(F)$ is a set by Axiom (A4) and therefore $\mathbf{D}(F)$ is an ordinal number. Set $f = F$, $\gamma = \mathbf{D}(f)$ and $g = f \cup \{\langle G'(f''\gamma), \gamma\rangle\}$. Then $f, g \in K$, $\mathbf{D}(g) = \gamma + 1$, hence $\gamma \in \mathbf{D}(f) = \gamma$ which contradicts 2108. Hence F is not a set and consequently K is not a set by 1427.

We now prove that $\mathbf{D}(F)$ is not a semiset. Set $X = \{\langle f, \alpha\rangle; f \in K \;\&\; \mathbf{D}(f) = \alpha\}$. Obviously X is a one-to-one mapping, $\mathbf{D}(X) = \mathbf{D}(F)$ and $\mathbf{W}(X) = K$. If $\mathbf{D}(F)$ is a semiset then K is a semiset and therefore a set since K is real. But we proved that K is not a set and so $\mathbf{D}(F)$ is not a semiset. Furthermore, F is not a semiset since $\mathrm{Sm}(F) \rightarrow \mathrm{Sm}(\mathbf{D}(F))$.

It follows by 1440 that F is real and by 2115 that $\mathbf{D}(F) = \mathbf{On}$. The proof of uniqueness is straightforward.

Remark. In the course of the above proof we gave the definition of F in terms of G. Hence we are justified stating the following matemathical version of Construction by Transfinite Recursion:

2125. METATHEOREM. There is a gödelian term $\mathbf{F}(G)$ with one class variable G such that

$$\mathbf{TSS} \vdash \mathrm{Real}(G) \;\&\; \mathrm{Un}(G) \,.\rightarrow.\, \mathrm{Real}(\mathbf{F}(G)) \;\&\; \mathrm{Un}(\mathbf{F}(G)) \;\&\; \mathbf{D}(\mathbf{F}(G)) = \mathbf{On} \;\&\; (\forall\alpha)(\mathbf{F}(F)'\,\alpha = G'(\mathbf{F}(G)''\,\alpha)) \,.$$

Demonstration. The formula $\mathrm{Un}(f) \;\&\; \mathbf{D}(f) \in \mathbf{On} \;\&\; (\forall\alpha)(f'\alpha = G'(f''\alpha))$ is normal in **TSS**, i.e. there is a NF $\varphi(f, G)$ equivalent to the former one in **TSS**. Hence, by Corollary 1122 there is a gödelian term $\mathbf{K}(G)$ such that $\mathbf{TSS} \vdash \mathbf{K}(G) = \{f;\ \mathrm{Un}(f) \;\&\; \mathbf{D}(f) \in \mathbf{On} \;\&\; (\forall\alpha \in \mathbf{D}(f))(f'\alpha = G'(f''\alpha))\}$. The formula $x \in \bigcup(\mathbf{K}(G))$ is equivalent to $(\exists f)(\varphi(f, G) \;\&\; x \in f)$ in **TSS** and is therefore normal; hence there is a gödelian term $\mathbf{F}(G)$ such that $\mathbf{TSS} \vdash \mathbf{F}(G) = \bigcup(\mathbf{K}(G))$. In the same way as Theorem 2124 was proved we can prove in **TSS**: If G is a real function then $\mathbf{F}(G)$ is a real function, $\mathbf{D}(\mathbf{F}(G)) = \mathbf{On}$ and $\mathbf{F}(G)'\,\alpha = G'(\mathbf{F}(G)''\,\alpha)$ for every α.

2126. DEFINITION (**TSS**). An ordinal number α is called *isolated* if $\alpha = 0$ or if there is a β such that $\beta + 1 = \alpha$. An ordinal number which is not isolated is called a *limit number*. The classes of all isolated and all limit numbers are denoted by $\mathbf{On_I}$ and $\mathbf{On_{II}}$ respectively.

2127. THEOREM (**TSS**). Both $\mathbf{On_I}$ and $\mathbf{On_{II}}$ are real classes and proper classes, i.e. they are not sets.

Proof. For any x we have $x \in \mathbf{On_I}$ iff $\mathrm{Ord}(x) \& (x = 0 \vee (\exists y \in x)(x = y \cup \{y\}))$. The last formula is evidently equivalent to a PUP-formula, hence, by Metatheorem 1438 $\mathbf{On_I}$ is real. Since $\mathbf{On_{II}} = \mathbf{On} - \mathbf{On_I}$, $\mathbf{On_{II}}$ is also real.

Suppose that the isolated numbers form a set and let $\gamma = \mathbf{Sup}(\mathbf{On_I})$. Then $\gamma + 1$ is again isolated, a contradiction.

Suppose that the limit numbers form a set and let $\gamma = \mathbf{Sup}(\mathbf{On_{II}})$. Then for each nonempty set x of ordinals, $x \subseteq \mathbf{On} - \gamma$ implies $\mathbf{Sup}(x) \in x$ and consequently x has a maximal element. Let a be a set satisfying the axiom C1 of infinity, i.e. suppose that $0 \in a$ and that $\{x\} \in a$ whenever $x \in a$. Let X be the class of all functions f such that $\mathbf{D}(f) = \beta - \gamma$ for some β and such that $f`\gamma = 0$ and $f`(\alpha + 1) = \{f`\alpha\}$ for each $\alpha$ such that $\alpha + 1 \in \mathbf{D}(f)$. It is easy to write the condition defining $X$ as a PUP-formula; hence $X$ is real. We first prove that every $f \in X$ is one-one and has its values in $a$. Let $\alpha$ be the least element of $\mathbf{D}(f)$ such that $f`\alpha = f`\beta$ for some $\beta \in \alpha$. There exist $\alpha_0$ and $\beta_0$ such that $\alpha = \alpha_0 + 1$ and $\beta = \beta_0 + 1$. Then $f`\alpha_0 = f`\beta_0$, a contradiction. Hence $f$ is one-one. Suppose that $\mathbf{W}(f)$ is not included in $a$ and let $\alpha$ be the least element of $\mathbf{D}(f)$ which is not in $a$. Then $f`\alpha = \{f`\alpha_0\}$, $\alpha = \alpha_0 + 1$, $f`\alpha_0 \in a$ and hence $f`\alpha \in a$, a contradiction. We can prove in a similar manner that if $\mathbf{D}(f) \subseteq \mathbf{D}(g)$ then $f \subseteq g$, whenver f and g are in X. Set $F = \bigcup(X)$. We prove that F is not a set. Suppose that F is a set and let α_0 be the maximal element of the domain of F. Set $\alpha = \alpha_0 + 1$, $g = F \cup \{\langle\{F`\alpha_0\}, \alpha\rangle\}$. Then we have $g \in X$ and $\mathbf{D}(g) = \alpha + 1$, a contradiction. Hence F is not a set and consequently X is not a set. Since X is real X is not a semiset. Analogously to the proof of 2124 we prove that neither F nor $\mathbf{D}(F)$ are semisets. But then F is real by 1440. F is one-to-one and its range is a set. Hence F is itself a set, contradiction.

2128. DEFINITION (**TSS**). The least limit number is denoted by $\boldsymbol{\omega}$ (or $\boldsymbol{\omega}_0$). The elements of $\boldsymbol{\omega}$ are *natural numbers* and are denoted by the variables m, n etc.

2129. THEOREM (**TSS**). (The induction principle.)

If a real class X contains 0 and with each natural number n also its successor $n + 1$, then X contains all natural numbers.

Proof. Suppose that $\omega - X \neq 0$ and let n be the least natural number not in X. The number n is a successor of a number $m \in X$ and hence is itself in X, a contradiction.

The following theorem is a variant of Theorem 2124:

2130. THEOREM (**TSS**). If G_1, G_2 are real mappings and a is a set then there is exactly one real mapping F such that

$$(1) \qquad F\text{'}0 = a\,,$$

$$(2) \qquad F\text{'}(\alpha + 1) = G_1\text{'}(F\text{'}\alpha)\,,$$

$$(3) \qquad F\text{'}\alpha = G_2\text{'}(F\text{''}\alpha) \quad \text{for } \alpha \text{ limit}\,.$$

The proof is analogous to the proof of Theorem 2124. We define K as the class of all functions f such that $\mathbf{D}(f) \in \mathbf{On}$ and f fulfils (1), (2), (3) for any α for which these equations make sense. K is real (because it is defined by a PUP-formula), $\bigcup(K)$ is real by Lemma 1440 and $F = \bigcup(K)$ fulfils (1)–(3).

2131. COROLLARY (**TSS**). (Construction by Recursion.)

Let a be a set and let G be a real mapping. Then there is exactly one mapping f such that $\mathbf{D}(f) = \omega$, $f\text{'}0 = a$ and $f\text{'}(n + 1) = G\text{'}(f\text{'}n)$ for each n.

(Set $G_1 = G$, and let G_2 be an arbitrary real mapping. (Take $f = F \restriction \omega$ where F fulfils (1)–(3). f is a set because F is a real mapping.)

The reader may formulate Metatheorems analogous to Metatheorem 2125 which express the fact that we may "describe" F in terms G_1, G_2, a (f in terms of G and a) by means of gödelian operations. In particular define:

2132. DEFINITION (**TSS**).

$$f_x\text{'}0 = x\,, \quad f_x\text{'}(n + 1) = \bigcup(f_x\text{'}n)\,; \quad \mathbf{Unv}(x) = \bigcup(\mathbf{W}(f_x))\,.$$

Roughly speaking $\mathbf{Unv}(x)$ is the "infinite union" $x \cup \bigcup(x) \cup \bigcup\bigcup(x) \cup \dots$ It follows from the definition of $\mathbf{Unv}(x)$ that $\mathbf{TSS} \vdash (\mathbf{Unv}(x)$ is a set). ($\mathbf{Unv}(x)$ is the union of the domain of values of some set function f_x.)

The following definition is a generalization of Definition 2132:

2133. DEFINITION (**TSS**). If R is a real regular relation then

$$\mathbf{Unv}_R(x) = \bigcup(\mathbf{W}(f)) \quad \text{where} \quad f\text{'}0 = \mathbf{Ext}_R(x),\ f\text{'}(n + 1) = R\text{''}(f\text{'}n)\,.$$

The definition is justified by the following easy

2134. LEMMA (**TSS**). If R is a real regular relation then, for every x, $\mathbf{Ext}_R(x)$ is a set.

Proof. Clearly $\mathbf{Ext}_R(x) = \mathbf{W}(R \cap (\mathbf{V} \times \{x\}))$; $R \cap (\mathbf{V} \times \{x\})$ is real and it is a semiset (indeed, $\mathbf{Ext}_R(x)$ is a semiset and if $\mathbf{Ext}_R(x) \subseteq a$ then $R \cap (\mathbf{V} \times \{x\}) \subseteq a \times \{x\}$), hence $R \cap (\mathbf{V} \times \{x\})$ is a set and therefore $\mathbf{W}(R \cap (\mathbf{V} \times \{x\}))$ is a set.

Note that **TSS** $\vdash \mathbf{Unv}(x) = \mathbf{Unv}_{\mathbf{E}}(x)$.

2135. Remark. The first place in this Section where axiom C1 was used is the proof of 2127. Hence if we denote by $C1^{bis}$ the assumption $\mathbf{On}_{\mathrm{II}} \neq 0$ ("limit numbers exist") and by $\mathbf{TSS}^{bis}$ the theory **TSS** with C1 replaced by $C1^{bis}$, then we can define ω in $\mathbf{TSS}^{bis}$ and prove Theorem 2131. Hence in $\mathbf{TSS}^{bis}$ define $f'0 = 0$, $f'(n + 1) = \{f'n\}$, $a = \mathbf{W}(f)$; it follows in $\mathbf{TSS}^{bis}$ that $0 \in a$ and that $\{y\} \in a$ for any $y \in a$. This means that C1 is provable in $\mathbf{TSS}^{bis}$. Consequently, C1 can be equivalently replaced by $C1^{bis}$ in the axioms of **TSS**.

In the rest of this Section, we shall consider the theory **TSS'** (i.e. **TSS** + + D1). Recall Metatheorem 1453 which enables to prove many classes to be real.

We shall now investigate well orderings.

2136. DEFINITION (**TSS'**). A relation R is a *well-ordering* (WOrdg (R)) if it has the following properties:

(0) R is a real class,

(1) R is a linear ordering,

(2) each nonempty real subclass of $\mathbf{C}(R)$ has a least element, i.e.

$$(\forall X)\,[0 \neq X \subseteq \mathbf{C}(R) \,\&\, \mathrm{Real}(X)\,.\rightarrow (\exists x \in X)\,[X \cap \mathbf{Ext}_R(x) = 0]]\,.$$

2137. DEFINITION (**TSS'**). A relation R is called a *regular well-ordering* (RWOrdg (R)) if it has the following properties:

(0) R is a real class,

(1) R is regular,

(2) R is a linear ordering,

(3) each nonempty subset of $\mathbf{C}(R)$ has a least element.

A real class P is *well-ordered* (*regularly well-ordered*) by a real relation R if $P \subseteq \mathbf{C}(R)$ and $R \cap P^2$ is a well-ordering (a regular well-ordering).

2138. LEMMA (**TSS'**). A real relation R is a regular well ordering if and only if it is regular and a well ordering.

Proof. A real regular relation which is a well ordering is obviously a regular well ordering; conversely, let R be a regular well ordering. Let X be a nonempty real subclass of its field. To find the least element of X, take an arbitrary $z \in X$ and set $u = X \cap \mathbf{Ext}_R(z)$. If $u = 0$ we are finished; otherwise, take the least element x of u. We assert that $X \cap \mathbf{Ext}_R(x) = 0$. Assume the contrary and take $y \in X \cap \mathbf{Ext}_R(x)$. We have yRx, xRz whence yRz. Since $y \in u$ we have $u \cap \mathbf{Ext}_R(x) \neq 0$, which contradicts to the minimality of x in u and our assertion is proved.

2139. Remark. The formula $\mathrm{RWOrdg}(R)$ is normal in **TSS'**.

2140. THEOREM (**TSS'**). The class of all ordinal numbers is regularly well ordered by $\mathbf{E}$ (the $\in$-relation).

Proof. The theorem is an immediate consequence of preceding lemmas.

2141. THEOREM (**TSS'**). Let P, R be real and suppose that P is regularly well ordered by R. Then there exists a unique real ordinal X and a unique real isomorphism F between P and X w.r.t. R and $\mathbf{E}$.

Proof. Let P be regularly well ordered by R and suppose, for simplicity, that $0 \notin P$. First we shall prove the existence of an ordinal X and an isomorphism F. Define the function G as follows:

$G\text{'}x = y$ if y is the least element of $P - x$, and $G\text{'}x = 0$ if $P - x = 0$. By Theorem 2124 there is a real function F such that $F\text{'}\alpha = G\text{'}(F\text{''}\alpha)$ for each α. Hence $F\text{'}\alpha$ is the least element of $P - (F\text{''}\alpha)$. Set $X = \{\alpha; F\text{'}\alpha \neq 0\}$. Then $F \upharpoonright X$ is the required isomorphism.

To prove the uniqueness suppose first that there are two real isomorphisms F_1 and F_2 between P and an ordinal X. Let α be the least ordinal in X such that $F_1\text{'}\alpha \neq F_2\text{'}\alpha$. Then $F_1\text{'}\beta = F_2\text{'}\beta$ whenever $\beta \in \alpha$. By trichotomy we have e.g. $\langle F_1\text{'}\alpha, F_2\text{'}\alpha\rangle \in R$. Then there exists $\gamma \in X$ such that $F_1\text{'}\alpha = F_2\text{'}\gamma$. Obviously $\gamma \in \alpha$ and hence $F_1\text{'}\gamma = F_2\text{'}\gamma$. It follows that $F_1\text{'}\alpha = F_1\text{'}\gamma$, a contradiction.

Now suppose that there is a real isomorphism F_1 between P and X_1 and a real isomorphism F_2 between P and X_2 where X_1 and X_2 are different ordinals. We have e.g. $X_1 \in X_2$. Define the function G on X_2 as follows: $G\text{'}\alpha = \beta \equiv F_1\text{'}\beta = F_2\text{'}\alpha$. G is obviously an isomorphism between X_1 and X_2 (w.r.t. $\mathbf{E}$). Since $G\text{'}X_1 \in X_1$, we may denote by γ the least element of X_2 such that $G\text{'}\gamma \in \gamma$. We have $G\text{'}(G\text{'}\gamma) \in G\text{'}\gamma$ and so by the minimality of γ, either $\gamma \in G\text{'}\gamma$ or $\gamma = G\text{'}\gamma$, contradicting $G\text{'}\gamma \in \gamma$.

We conclude our treatment of ordinal numbers with some metamathematical results concerning the possibility of definitions by induction in **TSS′** even in the case where the arguments are proper classes. Although we cannot deal with ordinal functions which assume proper classes as values, we can employ relations such that the extensions of ordinal numbers (e.g.) are proper classes with certain properties.

2142. METADEFINITION. Let **T** be a theory stronger than **TC**. Let X be a variable. A term $\mathbf{T}(X)$ is *local* in **T** if

$$\mathsf{T} \vdash \mathbf{T}(X) = \bigcup_{y \subseteq X} \mathbf{T}(y)$$

(i.e. if we can prove that $u \in \mathbf{T}(X)$ iff there is a subset y of X such that $u \in$ $\in \mathbf{T}(y)$).

Note that if $\mathbf{T}(X)$ is local in **T** then $\mathsf{T} \vdash X \subseteq Y \rightarrow \mathbf{T}(X) \subseteq \mathbf{T}(Y)$ (where X and Y are variables of the same sort).

2143. METALEMMA. A term $\mathbf{T}(X)$ is local in **T** iff the following is provable in **T**:

$$\mathbf{T}(X) = \bigcup_{y \in F} \mathbf{T}(y)$$

whenever $F \subseteq \mathbf{P}(X)$ contains for each $x \subseteq X$ a superset $y \supseteq x$ (F may be called a *⊆-cofinal system of subsets of* X).

2144. METADEFINITION. Let **T** be a theory stronger than **TSS′**. A gödelian term $\mathbf{T}(X)$ of **T** is *iterable* in **T** if

(a) $\mathbf{T}(X)$ has the *set images property* (i.e. $\mathsf{T} \vdash \mathbf{T}(x)$ is a set),
(b) $\mathbf{T}(X)$ is local.

2145. METATHEOREM. Let **T** be a stronger than **TSS′**. Let $\mathbf{T}(X)$ be a gödelian term of **T** iterable in **T**. Then there is a gödelian term $\mathbf{S}(X)$ (called the *class iteration of* $\mathbf{T}(X)$) such that the following is provable in **T**:

If $X \neq 0$ is real and $H = \mathbf{S}(X)$ then H is a real relation, $\mathbf{D}(H)$ is an ordinal, $H''\{0\} = X$ and $H''\{\alpha\} = \mathbf{T}(H''\alpha)$ for every $\alpha > 0$.

Demonstration. Put $\mathbf{K}(X) = \{r;\ \mathrm{Rel}(r)\ \&\ \mathbf{D}(r) \in \mathbf{On}\ \&\ (\exists z \subseteq X)(r''\{0\} = z)\ \&\ (\forall \alpha > 0)(\alpha \in \mathbf{D}(r) \rightarrow r''\{\alpha\} = \mathbf{T}(r''\alpha))$. The formula defining $\mathbf{K}(X)$ is normal (see 1126), hence the gödelian term $\mathbf{K}(X)$ can be constructed following 1122. Furthermore, by Metatheorem 1454 $\mathsf{T} \vdash \mathrm{Real}(X) \rightarrow$

$\rightarrow$ Real $(\mathbf{K}(X))$. Put $\mathbf{S}(X) = \bigcup(\mathbf{K}(X))$; we show that $\mathbf{S}(X)$ has the desired properties. We proceed in **T**. Denote $\mathbf{S}(X)$ by H. Then H is real relation and $\mathbf{D}(H)$ is an ordinal. Evidently $H``\{0\} = X$. Further, $H``\{\alpha\} = \bigcup_{r \in \mathbf{K}(X)} r``\{\alpha\}$ and $H``\alpha = \bigcup_{r \in \mathbf{K}(X)} r``\alpha$. We have $H``\{\alpha\} = \bigcup_{r \in \mathbf{K}(X)} r``\{\alpha\} = \bigcup_r \mathbf{T}(r``\alpha)$ and $\mathbf{T}(\bigcup_r r``\alpha) = \mathbf{T}(H``\alpha)$. If we prove $\bigcup_r \mathbf{T}(r``\alpha) = \mathbf{T}(\bigcup_r r``\alpha)$ we have $H``\{\alpha\} = \mathbf{T}(H``\alpha)$. It suffices to prove that $\{r``\alpha;\ r \in \mathbf{K}(X)\}$ is an $\subseteq$-cofinal system in $\bigcup_r r``\alpha$. Let $q \subseteq \bigcup_r r``\alpha$; we find $r_0 \in \mathbf{K}(X)$ such that $q \subseteq r_0``\alpha$. Define $\langle r, u\rangle \in Q \equiv .\ u \in q \,\&\, r \in \mathbf{K}(X) \,\&\, u \in r``\alpha$. Q is a relation and $\mathbf{D}(Q)$ is a set; by (D1), there is a semiset relation $\varrho \subseteq Q$ such that $\mathbf{D}(\varrho) = \mathbf{D}(Q)$. Put $\mathbf{W}(\varrho) = \sigma$; σ is a subsemiset of $\mathbf{K}(X)$. Let $\sigma \subseteq a$; we can suppose that $a \subseteq \mathbf{K}(X)$ (since $\mathbf{K}(X)$ is real). $\bigcup(a)$ is a relation and $\bigcup(a)``\{0\}$ is a subset $x_0$ of $X$. There exists $r_0 \in \mathbf{K}(X)$ such that $r_0``\{0\} = x_0$ and $0 < \beta \leqq \alpha \rightarrow r_0``\{\beta\} = \mathbf{T}(r_0``\beta)$; it follows that $r``\alpha \subseteq r_0``\alpha$ for any $r \in a$. (Prove $r, s \in \mathbf{K}(X) \,\&\, \alpha \in \mathbf{D}(r) \cap \mathbf{D}(s) \,\&\, r``\{0\} \subseteq s``\{0\} \,. \rightarrow r``\{\alpha\} \subseteq s``\{\alpha\}$ by transfinite induction.) Therefore $q \subseteq r_0``\alpha$.

2146. METATHEOREM. Let $\mathbf{T}(X)$ be iterable in **T** where **T** is stronger than **TSS′** and let $\mathbf{O}(X)$ be the range of the class iteration of $\mathbf{T}(X)$. Then the following is provable in **T**: For every real X, $Z = \mathbf{O}(X)$ is the least real class with the properties $X \subseteq Z$ and $\mathbf{T}(Z) \subseteq Z$.

Demonstration. We use the denotation from the proof of the preceding Metatheorem. In **T** we have

$$Z = \mathbf{W}(H)\,,\quad \mathbf{T}(\mathbf{W}(H)) = \bigcup_{\substack{r \in \mathbf{K}(X) \\ \alpha \in \mathbf{On}}} \mathbf{T}(r``\alpha) = \bigcup_{\substack{r \in \mathbf{K}(X) \\ 0 < \alpha \in \mathbf{On}}} r``\{\alpha\} \subseteq \mathbf{W}(H)\,;$$

furthermore, it is obvious that $X \subseteq Z$. If W is real, $X \subseteq W$ and $\mathbf{T}(W) \subseteq W$, then by induction $H``\{\alpha\} \subseteq W$ for each α; hence $Z \subseteq W$.

2147. Example. Define H as follows (for X real):

$$H``\{0\} = X\,,$$
$$H``\{n + 1\} = \bigcup(H``(n + 1)) \quad \text{for} \quad n \in \omega$$

and let

$$\mathbf{Unv}(X) = \mathbf{W}(H)\,.$$

Verify that the term $\mathbf{Unv}(X)$ is definable in **TSS′** by a normal formula, that in case X is a set this term coincides with the previously defined term $\mathbf{Unv}(X)$ (cf. Definition 2132), and that **TSS′** $\vdash$ Comp $(\mathbf{Unv}(X))$.

SECTION 2

Equivalence and Subvalence of Sets. Cardinal Numbers

Throughout the present Section we shall work within the theory **TSS′**. We shall define equivalence of sets (having the same power) and subvalence (having smaller than or equal power). We will not define "the power of x" for arbitrary sets x, because without assuming the axiom of choice or some stronger axiom of regularity we cannot define an operation assigning to each set its "power". However we shall prove in **TSS′** a number of useful statements concerning cardinalities which can be applied in those theories where the axiom of choice is not assumed (or even does not hold), namely a number of statements on finite sets. On the other hand, in **TSS′** we can define cardinal numbers as powers of well-orderable sets and prove a number of statements about them. The concept of equivalence will be defined for classes in general (not only for sets).

Equivalence of two objects (sets or classes) usually means the existence of a one-one mapping of one object onto the other. But observe that – at least as far as sets are concerned – we have two possible ways to define equivalence. Given two sets x and y we may ask whether there is a set f which is a one-one mapping of x onto y or whether there is a class with this property. (It must necessarily be a semiset.) More generally, for real classes X and Y, we may ask whether there is a real class which is a one-one mapping of X onto Y or whether there is simply a class with this property. This gives the following definitions:

2201. Definition (**TSS′**). (1) Let X, Y be real classes. X is *equivalent* to Y (or X *has the same power as* Y, $X \approx Y$) if there is a real class which is a one-one mapping of X onto Y. X is *subvalent* to Y ($X \preccurlyeq Y$) if there is a real class which is a one-one mapping of X into Y. X is *strictly subvalent* to Y ($X \prec Y$) if $X \preccurlyeq Y$ and not $X \approx Y$.

(2) Let X, Y be arbitrary classes. X is *absolutely equivalent* to Y ($X \hat{\approx} Y$) if there is a class which is a one-one mapping of X onto Y. X is *absolutely subvalent* to Y ($X \hat{\preccurlyeq} Y$) if there is a class which is a one-one mapping of X into Y. X is *absolutely strictly subvalent* to Y if $X \hat{\preccurlyeq} Y$ and not $X \hat{\approx} Y$.

2202. *Remark.* Sets x and y are equivalent iff there is a set which is a one-one mapping of x onto y; x is absolutely equivalent to y iff there is a semiset which is a one-one mapping of x onto y. Similarly for subvalence. Obviously, $X \approx Y \to X \hat{\approx} Y$ and $X \preccurlyeq Y \to X \hat{\preccurlyeq} Y$ are provable in **TSS'**; further, the reflexivity, symmetry and transitivity of $\approx$ are provable in **TSS'**; i.e. $X \approx X$, $X \approx Y \to Y \approx X$, $X \approx Y \,\&\, Y \approx Z \,.\!\to X \approx Z$ for real X, Y, Z; similarly for $\hat{\approx}$ instead of $\approx$ (and arbitrary X, Y, Z). Also $X \preccurlyeq X$ and $X \preccurlyeq Y \,\&\, Y \preccurlyeq \preccurlyeq Z \,.\!\to X \preccurlyeq Z$ are provable in **TSS'** and similarly for $\hat{\preccurlyeq}$ instead of $\preccurlyeq$.

2203. DEFINITION (**TSS'**). A relation R is an *equivalence* if the following hold for all $x, y, z \in \mathbf{C}(R)$:

(1) $$\langle x, x\rangle \in R\,,$$

(2) $$\langle x, y\rangle \in R \to \langle y, x\rangle \in R\,,$$

(3) $$\langle x, y\rangle \in R \,\&\, \langle y, z\rangle \in R \,.\!\to \langle x, z\rangle \in R\,.$$

Evidently, the relation **Eq** defined by $\langle x, y\rangle \in \mathbf{Eq} \equiv x \approx y$ is an equivalence.

(Note that the formula $x \approx y$ is normal in **TSS'**.)

In the present section we shall be interested in the notions of equivalence and subvalence ($\approx$, $\preccurlyeq$); we shall deal with absolute equivalence and absolute subvalence later. For the moment we shall only show that the Cantor-Bernstein theorem holds for both notions.

2204. THEOREM (**TSS'**). (a) $X \preccurlyeq Y \,\&\, Y \preccurlyeq X \,.\!\to X \approx Y$ for any real classes X, Y;

(b) $X \hat{\preccurlyeq} Y \,\&\, Y \hat{\preccurlyeq} X \,.\!\to X \hat{\approx} Y$ for any classes X, Y.

Proof. Given one-one mappings F of X into Y and G of Y into X we construct a one-one mapping H of X onto Y; then we show that H is real provided that F and G are. Obviously it is sufficient to construct a one-one mapping of X onto G"Y which is real provided that F and G are. Let K be the class of all one-one mappings f such that

(1) $$\mathbf{D}(f) \text{ is a natural number },$$

(2) $$f'0 \in X - G''Y,$$

(3) $$(\forall n)(n + 1 \in \mathbf{D}(f) \rightarrow f'(n + 1) = G'(F'(f'n)) .$$

For any $f, g \in K$, $n + 1 \in \mathbf{D}(f) \cap \mathbf{D}(g)$ implies $f'(n + 1) = g'(n + 1) \equiv \equiv f'n = g'n$. Define $x \in W \equiv (\exists f \in K)(\exists n)(x = f'n)$; $S = X - W (= = G''Y - W)$. Note that if F and G are real then K, S and W are also real. Now define $H'x = G'(F'x)$ for $x \in W$, $H'x = x$ for $x \in S$. We claim that H is the required mapping. First, H is real if F, G are. Further, $H \restriction S$ is a one-one mapping of S onto S and $H \restriction W$ is a one-one mapping of W onto $(G''Y - S)$; hence $H = (H \restriction W) \cup (H \restriction S)$ is a one-one mapping of X onto $G''Y$, q.e.d.

2205. LEMMA (**TSS**′). Every set x is strictly subvalent to $\mathbf{P}(x)$.

Proof. The mapping f on x which assigns to each $y \in x$ the set $\{y\}$ is one-one and into $\mathbf{P}(x)$. Hence $x \preccurlyeq \mathbf{P}(x)$. Suppose now that $x \approx \mathbf{P}(x)$. Let f be a one-one mapping of x onto $\mathbf{P}(x)$. Consider the set a of all $y \in x$ which do not belong to their images, i.e. $y \notin f'y$. Since $a \in \mathbf{P}(x)$ there is a $y_0 \in x$ such that $a = f'y_0$. If $y_0 \in a$, then $y_0 \notin f'y_0$, a contradiction. If $y_0 \notin a$, then $y_0 \in f'y_0 = a$, a contradiction. Hence $x \not\approx \mathbf{P}(x)$.

2206. DEFINITION (**TSS**′). X is a *cardinal* (Card (X)) if it is a real ordinal and is not equivalent to any of its elements. Ordinal numbers which are cardinals are called *cardinal numbers*. The class of all cardinal numbers is denoted by **Cn**.

2207. LEMMA (**TSS**′). The class **On** is a cardinal.

Proof. **On** cannot be equivalent to any ordinal number α, since it is a proper class.

2208. LEMMA (**TSS**′). The class **Cn** of all cardinal numbers is a proper class.

Proof. Suppose the contrary and set $\xi = \mathbf{Sup}(\mathbf{Cn})$. Clearly ξ is equivalent to each $\alpha > \xi$. Let the function F assign to each $\alpha > \xi$ the set of orderings of ξ which are isomorphic to the natural ordering of α (by **E**). The function F is defined for every $\alpha > \xi$ and assumes different values at different points, i.e. is one-one. Since all of its values $F'\alpha$ belong to $\mathbf{P}(\mathbf{P}(\xi \times \xi))$, the range of F is a set. Consequently the domain of F is a set too, a contradiction.

2209. LEMMA (**TSS'**). If X is a real class which can be regularly well-ordered then X is equivalent to a cardinal, and this cardinal is unique.

Proof. By Theorem 2141 every real class X which can be regularly well-ordered is equivalent to an ordinal. Each ordinal is equivalent to a cardinal. The uniqueness is obvious.

2210. LEMMA (**TSS'**). The supremum of any real class X of cardinal numbers is a cardinal.

Proof. The assertion holds whenever X is a proper class since then $\mathbf{Sup}\, X = \mathbf{On}$. Let X be a set and let $\mathbf{Sup}\, X \notin X$. Suppose that there is a $\beta \in \mathbf{Sup}\, X$ equivalent to $\mathbf{Sup}\, X$. Then there is a $\gamma \in X$ such that $\beta \in \gamma \in \mathbf{Sup}\, X$. We have then $\beta \approx \gamma$ by 2204, since γ is a cardinal number, this is a contradiction.

2211. LEMMA (**TSS'**). $x \subset n \rightarrow x \prec n$.

Proof. By induction. Let n be the least natural number equivalent to a proper subset, let x be such a subset and let f be a one-one mapping of n onto x. Since $n \neq 0$ let $n = n_0 + 1$. The mapping $f \restriction n_0$ is one-one and onto $x - \{f'n_0\}$. If $f'n_0 = n_0$ or if n_0 is not in the range of f, then $x - \{f'n_0\} \subset n_0$, which contradicts the induction hypothesis. If $n_0 = f'i$ and $i \in n_0$, define $\bar{f}$ on n_0 as follows: $\bar{f}'j = f'n_0$ if $j = i$ and $\bar{f}'j = f'j$ otherwise. The function $\bar{f}$ is then a one-one mapping of n_0 onto a proper subset, a contradiction.

2212. THEOREM (**TSS'**). a) Every natural number is a cardinal number.

b) The class of all infinite cardinal numbers is isomorphic to **On.**

Proof. a) follows from 2211 because $m \in n \rightarrow m \subset n$ and hence $m \prec n$, i.e. each n is a cardinal number.

b) follows from 2208, 2141, 2115.

2213. DEFINITION (**TSS'**). The unique enumeration of infinite cardinal numbers by ordinal numbers is denoted by $\aleph$. The α-th infinite cardinal number is denoted by $\aleph_\alpha$ or ω_α.

2214. Remark. Both ω_α and $\aleph_\alpha$ denote the same operation on ordinal numbers. The former notation is used if the ordinal number ω_α is involved, whereas the latter one is used of the cardinality of the number ω_α is under consideration; however, both symbols may be used interchangeably.

2215. LEMMA (**TSS'**). Each ω_α is a limit ordinal number.

Proof. Suppose $\omega_\alpha = \beta + 1$. Define f as follows: $f'\gamma = 0$ if $\gamma = \beta$, $f'\gamma = \gamma + 1$ if $\gamma \in \omega$ and $f'\gamma = \gamma$ otherwise. Then f is a one-one mapping of ω_α onto β, so that $\omega_\alpha \approx \beta$, a contradiction.

In what follows we investigate the properties of finite sets.

2216. DEFINITION (**TSS'**). A set x is *finite* (in the sense of Tarski), if every non-void subset of the power-class of x has a maximal element with respect to inclusion, i.e.

$$(\forall y \subseteq \mathbf{P}(x))\,(y \neq 0 \rightarrow (\exists z \in y)\,(\forall u \in y)\,(\neg z \subset u))\,.$$

A set x is *countable if* $x \approx \aleph_0$.

2217. LEMMA (**TSS'**). (a) A singleton is finite.

(b) The sum of two finite sets is finite.

Proof. (a) Obvious. (b) Let y be a non-void subset of $\mathbf{P}(x_1 \cup x_2)$. Set $u \in y_1 \equiv (\exists v \in y)\,(u = v \cap x_1)$. If $y_1 = \{0\}$ then y contains only subsets of x_2 and hence has a maximal element. If $y_1 \neq \{0\}$ then there exists $z_1 \in y_1$ such that $(\forall u \in y_1)\,(\neg z_1 \subset u)$. Set

$$u \in y_2 \equiv (\exists v \in y)\,(u = v \cap x_2 \;\&\; z_1 = v \cap x_1)\,.$$

If $y_2 = \{0\}$, then z_1 is maximal in y, i.e. $(\forall u \in y)\,(\neg z_1 \subset u)$.

If $y_2 \neq \{0\}$ take a maximal z_2 in y_2 and set $z = z_1 \cup z_2$; then z is maximal in y.

2218. LEMMA (**TSS'**). (a) If x_1 is subvalent to x_2 and x_2 is finite then x_1 is also finite.

(b) A set which is equivalent to a finite set is itself finite.

Proof. (a) Let f be a one-one mapping of x_1 into x_2 and let y be a non-void set of subsets of x_1. Set $\bar{y} = \{f''z;\, z \in y\}$. Obviously $\bar{y} \subseteq \mathbf{P}(x_2)$ and $\bar{y} \neq 0$ and so $\bar{y}$ has a maximal element z. Then the inverse image of z under f is maximal in y.

(b) follows from (a).

2219. LEMMA (**TSS'**). (a) Each natural number is finite.

(b) $\aleph_0$ is infinite.

Proof. (a) Denote by F the class of all finite sets. F contains the void set and, by 2217, if $x \in F$ then $x + 1 \in F$. Hence, by the Induction Principle, F contains all natural numbers.

(b) We have $\omega_0 \subseteq \mathbf{P}(\omega_0)$ and ω_0 has no maximal element with respect to inclusion.

2220. THEOREM (**TSS′**). (a) A set is finite if and only if it is equivalent to some natural number.

(b) If x contains a countable subset (i.e. if $\aleph_0 \preccurlyeq x$) then x is infinite.

Proof. (a) If $x \approx n$ then x is finite by preceding lemmas. If x is finite suppose that x is not equivalent to any natural number n and set $u \in y \equiv$ $\equiv u \subseteq x \,\&\, (\exists n)(u \approx n)$. We have $y \neq 0$ and $y \subseteq \mathbf{P}(x)$. Let z be a maximal element of y and denote by n_0 the natural number equivalent to z. Since $x - z \neq 0$, take $q \in x - z$ and set $z_1 = z \cup \{q\}$. We have $z_1 \supset z$ and $z_1 \approx$ $\approx n_0 + 1$, a contradiction.

(b) follows from 2219(b) and 2218(a).

2221. COROLLARY (**TSS′**). x is finite if and only if $x \prec \omega_0$.

2222. Remark. The converse of 2220(b) is not provable in **TSS′**, see 6217.

2223. LEMMA (**TSS′**). The power set of a finite set is finite.

Proof. It suffices to prove that $\mathbf{P}(n)$ is finite for each n. Let n be the least number with an infinite power set. Clearly $n \neq 0$ and thus $n = n_0 + 1$ for some n_0. We have $\mathbf{P}(n) = \mathbf{P}(n_0) \cup (\mathbf{P}(n) - \mathbf{P}(n_0))$. By the induction hypothesis, $\mathbf{P}(n_0)$ is finite. Define f by setting $f'x = x \cup \{n_0\}$ for $x \subseteq n_0$. The function f is a one-one mapping of $\mathbf{P}(n_0)$ onto $\mathbf{P}(n) - \mathbf{P}(n_0)$ and hence $\mathbf{P}(n)$ is a union of two finite sets; thus $\mathbf{P}(n)$ is finite, a contradiction.

2224. LEMMA (**TSS′**). If x is a non-void finite set of ordinal numbers then $\mathbf{Sup}(x) \in x$ (i.e. $\mathbf{Sup}(x) = \mathbf{Max}(x)$).

Proof. Suppose that $n \neq 0$ is the least natural number for which there is an n-element set $x \subseteq \mathbf{On}$ which has no maximum. Obviously $n \neq 1$. Set $n = n_0 + 1$. If f is a one-one mapping of n onto x then by the induction hypothesis the image of n_0 under f has a maximum α. The number $\gamma =$ $= \mathbf{Max}(\alpha, f'n_0)$ is obviously the maximum of x, a contradiction.

2225. DEFINITION (**TSS′**). $\mathbf{P}_{\mathrm{fin}}(X)$ is the class of all finite subsets of X.

2226. DEFINITION (**TSS′**). The relation **Sd** on the class of all finite sets of ordinals is defined as follows:

$$\langle x, y\rangle \in \mathbf{Sd} \equiv .\ x \neq y \,\&\, \mathbf{Max}((x - y) \cup (y - x)) \in y\,.$$

2227. LEMMA (**TSS**′). The relation **Sd** is regular and real.

Proof. Let $y \in \mathbf{P_{fin}}(\mathbf{On})$. If $y = 0$ then $\mathbf{Sd}''(y) = 0$. If $y \neq 0$ let $\alpha = \mathbf{Max}(y)$. Since $\langle x, y\rangle \in \mathbf{Sd}$ implies $x \subseteq \alpha + 1$, we have $\mathbf{Sd}''\{y\} \subseteq \mathbf{P}(\alpha + 1)$ and so **Sd** is regular. **Sd** is real by Metatheorem 1453.

2228. LEMMA (**TSS**′). $\mathbf{P_{fin}}(\mathbf{On})$ is linearly ordered by **Sd**.

Proof. The only non-trivial part of the proof is the proof of transitivity. Suppose that $\langle x, y\rangle \in \mathbf{Sd}$ and $\langle y, z\rangle \in \mathbf{Sd}$. Set $\alpha_1 = \mathbf{Max}((x - y) \cup (y - x))$, $\alpha_2 = \mathbf{Max}((y - z) \cup (z - y))$. We have $\alpha_1 \in y - x$, $\alpha_2 \in z - y$ and $\alpha_1 \neq \alpha_2$. Suppose that $\alpha_2 \in \alpha_1$. Then $\gamma \in y \equiv \gamma \in z$ whenever $\alpha_2 \in \gamma$ and hence $\alpha_1 \in z - x$. Let $\alpha_1 \in \gamma$ and $\gamma \in x - z$ for some γ. Then $\gamma \in x - y$, contradicting the definition of α_1. Similarly, if $\alpha_1 \in \gamma$ and $\gamma \in z - x$ then $\gamma \in y - x$, contradicting the definition of α_1. Hence $\alpha_1 = \mathbf{Max}((x - z) \cup (z - x))$, and since $\alpha_1 \in z$ we have $\langle x, z\rangle \in \mathbf{Sd}$. The case $\alpha_1 \in \alpha_2$ is treated analogously.

2229. THEOREM (**TSS**′). $\mathbf{P_{fin}}(\mathbf{On})$ is regularly well ordered by **Sd**.

Proof. It suffices to prove that each non-void subset of $\mathbf{P_{fin}}(\mathbf{On})$ has an **Sd**-least element. Each $u \subseteq \mathbf{P_{fin}}(\mathbf{On})$ is a subset of $\mathbf{P_{fin}}(\alpha)$ for some α; we take the least α such that there is a non-void $u \subseteq \mathbf{P_{fin}}(\alpha)$ having no **Sd**-least element. Set $\beta = \mathbf{Min}(\{\mathbf{Max}(x)\,;\, x \in u\})$; obviously $\beta \in \alpha$. Denote by u_0 the set of all $x \in u$ such that $\beta = \mathbf{Max}(x)$. The set u_0 has no least element, for otherwise the least element of u_0 would also be least in u. We have $u_0 \subseteq \mathbf{P_{fin}}(\beta + 1)$ and hence $\alpha = \beta + 1$. Let $x \in u_1 \equiv x \subseteq \beta \;\&\; x \cup \{\beta\} \in u_0$. Since u_0 is non-void, the set u_1 is also non-void. Hence, u_1 has a least element (by the induction hypothesis). Denote by z_1 the least element of u_1 and set $z = z_1 \cup \{\beta\}$; z is the **Sd**-least element of u_0, a contradiction.

2230. THEOREM (**TSS**′). (a) $\mathbf{P_{fin}}(\mathbf{On}) \approx \mathbf{On}$,

(b) $\mathbf{P_{fin}}(\boldsymbol{\omega}_\alpha) \approx \aleph_\alpha$.

Proof. (a) Since $\mathbf{P_{fin}}(\mathbf{On})$ is a proper class regularly well-ordered by **Sd**, there is a real isomorphism F between **On** and $\mathbf{P_{fin}}(\mathbf{On})$ with respect to **E** and **Sd**. Hence $\mathbf{P_{fin}}(\mathbf{On})$ and **On** are equivalent.

(b) Let F be the above-mentioned isomorphism. We shall prove that $F''\boldsymbol{\omega}_\alpha = \mathbf{P_{fin}}(\boldsymbol{\omega}_\alpha)$ for each $\boldsymbol{\omega}_\alpha$. Suppose the contrary and consider the least $\boldsymbol{\omega}_\alpha$ such that $F''\boldsymbol{\omega}_\alpha \neq \mathbf{P_{fin}}(\boldsymbol{\omega}_\alpha)$. Since $\beta \in \gamma$ implies $\langle\{\beta\}, \{\gamma\}\rangle \in \mathbf{Sd}$, $\mathbf{P_{fin}}(\boldsymbol{\omega}_\alpha)$ contains a subset isomorphic to $\boldsymbol{\omega}_\alpha$, and $\mathbf{Sd}''\{x\} \subseteq \mathbf{P_{fin}}(\boldsymbol{\omega}_\alpha)$ whenever $x \in \mathbf{P_{fin}}(\boldsymbol{\omega}_\alpha)$. Hence $F''\boldsymbol{\omega}_\alpha \subset \mathbf{P_{fin}}(\boldsymbol{\omega}_\alpha)$ and there exists a finite $y \subseteq \boldsymbol{\omega}_\alpha$

such that $y \notin F``\boldsymbol{\omega}_\alpha$. Let $\beta$ be the maximum of $y$. Then $F``\boldsymbol{\omega}_\alpha \subseteq \mathbf{P}_{\text{fin}}(\beta + 1)$. Since $\beta + 1 \prec \aleph_\alpha$, we have $\mathbf{P}_{\text{fin}}(\beta + 1) \prec \aleph_\alpha$, a contradiction.

2231. THEOREM (**TSS'**). If X is a cardinal such that $X \succcurlyeq \aleph_0$ then $X \approx X^2$.

Proof. Since $X^2 \subseteq \mathbf{P}_{\text{fin}}(\mathbf{P}_{\text{fin}}(X))$, we have $X^2 \preccurlyeq X$ by the preceding theorem; obviously $X \preccurlyeq X^2$ and so $X \approx X^2$.

2232. DEFINITION (**TSS'**). $\boldsymbol{\omega}_\alpha$ is *cofinal with* $\boldsymbol{\omega}_\beta$ $(\text{Conf}(\boldsymbol{\omega}_\alpha, \boldsymbol{\omega}_\beta))$ if $\boldsymbol{\omega}_\alpha \leqq \boldsymbol{\omega}_\beta$ and if there is a function f such that $\mathbf{D}(f) = \boldsymbol{\omega}_\alpha$, $\mathbf{W}(f) \subseteq \boldsymbol{\omega}_\beta$ and $\bigcup(\mathbf{W}(f)) = \boldsymbol{\omega}_\beta$.

2233. LEMMA (**TSS'**). If $\boldsymbol{\omega}_\alpha$ is cofinal with $\boldsymbol{\omega}_\beta$ and $\boldsymbol{\omega}_\alpha \leqq \boldsymbol{\omega}_\delta \leqq \boldsymbol{\omega}_\beta$ then $\boldsymbol{\omega}_\delta$ is cofinal with $\boldsymbol{\omega}_\beta$. (Obvious.)

2234. DEFINITION (**TSS'**). $\mathbf{cf}(\boldsymbol{\omega}_\beta)$ is the least cardinal number $\boldsymbol{\omega}_\alpha$ which is cofinal with $\boldsymbol{\omega}_\beta$.

2235. LEMMA (**TSS'**). If $\boldsymbol{\omega}_\alpha = \mathbf{cf}(\boldsymbol{\omega}_\beta)$ then there is a function f such that $\mathbf{D}(f) = \boldsymbol{\omega}_\alpha$, $\mathbf{W}(f) \subseteq \boldsymbol{\omega}_\beta$, $\bigcup(\mathbf{W}(f)) = \boldsymbol{\omega}_\beta$ and

$$(\forall \iota, \kappa \in \boldsymbol{\omega}_\alpha)(\iota < \kappa \equiv f`\iota < f`\kappa) .$$

Proof. Let g be a function from $\boldsymbol{\omega}_\alpha$ into $\boldsymbol{\omega}_\beta$ such that $\bigcup(\mathbf{W}(g)) = \boldsymbol{\omega}_\beta$. Define the set d by the following induction: $0 \in d$ and, for $0 < \lambda < \boldsymbol{\omega}_\alpha$, $\lambda \in d$ iff $g`\lambda > g`\gamma$ for all $\gamma \in d \cap \lambda$. Then d is a subset of $\boldsymbol{\omega}_\alpha$, $g \upharpoonright d$ is a $1-1$ function and $\iota < \kappa \equiv g`\iota < g`\kappa$ for all $\iota, \kappa \in d$. Clearly d is isomorphic with an ordinal $\delta \leqq \boldsymbol{\omega}_\alpha$; but δ cannot be less than $\boldsymbol{\omega}_\alpha$ because otherwise some cardinal less than $\boldsymbol{\omega}_\alpha$ would be cofinal with $\boldsymbol{\omega}_\beta$. Hence there is an isomorphism h of $\boldsymbol{\omega}_\alpha$ onto d; we set $f`\iota = g`(h`\iota)$ for $\iota \in \boldsymbol{\omega}_\alpha$.

2236. LEMMA (**TSS'**). $\mathbf{cf}(\mathbf{cf}(\boldsymbol{\omega}_\beta)) = \mathbf{cf}(\boldsymbol{\omega}_\beta)$.

Proof. Let $\boldsymbol{\omega}_\alpha = \mathbf{cf}(\boldsymbol{\omega}_\beta)$, $\boldsymbol{\omega}_\delta = \mathbf{cf}(\boldsymbol{\omega}_\alpha)$. Let f be a monotone mapping of $\boldsymbol{\omega}_\alpha$ into $\boldsymbol{\omega}_\beta$ such that $\bigcup(\mathbf{W}(f)) = \boldsymbol{\omega}_\beta$ and let g be a monotone mapping of $\boldsymbol{\omega}_\delta$ into $\boldsymbol{\omega}_\alpha$ such that $\bigcup(\mathbf{W}(g)) = \boldsymbol{\omega}_\alpha$. Define $h`\iota = f`(g`\iota)$ for $\iota \in \boldsymbol{\omega}_\delta$; then h is a (monotone) mapping of $\boldsymbol{\omega}_\delta$ into $\boldsymbol{\omega}_\beta$ such that $\bigcup(\mathbf{W}(h)) = \boldsymbol{\omega}_\beta$; hence $\boldsymbol{\omega}_\alpha \leqq \boldsymbol{\omega}_\delta$. Evidently $\boldsymbol{\omega}_\delta \leqq \boldsymbol{\omega}_\alpha$, so that $\boldsymbol{\omega}_\alpha = \boldsymbol{\omega}_\delta$.

2237. Remark (**TSS'**). There exist infinite cardinal numbers such that $\mathbf{cf}(\alpha) = \alpha$ (e.g. $\aleph_0$) and infinite cardinal numbers with $\mathbf{cf}(\alpha) \in \alpha$ (e.g. $\aleph_{\omega_0}$).

2238. DEFINITION (**TSS'**). (a) A cardinal number α is *regular* if it is infinite and $\mathbf{cf}(\alpha) = \alpha$.

(b) A cardinal number α is *singular* if it is infinite and $\mathbf{cf}(\alpha) \in \alpha$.

2239. LEMMA (**TSS'**). The class of all singular cardinal numbers is a proper class.

Proof. Suppose that the class of all singular cardinal numbers is a set. Since it is non-void, this set has a supremum γ; we have $\boldsymbol{\omega}_0 \subseteq \gamma$. Put $f'0 = \gamma$, $f'(n+1) =$ the first cardinal number greater than $f'n$, $x = f''\boldsymbol{\omega}$. We have $\mathbf{Sup}(x) \notin x$ and $\gamma \in \mathbf{Sup}(x)$; furthermore $\mathbf{Sup}(x)$ is a cardinal number. $\mathbf{Sup}(x)$ is singular, since it is cofinal with a countable subset; this is a contradiction.

2240. THEOREM (**TSS'**). For every limit number α, $\mathbf{cf}(\alpha)$ is a regular cardinal.

To close this Section we define a regular-ordering of $\mathbf{On}^2$ which will be useful later.

2241. DEFINITION (**TSS'**).

$$\begin{aligned} \langle\alpha, \beta\rangle \,\mathbf{Maxlex}\, \langle\gamma, \delta\rangle \equiv . \,\mathbf{Max}(\alpha, \beta) < \mathbf{Max}(\gamma, \delta) \vee \\ \vee \left[(\mathbf{Max}(\alpha, \beta) = \mathbf{Max}(\gamma, \delta) \,\&\, \alpha < \gamma)\right] \vee \\ \vee \left[(\mathbf{Max}(\alpha, \beta) = \mathbf{Max}(\gamma, \delta) \,\&\, \alpha = \gamma \,\&\, \beta < \delta)\right]. \end{aligned}$$

(the *maximo-lexicographical ordering*).

2242. LEMMA (**TSS'**). **Maxlex** is a regular well-ordering of $\mathbf{On}^2$; there is an isomorphism *Is* of $\mathbf{On}^2$, **On** w.r.t. **Maxlex, E**.

Proof. Exercise.

SECTION 3

The First Axiom of Choice

We shall study three forms of the axiom of choice E1, E2 and E3. E1 is weaker than E2 and E2 in turn is weaker than E3. The present section deals with the first of these which is a "set form" of the axiom of choice; it ensures the existence of a selector (choice-function) on each set. In addition, we define the exponentiation of cardinal numbers and prove the recurrence formulas for cardinal exponentiation. It turns out that problems concerning cardinal exponentiation can be reduced to the investigation of a certain unary operation $\gimel$ on cardinal numbers.

2301. DEFINITION (**TSS**′). A function f defined on x is called a *selector for* x, if $f\text{'}y \in y$ whenever $0 \neq y \in x$. **Sel** (x) denotes the set of all selectors for x.

2302. AXIOM (E1). Every set has a selector.

In the sequel we shall work in the theory (**TSS**′, E1).

2303. THEOREM (**TSS**′, E1). Every set can be well ordered.

Proof. Let f be a selector for $\mathbf{P}(x)$. Define the function F by transfinite recursion: $F\text{'}0 = f\text{'}x$, $F\text{'}\alpha = f\text{'}(x - F\text{''}\alpha)$ for $\alpha < 0$. Note that F is a real class by 2124. Since x is not a proper class, there exists $\alpha \in \mathbf{On}$ such that $x = F\text{''}\alpha$. If α is the least number with this property, then $F \restriction \alpha$ is a one-one mapping of α onto x. F induces a well ordering r on x, namely $\langle F(\beta), F(\gamma)\rangle \in r \equiv \beta \in \gamma$ for $\beta, \gamma \in \alpha$.

2304. LEMMA (**TSS**′, E1). For every set x there is a unique cardinal number equivalent to x.

2305. DEFINITION (**TSS**′, E1). $\overline{\overline{x}}$ is the unique cardinal number equivalent to x. (The *cardinality of x*.)

2306. LEMMA (**TSS′**, E1). Let $\overline{\overline{x}} \leqq \aleph_\alpha$ and suppose that $\overline{\overline{y}} \leqq \aleph_\alpha$ whenever $y \in x$. Then $\overline{\overline{\bigcup(x)}} \leqq \aleph_\alpha$. Moreover, if $\overline{\overline{x}} = \aleph_\alpha$ or if $\overline{\overline{y}} = \aleph_\alpha$ for some $y \in x$, then $\overline{\overline{\bigcup(x)}} = \aleph_\alpha$.

Proof. For each $z \in \bigcup(x)$ choose (by E1) some $y \in x$ such that $z \in y$ and denote it by $k'z$. For each $y \in x$ choose a one-one mapping f_y of y into $\boldsymbol{\omega}_\alpha$ and let g be a one-one mapping of x into $\boldsymbol{\omega}_\alpha$. Let h be the following function on $\bigcup(x) : h'z = \langle g'(k'z), f_{k'z}'z\rangle$. The function h is obviously one-one and into $\boldsymbol{\omega}_\alpha \times \boldsymbol{\omega}_\alpha$. This completes the proof by Theorem 2231.

2307. LEMMA (**TSS′**, E1). For every α, $\aleph_{\alpha+1}$ is a regular cardinal number.

Proof. Take the least α such that $\aleph_{\alpha+1}$ is singular. Then there is a subset x of $\boldsymbol{\omega}_{\alpha+1}$ such that $\overline{\overline{x}} \leqq \aleph_\alpha$ and $\bigcup(x) = \boldsymbol{\omega}_{\alpha+1}$. By 2306, $\mathbf{Sup}\,(x) = \bigcup(x)$ is strictly subvalent to $\aleph_{\alpha+1}$. This is a contradiction.

2308. COROLLARY (**TSS′**, E1). The class of all regular cardinals is a proper class.

2309. DEFINITION (**TSS′**, E1). The set of all sets which are mappings of the set y into the set x is denoted by $\mathbf{exp}\,(x, y)$.

From now on we shall be interested in cardinal exponentiation, i.e. in the cardinalities of the sets $\mathbf{exp}\,(x, y)$ where x, y are cardinal numbers at least one of which is infinite. From Theorem 2231 we immediately deduce that $\overline{\overline{\mathbf{exp}\,(\aleph_\alpha, n)}} = \aleph_\alpha$ for every α and every $n \neq 0$. Also, as we shall see later, $\overline{\overline{\mathbf{exp}\,(n, \aleph_\alpha)}} = \overline{\overline{\mathbf{exp}\,(2, \aleph_\alpha)}}$ for every $n \geqq 2$. Hence, we are interested in cardinalities of sets $\mathbf{exp}\,(\aleph_\alpha, \aleph_\beta)$ and $\mathbf{exp}\,(2, \aleph_\alpha)$. These cardinalities have traditionally been denoted by $\aleph_\alpha^{\aleph_\beta}$, $2^{\aleph_\alpha}$ respectively. This notation is very common but it is defective in that the operation symbol for exponentiation is absent. (Let us mention that for this reason it is necessary to write (e.g. $n ** m$ instead of n^m in programming languages.) We preserve this traditional notation in the present section for the sake of intelligibility; but we introduce simultaneously another notation which will turn out to be useful in the sequel; namely, we take the Hebrew letter $\beth$ for the sign of exponentiation. The reader may rewrite the recurrence formulas given in the sequel in the $\beth$ symbolism.

2310. DEFINITION (**TSS′**, E1). If x, y are cardinal numbers then $x^y = x \beth y = \overline{\overline{\mathbf{exp}\,(x, y)}}$.

We now define some important suprema. In each case we give two alternative notations; one of these is traditional and the other uses the $\beth$ symbolism.

2311. DEFINITION (**TSS′**, E1).

$$\aleph_\beta \overset{\aleph_\alpha}{\cup} = \aleph_\beta \sqsupset \breve{\aleph}_\alpha = \mathbf{Sup}\,\{\aleph_\beta^{\aleph_\gamma};\ \gamma \in \alpha\}\,,$$

$$\breve{\aleph}_\beta^{\aleph_\alpha} = \breve{\aleph}_\beta \sqsupset \aleph_\alpha = \mathbf{Sup}\,\{\aleph_\gamma^{\aleph_\alpha};\ \gamma \in \beta\}\,,$$

$$2 \overset{\aleph_\beta}{\cup} = 2 \sqsupset \breve{\aleph}_\beta = \mathbf{Sup}\,\{2^{\aleph_\gamma};\ \gamma \in \beta\}\,.$$

2312. LEMMA (**TSS′**, E1). If $x, y, s, t \in \mathbf{Cn}$ and $x \leqq y$, $s \leqq t$ then $x^s \leqq y^t$.

2313. LEMMA (**TSS′**, E1). $(\aleph_\alpha^{\aleph_\beta})^{\aleph_\gamma} = \aleph_\alpha^{\max(\aleph_\beta, \aleph_\gamma)}$.

Proof. Let f be a mapping of $\boldsymbol{\omega}_\gamma$ into $\mathbf{exp}\,(\boldsymbol{\omega}_\alpha, \boldsymbol{\omega}_\beta)$, i.e. $f`\xi$ is a mapping of $\boldsymbol{\omega}_\beta$ into $\boldsymbol{\omega}_\alpha$ for each $\xi \in \boldsymbol{\omega}_\gamma$. Define a mapping $\bar{f}$ of $\boldsymbol{\omega}_\beta \times \boldsymbol{\omega}_\gamma$ into $\boldsymbol{\omega}_\alpha$ by $\bar{f}`\langle \eta, \xi\rangle = (f`\xi)`\,\eta$. The function which assigns to each $f \in \mathbf{exp}\,(\mathbf{exp}\,(\boldsymbol{\omega}_\alpha, \boldsymbol{\omega}_\beta), \boldsymbol{\omega}_\gamma)$ the corresponding $\bar{f} \in \mathbf{exp}\,(\boldsymbol{\omega}_\alpha, \boldsymbol{\omega}_\beta \times \boldsymbol{\omega}_\gamma)$ is one-one and onto. Hence $\mathbf{exp}\,(\mathbf{exp}\,(\boldsymbol{\omega}_\alpha, \boldsymbol{\omega}_\beta), \boldsymbol{\omega}_\gamma)$ and $\mathbf{exp}\,(\boldsymbol{\omega}_\alpha, \boldsymbol{\omega}_\beta \times \boldsymbol{\omega}_\gamma)$ are equivalent. By 2231 the latter set has cardinality $\aleph_\alpha^{\max(\aleph_\beta, \aleph_\gamma)}$.

2314. LEMMA (**TSS′**, E1). $2^{\aleph_\alpha} = \overline{\overline{\mathbf{P}\,(\boldsymbol{\omega}_\alpha)}} > \aleph_\alpha$ for each α. If $x \in \mathbf{Cn}$ and $2 \leqq x \leqq \aleph_\alpha$ then $x^{\aleph_\alpha} = 2^{\aleph_\alpha}$.

Proof. If $f \in \mathbf{exp}\,(2, \boldsymbol{\omega}_\alpha)$ (i.e. if f is a function on $\boldsymbol{\omega}_\alpha$ assuming the values 0, 1) we set $F`f = \{\gamma, \gamma \in \boldsymbol{\omega}_\alpha \,\&\, f`\gamma = 1\}$; F is obviously a one-one mapping of $\mathbf{exp}\,(2, \boldsymbol{\omega}_\alpha)$ onto $\mathbf{P}\,(\boldsymbol{\omega}_\alpha)$. This proves the first assertion; now we prove the second. Let $x \leqq \aleph_\alpha$. We claim that $\overline{\overline{\mathbf{exp}\,(x, \boldsymbol{\omega}_\alpha)}} \leqq 2^{\aleph_\alpha}$. If $f \in \mathbf{exp}\,(x, \boldsymbol{\omega}_\alpha)$ then $f \subseteq x \times \boldsymbol{\omega}_\alpha$ so that $f \in \mathbf{P}\,(x \times \boldsymbol{\omega}_\alpha)$. Since $\overline{\overline{x \times \boldsymbol{\omega}_\alpha}} = \aleph_\alpha$ the assertion is proved.

2315. DEFINITION (**TSS′**, E1). $\mathbf{P}_{\aleph_\alpha}(x)$ is the set of all subsets of x having power at most $\aleph_\alpha$.

2316. LEMMA (**TSS′**, E1). If $\alpha \leqq \beta$ then $\aleph_\beta^{\aleph_\alpha} = \overline{\overline{\mathbf{P}_{\aleph_\alpha}(\boldsymbol{\omega}_\beta)}}$.

Proof. Let f be a mapping of $\boldsymbol{\omega}_\alpha$ into $\boldsymbol{\omega}_\beta$. Clearly f is a subset of $\boldsymbol{\omega}_\beta \times \boldsymbol{\omega}_\alpha$ and has cardinality $\aleph_\alpha$, so that $f \in \mathbf{P}_{\aleph_\alpha}(\boldsymbol{\omega}_\beta \times \boldsymbol{\omega}_\alpha)$. Since $\overline{\overline{\boldsymbol{\omega}_\beta \times \boldsymbol{\omega}_\alpha}} = \aleph_\beta$ we have $\aleph_\beta^{\aleph_\alpha} \leqq \overline{\overline{\mathbf{P}_{\aleph_\alpha}(\boldsymbol{\omega}_\beta)}}$. Conversely, if y is a subset of $\boldsymbol{\omega}_\beta$ having cardinality at most $\aleph_\alpha$, then there is a function f_y of $\boldsymbol{\omega}_\alpha$ into $\boldsymbol{\omega}_\beta$ whose range is exactly y. We choose such a function f_y for each $y \subseteq \boldsymbol{\omega}_\beta$ with $\overline{\overline{y}} \subseteq \aleph_\alpha$. Obviously if $y_1 \neq y_2$ then $f_{y_1} \neq f_{y_2}$. Thus the mapping associating $f_y \in \mathbf{exp}\,(\boldsymbol{\omega}_\beta, \boldsymbol{\omega}_\alpha)$ with each $y \in \mathbf{P}_{\aleph_\alpha}(\boldsymbol{\omega}_\beta)$ is one-one and hence $\overline{\overline{\mathbf{P}_{\aleph_\alpha}(\boldsymbol{\omega}_\beta)}} \leqq \aleph_\beta^{\aleph_\alpha}$.

2317. LEMMA (**TSS**′, E1). Let $\mathbf{cf}(\aleph_\beta) \leqq \aleph_\alpha < \aleph_\beta$. Then

$$\aleph_\beta^{\aleph_\alpha} = (\breve{\aleph}_\beta^{\aleph_\alpha})^{\mathbf{cf}(\aleph_\beta)}.$$

Proof. Denote $\mathbf{cf}(\boldsymbol{\omega}_\beta)$ by $\boldsymbol{\omega}_{\beta_0}$. There is an increasing function g from $\boldsymbol{\omega}_{\beta_0}$ into a cofinal subset of $\boldsymbol{\omega}_\beta$. We claim that $\mathbf{P}_{\aleph_\alpha}(\boldsymbol{\omega}_\beta)$ is subvalent to $\mathbf{exp}\,(\bigcup_{\delta \in \omega_{\beta_0}} \mathbf{P}_{\aleph_\alpha}(g'\delta), \boldsymbol{\omega}_{\beta_0})$. If $x \subseteq \boldsymbol{\omega}_\beta$ is of power at most $\aleph_\alpha$, set $g_x'\gamma = g'\gamma \cap x$ for each $\gamma \in \boldsymbol{\omega}_{\beta_0}$. The function associating g_x with each $x \in \mathbf{P}_{\aleph_\alpha}(\boldsymbol{\omega}_{\beta_0})$ maps the set $\mathbf{P}_{\aleph_\alpha}(\boldsymbol{\omega}_\beta)$ into $\mathbf{exp}\,([\bigcup_{\delta \in \omega_{\beta_0}} \mathbf{P}_{\aleph_\alpha}(g'\delta)], \boldsymbol{\omega}_{\beta_0})$ and is one-one. Since $\aleph_\alpha < \aleph_\beta$, there is a $\delta_0 \in \boldsymbol{\omega}_{\beta_0}$ such that $\boldsymbol{\omega}_\alpha \in g'\delta_0$. By 2316 we have $\mathbf{P}_{\aleph_\alpha}(g'\delta) \approx \mathbf{exp}\,(g'\delta, \boldsymbol{\omega}_\alpha)$ whenever $\delta_0 < \delta < \boldsymbol{\omega}_{\beta_0}$. Hence the cardinality of the set $\mathbf{exp}\,(\bigcup_{\delta \in \omega_{\beta_0}} \mathbf{P}_{\aleph_\alpha}(g'\delta), \boldsymbol{\omega}_{\beta_0})$ is $(\breve{\aleph}_\beta^{\aleph_\alpha})^{\aleph_{\beta_0}}$ and we have proved $\aleph_\beta^{\aleph_\alpha} \leqq (\breve{\aleph}_\beta^{\aleph_\alpha})^{\aleph_{\beta_0}}$.

As for the converse inequality $(\breve{\aleph}_\beta^{\aleph_\alpha})$ is at most $\aleph_\beta^{\aleph_\alpha}$ (since $\alpha < \beta$) and hence the right-hand side is at most $(\aleph_\beta^{\aleph_\alpha})^{\aleph_{\beta_0}}$. Since $\beta_0 \leqq \alpha$ we have $(\aleph_\beta^{\aleph_\alpha})^{\aleph_{\beta_0}} = \aleph_\beta^{\aleph_\alpha}$ by 2313.

2318. LEMMA (**TSS**′, E1). $2^{\aleph_\beta} = (2 \overset{\aleph_\beta}{\cup})^{\mathbf{cf}(\aleph_\beta)}$.

Proof. We again set $\boldsymbol{\omega}_{\beta_0} = \mathbf{cf}(\boldsymbol{\omega}_\beta)$. Let g be an increasing function with domain $\boldsymbol{\omega}_{\beta_0}$ and range cofinal with $\boldsymbol{\omega}_\beta$. As in the proof of the preceding lemma it can be shown that $\mathbf{P}(\boldsymbol{\omega}_\beta)$ is subvalent to $\mathbf{exp}\,([\bigcup_{\delta \in \omega_{\beta_0}} \mathbf{P}(g'\delta)], \boldsymbol{\omega}_{\beta_0})$ and hence $2^{\aleph_\beta} \leqq (2 \overset{\aleph_\beta}{\cup})^{\aleph_{\beta_0}}$.

The converse inequality is obvious; indeed we have $(2 \overset{\aleph_\beta}{\cup})^{\aleph_{\beta_0}} \leqq (2^{\aleph_\beta})^{\aleph_{\beta_0}} = 2^{\aleph_\beta}$.

2319. DEFINITION (**TSS**′, E1). $\gimel(\aleph_\beta) = \aleph_\beta^{\mathbf{cf}(\aleph_\beta)}$.

2320. THEOREM (**TSS**′, E1).

(1) If $\aleph_\beta \leqq \aleph_\alpha$ then $\aleph_\beta^{\aleph_\alpha} = 2^{\aleph_\alpha}$.

(2) If $\aleph_\alpha \leqq \aleph_\beta$ then $\aleph_{\beta+1}^{\aleph_\alpha} = \mathbf{Max}\,(\aleph_{\beta+1}, \aleph_\beta^{\aleph_\alpha})$ (Hausdorff's recurrence formula).

(3) If β is a limit number and $\aleph_\alpha < \mathbf{cf}(\aleph_\beta)$ then $\aleph_\beta^{\aleph_\alpha} = \breve{\aleph}_\beta^{\aleph_\alpha}$ (Tarski's recurrence formula).

(4) Let $\aleph_\beta$ be singular and let $\mathbf{cf}(\aleph_\beta) \leqq \aleph_\alpha < \aleph_\beta$. Denote by $\aleph_\delta$ the cardinal number $(\breve{\aleph}_\beta^{\aleph_\alpha})$. We distinguish two cases as follows:

(a) if $\aleph_\delta = \aleph_\gamma^{\aleph_\alpha}$ for some $\gamma < \beta$ then

$$\aleph_\beta^{\aleph_\alpha} = \breve{\aleph}_\beta^{\aleph_\alpha} = \aleph_\delta\,;$$

(b) if $\aleph_\delta > \aleph_\gamma^{\aleph_\alpha}$ for every $\gamma < \beta$ then

$$\aleph_\beta^{\aleph_\alpha} = \gimel\,(\breve{\aleph}_\beta^{\aleph_\alpha}) = \gimel\,(\aleph_\delta)\,.$$

Proof. (1) See Lemma 2314.

(2) and (3): Since $\aleph_\beta \leqq \breve{\aleph}_\beta^{\aleph_\alpha}$ whenever β is limit, and $\aleph_\beta^{\aleph_\alpha} = \breve{\aleph}_{\beta+1}^{\aleph_\alpha}$, both cases are included in the formula $\aleph_\beta^{\aleph_\alpha} = \mathbf{Max}\,(\aleph_\beta, \breve{\aleph}_\beta^{\aleph_\alpha})$ whenever $\aleph_\alpha < \mathbf{cf}\,(\aleph_\beta)$. The left-hand side is obviously at least as great as the right-hand side and hence it suffices to prove $\aleph_\beta^{\aleph_\alpha} \leqq \mathbf{Max}\,(\aleph_\beta, \breve{\aleph}_\beta^{\aleph_\alpha})$. Here we use 2306 and the fact that $\mathbf{exp}\,(\boldsymbol{\omega}_\beta, \boldsymbol{\omega}_\alpha) = \bigcup_{\gamma \in \boldsymbol{\omega}_\beta} \mathbf{exp}\,(\gamma, \boldsymbol{\omega}_\alpha)$; for if $f \in \mathbf{exp}\,(\boldsymbol{\omega}_\beta, \boldsymbol{\omega}_\alpha)$, then there is a $\gamma \in \boldsymbol{\omega}_\beta$ such that $\mathbf{W}\,(f) \subseteq \gamma$ since $\aleph_\alpha < \mathbf{cf}\,(\aleph_\beta)$. The cardinality of $\bigcup_{\gamma \in \boldsymbol{\omega}_\beta} \mathbf{exp}\,(\gamma, \boldsymbol{\omega}_\alpha)$ is just the right-hand side of the equality.

(4a) Suppose that $\aleph_\delta = \breve{\aleph}_\beta^{\aleph_\alpha}$ is equal to $\aleph_\gamma^{\aleph_\alpha}$ for some $\gamma < \beta$. By 2317 we have $\aleph_\beta^{\aleph_\alpha} = (\breve{\aleph}_\beta^{\aleph_\alpha})^{\mathbf{cf}(\aleph_\beta)} = (\aleph_\gamma^{\aleph_\alpha})^{\mathbf{cf}(\aleph_\beta)} = \aleph_\gamma^{\mathbf{Max}(\aleph_\alpha,\, \mathbf{cf}(\aleph_\beta))} = \aleph_\gamma^{\aleph_\alpha} = \breve{\aleph}_\beta^{\aleph_\alpha}$.

(4b) Suppose that there is no maximal number among the cardinals $\aleph_\gamma^{\aleph_\alpha}$ $(\gamma < \beta)$. We claim that $\mathbf{cf}\,(\boldsymbol{\omega}_\delta) = \mathbf{cf}\,(\boldsymbol{\omega}_\beta)$. Since $\mathbf{cf}\,(\boldsymbol{\omega}_\delta) \leqq \mathbf{cf}\,(\boldsymbol{\omega}_\beta)$ is obvious, it suffices to prove $\mathbf{cf}\,(\boldsymbol{\omega}_\delta) \geqq \mathbf{cf}\,(\boldsymbol{\omega}_\beta)$. Suppose that $\mathbf{cf}\,(\boldsymbol{\omega}_\delta) < \mathbf{cf}\,(\boldsymbol{\omega}_\beta)$ and let f be an increasing function of $\mathbf{cf}\,(\boldsymbol{\omega}_\delta)$ onto a cofinal subset of $\boldsymbol{\omega}_\delta$. For every $\xi \in \mathbf{D}\,(f)$ let $g'\xi$ be the first $\gamma < \beta$ such that $\aleph_\gamma^{\aleph_\alpha} \geqq f'\xi$; set $\gamma_0 = \bigcup(\mathbf{W}\,(g))$. Then $\aleph_{\gamma_0} < \aleph_\beta$ and $\aleph_{\gamma_0}^{\aleph_\alpha} \geqq f'\xi$ for all ξ, i.e. $\aleph_{\gamma_0}^{\aleph_\alpha} = \boldsymbol{\omega}_\delta$. Consequently, for each γ between γ_0 and β we have $\aleph_\gamma^{\aleph_\alpha} = \aleph_{\gamma_0}^{\aleph_\alpha}$ which contradicts the assumption. Now, we have $\aleph_\beta^{\aleph_\alpha} = \aleph_\delta^{\mathbf{cf}(\aleph_\beta)}$ by 2317 and hence $\aleph_\beta^{\aleph_\alpha} = \gimel(\aleph_\delta)$.

2321. Corollary (**TSS′**, E1). If β is a limit number and $2^{\aleph_\gamma} < \aleph_\beta$ for every $\gamma < \beta$ then

(i) $\aleph_\alpha^{\aleph_\gamma} < \aleph_\beta$ whenener $\alpha, \gamma < \beta$,

(ii) $\aleph_\beta^{\aleph_\beta} = \gimel\,(\aleph_\beta)\,.$

Proof. (i) Let $\delta = \mathbf{Max}\,(\alpha, \gamma)$. Then $\aleph_\alpha^{\aleph_\gamma} \leqq \aleph_\delta^{\aleph_\delta} = 2^{\aleph_\delta} < \aleph_\beta$.

(ii) By 2318 $\aleph_\beta^{\aleph_\beta} = 2^{\aleph_\beta} = (2 \overset{\aleph_\beta}{\cup})^{\mathbf{cf}(\aleph_\beta)}$. Using the assumption we have $2 \overset{\aleph_\beta}{\cup} = \aleph_\beta$ and hence $\aleph_\beta^{\aleph_\beta} = \gimel\,(\aleph_\beta)$.

2322. Theorem (**TSS′**, E1). (1) If $\aleph_\alpha$ is regular, then

$$2^{\aleph_\alpha} = \gimel\,(\aleph_\alpha)\,.$$

(2) If $\aleph_\alpha$ is singular, then we again distinguish two cases:

(a) if $2\overset{\aleph_\alpha}{\cup} = 2^{\aleph_\gamma}$ for some $\gamma < \alpha$, then

$$2^{\aleph_\alpha} = 2\overset{\aleph_\alpha}{\cup}\,;$$

(b) if $2\overset{\aleph_\alpha}{\cup} > 2^{\aleph_\gamma}$ for every $\gamma < \alpha$, then

$$2^{\aleph_\alpha} = \gimel\,(2\overset{\aleph_\alpha}{\cup})\,.$$

Proof. (1) follows from Theorem 2320 Case (1).

(2a) Let $2\overset{\aleph_\alpha}{\cup} = 2^{\aleph_\gamma}$ for some $\gamma < \alpha$; we may suppose that $\mathbf{cf}\,(\aleph_\alpha) \leqq \aleph_\gamma$. By 2318 we have $2^{\aleph_\alpha} = (2\overset{\aleph_\alpha}{\cup})^{\mathbf{cf}(\aleph_\alpha)} = (2^{\aleph_\gamma})^{\mathbf{cf}(\aleph_\alpha)} = 2^{\mathbf{Max}(\aleph_\gamma,\, \mathbf{cf}(\aleph_\alpha))} = 2^{\aleph_\gamma} = 2\overset{\aleph_\alpha}{\cup}$.

(2b) It suffices to prove $\mathbf{cf}\,(2\overset{\aleph_\alpha}{\cup}) = \mathbf{cf}\,(\aleph_\alpha)$ and then to use Lemma 2318. The proof is analogous to the proof of Theorem 2320 Case (4b).

Remark. Theorems 2320 and 2322 give the promised recurrence formulas for cardinal exponentiation.

2323. THEOREM (**TSS′**, E1). Properties of the function $\gimel$.

(1) $\aleph_\beta < \gimel\,(\aleph_\beta)$,

(2) $\mathbf{cf}\,(\aleph_\beta) < \mathbf{cf}\,(\gimel\,(\aleph_\beta))$,

(3) if $\aleph_\beta \leqq \gimel\,(\aleph_\alpha)$ and $\mathbf{cf}\,(\aleph_\beta) \leqq \mathbf{cf}\,(\aleph_\alpha)$, then $\gimel\,(\aleph_\beta) \leqq \gimel\,(\aleph_\alpha)$.

Proof. (1) If $\aleph_\beta$ is regular then $\gimel\,(\aleph_\beta) = 2^{\aleph_\beta}$ and the assertion follows from Lemma 2314. Suppose that $\aleph_\beta$ is singular. Let g be a function on $\mathbf{cf}\,(\boldsymbol{\omega}_\beta)$ whose range is cofinal with $\boldsymbol{\omega}_\beta$. Let a be the set of all pairs $\langle \delta, \gamma \rangle$ such that $\gamma < \mathbf{cf}\,(\boldsymbol{\omega}_\beta)$ and $\delta \in g\text{'}\gamma$. Obviously, $\overline{\overline{a}} = \aleph_\beta$. We have, of course, $\aleph_\beta \leqq \gimel\,(\aleph_\beta)$, so suppose $\aleph_\beta = \gimel\,(\aleph_\beta)$. Then there is a one-one mapping of a onto $\mathbf{exp}\,(\boldsymbol{\omega}_\beta, \mathbf{cf}\,(\boldsymbol{\omega}_\beta))$. Denote by $f_{\delta\gamma}$ the image of $\langle \delta, \gamma \rangle$ under this mapping and define a mapping h of $\mathbf{cf}\,(\boldsymbol{\omega}_\beta)$ into $\boldsymbol{\omega}_\beta$ as follows: $h\text{'}\gamma = \mathbf{Min}\,(\boldsymbol{\omega}_\beta - \{f\text{'}_{\delta\gamma}(\gamma);\ \delta \in g(\gamma)\})$. The function h differs from all $f_{\delta\gamma}$'s; for, if $h = f_{\xi\gamma}$ where $\langle \xi, \gamma \rangle \in a$, then $h\text{'}\gamma \neq f\text{'}_{\delta\gamma}(\gamma)$ whenever $\delta \in g\text{'}\gamma$ and hence $h\text{'}\gamma \neq f\text{'}_{\xi\gamma}(\gamma)$, a contradiction. But the functions $f_{\delta\gamma}$ were supposed to include all functions in $\mathbf{exp}\,(\boldsymbol{\omega}_\beta, \mathbf{cf}\,(\boldsymbol{\omega}_\beta))$; thus we have reached a contradiction.

(2) Suppose that $\mathbf{cf}(\gimel(\aleph_\beta)) = \mathbf{cf}(\aleph_\beta)$. Then

$$\gimel(\gimel(\aleph_\beta)) = \gimel(\aleph_\beta)^{\mathrm{cf}(\gimel(\aleph_\beta))} = (\aleph_\beta^{\mathrm{cf}(\aleph_\beta)})^{\mathrm{cf}(\gimel(\aleph_\beta))} = \aleph_\beta^{\mathrm{Max}\,(\mathrm{cf}(\aleph_\beta),\,\mathrm{cf}(\gimel(\aleph_\beta)))} = \\ = \aleph_\beta^{\mathrm{cf}(\aleph_\beta)} = \gimel(\aleph_\beta)\,,$$

contradicting (1).

(3) $\gimel(\aleph_\beta) = \aleph_\beta^{\mathrm{cf}(\aleph_\beta)} \leqq \gimel(\aleph_\alpha)^{\mathrm{cf}(\aleph_\alpha)} = (\aleph_\alpha^{\mathrm{cf}(\aleph_\alpha)})^{\mathrm{cf}(\aleph_\alpha)} = \aleph_\alpha^{\mathrm{cf}(\aleph_\alpha)} = \gimel(\aleph_\alpha)$.

2324. LEMMA (**TSS'**, E1). $\aleph_\beta < \mathbf{cf}(2^{\aleph_\beta})$.

Proof. Suppose $\mathbf{cf}(2^{\aleph_\beta}) \leqq \aleph_\beta$. Then

$$2^{\aleph_\beta} < \gimel(2^{\aleph_\beta}) \leqq (2^{\aleph_\beta})^{\aleph_\beta} = 2^{\aleph_\beta}\,,$$

a contradiction.

2325. The *Continuum Axiom*:

$$(\mathrm{Cont}) \quad (\forall\alpha)\,(2^{\aleph_\alpha} = \aleph_{\alpha+1})\,.$$

2326. THEOREM (**TSS'**, E1). The Continuum Axiom is equivalent to the statement $(\forall\alpha)\,(\gimel(\aleph_\alpha) = \aleph_{\alpha+1})$.

Proof. Since $\gimel(\aleph_\alpha) \leqq \aleph_\alpha^{\aleph_\alpha} = 2^{\aleph_\alpha}$, the Continuum axiom implies $(\forall\alpha)$ $(\gimel(\aleph_\alpha) = \aleph_{\alpha+1})$. Conversely, suppose that $(\forall\alpha)\,(\gimel(\aleph_\alpha) = \aleph_{\alpha+1})$ and consider the least α with $2^{\aleph_\alpha} > \aleph_{\alpha+1}$. If $\aleph_\alpha$ is regular then $\gimel(\aleph_\alpha) = 2^{\aleph_\alpha}$ by 2322, Case 1. If $\aleph_\alpha$ is singular then by the induction hypothesis $2 \overset{\aleph_\alpha}{\cup} = \aleph_\alpha$ and we have $2^{\aleph_\alpha} = \gimel(2 \overset{\aleph_\alpha}{\cup}) = \gimel(\aleph_\alpha) = \aleph_{\alpha+1}$ by 2322, Case (2b).

SECTION 4

Complete Boolean Algebras

This section deals with the basic properties of complete Boolean algebras, which play an important role throughout the book. A complete Boolean algebra is a class which contains the complement of each of its elements and the meet of each of its subsets and which satisfies certain additional conditions. Thus a complete Boolean algebra is determined by a class B, a function C mapping B into B and a function F of $\mathbf{P}(B)$ into B. It is convenient to consider only real B, C, F.

Thus, a Boolean algebra is formed by three classes fulfilling certain conditions. If these classes are sets we speak of a set Boolean algebra. In this case it is more convenient to consider the ordered triple of these sets rather than the three sets separately. To be able to speak of an ordered pair (triple) of classes which are not necessarily sets we introduce the notion of generalized pair (triple) as follows:

2401. Definition (**TSS′**).

$$[X, Y] = (X \times \{0\}) \cup (Y \times \{1\});$$

$$[X, Y, Z] = (X \times \{0\}) \cup (Y \times \{1\}) \cup (Z \times \{2\}).$$

2402. Definition (**TSS′**). $\boldsymbol{B} = [B, C, F]$ is a *complete Boolean algebra* (Cba $(\boldsymbol{B})$) if B, C, F are real classes, C is a mapping of B into B, F is a mapping of $\mathbf{P}(B)$ into B and the following conditions hold:

(1) $F'\{x, C'x\} = C'F'0$ for each $x \in B$,

(2) $F'\{x, C'y\} = C'F'0 \equiv F'\{x, y\} = x$ for each $x, y \in B$,

(3) $F'z = F'F''q$ provided $q \subseteq \mathbf{P}(B)$ and $z = \bigcup(q)$.

2403. Remark. We introduce the following notation:

$F'0 \;= 1_B$	(the *unit element*),
$C'1_B = 0_B$	(the *zero element*),
$F'\{x, y\} = x \wedge_B y$	(the *meet* of x and y),
$C'x \;= -_B x$	(the *complement* of x).

If $z \subseteq B$ we write

$F'z \;= \bigwedge_B z$	(the meet or Boolean product of z).

If f is a function with values in B we write

$$F'\mathbf{W}(f) = \bigwedge_{x \in \mathbf{D}(f)} f(x).$$

The subscript $_B$ will be omitted if there is no danger of misunderstanding.

The conditions (1)–(3) may now be rewritten as follows:

(1) $\quad x \wedge - x = 0_B$,

(2) $\quad x \wedge - y = 0_B \equiv x \wedge y = x$,

(3) $\quad$ if $z \subseteq B$ and $z = \bigcup(q)$, then $\bigwedge z = \bigwedge_{t \in q} (\bigwedge t)$.

(This is in fact the associative law.)

2404. LEMMA (**TSS'**). If $x, y, z \in B$ then

$$x \wedge y = y \wedge x,$$

$$x \wedge (y \wedge z) = (x \wedge y) \wedge z.$$

2405. DEFINITION (**TSS'**). (1) R is a *reflexive ordering* of A ($\mathrm{Ordg}_=(R, A)$) if for all $x, y, z \in A$ we have

(a) xRx,

(b) $xRy \,\&\, yRz \,.\rightarrow xRz$,

(c) $xRy \,\&\, yRx \,.\rightarrow x = y$.

(2) $\boldsymbol{A} = [A, R]$ is an *ordered class* if R is a reflexive ordering of A. If $\boldsymbol{A} = [A, R]$ is an ordered class then A is said to be the *field* of $\boldsymbol{A}$ and the reflexive ordering R is often denoted by $\leqq$.

2406. Remark. If $\mathbf{I} = \{\langle x, x\rangle; x \in \mathbf{V}\}$ and R is a reflexive ordering of A then $R - I$ is an (irreflexive) ordering of A.

2407. DEFINITION (**TSS'**). Let $\leqq$ be a reflexive ordering of A and let $z \subseteq A$, $x \in A$. The element x is the *infimum* of z ($\inf_{\leqq}(x, z)$) if

(a) $(\forall y \in z)(x \leqq y)$,

(b) $(\forall w)[(\forall y \in z)[w \leqq y] \rightarrow w \leqq x]$.

2408. THEOREM (**TSS'**). Let $\leqq$ be a complete Boolean algebra. For $x, y \in B$ define $x \leqq y$ by $x \wedge - y = 0_B$ (i.e. $x \leqq y$ iff $x \wedge y = x$). The relation $\leqq$ is then a (real) reflexive ordering of B and for each $z \subseteq B$, $\bigwedge z$ is the infimum of z. If $x \leqq y$ then $-y \leqq -x$; $--x = x$; $x \neq -x$ unless B contains the single element x.

Proof. Clearly $x \leqq x$, since $x \wedge -x = 0_B$ by (1). Suppose $x \leqq y \leqq z$, i.e. $x \wedge y = x$ and $y \wedge z = y$. Then $x \wedge z = (x \wedge y) \wedge z = x \wedge (y \wedge z) = x \wedge y = x$, i.e. $x \leqq z$. If $x \leqq y \leqq x$ then $x = x \wedge y = y$. Hence $\leqq$ is a reflexive ordering. Since $x \wedge -x = 0_B$ we have $x \wedge x = x$. Suppose $z \subseteq B$ and let $x = \bigwedge z = F\text{'}z$; we shall prove that $x \leqq y$ for every $y \in z$. We have

$$x \wedge y = \bigwedge z \wedge y = F\text{'}\{F\text{'}z, y\} = F\text{'}F\text{''}\{z, \{y\}\} = \\ = F\text{'}(z \cup \{y\}) = F\text{'}z = \bigwedge z = x\,.$$

Suppose that $w \leqq y$ for all $y \in z$; we shall prove $w \leqq x$.

If $y \in z$ then $w \wedge y = w$; hence for $\hat{z} = z \cup \{w\}$ we have $F\text{'}\hat{z} = \bigwedge z \wedge w = x \wedge w$. On the other hand,

$$\bigwedge \hat{z} = F\text{'} \bigcup_{y \in z} \{y, w\} = F\text{'}\{y \wedge w; y \in z\} = \bigwedge\{w\} = w\,.$$

Hence $w \wedge x = w$, i.e. $w \leqq x$, and $x = \bigwedge z$ is the infimum of z.

By (1), $-x \wedge --x = 0_B$; i.e. $--x \leqq x$. If $x \leqq y$, then $--x \leqq y$; i.e. $--x \wedge -y = 0_B$ and so $-y \leqq -x$. Since $--x \leqq x$ we deduce that $-x \leqq ---x$; on the other hand, substituting $-x$ for x in $--x \leqq x$, we get $---x \leqq -x$ and hence $---x = -x$. Using $x \wedge -x = 0_B$ we get $x \wedge ---x = 0_B$, i.e. $x \leqq --x$. Hence $x = --x$.

If $x = -x$ then $0_B = x \wedge -x = x \wedge x = x$ and $0_B = -0_B = 1_B$. Since $0_B \leqq y \leqq 1_B$ for each $y \in B$, we have $0_B = y = 1_B$ for each $y \in B$.

2409. DEFINITION (**TSS'**). (1) Let $\boldsymbol{B}$ be a complete Boolean algebra. The *canonical ordering* of the algebra $\boldsymbol{B}$ is the (reflexive) ordering defined by $x \leqq y \equiv x \wedge y = x$.

(2) For $x, y \in B$ define

$$x < y \equiv x \leqq y \,\&\, x \neq y\,,$$
$$x \vee y = -(-x \wedge -y)$$

(the *join* or the *Boolean sum* of x, y), for $z \subseteq B$ define

$$\bigvee z = C\text{'}(\bigwedge C\text{''}z)$$

(the *join* of z).

2410. DEFINITION (**TSS'**). Let $\leqq$ be a (reflexive) ordering of A and let $z \subseteq A$, $x \in A$. The element x is the *supremum* of z ($\sup_{\leqq}(x, z)$) if

(a) $(\forall y \in z)(y \leqq x)$,
(b) $(\forall w)[(\forall y \in z)[y \leqq w] \rightarrow x \leqq w]$.

2411. LEMMA (**TSS'**). Let $\boldsymbol{B}$ be a complete Boolean algebra and let $z \subseteq \boldsymbol{B}$. Let $\leqq$ be the canonical ordering of $\boldsymbol{B}$. Then $\bigvee z$ is the supremum of z and we have

$$C\text{'}(\bigwedge u) = \bigvee(C\text{''}u) \quad\text{and}\quad C\text{'}(\bigvee u) = \bigwedge(C\text{''}u) \quad \text{for each} \quad u \in B$$

(De Morgan rules); in particular, for $x, y \in B$,

$$-(x \wedge y) = -x \vee -y\,, \quad -(x \vee y) = -x \wedge -y\,.$$

2412. LEMMA (**TSS'**). Let $\boldsymbol{B}$ be a complete Boolean algebra. For any elements $x, y, z \in B$ we have

(1)	$x \wedge y = y \wedge x$	$x \vee y = y \vee x$
(2)	$x \wedge (y \wedge z) = (x \wedge y) \wedge z$	$(x \vee y) \vee z = x \vee (y \vee z)$
(3)	$(x \wedge y) \vee y = y$	$(x \vee y) \wedge y = y$
(4)	$1_{\boldsymbol{B}} \wedge x = x$	$0_{\boldsymbol{B}} \vee x = x$
(5)	$1_{\boldsymbol{B}} \vee x = 1_{\boldsymbol{B}}$	$0_{\boldsymbol{B}} \wedge x = 0_{\boldsymbol{B}}$

(6) $$(x \wedge y = 0_{\boldsymbol{B}} \;\&\; x \vee y = 1_{\boldsymbol{B}}) \equiv x = -y$$

(7) $$x \wedge (y \vee z) = (x \wedge y) \vee (x \wedge z)$$

(8) $$x \vee (y \wedge z) = (x \vee y) \wedge (x \vee z)$$

Proof. We shall only prove (6) and (7); the proofs of (1)–(5) are easy, while (8) is analogous to (7).

(6): If $y = -x$ we have $x \leqq -y$, i.e. $x \wedge y = 0_{\boldsymbol{B}}$. Similarly we have $-x \leqq y$, i.e. $-x \wedge -y = 0_{\boldsymbol{B}}$; hence $x \vee y = 1_{\boldsymbol{B}}$. On the other hand, assume that $x \wedge y = 0_{\boldsymbol{B}}$ and $x \vee y = 1_{\boldsymbol{B}}$. Then we have $x \wedge - - y = 0_{\boldsymbol{B}}$, i.e. $x \leqq -y$, and $-x \wedge -y = 0_{\boldsymbol{B}}$, i.e. $-y \leqq x$; hence $x = -y$.

(7): We have $x \wedge y \leqq x$ and $x \wedge z \leqq x$; hence $(x \wedge y) \vee (x \wedge z) \leqq x$. Similarly we have $x \wedge y \leqq y \vee z$ and $x \wedge z \leqq y \vee z$; i.e. $(x \wedge y) \vee (x \wedge z) \leqq y \vee z$. Hence $(x \wedge y) \vee (x \wedge z) \leqq x \wedge (y \vee z)$. It remains to prove the converse inequality. Set $b = y \vee z$ and $a = (x \wedge y) \vee (x \wedge z)$. Since $x \wedge y \leqq a$, we have $x \wedge y \wedge -a = 0_{\boldsymbol{B}}$. Similarly $x \wedge z \wedge -a = 0_{\boldsymbol{B}}$. Hence $x \wedge -a \leqq -y$ and $x \wedge -a \leqq -z$. Consequently $x \wedge -a \leqq -y \wedge -z = -(y \vee z) = -b$. Thus we have $x \wedge -a \wedge b = 0_{\boldsymbol{B}}$, i.e. $x \wedge b \leqq a$; hence $x \wedge (y \vee z) \leqq (x \wedge y) \vee (x \wedge z)$.

2413. Definition (**TSS'**). Let $\boldsymbol{B} = [B, C, F]$ be a complete Boolean algebra and let $x \in B$. The *partial algebra determined by* x is the algebra $\boldsymbol{B} \mid x$ formed by the following classes:

$$B \mid x = \{y; y \leqq x\}, \quad (C \mid x)\text{'}\, y = -y \wedge x \quad \text{for} \quad y \in B \mid x,$$
$$(F \mid x)\text{'}\, z = \bigwedge z \wedge x \quad \text{for} \quad z \subseteq B \mid x.$$

2414. Lemma (**TSS'**). If $\boldsymbol{B}$ is a complete Boolean algebra and if $x \in B$ then $\boldsymbol{B} \mid x$ is a complete Boolean algebra.

Proof. Condition (3) is satisfied trivially. We have $0_{\boldsymbol{B}|x} = 0_{\boldsymbol{B}}$. Since $y \wedge -y = 0_{\boldsymbol{B}}$ we have $y \wedge -y \wedge x = 0_{\boldsymbol{B}}$ hence (1) is satisfied. Finally, if $u, v \in B \mid x$, i.e. $u, v \leqq x$, then $u \wedge -v \wedge x = 0_{\boldsymbol{B}}$ means the same as $u \wedge -v = 0_{\boldsymbol{B}}$, i.e. $u \wedge v = u$. Hence (2) is satisfied.

2415. Definition (**TSS'**). Let $\boldsymbol{B}$ be a complete Boolean algebra and let $Z \subseteq B$.

(a) Z is a *filter* on $\boldsymbol{B}$ if

(1) $Z \neq 0$,

(2) $(\forall x, y \in Z)(x \wedge y \in Z)$,

(3) $(\forall x \in Z)(\forall y \in B)(x \leqq y \rightarrow y \in Z)$.

(b) Z is a *proper filter* if it satisfies (1), (2), (3) above and if

(4) $0_{\boldsymbol{B}} \notin Z$.

(c) Z is a *complete filter* if it satisfies (3) above and if

(5) $(\forall q \subseteq Z)(\bigwedge q \in Z)$.

(d) Z is an *ultrafilter* if it satisfies (1), (2), (4) above and if

(6) $(\forall x \in B)(x \in Z \vee -x \in Z)$.

2416. Lemma (**TSS'**). Every ultrafilter is a proper filter; every complete filter is a filter.

Proof. Let Z be an ultrafilter; we shall prove (3). Suppose that for some x, y we have $x \in Z$, $y \notin Z$ and $x \leqq y$. Then $-y \in Z$ and so $x \wedge -y \in Z$; i.e. $0_{\boldsymbol{B}} \in Z$, a contradiction.

Let Z be a complete filter; we shall prove (1), (2). We have $0 \subseteq Z$ and so $\bigwedge 0 = 1_{\boldsymbol{B}} \in Z$. (2) is a particular case of (5).

2417. *Remarks*. 1) A complete ultrafilter is an ultrafilter which is a complete filter. Obviously a class $Z \subseteq B$ is a complete ultrafilter iff it satisfies (4), (5) and (6) above.

2) A proper filter Z is an ultrafilter iff

(6') $(\forall x, y \in B)(x \vee y \in Z \rightarrow x \in Z \vee y \in Z)$

(obvious).

2418. Theorem (**TSS'**). Let $\boldsymbol{B}$ be a complete Boolean algebra and let Z be a proper filter on $\boldsymbol{B}$. Then Z is an ultrafilter iff there is no proper filter Z_1 on $\boldsymbol{B}$ such that $Z \subset Z_1$.

Proof. Let Z be an ultrafilter. If $u \in Z_1 - Z$ then $-u \in Z$ and so $-u \in Z_1$; thus $u \wedge -u \in Z_1$, a contradiction.

Now suppose that Z is not an ultrafilter. Then there exists $u \in B$ such that neither $u \in Z$ nor $-u \in Z$. We have $z \wedge u \neq 0_{\boldsymbol{B}}$ for each $z \in Z$; for, if $z \wedge u = 0_{\boldsymbol{B}}$ for some $z \in Z$ then $z \leqq -u$, so that $-u \in Z$, a contradiction. Now define $x \in Z_1 \equiv (\exists z \in Z)(x \geqq z \wedge u)$. Z_1 is a proper filter and $Z_1 \supseteq Z$; since $u \in Z_1 - Z$, we have $Z_1 \neq Z$.

2419. THEOREM (**TSS'**). Let Z_1 be a real proper filter on $\boldsymbol{B}$ and let H be a real one-one mapping of an ordinal onto $\boldsymbol{B}$. Then there exists a real ultrafilter Z on $\boldsymbol{B}$ such that $Z_1 \subseteq Z$.

Proof. By induction we define

$$H'\alpha \in Z \equiv (\forall u)\,(\mathrm{Fin}\,(u)\,\&\, u \subseteq Z_1 \cup (H''\alpha \cap Z) \rightarrow \bigwedge u \wedge H'\alpha \neq 0_{\boldsymbol{B}})\,.$$

We have $0_{\boldsymbol{B}} \notin Z$ and $Z_1 \subseteq Z$. If u is a finite subset of Z then $\bigwedge u \in Z$; hence Z is a proper filter. Suppose that there exists an $x \in B$ such that neither $x \in Z$ nor $-x \in Z$. There exist finite subsets u, v of Z such that $x \wedge \bigwedge u = -x \wedge \wedge \bigwedge v = 0_{\boldsymbol{B}}$. Set $w = u \cup v$; then $x \wedge \bigwedge w = -x \wedge \bigwedge w = 0_{\boldsymbol{B}}$ so that $(x \vee -x) \wedge \bigwedge w = 0_{\boldsymbol{B}}$, i.e. $\bigwedge w = 0_{\boldsymbol{B}}$. This contradicts $w \subseteq Z$, since w is finite. Hence Z is an ultrafilter.

2420. DEFINITION (**TSS'**). Let $\boldsymbol{B}$ be a complete Boolean algebra. An element u of B is an *atom* if it is non-zero and if there is no $v \in B$ such that $0_{\boldsymbol{B}} < v < u$. An algebra $\boldsymbol{B}$ is *atomic* if for each non-zero element $x \in B$ there exists an atom y such that $y \leqq x$. $\boldsymbol{B}$ is *atomless* if it has no atoms.

2421. LEMMA (**TSS'**). If u is an atom then $\{x; u \leqq x\}$ is a complete ultrafilter.

2422. DEFINITION (**TSS'**). Let $\boldsymbol{B}$ be a complete Boolean algebra. A subset u of B is a *partition* of $\boldsymbol{B}$ if

(1) $\bigvee u = 1_{\boldsymbol{B}}\,, \quad 0_{\boldsymbol{B}} \notin u$

(2) $(\forall x, y \in u)\,(x \neq y \rightarrow x \wedge y = 0_B)\,.$

Part $(\boldsymbol{B})$ (briefly **Part**) is the class of all partitions. If u and v are partitions, we define

$$u \lessdot v \equiv (\forall x \in u)\,(\exists y \in v)\,(x \leqq y)\,,$$

$$u \mathbin{\triangle} v = \{x \wedge y, x \in u \,\&\, y \in v \,\&\, x \wedge y \neq 0_{\boldsymbol{B}}\}\,.$$

2423. LEMMA (**TSS'**). The relation $\lessdot$ is a reflexive ordering of the class **Part**; $u \mathbin{\triangle} v$ is the infimum of $\{u, v\}$ in this ordering.

2424. DEFINITION (**TSS'**). A non-void real class $P \subseteq$ **Part** is called a *partition filter* (*on* $\boldsymbol{B}$) if

(1) $(\forall x, y \in P)\,(x \mathbin{\triangle} y \in P)\,,$

(2) $(\forall x \in P)\,(\forall y \in \mathbf{Part})\,(x \lessdot y \rightarrow y \in P)\,.$

2425. DEFINITION (**TSS'**). Let Z be a filter on $\boldsymbol{B}$. Set

$$\mathbf{Pt}(Z) = \{u \in \mathbf{Part},\ u \cap Z \neq 0\}$$

(the *partition filter induced by a filter*).

2426. LEMMA (**TSS'**). $\mathbf{Pt}(Z)$ is a partition filter.

2427. DEFINITION (**TSS'**) Let $\boldsymbol{B}$ and $\boldsymbol{B}_1$ be complete Boolean algebras.
(a) $\boldsymbol{B}_1$ is a *subalgebra* of $\boldsymbol{B}$ if

$$B_1 \subseteq B\,, \quad C_1 \subseteq C\,, \quad F_1 \subseteq F\,;$$

(b) A mapping H of B onto B_1 is a *homomorphism* of $\boldsymbol{B}$ onto $\boldsymbol{B}_1$ if H is real and

(1) $\quad (\forall x \in B)(C_1\text{'}H\text{'}x = H\text{'}C\text{'}x)\,,$

(2) $\quad (\forall z \subseteq B)(F_1\text{'}H\text{''}z = H\text{'}F\text{'}z)\,.$

If H is one-one it is called an *isomorphism*. An isomorphism of $\boldsymbol{B}$ onto itself is called an *automorphism*.

2428. LEMMA (**TSS'**). Let $\boldsymbol{B}_1$ be a subalgebra of $\boldsymbol{B}$. Then $0_{\boldsymbol{B}} = 0_{\boldsymbol{B}_1}$, $1_{\boldsymbol{B}} = 1_{\boldsymbol{B}_1}$ and for $z \subseteq B_1$ we have $\bigvee_1 z = \bigvee z$, where $\bigvee_1$ is the symbol for the Boolean sum in $\boldsymbol{B}_1$. If $x, y \in B_1$ then $x \leqq_1 y \equiv x \leqq y$, where $\leqq_1$ is the canonical ordering of B_1. (Obvious.)

2429. LEMMA (**TSS'**). Let $\boldsymbol{B}$ be a complete Boolean algebra; let $B_1 \subseteq B$ and suppose

$$(\forall x \in B_1)(-x \in B_1)\,,$$
$$(\forall z \subseteq B_1)(F\text{'}z \in B_1)\,.$$

Then $[B_1, C \upharpoonright B_1, F \upharpoonright \mathbf{P}(B_1)]$ is a complete Boolean algebra which is a subalgebra of $\boldsymbol{B}$. (Obvious.)

2430. LEMMA (**TSS'**). Let H be a homomorphism of $\boldsymbol{B}$ onto $\boldsymbol{B}_1$. Then

$$(\forall x, y \in B)(x \leqq y \rightarrow H\text{'}x \leqq H\text{'}y)\,,$$
$$H\text{'}0_{\boldsymbol{B}} = 0_{\boldsymbol{B}_1}\,, \quad H\text{'}1_{\boldsymbol{B}} = 1_{\boldsymbol{B}_1}\,,$$
$$H\text{'}\bigvee u = \bigvee H\text{''}u \quad \text{for} \quad u \subseteq B\,.$$

The class $\{x \in B, H'x = 1_{\boldsymbol{B}_1}\}$ is a complete filter on $\boldsymbol{B}$. If H is an isomorphism, then its inverse is an isomorphism of $\boldsymbol{B}_1$ onto $\boldsymbol{B}$. (Obvious.)

2431. THEOREM (**TSS′**). Let $\boldsymbol{B}$ be a complete Boolean algebra and let X be a real class such that $X \subseteq B$. Then there exists an algebra $\boldsymbol{B}_1$ which is the least subalgebra of $\boldsymbol{B}$ (with respect to inclusion) such that $X \subseteq B_1$.

Proof. By recursion we define $H''\{0\} = X$ and

$$H''\{\alpha\} = \{x; (\exists y \in H''\alpha)(x = -y) \vee (\exists z \subseteq H''\alpha)(x = F'z)\}$$

for $\alpha > 0$. Set $B_1 = \mathbf{W}(H)$. By 2429, $[B_1, C \upharpoonright B_1, F \upharpoonright \mathbf{P}(B_1)]$ is a complete Boolean subalgebra of $\boldsymbol{B}$; from 2146 it follows that $\boldsymbol{B}_1$ is the least subalgebra containing X.

2432. DEFINITION (**TSS′**). Let $\boldsymbol{B}$ be a complete Boolean algebra and let X be a real class such that $X \subseteq B$. The class B_1 of Theorem 2431 will be denoted by $\mathbf{Gen}_{\boldsymbol{B}}(X)$ (or simply $\mathbf{Gen}(X)$). If $B = \mathbf{Gen}_{\boldsymbol{B}}(X)$ then we say that X generates $\boldsymbol{B}$.

2433. DEFINITION (**TSS′**). Let $\boldsymbol{B}$ be a complete Boolean algebra; a real class $X \subseteq B$ is a *base* (for $\boldsymbol{B}$) if $0_{\boldsymbol{B}} \notin X$ and if for each non-zero $x \in B$ there exists $y \in X$ such that $y \leqq x$.

2434. In the remainder of this section, we restrict our attention to those complete Boolean algebras which are sets. We say that a triple $\langle b, c, f\rangle$ is a complete Boolean algebra, if $[b, c, f]$ is a complete Boolean algebra. The set b is then called the field of the complete Boolean algebra $\langle b, c, f\rangle$ etc.

2435. LEMMA (**TSS′**). Let $\boldsymbol{b} = \langle b, c, f\rangle$ be a complete Boolean algebra, and let $z \subseteq b$ be a complete filter. Then there exists $u \in b$ such that $z = \{x, u \leqq x\}$; z is an ultrafilter iff this u is an atom.

Proof. Take $u = \bigwedge z \in z$. If u is an atom then z is an ultrafilter (see above); if not, then there is some w such that $0_b < w < u$. Suppose z is an ultrafilter; since $w \notin z$ we have $-w \in z$; hence $w \leqq -w$, a contradiction since $w \neq 0$.

2436. LEMMA (**TSS′**). If z is a base for a complete Boolean algebra $\boldsymbol{b} = \langle b, c, f\rangle$ then for each $u \in b$ there exists some $t \subseteq z$ such that $u = \bigvee t$.

Proof. Set $t = \{v \in z, v \leqq u\}$. We have $\bigvee t \leqq u$; suppose $\bigvee t < u$. Then $u \wedge -\bigvee t \neq 0_B$ and there exists $v \in z$ such that $v \leqq u \wedge -\bigvee t$; i.e. $v \in t$ and $v \leqq \bigvee t$. Since $v \leqq -\bigvee t$ we have $v \leqq \bigvee t \wedge -\bigvee t = 0_{\boldsymbol{B}}$, a contradiction.

2437. COROLLARY. If z is a base for a complete Boolean algebra $\boldsymbol{b} = \langle b, c, f \rangle$ then z generates $\boldsymbol{b}$.

2438. THEOREM (**TSS'**). Let q be a base for a complete Boolean algebra $\boldsymbol{b} = \langle b, c, f \rangle$ and let q_1 be a base for a complete Boolean algebra $\boldsymbol{b}_1 = \langle b_1, c_1, f_1 \rangle$. Let h be a $1-1$ mapping of q onto q_1 such that

$$(\forall u, v \in q)\,(u \leqq v \equiv h'u \leqq_1 h'v)$$

($\leqq_1$ is the ordering of $\langle b_1, c_1, f_1 \rangle$). Then there is a unique isomorphism g of $\boldsymbol{b}$ onto $\boldsymbol{b}_1$ such that $g \restriction q = h$.

Proof. Define $g'x = \bigvee_1 \{h'u, u \leqq x\}$ for $x \in b$. Since this must be satisfied by each isomorphism, the uniqueness will follow from the existence.

(1) $x \leqq y \equiv g'x \leqq_1 g'y$ for $x, y \in b$. If $x \leqq y$ then we have $g'x \leqq_1 g'y$ immediately from the definition. To prove the other implication suppose $g'x \leqq_1 g'y$, i.e. $\bigvee_1 \{h'u,\ u \leqq x\} \leqq \bigvee_1 \{h'v,\ v \leqq y\}$. It follows that $u \leqq x$ implies $u \leqq y$ for each $u \in q$. Hence $\bigvee\{u, u \leq x\} \leq \bigvee\{v,\ v \leq y\}$, i.e. $x \leqq y$. This completes the proof of (1); moreover, we have $x = y \equiv g'x = g'y$ for $x, y \in b$ and hence g is one-to-one.

(2) g maps b onto b_1. For $y \in b_1$ set $x = \bigvee \{u, h'u \leqq y\}$. Then $g'x = y$. We also have $g'0_b = 0_{b_1}$ and $g'1_b = 1_{b_1}$.

(3) $g'\bigwedge z = \bigwedge_1 g''z$ for $z \subseteq b$. To prove the inequality $\leqq_1$, consider some $w \in z$; then $\bigwedge z \leqq w$ implies $g'\bigwedge z \leqq_1 g'w$. Hence $g'\bigwedge z \leqq_1 \bigwedge_1 \{g'w, w \in z\} = \bigwedge_1 g''z$. To prove the converse, set $\bigwedge_1 g''z = y$. Then $w \in z$ implies $y \leqq_1 g'w$; if $y = g'x$ then $x \leqq w$, and hence $x \leqq \bigwedge z$. It follows that $g'x = y = \bigwedge_1 g''z \leqq_1 g'\bigwedge z$. It can be proved analogously that $g'\bigvee z = \bigvee_1 g''z$ for $z \subseteq b$.

(4) $g'(-x) = -_1\, g'x$ for $x \in b$. We have $g'x \wedge_1 g'(-x) = g'(x \wedge -x) = g'0_b = 0_{b_1}$ and $g'x \vee_1 g'(-x) = g'(x \vee -x) = g'1_{\boldsymbol{b}} = 1_{b_1}$. Hence $g'(-x) = -_1\,(g'x)$ by Lemma 2412 (6).

This completes the proof of the Theorem.

2439. DEFINITION (**TSS'**). We say that an ordering $\leqq$ on A is *separative* if for each $x, y \in A$

$$(\forall z \leqq x)\,(\exists t)\,(t \leqq z\,\&\,t \leqq y)$$

implies $x \leqq y$.

2440. Remark. The condition can be restated as follows:

$$x \nleq y \to (\exists z \leqq x)(\forall t \leqq z)(t \nleq y).$$

2441. LEMMA (**TSS'**). Let q be a base for a complete Boolean algebra $\boldsymbol{b} = \langle b, c, f \rangle$. Then the ordering $\leqq$ of b is separative on q.

Proof. If $x, y \in q$ and $x \nleq y$, then $x \wedge -y < 0_{\boldsymbol{b}}$; hence there is some $z \in q$ with $z \leqq x \wedge -y$. Now z satisfies the condition

$$(\forall t \leqq z)(t \nleq y).$$

2442. THEOREM (**TSS'**). Let $\leqq_1$ be a reflexive separative ordering of a set q. Then there exist a complete Boolean algebra $\boldsymbol{b} = \langle b, c, f \rangle$ and a 1-1 function g which maps q onto some base for $\boldsymbol{b}$ and which satisfies the condition

$$(\forall u, v \in q)(u \leqq_1 v \equiv g'u \leqq g'v).$$

Proof. (1) We say that $u \subseteq q$ is *saturated* if the following conditions hold:

(a) if $x \leqq y$ and $y \in u$ then $x \in u$,

(b) if $(\forall y \leqq x)(\exists z \leqq y)(z \in u)$ then $x \in u$.

Denote by b the set of all saturated subsets of q.

(2) If z is a non-void subset of b then $\bigcap z$ is saturated. (Exercise.) Define $F'0 = q, \quad F'z = \bigcap z \quad \text{for} \quad 0 \neq z \subseteq b$.

(3) For $u \in b$ define $C'u = \{x \in q;\ \neg(\exists z \leqq x)(z \in u)\}$. Evidently $u \cap C'u = 0$.

(4) We prove that $C'u$ is saturated for each $u \in b$. If $x \leqq y \in C'u$ then $\neg(\exists z \leqq y)(z \in u)$; hence $\neg(\exists z \leqq x)(z \in u)$ and so $x \in C'u$. Further, if we suppose

$$(\forall s \leqq x)(\exists t \leqq s)(t \in C'u)$$

then we may deduce successively the formulas

$$(\forall s \leqq x)(\exists t \leqq s)\neg(\exists z \leqq t)(z \in u),$$

$$\neg(\exists s \leqq x)(\forall t \leqq s)(\exists z \leqq t)(z \in u),$$

$$\neg\,(\exists s \leqq x)\,(\forall t \leqq s)\,(t \in u)\,,$$
$$\neg\,(\exists s \leqq x)\,(s \in u)\,,$$
$$x \in C\text{'}u\,.$$

(5) Define $\boldsymbol{b} = \langle b, C, F\rangle$; we have $0_{\boldsymbol{b}} = 0$ and $u \wedge v = u \cap v$ for $u, v \in b$. Obviously $u \cap C\text{'}u = 0$. If $z \subseteq b$ and $z = \bigcup(g)$ then $\bigcap z = \bigcap_{t \in g} (\bigcap t)$. (Exercise.)

(6) We prove that $u \cap C\text{'}v = 0$ iff $u \cap v = u$ (i.e. $u \subseteq v$). If $u \subseteq v$ then $u \cap C\text{'}v = 0$, since $v \cap C\text{'}v = 0$ (see (3)). If $u \nsubseteq v$, choose $x \in u$ such that $x \notin v$. It follows (from saturatedness) that $\neg\,(\forall y \leqq x)\,(\exists z \leqq y)\,(z \in v)$, i.e. $(\exists y \leqq x)\,(\forall z \leqq y)\,(z \notin v)$. For any such y we have $y \in u$ and $y \in C\text{'}v$; hence $u \cap C\text{'}v \neq 0$. Thus $\boldsymbol{b}$ is a complete Boolean algebra.

(7) For $w \in q$ set

$$g\text{'}w = \{x \in q;\ x \leqq w\}\,;$$

$g\text{'}w$ is saturated by the separativity. Evidently, $w_1 \leqq w_2 \equiv g\text{'}w_1 \subseteq g\text{'}w_2$. Finally, $g\text{''}q$ is a base, which completes the proof.

2443. LEMMA (**TSS'** + E1). If q is a base for a c. B. a. $\boldsymbol{b}$ and if $u \in b$ and $u \neq 0_{\boldsymbol{b}}$ then there exists $a \subseteq q$ such that

1) $\bigvee a = u\,,$

2) $x \wedge y = 0_{\boldsymbol{b}}$ whenever x nad y are distinct elements of a.

Proof. Let f be a selector for the power-set of q and let G be defined as follows:

$$G\text{'}0 = f\text{'}\{v \in q; v \leqq u\}\,,$$
$$G\text{'}\alpha = f\text{'}\{v \in q; v \leqq u - \bigvee G\text{''}\alpha\} \quad \text{for} \quad \alpha > 0\,.$$

We may now let $a = G\text{''}\lambda$ where λ is the least ordinal number such that $G\text{'}\lambda = 0$, i.e. such that $u - \bigvee G\text{''}\lambda = 0_{\boldsymbol{b}}$.

2444. THEOREM (**TSS'**, E1). If p is a partition of b and if q is a base for b then there exists a partition $\bar{p} \lessdot p$ such that $\bar{p} \subseteq q$.

Proof. Straightforward.

SECTION 5

Ordered and separatively ordered sets

The significance of separatively ordered sets lies in the fact, established in the preceding section, that they are precisely the bases for complete Boolean algebras which are sets. In the present Section we shall study some characteristics of separatively ordered sets and certain methods for constructing such sets. We shall use the axiom of choice (E1) in several places. Some statements are formulated not only for separatively ordered sets but for ordered sets in general.

2501. DEFINITION (**TSS′**). If $\boldsymbol{a} = \langle a, \leqq_{\boldsymbol{a}} \rangle$ is an ordered set then a *segment* of a determined by x is a set of the form

$$\mathbf{Seg}_{\boldsymbol{a}}(x) = \{y \in a;\ y \leqq_{\boldsymbol{a}} x\} \quad \text{where} \quad x \in a\,.$$

If it is clear which ordered set is involved we may write $\hat{x}$ instead of $\mathbf{Seg}_{\boldsymbol{a}}(x)$.

A set $z \subseteq a$ is called an *exclusive* system (in $\boldsymbol{a}$) ($\mathrm{Ex}_{\boldsymbol{a}}(z)$), if

$$(\forall x, y \in z)(x \neq y \rightarrow \hat{x} \cap \hat{y}) = 0\,.$$

An ordered set is $\aleph_\alpha$-*multiplicative* ($\nu(\aleph_\alpha, \boldsymbol{a})$) if for any subset u of a which is linearly ordered by $\leqq$ and has cardinality at most $\aleph_\alpha$ there is an $x \in a$ such that $(\forall y \in u)(x \leqq y)$. We say that $\boldsymbol{a}$ is $\aleph_\alpha$-*partitionable* ($\mu(\aleph_\alpha, \boldsymbol{a})$) if there is an exclusive system $u \subseteq a$ which has cardinality at least $\aleph_\alpha$.

2502. LEMMA (**TSS′**). If $\aleph_\alpha \leqq \aleph_\beta$ then for any ordered set $\boldsymbol{a}$ we have

$$\nu(\aleph_\beta, \boldsymbol{a}) \rightarrow \nu(\aleph_\alpha, \boldsymbol{a})\,,$$

$$\mu(\aleph_\beta, \boldsymbol{a}) \rightarrow \mu(\aleph_\alpha, \boldsymbol{a})\,.$$

2503. THEOREM (**TSS′**). If $\aleph_\xi$ is a singular cardinal and if an ordered set $\boldsymbol{a}$ is $\aleph_\beta$-multiplicative for every $\beta < \xi$ then $\boldsymbol{a}$ is $\aleph_\xi$-multiplicative.

Proof. Let $u \subseteq a$ be linearly ordered by $\leqq_{\boldsymbol{a}}$ and suppose $u \approx \aleph_\xi$; let f be a one-one mapping of $\boldsymbol{\omega}_\xi$ onto u. We let $g'0 = f'0$ and for any $\alpha > 0$ we let $\bar{\alpha}$ be the least β such that $(\forall\gamma < \alpha)(g'\gamma >_{\boldsymbol{a}} f'\beta)$ and $g'\alpha = f'\bar{\alpha}$. Clearly $\mathbf{D}(g) \leqq \boldsymbol{\omega}_\xi$. If $\mathbf{D}(g) < \boldsymbol{\omega}_\xi$ we let $\bar{g} = g$; if $\mathbf{D}(g) = \boldsymbol{\omega}_\xi$ let $\{\tau_\alpha\}_{\alpha < \mathbf{cf}(\omega_\xi)}$ be a sequence whose supremum is $\boldsymbol{\omega}_\xi$ and let $\bar{g}'\alpha = g'\tau_\alpha$. It follows that $\mathbf{D}(\bar{g}) < \boldsymbol{\omega}_\xi$ and that $\mathbf{W}(\bar{g})$ is cofinal with u in $\geqq_{\boldsymbol{a}}$; i.e., $(\forall z \in u)(\exists w \in \mathbf{W}(\bar{g}))(z \geqq_{\boldsymbol{a}} w)$. Let $\aleph_\beta$ be the cardinality of $\mathbf{W}(\bar{g})$; clearly, $\boldsymbol{a}$ is $\aleph_\beta$-multiplicative and hence there is an $x \in a$ such that $(\forall y \in \mathbf{W}(\bar{g}))(x \leqq_{\boldsymbol{a}} y)$. By cofinality $(\forall y \in u)(x \leqq_{\boldsymbol{a}} y)$.

2504. THEOREM (**TSS′**, E1). Let $\boldsymbol{a}$ be a separatively ordered set and suppose that $\boldsymbol{a}$ has no minimal elements, i.e. that $(\forall x \in a)(\exists y \in a)(y <_{\boldsymbol{a}} x)$. If $\boldsymbol{a}$ is $\aleph_\alpha$-multiplicative then it is $\aleph_{\alpha+1}$-partitionable.

Proof. Using a selector for $\mathbf{P}(a)$ we construct a function f from $\boldsymbol{\omega}_{\alpha+1}$ into a such that $f'\beta >_{\boldsymbol{a}} f'\gamma$ for any $\beta < \gamma < \boldsymbol{\omega}_{\alpha+1}$. By separativity we have

$$y <_{\boldsymbol{a}} x \to (\exists z <_{\boldsymbol{a}} x)(\hat{z} \cap \hat{y} = 0),$$

and hence $(\exists z \leqq_{\boldsymbol{a}} x)(\forall t \leqq_{\boldsymbol{a}} z)(t \nleqq_{\boldsymbol{a}} y)$. If follows that for any $\beta \in \boldsymbol{\omega}_{\alpha+1}$ we can define a function $g'\beta$ such that $g'\beta \leqq_{\boldsymbol{a}} f'\beta$ and $\mathbf{Seg}_{\boldsymbol{a}}(g'\beta) \cap \mathbf{Seg}_{\boldsymbol{a}}(g'(\beta + 1)) = 0$. We have $g''\boldsymbol{\omega}_{\alpha+1} \approx \boldsymbol{\omega}_{\alpha+1}$ and $\mathrm{Ex}_{\boldsymbol{a}}(g''\boldsymbol{\omega}_{\alpha+1})$.

2505. LEMMA (**TSS′**). Let u be an exclusive system in an ordered set $\boldsymbol{a}$. If for all $x \in u$, $v_x \subseteq \hat{x}$ is an exclusive system in $\boldsymbol{a}$ then $\bigcup_{x \in u} v_x$ is an exclusive system in $\boldsymbol{a}$.
(Exercise.)

2506. DEFINITION (**TSS′**). If $\boldsymbol{a}$ is an ordered set then an element x of a is called μ-*saturated* ($\mu\mathrm{sat}_{\boldsymbol{a}}(x)$) if for every $\aleph_\alpha$ and every $y <_{\boldsymbol{a}} x$ the existence of an exclusive system $u \subseteq \hat{x}$ of cardinality $\aleph_\alpha$ implies the existence of an exclusive system of cardinality $\aleph_\alpha$ included in $\hat{y}$.

2507. LEMMA (**TSS′**). If x is μ-saturated and $y \leqq_{\boldsymbol{a}} x$ then y is μ-saturated.

2508. LEMMA (**TSS′**). If x is an element of an ordered set a then there is a μ-saturated $y \leqq_{\boldsymbol{a}} x$.

Proof. For any $y \leqq_{\boldsymbol{a}} x$ we let $\mathbf{M}(y)$ be the least α such that there is no exclusive system of power $\aleph_\alpha$ of elements of $\hat{y}$. We let $y_0 \leqq_{\boldsymbol{a}} x$ be such that $\mathbf{M}(y_0) \leqq \mathbf{M}(y)$ for every $y \leqq_{\boldsymbol{a}} x$. Clearly y_0 is μ-saturated.

2509. LEMMA (**TSS'**, E1). If $\boldsymbol{a}$ is an ordered set then there is an exclusive system u of elements of a which contains only μ-saturated elements of a and such that $(\forall x \in a)(\exists y \in u)(\hat{x} \cap \hat{y}) \neq 0$.

Proof. Let $\{x_\alpha\}_{\alpha<\lambda}$ be a sequence consisting of all μ-saturated elements and let

$$f\text{'}0 = x_0 ,$$

$$f\text{'}\alpha = \begin{cases} x_\alpha \text{ if } (\forall\beta < \alpha)(\hat{x}_\alpha \cap (f\text{'}\beta)^\wedge) = 0 , \\ x_0 \text{ otherwise} , \end{cases}$$

for any $\alpha > 0$. We also let $u = f\text{''}\lambda$. Clearly $\mathrm{Ex}_{\boldsymbol{a}}(u)$. If $x \in a$ then there is an α such that $x_\alpha \leqq_{\boldsymbol{a}} x$. If $x_\alpha \in u$ then we let $y = x_\alpha$. If $x_\alpha \notin u$ then there exists $\beta < \alpha$ such that $x_\beta \in u$ and $\hat{x}_\beta \cap \hat{x}_\alpha \neq 0$; we then let $y = x_\beta$. Clearly, y is μ-saturated.

2510. THEOREM (**TSS'**, E1). Let $\boldsymbol{a}$ be an ordered set, let $\aleph_\zeta$ be a singular cardinal and suppose that for every $\beta < \zeta$ there is an exclusive system $u \subseteq \boldsymbol{a}$ of cardinality $\aleph_\beta$. Then there is an exclusive system of cardinality $\aleph_\zeta$.

Proof. Let u be an exclusive system in $\boldsymbol{a}$ which satisfies the conditions of Lemma 2509. If this system has cardinality $\geqq \aleph_\zeta$ then we are finished; we therefore suppose its cardinality to be $< \aleph_\zeta$. If there exists $x \in u$ such that there is an exclusive system $v \subseteq \hat{x}$ of cardinality $\geqq \aleph_\zeta$ then we are also finished, since v is the required system. Put $\beta_x = \mathbf{Min}\,\{\beta$, there is no exclusive system on $\hat{x}$ of cardinality $\aleph_\beta\}$; we suppose $\beta_x \leqq \zeta$ for each $x \in u$. On the other hand, it is possible (and necessary, as we shall show) that for any $\beta < \zeta$ there exists $x \in u$ such that there is an exclusive system $v \subseteq \hat{x}$ of cardinality $\aleph_\beta$. Then we can find a set u_1 and exclusive systems $w_x \subseteq \hat{x}\,(x \in u_1)$ such that the supremum of $\overline{\overline{w}}_x\,(x \in u_1)$ is $\aleph_\zeta$. If we let $w = \bigcup_{x \in u_1} w_x$ then w is the required system. (If there is a $y \in u$ such that $\beta_y = \zeta$ then u_1 is an exclusive system on $\hat{y}$; otherwise u_1 is a subset of u.)

It remains to prove that there is no $\beta < \zeta$ such that, for any x, there is no exclusive system $v_x \subseteq \hat{x}$ of cardinailty $\aleph_\beta$. Suppose there is such a β and let $\delta < \zeta$ be such that $\aleph_\delta > \aleph_\beta$ and $\aleph_\delta > \aleph_\gamma$, where $\aleph_\gamma$ is the cardinality of u. Let v be an exclusive system of cardinality $\aleph_\delta$. For every $x \in v$ we choose a $y_x \in \hat{x}$ such that $y_x \leqq_{\boldsymbol{a}} z$ for some $z \in u$. If we denote by $\bar{v}$ the set of all y_x such that $x \in v$ then for every $z \in u$, $\bar{v} \cap \hat{z} \subseteq \hat{z}$ is an exclusive system and has cardinality less than $\aleph_\beta$. Hence the cardinality of $\bar{v}$, and therefore also of v, is at most $\mathbf{Max}\,(\aleph_\beta, \aleph_\gamma) < \aleph_\delta$; a contradiction.

This completes the proof.

We now define two important indices for complete Boolean algebras.

2511. DEFINITION (**TSS′**). If $\boldsymbol{b}$ is a complete Boolean algebra then the *calibre* of $\boldsymbol{b}$ ($\boldsymbol{\mu}(\boldsymbol{b})$), is the least cardinal $\aleph_\alpha$ such that there is no $q \subseteq b$ of cardinality $\aleph_\alpha$ such that

$$(\forall x, y \in q)(x \neq y \rightarrow x \wedge y = 0_b) .$$

2512. LEMMA (**TSS′**, E1). If an ordered set $\boldsymbol{a}$ is a base for a complete Boolean algebra $\boldsymbol{b}$ then $\boldsymbol{\mu}(\boldsymbol{b})$ is the least cardinal $\aleph_\alpha$ such that there is no exclusive system $\boldsymbol{u} \subseteq a$ of cardinality $\aleph_\alpha$.
(Exercise.)

2513. DEFINITION (**TSS′**). If $\boldsymbol{b}$ is a complete Boolean algebra then the *draft* of $\boldsymbol{b}$ ($\boldsymbol{\nu}(\boldsymbol{b})$), is the least cardinal $\aleph_\alpha$ such that there is no $\aleph_\alpha$-multiplicative base for $\boldsymbol{b}$.

We shall now consider products of ordered sets.

2514. DEFINITION (**TSS′**). (1) If s is a nonempty set then $l \subseteq \mathbf{P}(s)$ is called a *cut* in s if it has the following properties:

$$l \neq 0 , \quad (\forall x \in l)(\forall y \subseteq x)(y \in l) .$$

(2) If m is a cardinal number then a cut l is m-additive if

$$(\forall z \subseteq l)(z \preccurlyeq m \rightarrow \bigcup z \in l) .$$

(3) A cut is an *ideal* on s if it is 2-additive, i.e. if.

$$(\forall x, y \in l)(x \cup y \in l) .$$

(4) The *norm* of a cut l (**Norm** (l)) is the least cardinal number $\aleph_\alpha$ such that

$$\neg (\exists x \in l)(x \not\prec \aleph_\alpha) .$$

2515. DEFINITION (**TSS′**). Let l be a cut in s and let $\mathbf{b}$ be a function on s such that, for any $x \in s$, $\boldsymbol{b}'x$ is an ordered set; for convenience we write $\boldsymbol{b}_x = \langle b_x, \leqq_x \rangle$ instead of $\boldsymbol{b}'x$. The product of ordered sets $\boldsymbol{b}_x$ over l is the ordered set $\prod^l_{x \in s} \boldsymbol{b}_x = \langle a, \leqq \rangle$ where

$$a = \{f; \operatorname{Un}(f) \,\&\, \mathbf{D}(f) \in l \,\&\, (\forall x \in \mathbf{D}(f))(f'x \in b_x)\} ,$$
$$f \leqq g \equiv \mathbf{D}(f) \supseteq \mathbf{D}(g) \,\&\, (\forall x \in \mathbf{D}(g))(f'x \leqq_x g'x) .$$

2516. Remark. The reader may easily verify that $\prod_{x \in s}^{l} \boldsymbol{b}_x$ is an ordered set.

2517. LEMMA (**TSS'**). 1) If all $\boldsymbol{b}_x$ are separatively ordered sets without greatest element then $\prod_{x \in s}^{l} \boldsymbol{b}_x$ is also separatively ordered.

2) Denote $\prod_{x \in s}^{l} \boldsymbol{b}_x$ by $\langle a, \leqq \rangle$. If f and g are in a and if $f = g \restriction \mathbf{D}(f)$ then $g \leqq f$; if $\mathbf{D}(f) \cap \mathbf{D}(g) = 0$ then $f \cup g$ is the infimum of f and g.

3) The empty function 0 is the greatest element of $\prod_{x \in s}^{l} \boldsymbol{b}_x$.

2518. DEFINITION (**TSS'**). A cut l is said to have the *singleton property* on s if $z \in l$ and $x \in s$ imply $z \cup \{x\} \in l$.

2519. Remark (**TSS'**). 1) Clearly if an ideal on s contains all singletons then it has the singleton property.

2) Let s be infinite and suppose that l has the singleton property and $s \notin l$ (e.g. $s = \omega_0$ and $l = \mathbf{P}_{\text{fin}}(s)$). If, for any $x \in s$, the set b_x has at least one element, then there are no minimal elements in $\prod_{x \in s}^{l} \boldsymbol{b}_x$. (Exercise.) Thus the algebra whose base is $\prod_{x \in s}^{l} \boldsymbol{b}_x$ does not contain any atoms.

2520. DEFINITION (**TSS'**). For any set x, **at** (x) is the ordered set $\langle x, \mathbf{I} \restriction x \rangle$. (It follows that every two distinct elements of **at** (x) are incomparable.)

2521. DEFINITION (**TSS'**). If $\boldsymbol{b}_0$, $\boldsymbol{b}_1$ are ordered sets then $\boldsymbol{b}_0 \odot \boldsymbol{b}_1 = \langle b_0 \times b_1, \leqq \rangle$ where

$$\langle u_0, u_1 \rangle \leqq \langle v_0, v_1 \rangle \equiv u_0 \leqq_0 v_0 \,\&\, u_1 \leqq_1 v_1$$

for any $u_0, v_0 \in b_0$ and $u_1, v_1 \in b_1$.

We may prove (in **TSS**) that if b_0, b_1 are separatively ordered sets (not necessarily without greatest elements) then $\boldsymbol{b}_0 \odot \boldsymbol{b}_1$ is a separatively ordered set. Further, we have the following

2522. LEMMA (**TSS'**). Let $\boldsymbol{b}_0$, $\boldsymbol{b}_1$ be separatively ordered sets with greatest elements 1_0, 1_1 respectively. If $\boldsymbol{a}_i = \langle b_i - \{1_i\}, \leqq_i \rangle$ $(i = 0, 1)$ and $l = \mathbf{P}(\{0, 1\})$ then $\boldsymbol{b}_0 \odot \boldsymbol{b}_1$ and $\prod_{i=0,1}^{l} \boldsymbol{a}_i$ are isomorphic.

Proof. Exercise.

2523. THEOREM (**TSS'**, E1). If $\boldsymbol{b}_x$ is $\aleph_\alpha$-multiplicative for each $x \in s$ and if l is $\aleph_\alpha$-additive then $\prod_{x \in s}^{l} \boldsymbol{b}_x$ is $\aleph_\alpha$-multiplicative.

Proof. Let $\langle a, \leqq \rangle = \prod^{l}_{x \in s} \boldsymbol{b}_x$ and let u be a linearly ordered subset of a of power $\leqq \aleph_\alpha$. For every $x \in s$ the set $\{f'x; f \in u\}$ is linearly ordered by $\leqq_x$. By $\aleph_\alpha$-additivity we have $\bigcup_{f \in u} \mathbf{D}(f) = z \in l$. Let f be a function which associates with each $x \in z$ an element $f'x$ of b_x such that $f'x \leqq_x g'x$ for any $g \in u$. The element f of a satisfies the condition of $\aleph_\alpha$-multiplicativity.

2524. THEOREM (**TSS'**, E1). If l is an ideal, if $\mathbf{Norm}(l) \leqq \aleph_{\xi+1}$ and if no $\boldsymbol{b}_x$ is $\aleph_{\xi+1}$ partitionable then $\prod^{l}_{x \in s} \boldsymbol{b}_x$ is not $\aleph_\alpha$-partitionable for any cardinal $\aleph_\alpha > 2^{\aleph_\xi}$.

Proof. For any $f \in a$ we shall call the elements of $\mathbf{D}(f)$ *fixing points* of f. The set of all fixing points of f belongs to l and hence it is of power at most $\aleph_\xi$. For any $f \in a$ we choose some (not necessarily one-to-one) mapping of $\boldsymbol{\omega}_\xi$ onto the set of all fixing points of f; for each $\alpha < \boldsymbol{\omega}_\xi$ we may refer to the α-th fixing point to be denoted by x^f_α.

Suppose that $u \subseteq a$ is an exclusive system in a; we shall prove that $u \preccurlyeq 2^{\aleph_\xi}$. Let A be a selector for $\mathbf{P}(u)$; i.e. we have $A'v \in v$ for any non-empty $v \subseteq u$. We shall prove that every element of u can be constructed by iterated application of A. For any $f \in u$ and $\alpha < \boldsymbol{\omega}_\xi$ we let $G(f, \alpha) = \{g \in u; \mathrm{Ex}_{\boldsymbol{b}'x^f_\alpha}(\{f'x^f_\alpha, g'x^f_\alpha\})\}$; $G(f, \alpha)$ is the set of all functions in u whose value at the α-th fixing point of f determines a segment disjoint from the segment determined by the value of f at this point.

We let D be the (real) class of all functions h such that $\mathbf{D}(h) \in \mathbf{On}$ and $\mathbf{W}(h) \subseteq \boldsymbol{\omega}_\xi$. To each $h \in D$ we assign a sequence f^h_α of elements of u as follows:

$$f^h_0 = A'u\,,$$
$$f^h_\alpha = A' \bigcap_{\beta<\alpha} G(f^h_\beta, h'\beta) \quad (0 < \alpha \leqq \mathbf{D}(h))\,.$$

If $\beta < \alpha$ the values of f^h_α and f^h_β at the $(h'\beta)$-th fixing of f^h_β are exclusive (in the ordered set corresponding to the $(h'\beta)$-th fixing point of f^h_β). We let D_0 be the class of all $h \in D$ such that $\bigcap_{\beta<\alpha} G(f^h_\beta, h'\beta) \neq 0$ for all $\alpha \leqq \mathbf{D}(h)$; this means that all f^h_α for $\alpha \leqq \mathbf{D}(h)$ are defined as elements of u. If $\alpha < \beta \leqq \mathbf{D}(h)$ and if $h \in D_0$ then $f^h_\alpha \neq f^h_\beta$, since they differ at the $(h'\alpha)$-fixing point of f^h_α. We shall prove the following:

1) For every $h \in D_0$ we have $\mathbf{D}(h) < \boldsymbol{\omega}_{\xi+1}$.

2) For every $f \in u$ there exist $h \in D_0$ and $\alpha \leqq \mathbf{D}(h)$ such that $f = f^h_\alpha$.

This will complete the proof since we shall have $D_0 \preccurlyeq 2^{\aleph_\xi}$ by 1) and $u \preccurlyeq \preccurlyeq \aleph_\xi \,.\, 2^{\aleph_\xi} = 2^{\aleph_\xi}$ by 2).

Sub 1): Suppose there exists $h \in D_0$ such that $\mathbf{D}(h) = \omega_{\xi+1}$ and let $f = f^h_{\omega_{\xi+1}}$. For each $\alpha < \omega_{\xi+1}$ we let $p'\alpha$ be the $(h'\alpha)$-th fixing point of f^h_α. Each $p'\alpha$ is also a fixing point of f, since $f \in G(f^h_\alpha, h'\alpha)$. Hence we have $p''\omega_{\xi+1} \prec \aleph_{\xi+1}$ and so there exists $x \in s$ such that the set $q = \{\alpha < \omega_{\xi+1}; p'\alpha = x\}$ has power $\aleph_{\xi+1}$. The set $w = \{f^{h'}_\alpha x; \alpha \in q\}$ is an exclusive system of power $\aleph_{\xi+1}$ in $\boldsymbol{b}_x$, a contradiction. This proves 1).

Sub 2): If $f \in u$ then we construct $h \in D_0$ as follows.

If $f = A'u$ we are finished since $f^h_0 = A'u$ for all $h \in D_0$. Otherwise we let $h'0 = \gamma$, where γ is the least number such that $f \in G(f^h_0, \gamma)$; such a number exists since $f \neq f^h_0$ and both f and f^h_0 are elements of the exclusive system u. Suppose that we have already constructed $h'\beta$ for all $\beta < \alpha$; hence f^h_β for all $\beta \leqq \alpha$. If $f = f^h_\alpha$ then we are finished. Otherwise we have $f \in \bigcap_{\beta<\alpha} G(f^h_\beta, h'\beta)$ and $f \neq f^h_\alpha$; we let $h'\alpha = \gamma$ where γ is the least number such that $f \in G(f^h_\alpha, \gamma)$.

We have $f \in \bigcap_{\beta \leqq \alpha} G(f^h_\beta, h'\beta)$ and the induction may be continued. In this way we construct a function $h \in D_0$; hence we have $f^h_\alpha = f$ for $\alpha = \mathbf{D}(h)$.

The following theorem can be proved similarly:

2525. THEOREM (**TSS′**, E1). If $\mathbf{Norm}(l) \leqq \omega_0$ and if no $\boldsymbol{b}_x$ is $\aleph_0$-partitionable then every exclusive system in $\prod^l_{\alpha \in s} \boldsymbol{b}_x$ is at most countable.

Proof. We proceed as above; we construct the fixing points x^f_n for $f \in \prod^l_{x \in s} \boldsymbol{b}_x$ and $n \in \omega$, the function $G(f, n)$, the class D of all functions with $\mathbf{D}(h) \in \mathbf{On}$ and $\mathbf{W}(h) \subseteq \omega$ and the class D_0. For any $h \in D_0$ the set $\mathbf{D}(h)$ is finite; for any $f \in u$ there is some $h \in D_0$ and some n such that $f = f^h_n$; hence D_0 is countable which proved the Theorem.

2526. THEOREM (**TSS′**, E1). Let $\mathbf{Norm}(l) \leqq \aleph_\alpha$ where α is a limit number. If $\aleph_\beta < \aleph_\alpha$ and if no $\boldsymbol{b}_x$ is $\aleph_\beta$-partitionable then every exclusive system in $\prod^l_{x \in s} \boldsymbol{b}_x$ is of power at most $\mathbf{Sup}_{\delta<\alpha} 2^{\aleph_\delta}$.

Proof. We let $l(\gamma) = \{z \in l; z \prec \aleph_\gamma\}$ for any $\gamma < \alpha$; $l(\gamma)$ is an ideal and $l = \bigcup_{\gamma<\alpha} l(\gamma+1)$. We let $\boldsymbol{a}(\gamma) = \langle a(\gamma), \leqq(\gamma)\rangle = \prod^{l(\gamma)}_{x \in s} \boldsymbol{b}_x$ so that $a = \bigcup_{\gamma<\alpha} a(\gamma + 1)$. If $u \leqq a$ is an exclusive system in $\boldsymbol{a}$ then $u \cap a(\gamma)$ is an exclusive system in $\boldsymbol{a}(\gamma)$; we have $u = \bigcup_{\beta<\gamma<\alpha} u \cap a(\gamma+1)$.

By the preceding Theorem we have $u \cap a(\gamma+1) \preccurlyeq 2^{\aleph_\gamma}$; the result now follows immediately.

2527. DEFINITION (**TSS'**). If t is a nonempty subset of s and if l is a cut in s then the *restriction of the cut l to t* is the set $l/t = l \cap \mathbf{P}(t)$.

2528. *Remark* (**TSS'**). 1) The restriction of l to t is a cut in t.

2) If l is a cut then $l = l/\bigcup(l)$.

2529. THEOREM (**TSS'**). Let l be an ideal on s and suppose that, for any $x \in s$, $\boldsymbol{b}_x$ is an ordered set. If $t \subseteq s$, $t \neq 0$ and $s - t \neq 0$ then $\prod^l_{x \in s} \boldsymbol{b}_x$ is isomorphic to $\prod^{l/t}_{x \in t} \boldsymbol{b}_x \odot \prod^{l/(s-t)}_{x \in s-t} \boldsymbol{b}_x$.

Proof. Exercise.

We shall now study products of separatively ordered sets as bases for complete Boolean algebras. If $\boldsymbol{b}$ is a complete Boolean algebra then the set of all non-zero elements endowed with the canonical ordering is a base for $\boldsymbol{b}$. This leads us to the following

2530. DEFINITION (**TSS'**). If $\boldsymbol{b}$ is a complete Boolean algebra with more than two elements then

$$\dot{\boldsymbol{b}} = \langle b - \{0_{\boldsymbol{b}}, 1_{\boldsymbol{b}}\}, \leqq \rangle$$

is called the *canonical base* for $\boldsymbol{b}$.

$$\mathring{\boldsymbol{b}} = \langle b - \{0_{\boldsymbol{b}}\}, \leqq \rangle$$

is called the *canonical base* for $\boldsymbol{b}$ *with the unit*.

As a consequence of Lemma 2522 we obtain the following

2531. LEMMA (**TSS'**). Let $\boldsymbol{b}_0$, $\boldsymbol{b}_1$ be complete Boolean algebras and let l be the ideal $\mathbf{P}(\{0, 1\})$. Then the separatively ordered sets $\mathring{\boldsymbol{b}}_0 \odot \mathring{\boldsymbol{b}}_1$, $\prod^l_{x \in 2} \dot{\boldsymbol{b}}_x$ are isomorphic.

2532. *Remark*. We shall denote by $\boldsymbol{b}_0 \odot \boldsymbol{b}_1$ the complete Boolean algebra with the base $\mathring{\boldsymbol{b}}_0 \odot \mathring{\boldsymbol{b}}_1$.

We first prove a theorem on uniqueness.

2533. THEOREM (**TSS'**, E1). If l is a cut in s and if for any $x \in s$ both $\boldsymbol{c}_x$ and $\boldsymbol{d}_x$ are bases for a Boolean algebra $\boldsymbol{b}_x$ then the complete Boolean algebras $\boldsymbol{c}$ and $\boldsymbol{d}$ which have bases $\prod^l_{x \in s} \boldsymbol{c}_x$ and $\prod^l_{x \in s} \boldsymbol{d}_x$ respectively*) are isomorphic.

*) We say that an ordered set $\langle \boldsymbol{a}, \leqq \rangle$ is a base for a complete Boolean algebra $\boldsymbol{b}$ if $\boldsymbol{a}$ is a base for $\boldsymbol{b}$ and $\leqq$ is the canonical ordering of $\boldsymbol{b}$ restricted to $\boldsymbol{a}$.

Proof. We may suppose w. l. o. g. that $d_x = \mathring{b}_x$ for any x. By Theorem 2438 it suffices to prove that $\prod_{x \in s}^{l} \boldsymbol{c}_x$ is a base for the algebra $\boldsymbol{d}$. If u is an element of d then there exists an element f of the ordered set $\prod_{x \in s}^{l} (\mathring{\boldsymbol{b}}_x)$ such that $f \leqq u$ (in the canonical ordering of d). We let g be a function with the same domain as f such that $g'x \in c_x$ and $g'x \leqq_x f'x$ for all $x \in \mathbf{D}(g)$; there is such a g, since, for every x, $\boldsymbol{c}_x$ is a base for $\boldsymbol{b}_x$. This g belongs to $\prod^{l} \boldsymbol{c}_x$ and we have $g \leqq f \leqq u$, so that $\prod_{x \in s}^{l} \boldsymbol{c}_x$ is a base for $\boldsymbol{d}$.

2534. THEROEM (**TSS'**). If $\prod_{x \in s}^{l} \boldsymbol{b}_x$ is a base for an algebra $\bar{\boldsymbol{b}} = \langle \bar{b}, \bar{c}, \bar{f} \rangle$ and if $t \subseteq s$ then $\prod_{x \in s}^{l/t} \boldsymbol{b}_x$ is a base for some subalgebra $\bar{\boldsymbol{b}}_0$ of $\bar{\boldsymbol{b}}$.

Proof. We denote $\prod^{l} \boldsymbol{b}_x$ by $\langle a, \leqq \rangle$ and $\prod^{l/t} b_x$ by $\langle a/t, \leqq/t \rangle$. Let $\bar{b}_0$ be the set of all joins $\bigvee z$ such that $z \subseteq a/t$. We shall prove that $\langle \bar{b}_0, \bar{c} \upharpoonright \bar{b}_0, \bar{f} \upharpoonright \mathbf{P}(\bar{b}_0) \rangle$ is a subalgebra. The set $\bar{b}_0$ is clearly closed under joins, so it is sufficient to show that $\bar{b}_0$ is closed under complementation. For any $f \in a$ we denote by f/t the unique element g of a/t which assumes in s the same values as f. We have $f \leqq f/t$ and $f_1 \leqq f_2 \to f_1/t \leqq f_2/t$. On the other hand, for any $f \in a$ and $g \in a/t$ we have $f \leqq g \to f/t \leqq g$. It follows that

$$f \wedge g = 0_{\bar{\boldsymbol{b}}} \to (f/t) \wedge g = 0_{\bar{\boldsymbol{b}}} \quad \text{for} \quad f \in a \quad \text{and} \quad g \in a/t\,,$$

and

$$f \wedge (\bigvee q) = 0_{\bar{\boldsymbol{b}}} \to (f/t) \wedge (\bigvee q) = 0_{\bar{\boldsymbol{f}}} \quad \text{for} \quad f \in a \quad \text{and} \quad q \subseteq a/t\,.$$

If $u \in \bar{b}_0$ and $u = \bigvee z$ for some $z \subseteq a/t$ we let $\bar{z}$ be the set of all $g \in a/t$ such that $g \wedge u = 0_{\bar{\boldsymbol{b}}}$. We prove $-u = \bigvee \bar{z}$. We have $\bigvee z \vee \bigvee \bar{z} = 1_{\bar{\boldsymbol{b}}}$. Indeed, in the contrary case there exists $f \in a$ such that $f \wedge (\bigvee z \vee \bigvee \bar{z}) = 0_{\bar{\boldsymbol{b}}}$, so that $(f/t) \wedge (\bigvee z \vee \bigvee \bar{z}) = 0_{\bar{\boldsymbol{b}}}$; hence $(f/t) \wedge \bigvee z = 0_{\bar{\boldsymbol{b}}}$ and so $f/t \in \bar{z}$, a contradiction. Further we have $\bigvee z \wedge \bigvee \bar{z} = 0_{\bar{\boldsymbol{b}}}$; for, otherwise there exist $g \in z$ and $\bar{g} \in \bar{z}$ such that $g \wedge \bar{g} \neq 0_{\bar{\boldsymbol{b}}}$, which contradicts the definition of $\bar{z}$. Thus $\bar{b}_0$ determines a subalgebra which evidently has $\langle a/t, \leqq/t \rangle$ as a base.

2535. THEOREM (**TSS'**, E1). Let $\prod_{x \in s}^{l} \boldsymbol{b}_x$ be a base for a complete Boolean algebra $\bar{\boldsymbol{b}} = \langle \bar{b}, \bar{c}, \bar{f} \rangle$; suppose that $\mathbf{Norm}(l) \leqq \aleph_{\alpha+1}$ and that $\neg\, \boldsymbol{\mu}(\prod_{x \in s}^{l} \boldsymbol{b}_x, \aleph_{\alpha+1})$. If $u \subseteq \bar{b}$ is of power $\aleph_\alpha$ then there exists a set t of power $\aleph_\alpha$ such that the algebra determined by the base $\prod^{l/t} \boldsymbol{b}_x$ contains u as a subset.

Proof. We denote $\prod^l \boldsymbol{b}_x$ by $\langle a, \leqq \rangle$. For any $f \in u$ let a_f be an exclusive set of elements of a such that $\mathrm{V} a_f = f$ and for each $z \in a_f$ let $\mathbf{fix}(z)$ be the set of all fixing points of z. If we set $a_u = \bigcup_{f \in u} a_f$ and $t = \bigcup_{z \in a_u} \mathbf{fix}(z)$ then t has power $\leqq \aleph_\alpha$, since every set $\mathbf{fix}(z)$ has power at most $\aleph_\alpha$ and a_u is the union of at most $\aleph_\alpha$ sets of power at most $\aleph_\alpha$. It follows that $u \subseteq \bar{\boldsymbol{b}}_0$ where $\bar{\boldsymbol{b}}_0$ is the algebra whose base is $\prod_{x \in s}^{l/t} \boldsymbol{b}_x$.

2536. *Example.* Let $\boldsymbol{b}_x = \mathbf{at}(2)$ for $x \in \boldsymbol{\omega}$ and let $l = \mathbf{P}_{\mathbf{fin}}(\boldsymbol{\omega})$. Denote by **Cant** the algebra with the base $\boldsymbol{b} = \prod_{x \in \boldsymbol{\omega}}^{l} \boldsymbol{b}_x$. Then **Cant** has no atoms.

Remark. This construction can be generalized; generalized Cantor algebras will be studied in Chapt. 6 Sect. 1.

CHAPTER III

SECTION 1

The second and third axioms of regularity. The second axiom of choice. Axioms concerning urelements

We shall now present some axioms stronger than D1. The discussion takes place in **TSS**. Among other things we shall establish the consistency of D1 with **TSS**.

3101.

(D2) $(\exists D)\,(\mathrm{Real}\,(D)\ \&\ \mathrm{Reg}\,(D)\ \&\ \mathbf{D}\,(D) = \mathbf{On}\ \&\ \mathbf{W}\,(D) = \mathbf{V})$,

(E2) $(\exists F)\,(\mathrm{Real}\,(F)\ \&\ \mathrm{Un}\,(F)\ \&\ \mathbf{D}\,(F) = \mathbf{On}\ \&\ \mathbf{W}\,(F) = \mathbf{V})$.

Both axioms are similar; they postulate the existence of a certain real relation whose domain is **On** and whose range is the universe; axiom D2 requires this relation to be regular while E2 requires it to be a function.

3102. LEMMA (**TSS**). (a) Let D be a real regular relation; define $F = \{\langle y, x\rangle;\ y = \mathbf{Ext}_D\,(x)\}$. Then F is a real mapping, $\mathbf{D}\,(D) = \mathbf{D}\,(F)$ and and $F\text{'}x = \mathbf{Ext}_D\,(x)$ for every $x \in \mathbf{D}\,(D)$.

(b) Let F be a real mapping, define $D = \{\langle y, x\rangle;\ y \in F\text{'}x\}$. Then D is a real regular relation, $\mathbf{D}\,(F) = \mathbf{D}\,(D)$ and $\mathbf{Ext}_D\,(x) = F\text{'}x$ for every $x \in \mathbf{D}\,(F)$.

Proof. The only non-trivial part is the statement that the F defined in (a) and the D defined in (b) are real classes. (a) Let a be a set; we prove that $a \cap F$ is a set. $\mathbf{D}\,(a)$ is a set and so $D \upharpoonright \mathbf{D}\,(a)$ is a set (it is a real semiset). Put $d = D \upharpoonright \mathbf{D}\,(a)$ and define $f = \{\langle y, x\rangle;\ y = \mathbf{Ext}_d\,(x)\}$ (by Metatheorem 1438 f is a set; for $u \in f$ iff $(\exists y, x)\,(u = \langle y, x\rangle\ \&\ y = \mathbf{Ext}_d\,(x))$ and it is easy to see that the last formula is equivalent to a PUP-formula). Now $a \cap F = a \cap f$, so $a \cap F$ is a set.

(b) Similarly we prove that $a \cap D$ is a set for any set a. $F \upharpoonright \mathbf{D}(a)$ is a set, f say, and $d = \{\langle y, x\rangle; y \in f\text{'}x\}$ is a set; then $a \cap D = a \cap d$, so $a \cap D$ is a set.

3103. Theorem (**TSS**). (a) Axiom (E2) implies (D2) and (E1).

(b) Axiom (D2) is equivalent to the statement

$$(\exists F)(\mathrm{Real}(F) \,\&\, \mathrm{Un}(F) \,\&\, \mathbf{D}(F) = \mathbf{On} \,\&\, \bigcup(\mathbf{W}(F)) = \mathbf{V}).$$

Proof. (a) is obvious; (b) follows immediately by the preceding Lemma.

3104. Theorem (**TSS**). Axiom (D2) implies (D1).

Proof. Let X be an arbitrary relation whose domain is a semiset, and let $\mathbf{D}(X) = \varrho$ and let D be a relation satisfying (D2). We prove that there is a semiset relation σ such that $\sigma \subseteq X$ and $\mathbf{D}(X) = \mathbf{D}(\sigma)$. Define $Q = \{\langle \alpha, x\rangle; x \in \varrho \,\&\, \mathbf{Ext}_X(x) \cap D\text{''}\alpha = 0\}$, Q is regular; for if y is an arbitrary element of $\mathbf{Ext}_X(x)$ then there is a γ such that $y \in D\text{''}\gamma$, hence $\mathbf{Ext}_Q(x) \subseteq \gamma$. Further, $\mathbf{D}(Q) \subseteq \varrho$ so that $\mathbf{D}(Q)$ is a semiset. By the axiom (C2) it follows that $\mathbf{W}(Q)$ is a semiset, hence there is an $\alpha \in \mathbf{On}$ such that $\mathbf{W}(Q) \subseteq \alpha$. Let $a = D\text{''}(\alpha + 1)$ (a is a set by the regularity and realness of D). Put $\sigma = X \cap (a \times \varrho)$. Obviously σ is a semiset and $\sigma \subseteq X$. We claim that $\mathbf{D}(\sigma) = \varrho\ (= \mathbf{D}(X))$. Clearly $\mathbf{D}(\sigma) \subseteq \varrho$. If $x \in \varrho$ then $\mathbf{Ext}_Q(x) \subseteq \alpha$; hence there is a y such that $y \in D\text{''}(\alpha + 1)$ and $\langle y, x\rangle \in X$. Consequently, $\langle y, x\rangle \in \sigma$ and $x \in \mathbf{D}(\sigma)$.

3105. *Remark*. The main significance of this Theorem lies in the fact that it allows us to use Metatheorem 1453 as a test for realness of classes in (**TSS**, D2). In particular the model $\mathfrak{Real}$ can be studied as a model in (**TSS**, D2). The latter theory will be denoted by **TSS**″.

3106. Definition (**TSS**″). Let D be a relation satisfying the conditions (D2) (or of (E2)); we define the *rank* of a set as follows:

$$\mathbf{r}_D(x) = \alpha \equiv \langle x, \alpha\rangle \in D \,\&\, (\forall \beta < \alpha)(\langle x, \beta\rangle \notin D).$$

Under the assumption of (D2), the rank of a set x is uniquely determined; the collection of sets which have the same rank as x is a set (by regularity and realness of D). Moreover, under the assumption of (E2), different sets have different ranks. In the following paragraphs we shall work in the theory **TSS** + (D2) with a constant **D** satisfying Axiom (D2).

3107. DEFINITION (**TSS**″). For each real class X, set

$$\mathbf{A}(X) = \{x \in X; \neg (\exists y \in X)(\mathbf{r_D}(y) < \mathbf{r_D}(x))\}$$

(the set of elements of X of least rank).

3108. Remark. $\mathbf{A}(X)$ has the set images property in **TSS** + (D2); in **TSS** + (D2) one can prove

$$\text{Real}(X) \,\&\, X \neq 0 \,.\rightarrow \mathbf{A}(X) \neq 0\,.$$

3109. THEOREM (**TSS**″). If X is a real equivalence then there exists a real $P \subseteq \mathbf{C}(X)$ such that for any $x \in \mathbf{C}(X)$ the class $\{y; \langle y, x\rangle \in X \,\&\, y \in P\}$ is a nonempty set. (We say that P is a *regulator* of X.)

Proof. We define P as follows:

$$y \in P \equiv (\exists x \in \mathbf{C}(X))(y \in \mathbf{A}(\mathbf{Ext}_X(x)))\,.$$

P has the required properties; it selects the set of elements of least rank from each equivalence class.

3110. Remark. We have given instructions for constructing P from X; i.e. we have described an operation which assigns to each equivalence X a class P satisfying the conditions of the theorem.

3111. THEOREM (**TSS**″). For any real relation X there exists a real regular relation $Y \subseteq X$ such that $\mathbf{D}(Y) = \mathbf{D}(X)$.

Proof. We define

$$\langle y, x\rangle \in Y \equiv \langle y, x\rangle \in X \,\&\, y \in \mathbf{A}(\mathbf{Ext}_X(x))\,.$$

Y has the required properties; each $x \in \mathbf{D}(X)$ is related only to elements of least rank.

3112. Remark. We can again define an operation which assigns to each relation X a regular subrelation Y having the same domain. Since we shall need this operation in the sequel, we define it explicitly:

3113. DEFINITION.

$$[\langle y, x\rangle \in \mathbf{Rg}(X) \equiv \langle y, x\rangle \in X \,\&\, y \in \mathbf{A}(\mathbf{Ext}_X(x))] \,\&\, \text{Rel}(\mathbf{Rg}(X))\,.$$

3114. Remark. Notice that in **TSS** Theorem 3111 follows from Theorem 3109 without using (D2). (Exercise.)

3115. METALEMMA. The predicate WOrdg is normal in **TSS**″.

Demonstration. We prove in **TSS**″ that the predicate WOrdg(X) is equivalent to the formula

$$(*) \qquad \mathrm{Real}(X) \,\&\, \mathrm{LOrdg}(X) \,\&\, (\forall u \subseteq \mathbf{C}(X))\,(u \neq 0 \rightarrow \rightarrow (\exists z \in u)(\forall v \in u)(\langle v, z\rangle \notin X)).$$

The statement $(*)$ obviously follows from WOrdg(X). Conversely, suppose that X satisfies $(*)$. Let U be a nonempty real subclass of $\mathbf{C}(X)$ which has no least element; we construct a nonempty set without least element. Let x_0 be some element of U; if $x_0, \ldots, x_n$ are elements of U such that $\langle x_{i+1}, x_i\rangle \in \in X$ $(i < n)$, then we let $u_{n+1} = \mathbf{A}(\{x \in U; \langle x, x_n\rangle \in X\}$. Since x_n is not least in U we have $u_{n+1} \neq 0$. By $(*)$ u_{n+1} has a least element; we let x_{n+1} be the least element of u_{n+1}. The set $\{x_n; n \in \omega\}$ has no least element and hence $(*)$ implies WOrdg(X).

Consider now the theory **TSS** + (E2). Since we can prove D2 in this theory we can also prove all the theorems mentioned above. These theorems can be strengthened in the sense that one can prove in **TSS** + (E2) that the operation $\mathbf{A}(X)$ assigns to each nonempty real class a set having just one element.

3116. DEFINITION (**TSS** + E2).

$$\langle y, x\rangle \in \mathbf{Ac} \equiv x \neq 0 \,\&\, \{y\} = \mathbf{A}(x).$$

Ac is obviously a real function and for any $x \neq 0$ we have $\mathbf{Ac}'x \in x$. We call **Ac** the universal selector.

3117. THEOREM (**TSS** + E2). For any real equivalence X there exists $P \subseteq \mathbf{C}(X)$ such that every equivalence class has exactly one element in common with P.

3118. THEOREM (**TSS** + E2). For any real relation X there exists a real function $Y \subseteq X$ with the same domain as X.

Remark. The proofs of these theorems are similar to the proofs of Theorems 3109, 3111; we can define corresponding operations in a similar fashion.

3119. THEOREM (**TSS**). Axiom (E2) implies (E1).

Proof. Let x be a nonempty set; we shall construct a well-ordering of x using the real mapping F of **On** onto **V** given by E2. We let H be the real function which assigns to each $y \in x$ the least $\alpha \in \mathbf{On}$ such that $\langle y, \alpha \rangle \in F$; H is a one-one mapping of x into **On**. For $y_1, y_2 \in x$ we let $\langle y_1 y_2 \rangle \in r \equiv$ $\equiv (H\text{'}y_1 < H\text{'}y_2)$. The relation r is a well-ordering of x.

Another aim of the present Section is to demonstrate so-called Reflection Principles; this will be done both in **TSS** + D2 and (in a slightly different form) in **TSS** + E2. To do this we show first that the relation D given by Axiom (D2) can be assumed to have certain additional properties.

3120. DEFINITION (**TSS**). A *reflecting system* Q is a real function on **On** with the following properties:

(a) $\bigcup(\mathbf{W}(Q)) = \mathbf{V}$,
(b) $Q\text{'}\alpha$ is complete for any α,
(c) $\mathbf{P}(Q\text{'}\alpha) \subseteq Q\text{'}(\alpha + 1)$ for any α,
(d) $Q\text{'}\lambda = \bigcup_{\alpha<\lambda} Q\text{'}\alpha$ for any limit λ.

3121. LEMMA (**TSS**). If Q is a reflecting system then for any limit λ we have

(1) $Q\text{'}\lambda$ is **V**-like,
(2) if $z \subseteq \lambda$ and $\bigcup(z) = \lambda$ then $Q\text{'}\lambda = \bigcup_{\alpha \in z} Q\text{'}\alpha$.

Proof. The set $Q\text{'}\lambda$ is complete and closed by (b), (c) and (d); hence it is **V**-like. If $\alpha < \beta$ then obviously $Q\text{'}\alpha \subseteq Q\text{'}\beta$; this together with (d) implies (2).

3122. THEOREM (**TSS** + D2). Reflecting systems exist.

Proof. Let D be the relation given by Axiom (D2). We define Q by recursion:

$$Q\text{'}0 = \{0\},$$

$$Q\text{'}(\alpha + 1) = \mathbf{Unv}(\mathbf{P}(Q\text{'}\alpha) \cup D\text{''}\alpha),$$

$$Q\text{'}\lambda = \bigcup_{\alpha<\lambda} Q\text{'}\alpha \quad \text{for limit} \quad \lambda.$$

We have $D\text{''}\{\alpha\} \subseteq Q\text{'}(\alpha + 1)$ and hence $\bigcup(\mathbf{W}(Q)) = \mathbf{V}$. The remaining conditions are also satisfied.

3123. METADEFINITION. I. For any gödelian term $\mathbf{G}(X_1, \bullet)$ containing no constants we define in **TSS**:

$$\mathrm{Refl}_{\mathbf{G}}(Z, X_1, \bullet) \equiv . \mathrm{Vlk}(Z) \,\&\, \mathbf{G}(X_1, \bullet) \cap Z = \mathbf{G}(X_1 \cap Z, \bullet).$$

(There is one definition for each term; if $\mathrm{Refl}_{\mathbf{G}}(Z, X_1, \bullet)$ we say that Z is a **G**-*reflecting class* w. r. t. $X_1, \bullet$.)

II. If $\mathbf{G}_0, \ldots, \mathbf{G}_k$ are all the subterms*) of **G** we define in **TSS**:

$$\mathrm{HRefl}_{\mathbf{G}}(Z, X_1, \bullet) \equiv . \mathrm{Refl}_{\mathbf{G}_0}(Z, X_1, \bullet) \,\&\, \ldots \,\&\, \mathrm{Refl}_{\mathbf{G}_k}(Z, X_1, \bullet).$$

(Z is a *hereditarily* **G**-*reflecting class* w. r. t. $X_1, \bullet$.)

3124. Remarks. 1) If X is a variable then

$$\mathbf{TSS} \vdash \mathrm{HRefl}_X(Z, X) \equiv \mathrm{Vlk}(Z).$$

(Obvious.)

2) If **G** does not contain the operation **D** then

$$\mathbf{TSS} \vdash \mathrm{HRefl}_{\mathbf{G}}(Z, X, \bullet) \equiv \mathrm{Vlk}(Z).$$

(Cf. 1304, 1319.)

To illustrate the meaning of the notion of reflecting classes we demonstrate the following

3125. METATHEOREM. Let $\varphi(x, X, \bullet)$ be a normal formula not containing any constants. By Metatheorem 1122 there is a gödelian term $\mathbf{H}(X, \bullet, Y)$ without constants such that $\mathbf{H}(X, \bullet, \mathbf{V})$ represents φ in **TC** w.r.t. x, i.e. $\mathbf{TC} \vdash x \in \mathbf{H}(X, \bullet, \mathbf{V}) \equiv \varphi(x, X, \bullet)$.

The following is provable in **TSS**, $\mathrm{Vlk}(\mathbf{M})$, $\partial\mathbf{TC}/\partial\mathfrak{S}\mathrm{t}(\mathbf{M})$:

$$(X^* = X \cap \mathbf{M} \,\&\, \bullet) \rightarrow [\mathrm{Refl}_{\mathbf{H}}(\mathbf{M}, X, \bullet, \mathbf{V}) \equiv$$
$$\equiv \{x^*; \varphi^*(x^*, X^*, \bullet)\}^* = \{x; \varphi(x, X, \bullet)\} \cap \mathbf{M}].$$

*) It is clear what is meant by "subterms of a term"; the precise recursive definition reads as follows:

(a) If **G** is an atomic term then **G** is its only subterm.

(b) If **F** is an operation and $\mathbf{t}_1, \ldots, \mathbf{t}_n$ are terms then the subterms of $\mathbf{F}(\mathbf{t}_1, \ldots, \mathbf{t}_n)$ are all subterms of all terms $\mathbf{t}_1, \ldots, \mathbf{t}_n$ together with the term $\mathbf{F}(\mathbf{t}_1, \ldots, \mathbf{t}_n)$ itself.

Demonstration. We proceed in the theory indicated, assuming $X^* = X \cap \mathbf{M}, \bullet$. In fact, $\{x; \varphi(x, X, \bullet)\} \cap \mathbf{M}$ equals $\mathbf{H}(X, \bullet\, \mathbf{V}) \cap \mathbf{M}$ and $\{x^*; \varphi^*(x^*, X^*, \bullet)\}^*$ equals $\mathbf{H}^*(X^*, \bullet, \mathbf{M})$ (since $\mathbf{M} = \mathbf{V}^*$). $\mathrm{Refl}_{\mathbf{H}}(X, \bullet, \mathbf{V})$ is equivalent to $\mathbf{H}(X, \bullet, \mathbf{V}) \cap \mathbf{M} = \mathbf{H}(X \cap \mathbf{M}, \bullet, \mathbf{M}) = \mathbf{H}(X^*, \bullet, \mathbf{V}^*)$. But, by 1322 (b), $\mathbf{H}^*(X^*, \bullet, \mathbf{V}^*) = \mathbf{H}(X^*, \bullet, \mathbf{V}^*)$. This completes the proof.

3126. The preceding metatheorem has some consequences for restricted formulas. Suppose that $\varphi(x, X, \bullet)$ is a restricted **TC**-formula without constants and that $\mathbf{H}(X, \bullet, Y)$ is a gödelian term without constants such that $\mathbf{H}(X, \bullet, \mathbf{V})$ represents $\varphi(x, X, \bullet)$ w.r.t. x in **TC**.

1) $\mathrm{Refl}_{\mathbf{H}}(\mathbf{M}, X, \bullet, \mathbf{V})$ is provable in **TSS**, Vlk ($\mathbf{M}$) & Comp ($\mathbf{M}$), $\partial\mathbf{TC}/\partial\mathfrak{St}(\mathbf{M})$. This follows by the preceding Metatheorem and Metatheorem 1327.

2) Since the theory considered sub 1) is an conservative extension of **TSS** we have the following result:

$$\mathbf{TSS} \vdash (\forall Z)(\mathrm{Vlk}(Z) \,\&\, \mathrm{Comp}(Z) \,.\rightarrow \mathrm{Refl}_{\mathbf{H}}(Z, X, \bullet, \mathbf{V}))\,.$$

In particular we have the following

3127. Metacorollary. Suppose that $\varphi(x, X, \bullet)$ is a restricted **TC**-formula and that $\mathbf{H}(X, \bullet, Y)$ is a gödelian term without constants such that $\mathbf{H}(X, \bullet, \mathbf{V})$ represents $\varphi(x, X, \bullet)$ w.r.t. x in **TC**. Then

$$\mathbf{TSS} \vdash \mathrm{Vlk}(Z) \,\&\, \mathrm{Comp}(Z) \,.\&\, (X \subseteq Z \,\&\, \bullet) \,.\rightarrow$$
$$\rightarrow Z \cap \mathbf{H}(X, \bullet, \mathbf{V}) = \mathbf{H}(X, \bullet, Z)\,.$$

Let us now consider arbitrary gödelian terms. Our aim is to demonstrate metatheorems of the following form: If $\mathbf{G}(X, \bullet)$ is a gödelian term without constants then it is provable that for arbitary X, $\bullet$ there is a "nice" Z such that Vlk (Z) and $\mathrm{Refl}_{\mathbf{G}}(Z, X, \bullet)$. By "nice" we mean that Z is big enough, i.e. contains an arbitrary given infinite set x and either is complete (this will be proved in **TSS**″) or is not too big i.e. has the same cardinality as x (this will be proved in **TSS**, E2).

3128. Metatheorem (Reflection Principle). Let $\mathbf{G}(X_1, \bullet)$ be a gödelian term without constants. Then we can prove the following in **TSS** + D2: Let Q be a reflecting system and let X_1, $\bullet$ be real classes.

1) For any x, there is a limit number λ such that $x \in Q'\lambda$ and $\mathrm{HRefl}_{\mathbf{G}}(Q'\lambda, X_1, \bullet)$.

2) If $\{\lambda_\alpha\}_{\alpha\in\gamma}$ is an increasing sequence of limit numbers and if $(\forall\alpha\in\gamma)\ \mathrm{HRefl}_{\mathbf{G}}(Q'\lambda_\alpha, X_1, \bullet)$ then $\mathrm{HRefl}_{\mathbf{G}}(Q'\lambda, X_1, \bullet)$ where $\lambda = \bigcup\lambda_\alpha$.

Demonstration. If **G** is an atomic term then both statements are obvious, since in **TSS** + D2 we can prove that $\mathrm{Refl}_{\mathbf{G}}(Q'\lambda, X_1, \bullet)$ for any limit number λ. If $\mathbf{G}_1$ and $\mathbf{G}_2$ are gödelian terms then in **TSS** + D2 we prove: if $\lambda\in Z_i \equiv \mathrm{HRefl}_{\mathbf{G}_i}(Q'\lambda, X_1, \bullet)$ then

$$\lambda\in Z_1\cap Z_2 \equiv \mathrm{HRefl}_{\mathbf{G}_1-\mathbf{G}_2}(Q'\lambda, X_1, \bullet) \equiv \mathrm{HRefl}_{\mathbf{G}_1\upharpoonright\mathbf{G}_2}(Q'\lambda, X_1, \bullet).$$

Only the implications from left to right are non-trivial; we use here the equalities from 1304, 1319 and the fact that Vlk (Z) implies $\mathbf{D}(Z) = Z$. If λ_0 is an ordinal number we let λ_{2n+1} be the least number such that $\lambda_{2n+1} > \lambda_{2n}$ and $\mathrm{HRefl}_{\mathbf{G}_1}(Q'\lambda_{2n+1}, X_1, \bullet)$ and we let λ_{2n+2} be the least number such that $\lambda_{2n+2} > \lambda_{2n+1}$ and $\mathrm{HRefl}_{\mathbf{G}_2}(Q'\lambda_{2n+2}, X_1, \bullet)$; we then set $\lambda = \bigcup_n \lambda_n$. We have $\mathrm{HRefl}_{\mathbf{G}_1}(Q'\lambda, X_1, \bullet)$ and $\mathrm{HRefl}_{\mathbf{G}_2}(Q'\lambda, X_1, \bullet)$. Hence $Z_1\cap Z_2$ is a proper class.

Further we have

$$\lambda\in Z_1 \equiv \mathrm{HRefl}_{\mathbf{E}(\mathbf{G}_1)}(Q'\lambda, X_1, \bullet) \equiv$$
$$\equiv \mathrm{HRefl}_{\mathbf{Cnv}(\mathbf{G}_1)}(Q'\lambda, X_1, \bullet) \equiv \mathrm{HRefl}_{\mathbf{Cnv_3}(\mathbf{G}_1)}(Q'\lambda, X_1, \bullet).$$

We again use the equalities from 1304, 1319; in the case of $\mathbf{Cnv}(X)$ we have $Z\cap\mathbf{Cnv}(X) = \mathbf{Cnv}(X\cap Z')$ where $Z' = \mathbf{Cnv}(Z)\cup\{y\in Z; y$ is not an ordered pair$\}$. If Z is **V**-like then $Z' = Z$. The case of $\mathbf{Cnv}_3$ is treated similarly.

Suppose now that $\lambda\in Z \equiv \mathrm{HRefl}_{\mathbf{D}(\mathbf{G}_1)}(Q'\lambda, X_1, \bullet)$. Obviously:

(1) Let $S\subseteq\mathbf{G}_1(X_1, \bullet, \mathbf{V})$ be a real regular relation with the same domain as $\mathbf{G}_1$. If λ_0 is the least λ such that $x\in Q'\lambda$ then we let λ_{2n+1} be the least λ greater than λ_{2n} and belonging to Z_1 and we let λ_{2n+2} be the least λ greater than λ_{2n+1} such that $S\upharpoonright Q'\lambda_{2n+1}\subseteq Q'\lambda_{2n+2}$; we then set $\lambda = \bigcup_n\lambda_n$. If $\mathbf{G}_1(X_1, \bullet)$ is denoted by R we have

$$\mathbf{D}(R\cap Q'\lambda) = \mathbf{D}(R)\cap Q'\lambda.$$

The inclusion $\subseteq$ is obvious; we prove the converse. If $u\in\mathbf{D}(R)\cap Q'\lambda$ then $u\in\mathbf{D}(S)\cap Q'\lambda$, i.e. $u\in\mathbf{D}(S)\cap Q'\lambda_{2n+1}$ for some n; hence there is a v such that $\langle v, u\rangle\in S\cap Q'\lambda_{2n+2}$ and so $\langle v, u\rangle\in R\cap Q'\lambda$ and $u\in\mathbf{D}(R\cap$

$\cap Q'\lambda$). Thus we have

$$Q'\lambda \cap \mathbf{D}(\mathbf{G}_1(X_1, \bullet)) = \mathbf{D}(Q'\lambda \cap \mathbf{G}_1(X_1, \bullet)) =$$
$$= \mathbf{D}(\mathbf{G}_1(X_1 \cap Q'\lambda, \bullet))$$

and hence $\mathrm{HRefl}_{\mathbf{D}(\mathbf{G}_1)}(Q'\lambda, X_1, \bullet)$ and $\lambda \in Z$.

(2) Suppose that λ_α and λ satisfy the conditions of the assertion; denote $Q'\lambda_\alpha$ and $Q'\lambda$ by x_α and x respectively. We want to prove

$$x \cap \mathbf{D}(\mathbf{G}_1(X_1, \bullet)) = \mathbf{D}(\mathbf{G}_1(x \cap X_1, \bullet)).$$

The right-hand side is equal to $\mathbf{D}(x \cap \mathbf{G}_1(X_1, \bullet))$ and is therefore contained in the left-hand side. If $u \in x \cap \mathbf{D}(\mathbf{G}_1(X_1, \bullet))$ then there is an α such that $u \in x_\alpha \cap \mathbf{D}(\mathbf{G}_1(X_1, \bullet)) = \mathbf{D}(x_\alpha \cap \mathbf{G}(X_1, \bullet)) \subseteq \mathbf{D}(x \cap \mathbf{G}_1(X_1, \bullet))$.

Let us consider Axiom (E2) again. In addition to Metatheorem 3128 which of course also holds for **TSS** + (E2), we shall prove another slightly different Reflection Principle.

3129. Metatheorem. The following may be proved in **TSS** + E2 for any godelian term $\mathbf{G}(X_1, \bullet)$ without constants: Let $X_1, \bullet$ be real classes.

(1) For any x, there exists $y \supseteq x$ such that

$$\overline{\overline{y}} = \max(\overline{\overline{x}}, \aleph_0) \quad \text{and} \quad \mathrm{HRefl}_{\mathbf{G}}(y, X_1, \bullet).$$

(2) If $\{x_\alpha\}_{\alpha\in\lambda}$ is a sequence of sets such that $x_\alpha \subseteq x_\beta$ whenever $\alpha < \beta$ and such that $\mathrm{HRefl}_{\mathbf{G}}(x_\alpha, X_1, \bullet)$ for any $\alpha < \lambda$, then $\mathrm{HRefl}_{\mathbf{G}}(\bigcup_\alpha x_\alpha, X_1, \bullet)$.

Demonstration. In the case of atomic terms we prove the following in **TSS** + E2: $(\forall x)(\exists y)(\mathrm{Vlk}(y) \,\&\, x \subseteq y \,\&\, \overline{\overline{y}} = \mathbf{Max}(\overline{\overline{x}}, \aleph_0))$. Let A be a universal selector. We set

$u_0 = x,\ u_{n+1} = u_n \cup \{z; (\exists p, q \in u_n)(z = \{p, q\} \vee z = p - q \vee z = A'p\}$ and $y = \bigcup_n u_n$; obviously y is V-like and $\overline{\overline{y}} = \mathbf{Max}(\overline{\overline{x}}, \aleph_0)$. The induction step is treated in the same way as in the preceding Metatheorem. In the case of $\mathbf{G}_1 - \mathbf{G}_2$ and $\mathbf{G}_1 \upharpoonright \mathbf{G}_2$ we proceed as follows. Suppose that $\overline{\overline{x}} = \aleph_\alpha$. Set

$$y_0 = x\,,\ y_{2n+1} = \mathbf{A}(\{z \supseteq y_{2n}; \mathrm{HRefl}_{\mathbf{G}_1}(z, X_1, \bullet) \,\&\, \overline{\overline{z}} = \aleph_\alpha\}),$$
$$y_{2n+2} = \mathbf{A}(\{z \supseteq y_{2n+1}; \mathrm{HRefl}_{\mathbf{G}_2}(z, X_1, \bullet) \,\&\, \overline{\overline{z}} = \aleph_\alpha\}).$$

If $y = \bigcup_n y_n$ and if $Z_i = \{z; \mathrm{HRefl}_{\mathbf{G}_i}(z, X_1 \bullet)\}$ then we have $y \in Z_1 \cap Z_2$. In the case of $\mathbf{D}(\mathbf{G}_1)$ we proceed as follows. Suppose that $\overline{\overline{x}} = \aleph_\alpha$ and that $S \subseteq \mathbf{G}_1(X_1, \bullet)$ is a real function with the same domain as $\mathbf{G}_1(X_1, \bullet)$. We let

$$y_0 = x\,,\ y_{2n+1} = \mathbf{A}(\{z \supseteq y_{2n}; \mathrm{HRefl}_{\mathbf{G}_1}(z, X_1, \bullet) \,\&\, \overline{\overline{z}} = \aleph_\alpha\})\,,$$

$y_{2n+2} = y_{2n+1} \cup S \upharpoonright y_{2n+1}$. If $y = \bigcup_n y_n$ then we have $\mathrm{HRefl}_{\mathbf{D}(\mathbf{G}_1)}(y, X_1, \bullet)$.

Remark. We may impose additional conditions on the reflecting set. As an example which will be useful in Section 5 we demonstrate here the following modification of the Reflection Principle:

3130. METATHEOREM. The following can be proved in **TSS** + E2 for any gödelian term $\mathbf{G}(X_1, \bullet)$: Let $X_1, \bullet$ be real classes. If $\aleph_\alpha$ is a regular cardinal then, for any x such that $\overline{\overline{x}} < \aleph_\alpha$ there exists $y \supseteq x$ such that $\overline{\overline{y}} < \aleph_\alpha$, $\mathrm{Refl}_{\mathbf{G}}(y, X_1, \bullet)$ and $y \cap \omega_\alpha$ is a complete set (a fortiori, an ordinal number).

Demonstration. In **TSS** + E2, we let $y_0 = x$, $y_{2n+1} = \mathbf{A}(\{z \supseteq y_{2n}; \mathrm{HRefl}_{\mathbf{G}}(z, X_1, \bullet) \,\&\, \overline{\overline{z}} = \overline{\overline{y_{2n}}}\})$, $y_{2n+2} = y_{2n+1} \cup \bigcup(y_{2n+1} \cap \omega_\alpha)$ and $y = \bigcup y_n$; we have $\overline{\overline{y}} < \aleph_\alpha$, $\mathrm{HRefl}_{\mathbf{G}}(y, X_1, \bullet)$ and $\mathrm{Comp}(y \cap \omega_\alpha)$.

We have seen that Axiom D2 is equivalent to the existence of reflecting systems. We shall now present the third axiom of regularity, D3, which is stronger than D2 and is equivalent to the assumption that a certain well-defined class is a reflecting system.

3131. DEFINITION (**TSS**).

$$\mathscr{P}'0 = 0\,,$$

$$\mathscr{P}'(\alpha + 1) = \mathbf{P}(\mathscr{P}'\alpha)\,,$$

$$\mathscr{P}'\lambda = \bigcup_{\beta < \lambda} \mathscr{P}'\beta \text{ if } \lambda \text{ is a limit number}\,,$$

$$\mathbf{Ker} = \bigcup_{\alpha \in \mathbf{On}} \mathbf{P}'\alpha = \bigcup(\mathbf{W}(\mathscr{P}))\,.$$

(The class **Ker** is called the *regular kernel.*)

Remark. This definition is correct by Metatheorem 2130. We take $G_1'x = \mathbf{P}(x)$, $G_2'x = \bigcup(x)$. (Notice that G_1, G_2 are definable by PUP-formulas.) By the Metatheorem $\mathscr{P}$ is a real class.

3132.

(D3) $$\mathbf{Ker} = \mathbf{V}\,.$$

3133. THEOREM (**TSS**). Axiom (D3) implies (D2).

Proof. By Theorem 3103 (the second formulation of (D2)).

Thus (D3) also implies (D1).

More generally we define:

3134. DEFINITION (**TSS**). $\mathscr{P}_x'0 = x$, $\mathscr{P}_x'(\alpha + 1) = \mathbf{P}(\mathscr{P}_x'\alpha)$, $\mathscr{P}_x'\lambda = \bigcup_{\alpha<\lambda} \mathscr{P}_x'\alpha$ for λ limit; $\mathbf{Ker}\,(x) = \bigcup_{\alpha\in\mathbf{On}} \mathscr{P}_x'\alpha$.

Thus, $\mathbf{Ker} = \mathbf{Ker}\,(0)$. We shall write p_α^x instead of $\mathscr{P}_x'\alpha$ and p_α instead of p_α^0. In order to prove the consistency of (D3) with **TSS** we shall investigate the sets p_α^x.

3135. LEMMA (**TSS**). Let x be a complete set.

(a) p_α^x is complete for every α, and the class $\mathbf{Ker}\,(x)$ is complete.

(b) $\alpha \leqq \beta$ implies $p_\alpha^x \subseteq p_\beta^x$ for all α, β.

(c) $\mathbf{Ker}\,(x)$ is an almost universal class; $\mathbf{Ker}\,(x)$ is not a semiset; $\mathbf{Ker}\,(x)$ is a real class.

(d) For every $u \in \mathbf{Ker}\,(x)$ there is a minimal α such that $u \in p_\alpha^x$; we denote this α by $\tau(u)$ and call the *rank* of u. (Note that no rank is a limit number.)

(e) For any $u, v \in \mathbf{Ker}\,(x)$, $\tau(v) > 0$ and $u \in v$ implies $\tau(u) < \tau(v)$.

(f) For every non-void $z \subseteq \mathbf{Ker}\,(x)$ there is a $u \in z$ with the minimal rank.

(g) For any u, $u \in \mathbf{Ker}\,(x)$ iff $u \subseteq \mathbf{Ker}\,(x)$.

Proof. (a) Let α be an arbitrary ordinal number. Suppose that p_α^x is not complete. Let $A = \{\beta \leqq \alpha;\ p_\beta^x \text{ is not complete}\}$. Note that $\boldsymbol{\mathscr{P}}_x$ is a real mapping and so $\boldsymbol{\mathscr{P}}_x \restriction (\alpha + 1)$ is a set; hence A is a set. Let γ be the minimum of A; then $\gamma > 0$. Either γ is limit and $p_\gamma^x = \bigcup_{\delta<\gamma} p_\delta^x$ or $\gamma = \delta_0 + 1$ and $p_\gamma^x = \mathbf{P}\,(p_{\delta_0}^x)$; in either case we obtain a contradiction.

(b) Let $\alpha < \beta$ and suppose $p_\alpha^x \nsubseteq p_\beta^x$. Let $A = \{\delta \leqq \beta; (\exists\gamma < \delta)\,(p_\gamma \nsubseteq p_\delta)\}$. A is a set; let δ be minimal in A and let $\gamma < \delta$ be such that $p_\gamma^x \nsubseteq p_\delta^x$. If $\delta = \delta_0 + 1$ then $p_\gamma^x \subseteq p_{\delta_0}^x$ and $\mathbf{P}\,(p_\gamma^x) \subseteq \mathbf{P}\,(p_{\delta_0}^x) = p_\delta^x$; but $p_\gamma^x \subseteq \mathbf{P}\,(p_\gamma^x)$ by the completeness of p_γ^x, hence $p_\gamma^x \subseteq p_\delta^x$. If δ is a limit then $p_\delta^x = \bigcup_{\eta<\delta} p_\eta^x$, hence $p_\gamma^x \subseteq p_\delta^x$. In either case we have a contradiction.

(c) Let $\sigma \subseteq \mathbf{Ker}\,(x)$ and put $R = \{\langle\alpha, u\rangle; u \in \sigma - p_\alpha^x\}$. R is regular and $\mathbf{D}\,(R)$ is a semiset, hence $\mathbf{W}\,(R)$ is a semiset of ordinal numbers. Let γ be such that $\mathbf{W}(R) \subseteq \gamma$; then $\sigma \subseteq p_\gamma^x$ and $p_\gamma^x \in p_{\gamma+1}^x \subseteq \mathbf{Ker}(x)$. Hence $\mathbf{Ker}(x)$ is an almost universal class. If $\mathbf{Ker}(x)$ were a semiset then there would be a γ such that $\mathbf{Ker}(x) \subseteq p_\gamma^x$, hence $\mathbf{Ker}(x) = p_\gamma^x$. But $p_{\gamma+1}^x = \mathbf{P}(p_\gamma^x) \succ p_\gamma^x$ and $p_{\gamma+1}^x \subseteq \mathbf{Ker}(x)$, a contradiction. Real $(\mathbf{Ker}(x))$ follows by 1440.

(d) Let $u \in \mathbf{Ker}(x)$, $u \in p_\alpha^x$. Put $A = \{\beta \leqq \alpha; u \in p_\beta^x\}$. Obviously $\beta \in A$ iff $(\exists q \in \mathscr{P}_x \restriction (\alpha + 1))(\exists v)(q = \langle v, \beta\rangle \mathbin{\&} u \in v)$ and $\mathscr{P}_x \restriction (\alpha + 1)$ is a set; hence A is a set (by Metatheorem 1431). Let α_0 be minimal in A; then $\tau(x) = \alpha_0$ has the required property.

(e) If $\tau(v) > 0$ and $u \in v$ then $\tau(v) = \gamma + 1$ for some γ, hence $v \in \mathbf{P}(p_\gamma^x)$, $v \subseteq p_\gamma^x$ and $u \in p_\gamma^x$. Consequently, $\tau(u) \leqq \gamma < \tau(v)$.

(f) Let $z \subseteq \mathbf{Ker}(x)$. Then $z \subseteq p_\gamma^x$ for some γ. Put $A = \{\alpha; (\exists q \in \mathscr{P}_x \restriction (\gamma + 1))(\exists u \in z)(\exists v)(q = \langle v, \alpha\rangle \mathbin{\&} u \in v)\}$. A is a set and has a minimal element α; there is a $u \in z$ such that $u \in p_\alpha^x$; then $\tau(u) = \alpha$ and no element of z has a smaller rank.

(g) The implication $\rightarrow$ is the completeness, see (a). If $u \subseteq \mathbf{Ker}(x)$ then $u \subseteq p_\gamma^x$ for some γ (by the proof of (c)), hence $u \in p_{\gamma+1}^x$, $p_{\gamma+1}^x \subseteq \mathbf{Ker}(x)$.

3136. Theorem (**TSS**). Axiom (D3) is equivalent to either of the following statements:

(a) For every non-void x, there is a $y \in x$ such that $y \cap x = 0$.

(b) For every non-void real class X, there is a $y \in X$ such that $y \cap X = 0$.

Proof. Suppose (D3). Let y be an element of x with the minimal rank. $\tau(y)$ cannot be 0 because there are no elements, of the rank 0. Hence, for any $z \in y$, $\tau(z) < \tau(y)$ and therefore $z \notin x$.

We show that (a) implies (b). Suppose that X is a real class such that $u \cap X \neq 0$ for every $u \in X$. Let u be an arbitrary element of X and set $x = \mathbf{Unv}(u) \cap X$. Then for any $y \in x$ we have $y \cap x \neq 0$, which contradicts (a).

Suppose (b). If $\mathbf{V} - \mathbf{Ker} \neq 0$ let $y \in (\mathbf{V} - \mathbf{Ker})$ be such that $y \cap (\mathbf{V} - \mathbf{Ker}) = 0$; we have $y \subseteq \mathbf{Ker}$ and hence $y \in \mathbf{Ker}$, a contradiction.

3137. Theorem (**TSS**). If x is complete then $\mathbf{Ker}(x)$ is a model class.

Proof. $\mathbf{Ker}(x)$ is a complete and almost universal class. Further more $\mathbf{Ker}(x)$ is a closed class. Indeed, suppose $z, y \in \mathbf{Ker}(x)$; then there is a γ such that $z, y \in p_\gamma^x$; moreover we may suppose γ to be a limit number; hence $\mathbf{E}(z)$, $z - y$, $\mathbf{D}(z)$, $z \restriction y$, $\mathbf{Cnv}(z)$, $\mathbf{Cnv}_3(z) \in p_\gamma^x$.

3138. Metatheorem. The specification of $\mathfrak{St}(\mathbf{M})$ by $\mathbf{M} = \mathbf{Ker}$ determines an essentially faithful model of **TSS** + (D3) in **TSS**.

Demonstration. Consider $\mathbf{M} = \mathfrak{Dir}(\partial\mathbf{TSS}/\partial\mathfrak{St}(\mathbf{M}))$ in **TSS**, $\mathbf{M} = \mathbf{Ker}$, $\partial\mathbf{TSS}/\partial\mathfrak{St}(\mathbf{M})$. By Theorem 3137 and Metatheorem 1445 $\mathfrak{M}$ is a model of $\mathbf{TSS}_{-\infty}$ in **TSS**. By 1327 every RF is absolute. We prove (C1)*.

Let a be a set satisfying (C1) and let $a' = a \cap \mathbf{Ker}$ (a' is a set because **Ker** is real). We have $a' \subseteq \mathbf{Ker}$, hence $a' \in \mathbf{Ker}$, $0 \in a'$ and if $x \in a'$ then $x \in p_\gamma$ for some γ; hence $\{x\} \in a$, $\{x\} \in p_{\gamma+1}$, $\{x\} \in a'$. Denote the theory **TSS**, $\mathbf{M} = \mathbf{Ker}$, $\partial\mathbf{TSS}/\partial\mathfrak{St}(\mathbf{M})$ by $\mathbf{TSS}_1$. We now show that

$$\mathbf{TSS}_1 \vdash (\forall x^*)(x^* \neq 0^* \rightarrow (\exists y^* \in^* x^*)(y^* \cap^* x^* = 0^*)).$$

The last formula is equivalent to $(\forall x \in \mathbf{Ker})(x \neq 0 \rightarrow (\exists y \in x)(y \cap x = 0))$ in $\mathbf{TSS}_1$, and this formula can be proved in $\mathbf{TSS}_1$ in the same way as the implication (D3) $\rightarrow$ (a) in Theorem 3136. Consequently (D3) holds in the model and $\mathfrak{M}$ is a model of **TSS** + (D3) in $\mathbf{TSS}_1$.

Finally, consider $\mathfrak{M}$ as a direct model in **TSS**, D3, $\mathbf{M} = \mathbf{Ker}$, $\partial\mathbf{TSS}/\partial\mathfrak{St}(\mathbf{M})$ and call it $\mathfrak{M}$'. This model is evidently equivalent to the identical model and is therefore essentially faithful; consequently $\mathfrak{M}$ is also essentially faithful.

3139. Corollary. (D3) is consistent with **TSS**. (A fortiori, (D2) and (D1) are consistent with **TSS**.)

We shall sometimes make use of the so-called transfinite non-empty powers defined as follows:

3140. (1) Definition (**TSS**).

$$\bar{p}_0^x = x,\ \bar{p}_{\alpha+1}^x = \mathbf{P}\big(\bigcup_{\beta \leqq \alpha} \bar{p}_\beta^x\big) - \{0\},\quad \bar{p}_\lambda^x = \bigcup_{\alpha<\lambda} \bar{p}_\alpha^x$$

if λ is a limit;

$$\mathbf{Ker}'(x) = \bigcup_{\alpha \in \mathbf{On}} \bar{p}_\alpha^x$$

(2) Lemma (**TSS**). (a) $\mathbf{Ker}'(0) = 0$.

(b) $0 \in x \rightarrow \mathbf{Ker}'(x) = \mathbf{Ker}(x)$.

(c) $0 \notin x \rightarrow 0 \notin \mathbf{Ker}'(x)$.

Proof. (a) Every $\bar{p}_\alpha^0$ is empty. (b) For every α, $p_\alpha^x = \bigcup_{\beta \leqq \alpha} \bar{p}_\beta^x$. (c) 0 is not an element of any $\bar{p}_\alpha^x$.

We shall now present a generalization of Axiom D3. We first give the following

3141. DEFINITION (**TSS**). A set x is an *urelement* (Urel (x)) if $x = \{x\}$. We denote by **Ur** the class of all urelements.

3142. Remark. **Ur** is a real class; for $x \in \mathbf{Ur}$ iff $(\forall y \in x)(y = x) \,\&$ $\& \,(\exists y \in x)(y = x)$ and the last formula is restricted.

3143. Axioms concerning urelements.

(U1) $\mathbf{V} = \bigcup_{x \subseteq \mathbf{Ur}} \mathbf{Ker}(x)$,

(U2) $\mathbf{V} = \bigcup_{x \subseteq \mathbf{Ur}} \mathbf{Ker}(x) \,\&\, (\mathrm{D2})$,

(U3) $\mathrm{M}(\mathbf{Ur}) \,\&\, \mathbf{V} = \mathbf{Ker}(\mathbf{Ur})$.

3144. Remark. U1 states that for every set y there is a set x of urelements such that $y \in \mathbf{Ker}(x)$. Note that the following can be proved in **TSS**:

$$(\mathrm{U2}) \equiv .\, (\mathrm{U1}) \,\&\, (\exists R)(\mathrm{Real}(R) \,\&\, \mathrm{Reg}(R) \,\&\, \mathbf{D}(R) \subseteq \mathbf{On} \,\&\, \mathbf{W}(R) = \mathbf{Ur}).$$

3145. LEMMA (**TSS**). (a) (U3) $\rightarrow$ (U2) $\rightarrow$ (U1).

(b) (D3) $\rightarrow$ (U3).

Obvious.

3146. METATHEOREM. The formula Ord (x) is a RSF in **TSS** + (U1).

Demonstration. We can prove in **TSS** + (U1) that

Ord $(x) \equiv \mathrm{Comp}(x) \,\&\, (\forall u, v \in x)(u \in v \vee u = v \vee v \in u \,\&\, x \cap \mathbf{Ur} = 0)$. For, if we assume the right-hand side and if $0 \neq y \subseteq x$ and $x \in \mathbf{Ker}(z)$ then there exists an element of y of least rank in $\mathbf{Ker}(z)$ which is disjoint from y. The formula on the right is restricted since

$$x \cap \mathbf{Ur} = 0 \equiv \neg\, (\exists y \in x)(y = \{y\}).$$

3147. METATHEOREM. Axiom (C1) holds in $\mathfrak{Dir}(\partial\mathbf{TSS}/\partial\mathfrak{St}(\mathbf{M}))$ as a model in **TSS**, U1, Mcl $(\mathbf{M})$, $\partial\mathbf{TSS}/\partial\mathfrak{St}(\mathbf{M})$.

Demonstration. By 1327 all RF's are absolute and therefore, by the preceding Metatheorem, Ord* $(x) \equiv$ Ord (x) whenever $x \in \mathbf{M}$. But if $\alpha \in \mathbf{On}$ then $\alpha \in \mathbf{M}$; for $\alpha \notin \mathbf{M}$ implies $(\forall \beta \geqq \alpha)(\beta \notin \mathbf{M})$ by the completeness of **M**. Hence $\mathbf{On} - \mathbf{M} \neq 0$ implies $\mathrm{M}(\mathbf{On}^*)$. But this implies $\mathrm{M}^*(\mathbf{On}^*)$ by the almost-universality of **M**, which is a contradiction since $\neg\, \mathrm{M}(\mathbf{On})$ is provable in **TSS** without using (Cl). Hence $\boldsymbol{\omega} \in \mathbf{M}$, Ord* $(\boldsymbol{\omega})$, $\boldsymbol{\omega} \in \mathbf{On}_{\mathrm{II}}^*$ (obvious). We have $(\mathbf{On}_{\mathrm{II}} \neq 0)^*$ which implies (Cl)* by 2135.

In Section 4 we shall strengthen the axiom (E2) to (E3) in such a way as we strengthened (D2) to (D3) in the present Section. We complete this Section with the following Figure depicting the interdependence of the axioms of groups D, E and U:

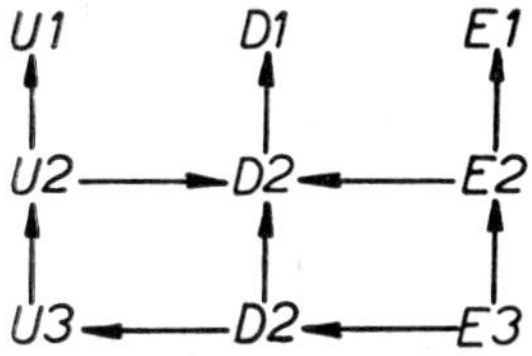

SECTION 2

Model-classes and set-universal relations. Independence of Axiom (D3)

The remaining Sections of Chapter III will be devoted to model-classes and some relations in **TSS′** and certain stronger theories. (In the rest of this book we shall seldom work with theories weaker than **TSS′**.) In the present Section, in addition to giving several general theorems on model-classes, we shall show how to construct certain types of model-relations; in this way we shall prove the independence of (D3) and the consistency of various strengthenings of (U1) concerning the size of the class **Ur**.

First we generalize the operation **Ker** (x); we define **Ker** (X) also for real classes.

3201. Lemma (**TSS′**). $x \subseteq y \rightarrow p_\alpha^x \subseteq p_\alpha^y$.

Proof. Suppose not; let $\alpha \in A \equiv p_\alpha^x \nsubseteq p_\alpha^y$. A is a real class so let α be the smallest element of A. Clearly α is not 0. If $\alpha = \beta + 1$ then $p_\beta^x \subseteq p_\beta^y$ and $p_\alpha^x = \mathbf{P}(p_\beta^x) \subseteq \mathbf{P}(p_\beta^y) = p_\alpha^y$. If α is limit then $p_\alpha^x = \bigcup_\alpha p_\beta^x \subseteq \bigcup_\alpha p_\beta^y = p_\alpha^y$. In either case we obtain a contradiction.

3202. Corollary (**TSS′**). (1) $x \subseteq y \rightarrow \mathbf{Ker}(x) \subseteq \mathbf{Ker}(y)$,

(2) $\mathbf{Ker}(x) = \bigcup_{y \subseteq x} \mathbf{Ker}(y)$.

3203. Definition (**TSS′**). For any real class X, $\mathbf{Ker}(X) = \bigcup_{y \subseteq X} \mathbf{Ker}(y)$.

Remark (**TSS′**). It follows by the preceding corollary that in case X is a set this definition coincides with Definition 3134.

3204. Theorem (**TSS′**). Let X be a real class. Then **Ker** (X) is the smallest real class Z such that $X \subseteq Z$ and $\mathbf{P}(Z) \subseteq Z$.

Proof. Suppose $X \subseteq Z \,\&\, \mathbf{P}(Z) \subseteq Z \,\&\, \mathbf{Real}(Z)$. Then by induction we have $x \subseteq X \rightarrow p_\alpha^x \subseteq Z$ so $\mathbf{Ker}(X) \subseteq Z$. On the other hand $X \subseteq \mathbf{Ker}(X)$ and we show that $\mathbf{P}(\mathbf{Ker}(X)) \subseteq \mathbf{Ker}(X)$. Indeed, let $z \subseteq \mathbf{Ker}(X)$ and set

$$Q = \{\langle u, v\rangle; u \subseteq X \,\&\, v \in z \,\&\, v \in \mathbf{Ker}(u)\}.$$

By (D1), there is a semiset relation Q_0 such that $Q_0 \subseteq Q$ and $\mathbf{D}(Q_0) = \mathbf{D}(Q)$; then $\bigcup \mathbf{W}(Q_0)$ is a subsemiset of X and there is a $\bar{u} \subseteq X$ such that $\bigcup \mathbf{W}(Q_0) \subseteq \bar{u}$. It follows that $z \subseteq \mathbf{Ker}(\bar{u})$, hence $z \in \mathbf{Ker}(\bar{u}) \subseteq \mathbf{Ker}(X)$. Thus $\mathbf{P}(\mathbf{Ker}(X)) \subseteq \mathbf{Ker}(X)$.

3205. Corollary (**TSS′**). (1) For any real X,

$$\mathbf{V} = \mathbf{Ker}(X) \equiv (\forall Z)(\mathrm{Real}(Z) \,\&\, X \subseteq Z \,\&\, \mathbf{P}(Z) \subseteq Z \,.\!\rightarrow Z = \mathbf{V}).$$

(2) $\mathrm{Real}(X) \rightarrow \mathbf{Ker}(X)$ is almost universal.

Proof. (1) is obvious. (2) If $\sigma \subseteq \mathbf{Ker}(X)$ and $\sigma \subseteq a$, then $\sigma \subseteq \mathbf{Ker}(X) \cap a$ and $\mathbf{Ker}(X) \cap a$ is a set, b say. ($\mathbf{Ker}(X)$ is real.) Hence it suffices to show that every subset of $\mathbf{Ker}(X)$ is (a subset of) an element of $\mathbf{Ker}(X)$. But this was shown in the preceding theorem.

3206. Theorem (**TSS′**). For every real complete class X, $\mathbf{Ker}(X)$ is a model class.

The proof is by the preceding corollary and by 3137.

Thus, the fixing of constants $\mathbf{X}, \mathbf{M}$ by $\mathrm{Real}(\mathbf{X}) \,\&\, \mathrm{Comp}(\mathbf{X}) \,\&\, \mathbf{M} = \mathbf{Ker}(\mathbf{X})$ is a specification for $\mathfrak{St}(\mathbf{M})$ in **TSS′**.

3207. Metatheorem. Consider the model $\mathfrak{Dir}(\partial\mathbf{TSS}/\partial\mathfrak{St}(\mathbf{M})$ as a model in **TSS′**, $(\mathrm{Real}(\mathbf{X}) \,\&\, \mathrm{Comp}(\mathbf{X}) \,\&\, \mathbf{M} = \mathbf{Ker}(\mathbf{X}))$, $\partial\mathbf{TSS}/\partial\mathfrak{St}(\mathbf{M})$.

(1) If φ is a PUP-formula (cf. 1437) then φ is absolute in this model.

(2) Moreover, the formulas $\mathrm{M}(X)$ and $\mathrm{Sm}(X)$ are absolute.

Demonstration. (1) It suffices to show (in the theory indicated) that $\mathbf{P}^*(x^*) = \mathbf{P}(x^*)$, $\bigcup^* x^* = \bigcup x^*$ and $\{x^*, y^*\}^* = \{x^*, y^*\}$ for all x^*, y^*. The last two equalities are evident (cf. 1327); the first follows from the fact that $z \subseteq \mathbf{Ker}(X) \rightarrow z \in \mathbf{Ker}(X)$ is provable in **TSS′**.

(2) With regard to $\mathrm{Sm}(X)$ see 1445; $\mathrm{M}^*(X^*) = \mathrm{M}(X^*)$ follows from the provability of $z \subseteq \mathbf{Ker}(X) \equiv z \in \mathbf{Ker}(X)$.

3208. Corollary. $\mathfrak{St}(\mathbf{M})$ with $\mathrm{Real}(\mathbf{X}) \,\&\, \mathrm{Comp}(\mathbf{X}) \,\&\, \mathbf{M} = \mathrm{Ker}(\mathbf{X})$ determines a model of (**TS**, D1) in (**TS**, D1).

Demonstration. The formula $\mathrm{Sm}(X) \rightarrow \mathrm{M}(X)$ is provable in (**TS**, D1); hence by the preceding Metalemma, $\mathrm{Sm}^*(X^*) \rightarrow \mathrm{M}^*(X^*)$ is also provable.

3209. Remarks. (1) The same discussion of **Ker** (X) is possible in **TSS** (without D1); but proofs are longer. As we have already mentioned, we shall in future always assume D1; nevertheless the reader may prove all preceding theorems in **TSS** and demonstrate all the preceding metatheorems for **TSS**.

(2) The reader may give a definition for **Ker** (X) (in **TSS′**) as a class iteration of the operation **P** (X) using Metatheorem 2145.

3210. METATHEOREM on $\in$-induction. The following can be proved in **TSS′** for any normal formula $\varphi(x, X_1, \bullet)$:

$$\text{Real}(X) \,\&\, (\text{Real}(X_1) \,\&\, \bullet) \,\&\, (\forall z \in X)\, \varphi(z, X_1, \bullet) \,\&$$
$$\&\, (\forall x)\, [(\forall y \in x)\, \varphi(y, X_1, \bullet) \to \varphi(x, X_1, \bullet)] \,.\to (\forall z \in \mathbf{Ker}(X))\, \varphi(z, X_1, \bullet)\,.$$

Demonstration. Use Theorem 3204.

3211. THEOREM (**TSS′**) (Construction by $\in$-recursion). If F is a real function on a complete real class X and if K is a real function then there exists a unique real function G such that $\mathbf{D}(G) = \mathbf{Ker}(X)$, $G \upharpoonright X = F$ and $G'x = K'(G''x)$ for any $x \in \mathbf{Ker}(X) - X$.

Proof. We fix F, K and X. We let $H''\{0\} = F$ and

$$H''\{\alpha\} = \{\langle z, x\rangle;\, x \subseteq \mathbf{D}(H''\alpha) \,\&\, x \notin \mathbf{D}(F) \,\&\, z = K'[(H''\alpha)''\, x]\}\,.$$

It may be shown that the operation

$$\mathbf{T}(Y - X) = \{\langle z, x\rangle;\, x \subseteq \mathbf{D}(Y - X) \,\&\, z = K'(Y''x)\}$$

is iterable, Y being e.g. a variable for a regular real relation. (Exercise.) Further we let $G = \mathbf{W}(H)$, so that $\text{Un}(G, \mathbf{Ker}(X), \mathbf{V})$, $G \upharpoonright X = F$ and $G'x = K'(G''x))$ for every $x \in \mathbf{Ker}(X) - X$.

3212. DEFINITION (**TSS′**). A real relation R is *well-founded* (WF (R)) if for any nonempty $x \subseteq \mathbf{C}(R)$ there is some $y \in x$ such that $\mathbf{Ext}_R(y) \cap x = 0$.

3213. Remark (**TSS′**). The relation $\mathbf{E} \cap \mathbf{Ker}(0)$ is well-founded.

3214. LEMMA (**TSS′**). If R is a real regular well-founded relation then for any real nonempty $X \subseteq \mathbf{C}(R)$ there is a $y \in X$ such that $\mathbf{Ext}_R(y) \cap X = 0$.

Proof. Suppose there is a real nonempty $X \subseteq \mathbf{C}(R)$ such that $(\forall y \in X)$ $(\mathbf{Ext}_R(y) \cap X \neq 0)$. If $y \in X$ then, by the regularity and realness of R,

$\mathbf{Unv}_R(y)$ is a set. If we let $\mathbf{Unv}_R(y) \cap X = x$ then $x \neq 0$ and for any $u \in x$ there is a $v \in \mathbf{Ext}_R(u) \cap X$. Hence $v \in x$ and so $v \in \mathbf{Ext}_R(u) \cap x$. The existence of such an x contradicts well-foundedness.

3215. THEOREM (**TSS′**). If R is a real regular, strongly extensional and well-founded relation, then there is a unique complete real class $X \subseteq \mathbf{Ker}(0)$ and a real isomorphism G between X and $\mathbf{C}(R)$ w.r.t. $\mathbf{E}$ and R.

Proof. There exists a unique $u \in \mathbf{C}(R)$ such that $\mathbf{Ext}_R(u) = 0$; we let $u = 0_R$ and define

$$H\text{''}\{0\} = \{\langle 0, 0_R\rangle\},$$
$$H\text{''}\{\alpha\} = \{\langle x, y\rangle; y \in \mathbf{C}(R) \,\&\, \mathbf{Ext}_R(y) \subseteq \subseteq \mathbf{D}(H\text{''}\alpha) \,\&\, x = (H\text{''}\alpha)\text{''}\, \mathbf{Ext}_R(y)\}$$

for $\alpha > 0$ and

$$G = H\text{''}\, \mathbf{On}.$$

G is a morphism w.r.t. R and $\mathbf{E}$; G is a function and is one-to-one by strong extensionality. We have $\mathbf{D}(G) = \mathbf{C}(R)$; for if $u \in \mathbf{C}(R) - \mathbf{D}(G)$ is such that $\mathbf{Ext}_R(u) \cap (\mathbf{C}(R) - \mathbf{D}(G)) = 0$ then $\mathbf{Ext}_R(u) \subseteq \mathbf{D}(G)$ and, since $\mathbf{Ext}_R(u)$ is a set, there is an α such that $\mathbf{Ext}_R(u) \subseteq \mathbf{D}(H\text{''}\alpha)$ and $u \in \mathbf{D}(H\text{''}\{\alpha\})$, a contradiction. Hence G is an isomorphism between $\mathbf{W}(G)$ and $\mathbf{C}(R)$ w.r.t. R and $\mathbf{E}$. If $\hat{G} \neq G$ is another real isomorphism then there is a least α such that $\hat{G} \restriction \mathbf{D}(H\text{''}\{\alpha\}) \neq G \restriction \mathbf{D}(H\text{''}\{\alpha\})$ and there is an $x \in \mathbf{D}(H\text{''}\{\alpha\})$ such that $G\text{'}x \neq \hat{G}\text{'}x$. Since $\mathbf{Ext}_R(x) \subseteq H\text{''}\alpha$, we have $G\text{'}x = G\text{''}(\mathbf{Ext}_R(x)) = \hat{G}\text{''}(\mathbf{Ext}_R(x)) = \hat{G}\text{'}x$, a contradiction. Hence G is unique.

3216. COROLLARY (**TSS′**). If R is a real well-founded $\mathbf{E}$-like relation then there is a unique complete class $A \subseteq \mathbf{Ker}(0)$ such that $\mathbf{E} \cap A$ and R are isomorphic.

We shall now consider the independence of Axiom D3. We shall do this in a rather general way; to speak imprecisely, we show that if the existence of some regular and strongly extensional relation is consistent then it is consistent to assume that this relation describes the membership relation between certain sets in an exhaustive way. The following consideratioins take place in **TSS′** (or in an arbitrary extension of **TSS′**).

3217. DEFINITION (**TSS′**). If R and S are relations we say that R is a *complete subrelation* of S ($R \subseteq \subseteq S$) if $R \subseteq S$ and $\langle x, y\rangle \in S$ implies $\langle x, y\rangle \in R$ for any $x \in \mathbf{C}(S)$ and $y \in \mathbf{C}(R)$; hence $\mathbf{Ext}_R(y) = \mathbf{Ext}_S(y)$ for any $y \in \mathbf{C}(R)$.

3218. Example. If P is complete then $(\mathbf{E} \cap P^2) \subseteq \subseteq \mathbf{E}$.

3219. LEMMA (**TSS′**). (a) If R is a real regular relation and if $x \in \mathbf{C}(R)$ then there exists $r \subseteq \subseteq R$ such that $x \in \mathbf{C}(r)$.

(b) If $R \subseteq \subseteq S$ and $S \subseteq \subseteq T$ the $R \subseteq \subseteq T$.

(c) If $r \subseteq \subseteq R$ for all $r \in Z$ then $\bigcup(Z) \subseteq \subseteq R$.

Proof. (a) Let $r = R \restriction \mathbf{Unv}_R(x)$;

(b) and (c): Exercise.

3220. DEFINITION (**TSS′**). A relation R is said to be *set-universal* (denotation: SetUnvr (R)) if for each $x \subseteq \mathbf{C}(R)$ there is a $u \in \mathbf{C}(R)$ such that $x = \mathbf{Ext}_R(u)$ (every subset of $\mathbf{C}(R)$ has an R-code).

Note that (1) $\mathbf{E}$ is a real set-universal relation and (2) if R is a real set-universal relation then R is almost universal (cf. 1415). (Both remarks are provable in **TSS′**.)

3221. METATHEOREM. (1) The F-definition $\mathfrak{N}(\mathbf{R})$ with the specification

$$\text{Real}(\mathbf{R}) \,\&\, \text{sExtl}(\mathbf{R}) \,\&\, \text{Reg}(\mathbf{R}) \,\&\, \text{SetUnvr}(\mathbf{R})$$

determines a model of **TSS′** in **TSS′**. Moreover the following is provable in **TSS′**, (Real $(\mathbf{R})$ & sExtl $(\mathbf{R})$ & Reg $(\mathbf{R})$ & SetUnvr $(\mathbf{R})$), ∂**TSS**$/\partial\mathfrak{N}(\mathbf{R})$:

(2) There is a real isomorphism Is_1 of **On**, **On*** w.r.t. **E**, **R**.

(3) There is a real isomorphism Is_2 of **Ker**, **Ker*** w.r.t. **E**, **R**.

(4) $\mathbf{M}(X^*) \equiv \mathbf{M}^*(X^*)$, Real $(X^*) \equiv$ Real* (X^*).

(5) If X^*, Y^* are real subclasses of $\mathbf{C}(R)$ then the following are equivalent:

(i) $(\exists F)(\text{Real}(F) \,\&\, \text{Un}_2(F, X^*, Y^*))$,

(ii) $(\exists F^*)(\text{Real}^*(F^*) \,\&\, \text{Un}_2^*(F^*, X^*, Y^*))$.

(6) Consequently if X^* is a real subclass of $\mathbf{C}(R)$, Λ is a real ordinal (i.e. either an ordinal number or **On**) and $\Lambda^* = Is_1^{*}{}''\Lambda$, then the following are equivalent:

(i) $(\exists F)(\text{Real}(F) \,\&\, \text{Un}_2(F, X^*, \Lambda))$,

(ii) $(\exists F^*)(\text{Real}^*(F^*) \,\&\, \text{Un}_2^*(F^*, X^*, \Lambda^*))$.

Demonstration. (1) By 1416 it suffices to prove Axioms (C1) and (A2) to (A7) in the sense of $\mathfrak{Dir}(\partial\mathbf{TSS}/\partial\mathfrak{N}(\mathbf{R}))$. We proceed in **TSS′**, (Real $(\mathbf{R})$ &

& sExtl $(\mathbf{R})$ & Reg $(\mathbf{R})$ & SetUnvr $(\mathbf{R})$), $\partial\mathbf{TSS}/\partial\mathfrak{R}(\mathbf{R})$. Denote $\mathbf{C}(\mathbf{R})$ by P. There is a unique $x_0 \in P$ such that $\mathbf{Ext_R}(x_0) = 0$. For any $n \in \boldsymbol{\omega}$ there is a unique x_{n+1} such that $\mathbf{Ext_R}(x_{n+1}) = \{x_n\}$. Hence there is a unique $x \in P$ such that $\mathbf{Ext_R}(x) = \{x_n; n \in \boldsymbol{\omega}\}$. It follows that (C1)* holds.

We now prove (A2)*–(A7)*. We let $\omega_i(u, x_1, \bullet)$ be a SF equivalent to $u \in \mathbf{F}_i(x_1, \bullet)$ $(i = 1, \ldots, 7)$; for $x_1, \bullet \in P$ we let $Z = \{u \in P; \omega_i^*(\mathbf{Ext_R}(u), \mathbf{Ext_R}(x_1), \bullet)\}$. Z is included in P and is a set; for, if $i = 1$ then Z has at most two elements, if $i = 2, 3, 5$ then $Z \subseteq \mathbf{Ext_R}(x_1)$ and if $i = 4, 6, 7$ then we can define a set-mapping of a subset of $\mathbf{Ext_R}(x_1)$ onto Z. By set-universality there is a $w \in \mathbf{R}$ such that $Z = \mathbf{Ext_R}(w)$ and hence we have $\mathrm{M}^*(Z)$. We have $x^* \in^* Z \equiv \omega_i^*(x^*, \mathbf{Ext_R}(x_1), \mathbf{Ext_R}(x_2))$ which completes the proof of (1).

(2) The class $\mathbf{On}^*$ is ordered by $\mathbf{R}$; by regularity and realness of $\mathbf{R}$ each segment of $\mathbf{On}^*$ is a set. Let x be a nonempty subset of $\mathbf{On}^*$; by universality $x = \mathbf{Ext_R}(z)$ for some z. Using properties of ordinals we have $(\exists y \in x)$ $(\forall z \in x)(\langle z, y\rangle \notin \mathbf{R})$, so that $\mathbf{On}^*$ is well-ordered by $\mathbf{R}$; $\mathbf{On}^*$ is obviously a proper class. Hence $\mathbf{On}^*$ is isomorphic to $\mathbf{On}$.

(3) Let $F = \{\langle u_0, 0\rangle\}$ where $u_0 \in \mathbf{C}(\mathbf{R})$ is such that $\mathbf{Ext_R}(u_0) = 0^* = 0$, and let $y = K'x \equiv x = \mathbf{Ext_R}(y)$ for and $x \subseteq \mathbf{C}(\mathbf{R})$. By 3211, there is a real function G on $\mathbf{Ker}$ such that $G \upharpoonright \{0\} = F$ and $G'x = K'(G''x)$ for any $x \in \mathbf{Ker} - \{0\}$. The function G is obviously an isomorphism w.r.t. $\mathbf{E}$ and $\mathbf{R}$; we prove that $\mathbf{W}(\mathbf{G}) = \mathbf{Ker}^*$. If $x \in \mathbf{Ker}$ is an element of least rank such that $G'x \notin \mathbf{Ker}^*$ then $G''x \subseteq \mathbf{Ker}^*$ and $\mathrm{M}^*(G''x)$ and we have $G''x = \mathbf{Ext_R}(G'x) \in^* \mathbf{Ker}^*$; hence $G'x \in \mathbf{Ker}^*$, a contradiction. If x is in $\mathbf{Ker}^*$ and if $x^* = \mathbf{Ext_R}(x)$ is an element of least $*$-rank such that $x \notin \mathbf{W}(G)$ then $x^* \subseteq \mathbf{W}(G)$ and $x^* = G''y$ foı some y; hence $x = G'y$, a contradiction.

(4) $\mathrm{M}^*(X^*) \to \mathrm{M}(X^*)$ follows by the regularity of $\mathbf{R}$, the converse implication follows by the set-universality of $\mathbf{R}$. Suppose Real* (X^*) and consider $x \cap X^*$. This semiset equals $(x \cap \mathbf{C}(\mathbf{R})) \cap X^*$ but $\mathbf{C}(\mathbf{R})$ is a real class, hence $x \cap \mathbf{C}(\mathbf{R})$ is a subset of $\mathbf{C}(\mathbf{R})$; by set-universality, there is a y^* such that $x \cap \mathbf{C}(\mathbf{R}) = y^*$. It follows that $\mathrm{M}^*(y^* \cap X^*)$ and hence $\mathrm{M}(x \cap X^*)$. Thus we have Real(X^*). Conversely, if X^* is a real class then, for any y^*, y^* is a set, $X^* \cap y^*$ is a set and $X^* \cap y^* \subseteq \mathbf{C}(\mathbf{R})$; by set-universality, there is an x^* such that $y^* \cap X^* = x^*$ and so we have Real* (X^*).

(5) If F is a real one-one mapping of X^* onto Y^* then we let H be such that $F = \mathbf{Dec_R}(H)$ (cf. 1308); H is a one-one real mapping of X^* onto Y^* in the sense of $\mathfrak{R}(\mathbf{R})$. Conversely, if Real* (F^*) & $\mathrm{Un}_2^*(F^*, X^*, Y^*)$ then Real $(\mathbf{Dec_R}(F^*))$ & $\mathrm{Un}_2(\mathbf{Dec_R}(F^*), X^*, Y^*)$.

(6) follows from (5) because $(\exists F)(\mathrm{Real}(F) \,\&\, \mathrm{Un}_2(F, X^*, \Lambda))$ is equivalent to $(\exists F)(\mathrm{Real}(F) \,\&\, \mathrm{Un}_2(F, X^*, \Lambda^*))$.

3222. COROLLARY.

$$(\mathbf{TSS}', (\mathrm{Real}(\mathbf{R}) \,\&\, \mathrm{sExtl}(\mathbf{R}) \,\&\, \mathrm{Reg}(\mathbf{R}) \,\&\, \mathrm{SetUnvr}(\mathbf{R})), \partial\mathbf{TSS}/\partial\mathfrak{N}(\mathbf{R})) .$$

(a) (E2) implies (E2)* ;

(b) (D2) implies (D2)* .

This follows easily from (6).

The following Metatheorem says that we can define various real, regular, strongly extensional set-universal relations in **TSS′**; in fact, we show how to construct such a relation starting from an arbitrary real regular strongly extensional relation and extending it "upwards".

3223. METATHEOREM. There are gödelian terms $\mathbf{RUpw}(Q)$, $\mathbf{Imb}(Q)$ such that

(1) the following is provable in **TSS′**: If Q is a real regular strongly extensional relation then $\mathbf{RUpw}(Q)$ is a real regular strongly extensional set-universal relation;

(2) the following is provable in the theory

$\mathbf{TSS}'$, $\mathrm{Real}(\mathbf{Q}) \,\&\, \mathrm{Reg}(\mathbf{Q}) \,\&\, \mathrm{sExtl}(\mathbf{Q}) \,\&\, \mathbf{R} = \mathbf{RUpw}(\mathbf{Q})$, $\partial\mathbf{TSS}/\partial\mathfrak{N}(\mathbf{R})$: One can define a real complete subrelation $\overline{\mathbf{Q}}$ of $\mathbf{R}$ such that $\mathbf{Imb}(\mathbf{Q})$ is an isomorphism of $\mathbf{Q}, \overline{\mathbf{Q}}$, and such that $\mathbf{V}^* = \mathbf{Ker}^*(\mathbf{C}(\overline{\mathbf{Q}}))$.

$\mathbf{RUpw}(Q)$ is called the *real upward extension* of Q. (Cf. Chapter V Section 1.)

Demonstration. In **TSS′** we define $\mathbf{Imb}(Q)\text{'}\, x = \langle 0, x\rangle$ for $x \in \mathbf{C}(Q)$. Set $\bar{Q} = \{\langle\langle 0, x\rangle, \langle 0, y\rangle\rangle; \langle x, y\rangle \in Q\}$. Evidently, $\mathbf{Imb}(Q)$ is an isomorphism of $Q, \bar{Q}$. Further, define a real relation $R_Q \subseteq [\{0, 1\} \times \mathbf{V}]^2$ as follows:

$$\langle\langle i, x\rangle, \langle j, y\rangle\rangle \in R_Q \equiv .\, (i = j = 0 \,\&\, \langle\langle 0, x\rangle, \langle 0, y\rangle\rangle \in \bar{Q}) \vee$$
$$\vee\, (j = 1 \,\&\, i \in \{0, 1\} \,\&\, \langle i, x\rangle \in y) .$$

We shall find a real class $P_Q \subseteq \mathbf{C}(R)$ such that $\mathbf{RUpw}(Q) = R_Q \cap P_Q^2$ is a relation with the required properties.

For $x \subseteq \{0, 1\} \times \mathbf{V}$ we define $\mathbf{Cod}_Q(x)$ (the *code* of x) as follows:

(a) $\mathbf{Cod}_Q(x) = y$, if $x = \mathbf{Ext}_{\bar{Q}}(y)$ and $y \in \mathbf{C}(\bar{Q})$,

(b) $\mathbf{Cod}_Q(x) = \langle 1, x\rangle$ otherwise.

The term $\mathbf{T}(X) = \{\mathbf{Cod}_Q(x); x \subseteq X\}$ is iterable; hence we can define a relation H as follows:

$$H''\{0\} = \mathbf{C}(\bar{Q}),$$
$$H''\{\alpha\} = \{\mathbf{Cod}_Q(x); x \subseteq H''\alpha\} \quad \text{for} \quad \alpha > 0,$$
$$P_Q = \mathbf{W}(H), \quad \mathbf{RUpw}(Q) = R_Q \cap P_Q^2.$$

Claim 1. $\bar{Q}$ is a complete subrelation of $\mathbf{RUpw}(Q)$. Indeed, $\bar{Q}$ is a subrelation of $\mathbf{RUpw}(Q)$ and if $x \in \mathbf{C}(\bar{Q})$ then $\mathbf{Ext}_{\bar{Q}}(x) = \mathbf{Ext}_{R_Q}(x) = \mathbf{Ext}_{\mathbf{RUpw}(Q)}(x)$; hence $\bar{Q} \subseteq \subseteq \mathbf{RUpw}(Q)$.

Claim 2. The relation $\mathbf{RUpw}(Q)$ is real and regular. Indeed, $\mathbf{RUpw}(Q)$ is real by 1454. We show that R_Q is regular. If $y = \langle 0, x\rangle$ then $\mathbf{Ext}_{R_Q}(y) = \mathbf{Ext}_{\bar{Q}}(y)$; since Q is regular, $\bar{Q}$ is regular and so $\mathbf{Ext}_{R_Q}(y)$ is a semiset. If $y = \langle 1, x\rangle$ then $\mathbf{Ext}_{R_Q}(y) = x$.

Claim 3. The relation $\mathbf{RUpw}(Q)$ is strongly extensional. Indeed, denote $\mathbf{RUpw}(Q)$ by S and suppose that $\langle i, x\rangle$ and $\langle j, y\rangle$ have the same extension; we prove $\langle i, x\rangle = \langle j, y\rangle$. If $i = j = 0$ then $\mathbf{Ext}_S(\langle 0, x\rangle) = \mathbf{Ext}_{\bar{Q}}(\langle 0, x\rangle)$ and $\mathbf{Ext}_S(\langle 0, y\rangle) = \mathbf{Ext}_{\bar{Q}}(\langle 0, y\rangle)$, hence $\langle 0, x\rangle = \langle 0, y\rangle$. It follows by construction of P_Q that $i \neq j$ is impossible. Hence let $i = j = 1$. Then $\mathbf{Ext}_S(\langle 1, x\rangle) = x$ and $\mathbf{Ext}_S(\langle 1, y\rangle) = y$ which implies $x = y$ and $\langle 1, x\rangle = \langle 1, y\rangle$.

Claim 4. The relation $\mathbf{RUpw}(Q)$ is set-universal. Indeed, using properties of iterable terms we get $x \subseteq P_Q \to \mathbf{Cod}_Q(x) \in P_Q$. If $x \subseteq P_Q$ then $x = \mathbf{Ext}_{\mathbf{RUpw}(Q)}(\mathbf{Cod}_Q(x))$.

Thus (1) is demonstrated. Now let us proceed in the theory described in (2). We already know that $\mathbf{Imb}(\mathbf{Q})$ is an isomorphism of $\mathbf{Q}$, $\bar{\mathbf{Q}}$ and that $\bar{\mathbf{Q}}$ is a complete subrelation of $\mathbf{RUpw}(\mathbf{Q}) = \mathbf{R}$. $\mathbf{C}(\bar{\mathbf{Q}})$ is trivially a class in the sense of $\mathfrak{N}(\mathbf{R})$. It remains to prove $\mathbf{V}^* = \mathbf{Ker}^*(\mathbf{C}(\bar{\mathbf{Q}}))$. For this purpose suppose $\mathbf{C}(\bar{\mathbf{Q}}) \subseteq X^*$, $\mathrm{Real}^*(X^*)$ and $(\forall x^*)(x^* \subseteq X^* \to x^* \in^* X^*)$; we shall prove $X^* = P_{\mathbf{Q}}$. If α is the least ordinal number such that $H''\{\alpha\} \nsubseteq X^*$ then we let $x \in H''\{\alpha\} - X^*$. We have $\alpha > 0$ and $\mathbf{Ext}_{\mathbf{R}}(x) \subseteq X^*$. Hence $\mathbf{Ext}_{\mathbf{R}}(x) \in^* X^*$ and $\mathbf{Cod}_{\mathbf{Q}}(\mathbf{Ext}_{\mathbf{R}}(x)) = x \in X^*$, a contradiction. It follows that $X^* = \mathbf{V}^*$. This proof completes the demonstration of the whole Metatheorem.

The preceding Metatheorem enables us to demonstrate the consistency of several statements contradicting (D3). As an important example we demonstrate the following

3224. METATHEOREM. Let $\mathbf{T}(X)$ be a gödelian term with one class variable and without constants such that

(1) $\mathbf{TSS}' \vdash \mathbf{T}(\mathbf{V})$ is an ordinal,

(2) $\mathbf{TSS}' \vdash \mathbf{T}(\mathbf{V}) = \mathbf{T}(\mathbf{Ker})$.

Then the following axiom is consistent with $\mathbf{TSS}'$:

$$(\exists F)(\mathrm{Real}(F) \& \mathrm{Un}_2(F, \mathbf{Ur}, \mathbf{T}(\mathbf{V})) \& \mathbf{V} = \mathbf{Ker}(\mathbf{Ur}).$$

Demonstration. We proceed in the theory $\mathbf{TSS}'$. In this theory we define $\mathbf{Q} = \{\langle \alpha, \alpha \rangle; \alpha \in \mathbf{T}(\mathbf{V})\}$ and consider $\mathfrak{Dir}(\partial\mathbf{TSS}/\partial\mathfrak{N}(\mathbf{RUpw}(\mathbf{Q}))$ as a model in $\mathbf{TSS}'$, $\partial\mathbf{TSS}/\partial\mathfrak{N}(\mathbf{RUpw}(\mathbf{Q}))$. We proceed in the latter theory.

The relation $\overline{\mathbf{Q}}$ is isomorphic to $\mathbf{Q}$ and we know that $\mathbf{V}^* = \mathbf{Ker}^*(\mathbf{C}(\overline{\mathbf{Q}}))$ holds. We prove $\mathbf{Ur}^* = \mathbf{C}(\overline{\mathbf{Q}})$. Evidently, $x \in \mathbf{C}(\overline{\mathbf{Q}}) \to x \in \mathbf{Ur}^*$ holds because $\mathbf{Ext}_{\mathbf{RUpw(Q)}}(x) = \mathbf{Ext}_{\overline{\mathbf{Q}}}(x) = \{x\}$. Hence $\mathbf{C}(\overline{\mathbf{Q}}) \subseteq \mathbf{Ur}^*$. On the other hand,

$$\mathbf{TSS}' \vdash \mathbf{V} = \mathbf{Ker}(X) \& \mathrm{Real}(X) \& \mathrm{Comp}(X) . \to \mathbf{Ur} \subseteq X$$

and hence the following is provable in the theory in question:

$$\mathbf{V}^* = \mathbf{Ker}^*(X^*) \& \mathrm{Real}^*(X^*) \& \mathrm{Comp}^*(X^*) . \to \mathbf{Ur}^* \subseteq X^* .$$

Thus we have $\mathbf{Ur}^* = \mathbf{C}(\overline{\mathbf{Q}})$. Furthermore, $\mathbf{Ker}$ and $\mathbf{Ker}^*$ are isomorphic w.r.t. $\mathbf{E}$, $\mathbf{RUpw}(\mathbf{Q})$. Denote by Λ^* the image of $\mathbf{T}(\mathbf{Ker})$ in the isomorphism of $\mathbf{On}$, $\mathbf{On}^*$ ($\Lambda^* = Is_1{}''\mathbf{T}(\mathbf{Ker})$, cf. 3221 (2)). Thus we have $(\exists F^*)(\mathrm{Real}^*(F^*) \& \mathrm{Un}_2^*(F^*, \mathbf{Ur}^*, \Lambda^*)$ by $(\exists F)(\mathrm{Real}(F) \& \mathrm{Un}_2(F, \mathbf{C}(\overline{\mathbf{Q}}), \mathbf{T}(\mathbf{V}))$ and by 3221 (6). By (2) we obtain $\Lambda^* = \mathbf{T}^*(\mathbf{Ker}^*) = \mathbf{T}^*(\mathbf{V}^*)$ and the Metatheorem is demonstrated.

Model-classes in $\mathbf{TSS}'$, U1.

We shall consider the model $\mathfrak{Dir}(\partial\mathbf{TSS}/\partial\mathfrak{St}(\mathbf{M}))$ in the theory $\mathbf{TSS}'$, U1, $\mathrm{Mcl}(\mathbf{M})$, $\partial\mathbf{TSS}/\partial\mathfrak{St}(\mathbf{M})$.

3225. THEOREM ($\mathbf{TSS}'$, U1, $\mathrm{Mcl}(\mathbf{M})$, $\partial\mathbf{TSS}/\partial\mathfrak{St}(\mathbf{M})$).

(1) $\mathbf{On} \subseteq \mathbf{M} \& \mathbf{On}^* = \mathbf{On}$;

(2) $\mathbf{V}^* = \mathbf{Ker}^*(\mathbf{Ur}^*)$;

(3) $p_\alpha^* = p_\alpha \cap \mathbf{M}$.

Proof. (1) First note that $\mathrm{Ord}(x)$ is a formula restricted in **TSS**$'$, U1, and hence $\mathrm{Ord}^*(x^*) \equiv \mathrm{Ord}(x^*)$. Since $\mathrm{Sm}^*(X^*) \equiv \mathrm{Sm}(X^*)$, $\mathbf{On}^*$ is not a semiset.

If $x^* \in \mathbf{On}^*$ then $x^* \in \mathbf{On}$ by absoluteness (cf. Section 1); hence $\mathbf{On}^* \subseteq \mathbf{On}$. If $\alpha \in \mathbf{On}$ then there is a $\beta > \alpha$ such that $\beta \in \mathbf{On}^*$; by completeness we have $\alpha \in \mathbf{M}$ and hence $\mathbf{On} \subseteq \mathbf{On}^*$.

(2) Since the formula $\mathrm{Urel}(x)$ is restricted in **TSS**$'$ we have $\mathrm{Urel}^*(x^*) \equiv \equiv \mathrm{Urel}(x^*)$ and hence $\mathbf{Ur}^* = \mathbf{Ur} \cap \mathbf{M}$. Suppose now that $\mathrm{Real}^*(X^*)$, that $X^* \supseteq \mathbf{Ur}^*$ and that $x^* \subseteq X^* \to x^* \in X^*$; we prove $X^* = \mathbf{V}^*$; if x^* is an element such that $x^* \notin X^*$ then $x^* \nsubseteq X^*$; $x^* \in \mathbf{Ker}(u)$ for some u and we may suppose that for each $y^* \in \mathbf{Ker}(u)$ of a smaller rank (w.r.t. $\mathbf{Ker}(u)$) we have $y^* \in X^*$. But then $x^* \subseteq X^*$, a contradiction. Hence $X^* = \mathbf{M}$, i.e. $X^* = \mathbf{V}^*$.

(3) We have $p_0 = 0 = 0^* = p_0^*$. If α is the least number such that $p_\alpha^* \neq \neq p_\alpha \cap \mathbf{M}$ then α is not a limit since $\bigcup$ is absolute; hence $\alpha = \beta + 1$ and we have $p_\beta^* = p_\beta \cap \mathbf{M}$ and $p_\alpha^* = \mathbf{P}^*(p_\beta^*) = \mathbf{P}(p_\beta^*) \cap \mathbf{M} = \mathbf{P}(p_\beta \cap \mathbf{M}) \cap \cap \mathbf{M} = \mathbf{P}(p_\beta) \cap \mathbf{M} = p_\alpha \cap \mathbf{M}$, a contradiction.

3226. Definition (**TSS**$'$, U1).

$$\mathrm{Mcl}_{\mathbf{cn}}(X) \equiv .\ \mathrm{Mcl}(X) \,\&\, (\forall \alpha, \beta)\,(\alpha \approx \beta \to (\exists f \in X)\,(\mathrm{Un}_2(f, \alpha, \beta)))\,.$$

(X is a *model-class with absolute cardinals.*)

Note that **TSS**$'$, U1 $\vdash \mathrm{Mcl}_{\mathbf{cn}}(\mathbf{V})$. We defined the predicate $\mathrm{Mcl}_{\mathbf{cn}}$ in **TSS**$'$, U1 because in this theory we know that **On** is a subclass of every model-class. (This cannot be proved in **TSS**$''$.) The name "model-class with absolute cardinals" is justified by the following

3227. Lemma (**TSS**', U1, $\mathrm{Mcl}(\mathbf{M})$, $\partial\mathbf{TSS}/\partial\mathfrak{St}(\mathbf{M})$).

$$\mathrm{Mcl}_{\mathbf{cn}}(\mathbf{M}) \equiv \mathbf{Cn} = \mathbf{Cn}^{\mathfrak{St}(\mathbf{M})}\,.$$

Proof. The right-hand side is equivalent to $(\forall\alpha)\,(\mathrm{Card}(\alpha) \equiv \mathrm{Card}^*(\alpha))$. The lemma follows by absoluteness of $\mathrm{Un}_2(f, x, y)$.

We shall consider model-classes with absolute cardinals on various places in this book.

Model-classes in **TSS**, D3.

Finally we consider $\mathfrak{Dir}\,(\partial\mathbf{TSS}/\partial\mathfrak{St}(\mathbf{M}))$ in the theory **TSS**, D3, $\mathrm{Mcl}(\mathbf{M})$,

∂**TSS**$/\partial\mathfrak{St}(\mathbf{M})$. Since (D3) is stronger than (U1 & D1), Theorem 3225 is provable in this theory. Moreover, we have the following

3228. THEOREM (**TSS**, D3, Mcl ($\mathbf{M}$), ∂**TSS**$/\partial\mathfrak{St}(\mathbf{M})$).

(1) (D3)*.

(2) For any x, x is an element of $\mathbf{M}$ iff there is an f and a relation $r \in \mathbf{M}$ such that f is an isomorphism of r and $\mathbf{E} \restriction \mathbf{Unv}(\{x\})$.

(3) $\mathrm{Mcl}^*(X^*) \equiv \mathrm{Mcl}(X^*)$.

Proof. (1) is obvious.

(2) Suppose first that $x \in \mathbf{M}$; if we let $r = \mathbf{E} \restriction \mathbf{Unv}(\{x\})$ and $f = \mathbf{I} \restriction \restriction \mathbf{Unv}(\{x\})$ then the condition obviously holds. Suppose conversely that the condition holds. The relation r is regular, strongly extensional and well-founded; hence r is regular etc. in $\mathfrak{St}(\mathbf{M})$. By 3215, there is a complete real $X^* \subseteq \mathbf{Ker}^*$ and a real isomorphism G^* of $\mathbf{E} \restriction X^*$ and r. It follows that the composition of G^* and f is an isomorphism of X^* and $\mathbf{Unv}(\{x\})$ w.r.t. $\mathbf{E}$, $\mathbf{E}$. By completeness, $X^* = \mathbf{Unv}(\{x\})$ and $x \in \mathbf{M}$.

(3) Suppose first that $\mathrm{Mcl}^*(X^*)$; then $\mathrm{Comp}^*(X^*)$ implies $\mathrm{Comp}(X^*)$ and $\mathrm{Clos}^*(X^*)$ implies $\mathrm{Clos}(X^*)$. It remains to prove $\mathrm{AUncl}(X^*)$. Suppose $\sigma \subseteq X^*$. Then $\mathrm{Sm}^*(\sigma)$ and, by $\mathrm{AUncl}^*(X^*)$, there is an $x^* \in X^*$ such that $\sigma \subseteq x^*$. Thus we have proved $\mathrm{Mcl}(X^*)$. Suppose conversely that X^* is a model-class. Then we have $\mathrm{Comp}^*(X^*)$ and $\mathrm{Clos}^*(X^*)$ since $\mathrm{Comp}(X^*)$ and $\mathrm{Clos}(X^*)$. $\mathrm{AUncl}(X)$ is evidently absolute from above and hence $\mathrm{AUncl}^*(X^*)$ follows from $\mathrm{AUncl}(X^*)$.

3229. COROLLARY (**TSS**, D3). If X is a model-class then $x \in X$ iff there is an f and a relation $r \in X$ such that f is an isomosphism of $X \restriction \mathbf{Unv}(x)$ and r.

Indeed, this is provable in **TSS**, D3 iff the following is provable in **TSS**, D3, Mcl ($\mathbf{M}$) ($\mathbf{M}$ being a constant):

$$x \in \mathbf{M} \equiv (\exists f)(\exists r \in \mathbf{M}) \quad (f \text{ is an isomorphism of } \mathbf{E} \restriction \mathbf{Unv}(x) \text{ and } r)\,.$$

But this is equivalent to the provability of (2) in Theorem 3228 (in the theory indicated in this theorem).

3230. DEFINITION. X is a *model-class with* E1 (denotation: $\mathrm{Mcl}_{\mathrm{E1}}(X)$)

$$\mathrm{Mcl}(X) \,\&\, (\forall x \in X)(x \neq 0 \rightarrow \mathbf{Sel}(x) \cap X \neq 0)\,.$$

3231. LEMMA **TSS**, D3, Mcl (**M**), ∂**TSS**/$\partial\mathfrak{St}$ (**M**).

$$(\mathrm{E1})^* \equiv \mathrm{Mcl}_{\mathrm{E1}}(\mathbf{M}).$$

Obvious by absoluteness.

We conclude this section with a theorem showing that different model-classes with E1 have different sets of ordinals.

This theorem is interesting but will not be used anywhere in the book.

3232. THEOREM (**TSS**, D3). If $\mathrm{Mcl}_{\mathrm{E1}}(X)$ and $\mathrm{Mcl}(Y)$ and $\mathbf{P}(\mathbf{On}) \cap X = \mathbf{P}(\mathbf{On}) \cap Y$ then $X = Y$.

Proof. Let us proceed in **TSS**, D3, Mcl (**A**), ∂**TSS**/$\partial\mathfrak{St}$ (**A**), Mcl (**B**), ∂**TSS**/$\partial\mathfrak{St}$ (**B**). This is a conservative extension of **TSS**, (D3.) Denote the notions defined if by ∂**TSS**/$\partial\mathfrak{St}$ (**A**) by superscript **A** and similarly for **B**. Suppose $(\mathrm{E1})^{\mathbf{A}}$.

Notice first that **A**, **B** are real, $\mathbf{On}^{\mathbf{A}} = \mathbf{On} = \mathbf{On}^{\mathbf{B}}$, $\mathbf{On} \times^{\mathbf{A}} \mathbf{On} = \mathbf{On} \times \mathbf{On} = \mathbf{On} \times^{\mathbf{B}} \mathbf{On}$, $\mathbf{P}^{\mathbf{A}}_{\mathbf{fin}}(\mathbf{On}) = \mathbf{P}_{\mathbf{fin}}(\mathbf{On}) = \mathbf{P}^{\mathbf{B}}_{\mathbf{fin}}(\mathbf{On})$ and $\mathbf{Sd}^{\mathbf{A}} = \mathbf{Sd} = \mathbf{Sd}^{\mathbf{B}}$ (cf. 2226) We define a well-ordering **R** of $\mathbf{On}^2$ as follows:

$$\langle\langle\alpha, \beta\rangle \langle\gamma, \delta\rangle\rangle \in \mathbf{R} \equiv \{\alpha, \beta\}\, \mathbf{Sd}\, \{\gamma, \delta\} \vee (\{\alpha, \beta\} = \{\gamma, \delta\} \,\&\, \alpha < \gamma)\,;$$

we have $\mathbf{R}^{\mathbf{A}} = \mathbf{R} = \mathbf{R}^{\mathbf{B}}$ and every segment is a set (the same holds also in the sense of both **A** and **B**). Hence there is a unique real isomorphism F between $\mathbf{On}^2$ and **On** w.r.t. **R** and **E**; we have $F^{\mathbf{A}} = F = F^{\mathbf{B}}$. It follows that $\mathbf{P}^{\mathbf{A}}(\mathbf{On}^2) = \mathbf{P}^{\mathbf{B}}(\mathbf{On}^2)$; for, if $x \in \mathbf{P}^{\mathbf{A}}(\mathbf{On}^2)$ then $F\text{''}x \subseteq \mathbf{On}$, $F\text{''}x \in \mathbf{A} \cap \mathbf{B}$, $(\mathbf{Cnv}(F))\text{''}(F\text{''}x) \in \mathbf{B}$ and hence $x \in \mathbf{P}^{\mathbf{B}}(\mathbf{On}^2)$ (the converse is treated similarly).

We shall now prove $\mathbf{A} \subseteq \mathbf{B}$. If $x \in \mathbf{A}$ then by $(\mathrm{E1})^{\mathbf{A}}$ there exist $\alpha \in \mathbf{On}$ and a one-one mapping $f \in \mathbf{A}$ of α onto $\mathbf{Unv}(x)$. We define $r \subseteq \alpha^2$ such that $\langle\gamma, \delta\rangle \in r \equiv f\text{'}\gamma \in f\text{'}\delta$; f is an isomorphism between α and $\mathbf{Unv}(x)$ w.r.t. r and **E**. Since $r \in \mathbf{P}(\mathbf{On}^2) \cap \mathbf{A}$ we have $r \in \mathbf{B}$ and hence, by 3229, $x \in \mathbf{B}$, which completes the proof that $\mathbf{A} \subseteq \mathbf{B}$. Moreover, we have proved that $f \in \mathbf{B}$; for f is the unique isomorphism between $\mathbf{E} \restriction \mathbf{Unv}(\{x\})$ and r, and since both sets are in **B**, f is also in **B**.

We shall now prove that $\mathbf{B} \subseteq \mathbf{A}$. Suppose that this is not the case and let α be the least number such that $p^{\mathbf{A}}_{\alpha} \subset p^{\mathbf{B}}_{\alpha}$. It is obvious that $\alpha > 0$ and that α is not a limit; hence $\alpha = \beta + 1$ for some β and $p^{\mathbf{A}}_{\beta} = p^{\mathbf{B}}_{\beta}$. There exists $\xi \in \mathbf{On}$ and a one-one mapping $f \in \mathbf{A}$ of $p^{\mathbf{B}}_{\beta}$ onto ξ; by the first part of the proof we have $f \in \mathbf{B}$. If $x \in \mathbf{B}$ and $x \subseteq p^{\mathbf{B}}_{\beta}$ then we let $y = f\text{''}x$; we have $y \in \mathbf{P}(\mathbf{On}) \cap \mathbf{B}$ and hence $y \in \mathbf{A}$; thus $x = (\mathbf{Cnv}(f))\text{''}\, y \in \mathbf{A}$ and $p^{\mathbf{A}}_{\alpha} = p^{\mathbf{B}}_{\alpha}$, a contradiction.

SECTION 3

Symmetric sets and the independence of Axiom (E1) in **TSS**′

We shall now present a classical method for constructing model-classes in the theory **TSS**, U3; this method enables us to construct various models in which Axiom (E1) is false. The method does not enable us to demonstrate the consistency of $\neg$ (E1) with **TSS** + (D3); however, it will appear later, when we have a method for proving the consistency of (E1), that we can automatically obtain various consistency results for **TSS** + (D3) using only the consistency results for **TSS** + (U3) (proved in the present Section). The basic idea is that a set which cannot be well-ordered is in a sense symmetric. The notion of symmetry will be precisely defined. Throughout this Section we shall work in **TSS** + (U3).

3301. Definition (**TSS** + U3). A one-to-one mapping of a set a onto itself is called a *permutation* of a. The set of all permutations of a is denoted by $\mathbf{g}(a)$. The *identical permutation* $\mathbf{e}_a$ is the permutation $\mathbf{I} \upharpoonright a$.

3302. Definition (**TSS** + U3). If p and q are permutations of a then $p \,.\, q$ (the *composition* of p and q) is the permutation defined by $y = (p \,.\, q)\text{'}\, x \equiv (\exists z)\,(y = p\text{'}z \;\&\; z = q\text{'}x)$ for $x, y \in a$; in other words, $(p \,.\, q)\text{'}\, x = p\text{'}(q\text{'}x)$. The converse of p is denoted by p^{-1}.

3303. Lemma (**TSS** + U3). The following identities hold for any permutations p, q and r of a:

(a) $(p \,.\, q) \,.\, r = p \,.\, (q \,.\, r)$,

(b) $p \,.\, \mathbf{e}_a = \mathbf{e}_a \,.\, p = p$,

(c) $p \,.\, p^{-1} = p^{-1} \,.\, p = \mathbf{e}_a$,

(d) $(p \,.\, q)^{-1} = q^{-1} \,.\, p^{-1}$.

3304. DEFINITION (**TSS** + U3).

(a) A set $h \subseteq \mathbf{g}(a)$ is called a *group* (of *permutations of* a) if $\mathbf{e}_a \in h$ and if $p . q \in h$ and $p^{-1} \in h$ whenever $p, q \in h$.

(b) A group $h_2 \subseteq \mathbf{g}(a)$ is called a *subgroup* of a group $h_1 \subseteq \mathbf{g}(a)$ if $h_2 \subseteq h_1$.

(c) If p is a permutation and h is a group then we write

$$p . h . p^{-1} = \{q \in \mathbf{g}(a); (\exists r \in h)(q = p . r . p^{-1})\}.$$

(d) The *commutator of a permutation* p is the set

$$[p] = \{q \in \mathbf{g}(a); q . p . q^{-1} = p\};$$

the *commutator of a group* h is the set

$$[h] = \{q \in \mathbf{g}(a); q . h . q^{-1} = h\}.$$

3305. LEMMA (**TSS** + U3).

(a) The set $\mathbf{g}(a)$ is a group.

(b) If $h \subseteq \mathbf{g}(a)$ is a group and p is a permutation then $p . h . p^{-1}$ is a group. The group $p . h . p^{-1}$ is said to be *conjugate* to h.

(c) The commutator of a permutation or group is a group; every group is a subgroup of its commutator.

(d) The intersection of a non-empty set of groups is a group.

Proof. Obvious.

3306. DEFINITION (**TSS** + U3). A non-empty set z of groups of permutations of a is called a *group-filter* (or a **g**-*filter*) on a if it satisfies the following conditions:

(i) $h_1, h_2 \in z \rightarrow h_1 \cap h_2 \in z$,

(ii) $h_1 \in z \;\&\; h_1 \subseteq h_2 . \rightarrow h_2 \in z$

for any groups h_1 and h_2 of permutations of a. ("**g**-filter" means "**g**-filter on **Ur**".)

3307. DEFINITION (**TSS** + U3). A function F is called an *automorphism* of a real class X (Aut (F, X)) if F is a real one-to-one mapping of X onto X and if $x \in y \equiv F'x \in F'y$ for all $x, y \in X$.

3308. LEMMA (**TSS** + U3). A real function F is an automorphism of the universal class if and only if F is a one-to-one mapping of $\mathbf{V}$ onto $\mathbf{V}$ and $F'x = F''x$ for all x.

3309. LEMMA (**TSS** + U3). If F is an automorphism of $\mathbf{V}$ then

a) $\tau(F'x) = \tau(x)$ for every x,

b) $F \upharpoonright \mathbf{Ur} \in \mathbf{g}(\mathbf{Ur})$.

3310. METATHEOREM. For any **TC**-formula $\varphi(x_1, \bullet, X_1, \bullet)$ without constants the following formula is provable in **TSS** + U3:

$$(\forall F)(\mathrm{Aut}(F, \mathbf{V}) \to (\forall x_1, \bullet, X_1, \bullet)(\varphi(x_1, \bullet, X_1, \bullet) \equiv \varphi(F'x_1, \bullet, F''X_1, \bullet)).$$

Demonstration by induction.

3311. COROLLARY. For any Gödelian term $\mathbf{F}(X_1, \bullet)$ without constants the following formula is provable in **TSS** + U3:

$$\mathrm{Aut}(F, \mathbf{V}) \to (\forall X_1, \bullet)(\mathbf{F}(F''X_1, \bullet) = F''\,\mathbf{F}(X_1, \bullet)).$$

We shall now be interested mainly in permutations of **Ur** which we shall refer to simply as *permutations.*

We write **g** instead of **g** (**Ur**).

3312. THEOREM (**TSS** + U3). For every permutation p there exists a unique automosphism G of $\mathbf{V}$ such that $G \upharpoonright \mathbf{Ur} = p$.

Proof. This follows by Theorem 3211 on setting $F = p$ and $K = \mathbf{I}$.

3313. DEFINITION (**TSS** + U3). For any $p \in \mathbf{g}$, we denote by **Aut** (p), or simply by $\check{p}$, the unique automorphism of $\mathbf{V}$ which extends p.

3314. LEMMA (**TSS** + U3). For any x we have $((p \,.\, q)^{\vee})'\,x = \check{p}'(\check{q}'x)$ and $((p^{-1})^{\vee})'\,x = (\mathbf{Cnv}(\check{p}))'\,x$.

Proof. By induction.

3315. DEFINITION (**TSS** + U3).

$\mathbf{Inv}(x) = \{p \in \mathbf{g};\, \check{p}'x = x\}$ (the *invariant* of x), $\mathbf{PInv}(x) = \bigcap_{y \in x} \mathbf{Inv}(y)$.

3316. LEMMA (**TSS** + U3). a) For every x, **Inv** (x) and **PInv** (x) are groups;

(b) For any group h we have $h \subseteq \mathbf{Inv}(x) \equiv (\forall p \in h)(\check{p}''x \subseteq x)$.

3317. LEMMA (**TSS** + U3).

(a) If $x \in \mathbf{Ker}$ then $\mathbf{Inv}(x) = \mathbf{g}$,

(b) $\mathbf{Inv}(\mathbf{E}(x)), \mathbf{Inv}(\mathbf{D}(x)), \mathbf{Inv}(\mathbf{Cnv}(x)), \mathbf{Inv}(\mathbf{Cnv}_3(x)) \supseteq \mathbf{Inv}(x)$,
$\mathbf{Inv}(x - y), \mathbf{Inv}(x \upharpoonright y), \mathbf{Inv}(\{x, y\}) \supseteq \mathbf{Inv}(x) \cap \mathbf{Inv}(y)$,

(c) $\mathbf{Inv}(\check{\mathbf{p}}\text{'}x) = p \,.\, \mathbf{Inv}(x) \,.\, p^{-1}$, for every $p \in \mathbf{g}$.

(d) $\mathbf{Inv}(p) = [p]$, $\mathbf{Inv}(h) = [h]$ for every permutation p and every group h.

(e) For any x and each $p \in \mathbf{g}$ we have $\mathbf{Inv}(\check{\mathbf{p}}\text{'}x) = \check{\mathbf{p}}\text{'}\, \mathbf{Inv}(x)$.

Proof. (a) Since $\check{\mathbf{p}}\text{'}0 = 0$ for every $p \in \mathbf{g}$, we obtain by transfinite induction $\check{\mathbf{p}}\text{'}x = x$ for every $p \in \mathbf{g}$ and $x \in \mathbf{Ker}$.

(b) By corollary 3311.

(c) We shall prove $\check{q}\text{'}p = q \,.\, p \,.\, q^{-1}$ for any $p, q \in \mathbf{g}$.
We have $\langle u, v\rangle \in p \equiv \langle q\text{'}u, q\text{'}v\rangle \in \check{q}\text{'}p$. Similarly we have

$$\langle u, v\rangle \in p \equiv \langle q\text{'}u, v\rangle \in q \,.\, p \equiv \langle v, q\text{'}u\rangle \in p^{-1} \,.\, q^{-1} \equiv$$
$$\equiv \langle q\text{'}v, q\text{'}u\rangle \in q \,.\, p^{-1} \,.\, q^{-1} \equiv \langle q\text{'}u, q\text{'}v\rangle \in qpq^{-1} \,.$$

This gives both c) and d).

3318. LEMMA (**TSS** + U3). If $\mathbf{Ur} \approx a$ for some $a \in \mathbf{Ker}$ then for every group h there exists x such that $h = \mathbf{Inv}(x)$.

Proof. Let f be a one-to-one mapping of a onto $\mathbf{Ur}$ and for any $p \in \mathbf{g}$ let $\bar{p}$ be the composition of p and f; we have $\bar{p}\text{'}u = p\text{'}(f\text{'}u)$ for every $u \in a$. If we define $\bar{h} = \{\bar{p}; p \in h\}$ then since $\overline{q \,.\, p} = \check{q}\text{'}\bar{p}$ we have $q \in \mathbf{Inv}(\bar{h})$ whenever $q \in h$; for, if $q \in h$ and $\bar{p} \in \bar{h}$ then $\check{q}\text{'}\bar{p} = \overline{q \,.\, p} \in \bar{h}$. Conversely, if $\check{q}\text{''}\bar{h} = \bar{h}$ then $\check{q}\text{'}\bar{\mathbf{e}} \in \bar{h}$; i.e. $\check{q}\text{'}\mathbf{e} = \overline{q \,.\, \mathbf{e}} = \bar{q}$ and hence $q \in h$. Thus we have $\mathbf{Inv}(\bar{h}) = h$.

3319. DEFINITION (**TSS** + U3). For any **g**-filter z we define the class $\mathbf{S}(z)$ of all *symmetric sets* and the class $\mathbf{HS}(z)$ of all *hereditarily symmetric sets* as follows:

$$\mathbf{S}(z) = \{x; \mathbf{Inv}(x) \in z\}\,,$$
$$\mathbf{HS}(z) = \{x; x \in \mathbf{S}(z) \,\&\, \mathbf{Unv}(x) \subseteq \mathbf{S}(z)\}\,.$$

3320. THEOREM (**TSS** + U3). For any **g**-filter z the class $\mathbf{HS}(z)$ is complete, closed and real.

Proof. Completeness follows immediately by definition; the fact that **HS** (z) is closed follows by Lemma 3317, (b). **HS** (z) is obviously real.

3321. Lemma (**TSS** + U3). If z_1 and z_2 are **g**-filters such that $z_1 \subseteq z_2$ then **HS** $(z_1) \subseteq$ **HS** (z_2).

3322. Definition (**TSS** + U3). A **g**-filter z is said to be *symmetric* (Sym (z)) if **Inv** $(z) \in z$.

3323. Theorem (**TSS** + U3). If a **g**-filter z is symmetric then **HS**(z) is a model-class.

Proof. By 3320 it suffices to prove that **HS** (z) is almost universal. If $x \subseteq$ **HS** (z) then we let $y = \{\breve{\mathbf{p}}\text{'}u;\ u \in x \ \&\ p \in \mathbf{Inv}\,(z)\}$. We have $y \subseteq$ **HS** (z): for, if $p \in \mathbf{Inv}\,(z)$ then $\breve{\mathbf{p}}\text{''}\mathbf{S}\,(z) = \mathbf{S}\,(z)$ and $\breve{\mathbf{p}}\text{''}\mathbf{HS}\,(z) = \mathbf{HS}\,(z)$. Moreover, we have **Inv** $(y) \supseteq$ **Inv** (z) and hence $y \in \mathbf{S}\,(z)$; thus $y \in$ **HS** (z) and $y \supseteq x$.

3324. Theorem (**TSS** + U3). If P is a model-class then there exists a **g**-filter z such that

(i) z is symmetric,

(ii) $P \subseteq$ **HS** (z),

(iii) if z_1 is a **g**-filter such that $P \subseteq$ **HS** (z_1) then $z \subseteq z_1$.

Proof. Let z_0 be the set $\{h;\ (\exists x \in P)\,(h = \mathbf{Inv}\,(x))\}$ and let z be the set $\{h;\ (\exists h_0 \in z_0)\,(h_0 \subseteq h)\}$. The set z contains the intersection of any two groups in z; for, if $x, y \in P$ then $\mathbf{Inv}\,(x) \cap \mathbf{Inv}\,(y) = \mathbf{Inv}\,(\langle x, y\rangle)$ and $\langle x, y\rangle \in$ $\in P$. Hence z is a **g**-filter. For every **g**-filter z_1 we have $P \subseteq \mathbf{HS}\,(z_1) \equiv z_0 \subseteq$ $\subseteq z_1$ and hence $P \subseteq \mathbf{HS}\,(z_1) \equiv z \subseteq z_1$; in particular, $P \subseteq$ **HS** (z). This proves (ii) and (iii). It remains to prove that z is symmetric. Since P is almost universal there is a set $x \in P$ such that $(\forall h \in z_0)\,(\exists u \in x)\,(h = \mathbf{Inv}\,(u))$. We have $\mathbf{Inv}\,(x) \in z_0$; we show that $\mathbf{Inv}\,(x) \subseteq \mathbf{Inv}\,(z)$. Suppose that $p \in$ $\in \mathbf{Inv}\,(x)$ and $h \in z$, we shall prove $\breve{\mathbf{p}}\text{'}h \in z$. There is an $h_0 \in z_0$ such that $h_0 \subseteq h$; we choose $u \in x$ such that $\mathbf{Inv}\,(u) = h_0$. We have $\breve{\mathbf{p}}\text{'}u \in x$ and $\mathbf{Inv}\,(\breve{\mathbf{p}}\text{'}u) \in z_0$; since $\mathbf{Inv}\,(\breve{\mathbf{p}}\text{'}u) = p \,.\, h_0 \,.\, p^{-1} = \breve{\mathbf{p}}\text{'}h_0$, we have $\breve{\mathbf{p}}\text{'}h \supseteq \breve{\mathbf{p}}\text{'}h_0$ and $\breve{\mathbf{p}}\text{'}h \in z$. This completes the proof.

3325. Corollary (**TSS** + U3). If **HS** (z) is a model-class then there exists a symmetric **g**-filter $z_1 \subseteq z$ such that $\mathbf{HS}\,(z_1) = \mathbf{HS}\,(z)$; if $\mathbf{HS}\,(z_2) =$ $= \mathbf{HS}\,(z)$ for some **g**-filter z_2 then $z_1 \subseteq z_2$.

3326. Definition (**TSS** + U3). A symmetric **g**-filter z is said to be *well-symmetric* (WSym (z)) iff $\mathbf{Inv}\,(x) \in z$ for each $x \in$ **Ur**.

3327. LEMMA (**TSS** + U3). A symmetric **g**-filter z is well-symmetric iff for each $e \in \mathbf{P}_{\text{fin}}(\mathbf{Ur})$ we have $\mathbf{PInv}(e) \in z$.

(Obvious.)

3328. LEMMA (**TSS**, U3, Sym (**z**), ∂**TSS**/$\partial\mathfrak{St}$ (**HS** (**z**))).

$\mathbf{Ur}^* = \mathbf{Ur}$ iff **z** is well-symmetric.

Proof. If $\mathbf{Ur}^* = \mathbf{Ur}$ then, for each $x \in \mathbf{Ur}$, $x \in \mathbf{HS}(\mathbf{z})$, hence $\mathbf{Inv}(x) \in \mathbf{z}$. On the other hand, if $\mathbf{Inv}(x) \in \mathbf{z}$ for each $x \in \mathbf{Ur}$ then $\mathbf{M}^*(x)$ for each $x \in \mathbf{Ur}$; hence $(\forall x)(\text{Urel}^*(x) \equiv \text{Urel}(x))$ (since Urel (x) is RF).

We now state several theorems which concern the possibility of well-ordering elements of the model-class **HS** (**z**) in the sense of **HS** (**z**). Consider a fixed well-symmetric **g**-filter **z** and denote **HS** (**z**) by P.

3329. THEOREM (**TSS**, U3, Sym (**z**), ∂**TSS**/$\partial\mathfrak{St}$ (**HS** (**z**))). If an element x of P is equivalent to an element y of **Ker** then

$$x \approx^* y \equiv \mathbf{PInv}(x) \in \mathbf{z}\,.$$

Proof. Since the concept of a mapping is absolute the condition $x \approx^* y$ is equivalent to $(\exists f \in P)(\text{Un}_2(f, x, y))$. Suppose first that $x \approx^* y$ and let $f \in P$ be a one-to-one mapping of x onto y. We have $\mathbf{Inv}(f) \in \mathbf{z}$; we shall prove $\mathbf{Inv}(f) \subseteq \mathbf{PInv}(x)$. If $p \in \mathbf{Inv}(f)$ and $u \in x$ we show that $\check{\mathbf{p}}\text{'}u = u$. There is a $v \in y$ such that $\langle v, u\rangle \in f$. Since $v \in \mathbf{Ker}$ we have $\check{\mathbf{p}}\text{'}v = v$ and $\check{\mathbf{p}}\text{'}(\langle v, u\rangle) = \langle \check{\mathbf{p}}\text{'}v, \check{\mathbf{p}}\text{'}u\rangle = \langle v, \check{\mathbf{p}}\text{'}u\rangle$. Since $p \in \mathbf{Inv}(f)$ we have $\check{\mathbf{p}}\text{'}f = f$ and $\langle v, \check{\mathbf{p}}\text{'}u\rangle \in f$. The function f is one-to-one so that $\check{\mathbf{p}}\text{'}u = u$. Suppose conversely that $\mathbf{PInv}(x) \in z$ and let f be a one-to-one mapping of x onto y. We shall prove $\mathbf{Inv}(f) \supseteq \mathbf{PInv}(x)$. If $p \in \mathbf{PInv}(x)$ and $\langle v, u\rangle \in f$ then $\check{\mathbf{p}}\text{'}\langle v, u\rangle = \langle v, u\rangle$; hence $\check{\mathbf{p}}\text{'}f = f$. This completes the proof.

3330. THEOREM (**TSS**, U3, Sym (**z**), ∂**TSS**/$\partial\mathfrak{St}$ (**HS** (**z**))).

$$(\forall x^*)\,[(\exists r^*)(\text{WOrdg}^*(r^*)\,\&\,\mathbf{C}^*(r^*) = x^*) \equiv$$
$$\equiv (\exists r)(\text{WOrdg}(r)\,\&\,\mathbf{C}(r) = x^*\,\&\,\mathbf{PInv}(x^*) \in \mathbf{z})]\,.$$

Proof. A set can be well-ordered iff it is equivalent to some ordinal. Since the concepts of ordinal and one-to-one mapping are absolute, the result follows from the preceding Theorem.

3331. THEOREM (**TSS**, U3, WSym (**z**), ∂**TSS**/$\partial\mathfrak{St}$ (**HS** (**z**))).

$$(\text{E1})^* \equiv (\text{E1})\,\&\,\{\mathbf{e}\} \in \mathbf{z}\,.$$

Proof. We again denote **HS** (**z**) by P. If Axiom (E1) holds and if $\{\mathbf{e}\} \in \mathbf{z}$ then the **g**-filter **z** contains all groups; hence $P = \mathbf{V}$ and (E1)* holds. Conversely, if (E1)* holds then there is an α such that $\mathbf{Ur} \approx^* \alpha$. By 3329, $\{\mathbf{e}\} = \mathbf{PInv}(\mathbf{Ur}) \in \mathbf{z}$ which gives $\mathbf{z} = \mathbf{g}$ and hence $\mathbf{V} = P$; (E1) follows by (E1)*.

3332. DEFINITION (**TSS**, U3). If l is an ideal on the set **Ur** then we call the set of all groups h such that $h \supseteq \mathbf{PInv}(u)$ for some $u \in l$ the *g-filter determined by l*.

3333. LEMMA (**TSS**, U3). If z is the **g**-filter determined by an ideal l on the set **Ur** then

(i) z is a **g**-filter;

(ii) $\{\mathbf{e}\} \in z$ iff $\mathbf{Ur} \in l \vee (\exists x \in \mathbf{Ur})(\mathbf{Ur} - \{x\} \in l)$;

(iii) z is symmetric iff $\mathbf{Inv}(l) \in z$.

Proof. (i) and (ii) are easy. (iii) If $\{\mathbf{e}\} \in z$ then the equivalence trivially holds. If $\{\mathbf{e}\} \notin z$ we show that $\mathbf{Inv}(l) = \mathbf{Inv}(z)$. For $u \subseteq \mathbf{Ur}$ let $h_u = \mathbf{PInv}(u)$. We have $\breve{p}\text{'}h_u = h_{\breve{p}\text{'}u}$ for any permutation p. Let $p \in \mathbf{Inv}(l)$ and $h \in z$ so that $h \supseteq h_u$ for some $u \in l$. We have $\breve{p}\text{'}h \supseteq \breve{p}\text{'}h_u = h_{\breve{p}\text{'}u}$ and $\breve{p}\text{'}u \in l$. Hence $\breve{p}\text{'}h \in z$. Conversely, let $p \in \mathbf{Inv}(z)$ and $u \in l$, so that $h_{\breve{p}\text{'}u} = \breve{p}\text{'}h_u \in z$. There is a $v \in l$ such that $h_{\breve{p}\text{'}u} \supseteq h_v$. We have $\breve{p}\text{'}u \subseteq v$; for if $\breve{p}\text{'}u - v \neq 0$ then there are $x \in \breve{p}\text{'}u - v$ and $y \neq x$ such that $y \notin v$. Consequently, if p is the transposition of x and y then $p \in h_v - h_{\breve{p}\text{'}u}$, a contradiction. Hence $\breve{p}\text{'}u \in l$.

3334. METATHEOREM. Axiom (E1) is not provable from the axioms of (**TSS** + U3).

Demonstration. By Section 2, the axiom $\mathbf{Ur} \approx \omega_0$ is consistent with **TSS**, (U3). In the theory **TSS**, (U3), ($\mathbf{Ur} \approx \omega_0$) we can prove that the **g**-filter $\mathbf{z}(\mathbf{P}_{\text{fin}}(\mathbf{Ur}))$ determined by the ideal $\mathbf{P}_{\text{fin}}(\mathbf{Ur})$ (which is obviously well-symmetric) does not contain the group $\{\mathbf{e}\}$. (Obvious.) Hence the F-definition $\mathfrak{St}(\mathbf{M})$ with the specification $\mathbf{M} = \mathbf{HS}(\mathbf{z}(\mathbf{P}_{\text{fin}}(\mathbf{Ur}))$ determines a model of **TSS**, (U3), $\neg$ (E1) in **TSS**, (U3), ($\mathbf{Ur} \approx \omega_0$) (more generally in **TSS**, (U3), $\neg$ Fin (**Ur**)).

We have the following more precise result:

3335. THEOREM (**TSS**, U3, Sym (**z**), ∂**TSS**/$\partial\mathfrak{St}$(**HS** (**z**))). If $\mathbf{Ur} \approx \aleph_\beta$, if l is the ideal of all sets of urelements of power less than $\aleph_\beta$ and if **z** is determined by l then the following holds in the sense of $\mathfrak{St}$(**HS** (**z**)):

(1) **Ur** cannot be linearly ordered by any set relation.

(2) If $x \subseteq \mathbf{Ur}$ and $\neg (x \prec \aleph_\beta)$ then $(\mathbf{Ur} - x) \prec \aleph_\beta$ and x cannot be linearly ordered.

(3) $\aleph_\alpha \prec \aleph_\beta \to \aleph_\alpha \prec \mathbf{Ur}$;

(4) neither $\aleph_\beta \preccurlyeq \mathbf{Ur}$ nor $\mathbf{Ur} \preccurlyeq \aleph_\beta$.

Proof. (1) follows from (2) since $\mathbf{Ur} \prec \aleph_\beta$ cannot hold in the model. (2) We first prove that for each $x \subseteq \mathbf{Ur}$, x is in the model iff $x \in l$ or $(\mathbf{Ur} - x) \in l$. If $x \subseteq \mathbf{Ur}$ and $x \in \mathbf{HS}(\mathbf{z})$ then $\mathbf{Inv}(x) \supseteq \mathbf{PInv}(u)$ for some $u \in l$ and hence either $x \subseteq u$ or $\mathbf{Ur} - x \subseteq u$. (Otherwise let $v \in x - u$ and $w \in \mathbf{Ur} - (x \cup u)$; the transposition of v and w is in $\mathbf{PInv}(u)$ but not in $\mathbf{Inv}(x)$, a contradiction.) The converse is obvious. Hence $x \prec \aleph_\beta$ or $\mathbf{Ur} - x \prec \aleph_\beta$ whenever $x \subseteq \mathbf{Ur}$ and $x \in \mathbf{HS}(\mathbf{z})$. By 3329, $x \prec \aleph_\beta \vee \mathbf{Ur} - x \prec \aleph_\beta$ holds in the sense of $\mathfrak{St}(\mathbf{HS}(z))$. To prove (2) it remains to show that if $(\mathbf{Ur} - x) \prec \aleph_\beta$ then no linear ordering r of x is symmetric. Suppose, on the contrary, that there is some $e \in l$ such that $\mathbf{PInv}(e) \subseteq \mathbf{Inv}(r)$. Let u, v be two different elements of $x - e$; we may suppose that $\langle u, v\rangle \in r$. If p is the transposition of u and v then $p \in \mathbf{Inv}(r)$ and $p'\langle u, v\rangle = \langle v, u\rangle \in r$ which contradicts the fact that r is an ordering. (3) follows from the fact that there is some $x \in l$ such that $\aleph_\alpha \approx x$.

Finally we prove (4). Suppose that f is a one-to-one mapping of $\boldsymbol{\omega}_\beta$ into **Ur** or a one-to-one mapping of a subset of $\boldsymbol{\omega}_\beta$ onto **Ur**. Then the range of f is included in **Ur** and is equivalent to $\aleph_\beta$; since it can be well-ordered (via f), it follows by (2) that f cannot be symmetric.

We shall now define and investigate some notions which will be used later on when we construct models where Axiom D3 holds but not E1. These results will be given in the last chapter of the book, and so the reader may postpone the reading of the remainder of this Section until Chapter VI, Section 2. In view of the applications, we shall be interested in the consistency of formulas of the form $(\exists x)(\varphi(x))$ where $\varphi(x)$ refers for example only to elements of x, subsets of x and sets of subsets of x but not to elements of elements of x. We shall define what it means for a formula of this form to have a permutation model and construct two particular models for such formulas. We shall also investigate some useful properties of complete Boolean algebras (or in general, of separatively ordered sets) in permutation models.

3336. METADEFINITION. Let $\mathbf{T}(X)$ be a gödelian term without constants and with exactly one variable X. The term $\mathbf{T}(\mathbf{V})$ is a *good definition of an ordinal number* in a theory **T** (stronger than **TSS**, U3) if

(1) $\mathbf{T} \vdash \mathbf{T}(\mathbf{V})$ is an ordinal number,

(2) $\mathbf{T} \vdash (\forall X)(\mathrm{Mcl}_{\mathbf{cn}}(X) \rightarrow \mathbf{T}(\mathbf{V}) = \mathbf{T}(X))$.

$\mathbf{T}(\mathbf{V})$ is a *good definition of a cardinal number* if instead of (1) we have

(1′) $\mathbf{T} \vdash \mathbf{T}(\mathbf{V})$ is a cardinal number.

Remark. (2) is equivalent to the following:

$$\mathbf{T}, \mathrm{Mcl}_{\mathbf{cn}}(\mathbf{M}), \partial\mathbf{TSS}/\partial\mathfrak{St}(\mathbf{M}) \vdash \mathbf{T}(\mathbf{V}) = \mathbf{T}^{\mathfrak{St}(\mathbf{M})}(\mathbf{V}^{\mathfrak{St}(\mathbf{M})}).$$

3337. METADEFINITION. (Relativizing quantifiers.) Let $\varphi(x, \bullet)$ be a SF and let **T** be a term such that no variable occurring in **T** is bound in $\varphi(x, \bullet)$. We define $\varphi^{\mathbf{T}}(x, \bullet)$ (*φ with the quantifiers relativized to* **T**) by the following induction:

(a) $(x \in y)^{\mathbf{T}}$ is $(x \in y)$, $(x = y^{\mathbf{T}})$ is $x = y$,

(b) $(\varphi \& \psi)^{\mathbf{T}}$ is $\varphi^{\mathbf{T}} \& \psi^{\mathbf{T}}$, $(\neg\varphi)^{\mathbf{T}}$ is $\neg(\varphi^{\mathbf{T}})$,

(c) $((\exists x)\,\varphi)^{\mathbf{T}}$ is $(\exists x \in \mathbf{T})\,\varphi^{\mathbf{T}}$.

Note that if **T** is a constant, **M** say, then one can prove the following for any closed SF φ:

$$\mathbf{TSS}', \mathrm{Mcl}(\mathbf{M}), \partial\mathbf{TSS}/\partial\mathfrak{St}(\mathbf{M}) \vdash (x^* = x \& \bullet) \rightarrow (\varphi^{\mathbf{M}}(x, \bullet) \equiv \varphi^*(x^*, \bullet)).$$

But we shall not make use of this fact. We need $\varphi^{\mathbf{T}}$ for the following

3338. METADEFINITION. Let $\varphi(x)$ be a SF with one free variable x; suppose that $\boldsymbol{\alpha}$ is a constant which is defined in a theory **T** by a good definition of an ordinal number. We say that $\varphi(x)$ is *$\boldsymbol{\alpha}$-boundable* if $\mathbf{T} \vdash \varphi(x) \equiv \equiv \varphi^{\bar{p}_{\alpha}^{x}}(x)$. (For the definition of $\bar{p}_{\alpha}^{x}$ see 3140.)

Until now we have mainly been concerned with permutations of urelements; this can however be generalized.

3339. DEFINITION (**TSS** + U3). A set x is called a *set of individuals* ($\mathrm{Ind}(x)$) if all elements of x have the same rank.

3340. LEMMA (**TSS** + U3). If x is a set of individuals and if p is a permutation of x then there exists a unique automorphism F of $\mathbf{Ker}'(x)$ such that $F \upharpoonright x = p$.

The proof is left to the reader.

If, for any $y \in \mathbf{Ker}'(x)$, we denote by $\mathbf{Inv}_x(y)$ the set of all permutations of x such that the corresponding automorphism keeps y fixed (i.e. $F'y = y$), then we can prove in a similar way that $\mathbf{Inv}_x(y)$ is a group of permutations of x. If x is a set of individuals and if z is a **g**-filter on x then we define:

$$\bar{p}_0^x(z) = \{y \in x; \mathbf{Inv}_x(y) \in z\},$$
$$\bar{p}_\alpha^x(z) = \{y \subseteq \bigcup_{\beta<\alpha} \bar{p}_\beta^x(z); y \neq 0 \,\&\, \mathbf{Inv}_x(y) \in z\} \quad (\alpha > 0),$$
$$\mathbf{HS}'(z, x) = \bigcup_{\alpha \in \mathbf{On}} \bar{p}_\alpha^x(z).$$

3341. Remark. **Ur** is a set of individuals; the set of all infinite subsets of $\boldsymbol{\omega}$ is a set of individuals etc. Notice that if x and y are sets of individuals, if $x \approx y$ and if neither x nor y contains urelements, then for each α the sets $\bar{p}_\alpha^x$ and $\bar{p}_\alpha^y$ are isomorphic w.r.t. **E** and **E**. Urelements must be excluded because the singleton of an urelement is this urelement itself. In what follows we specify a certain rather general kind of description for **g**-filters.

3342. METADEFINITION. Let $\omega(x, z)$ be a SF with two free variables and let $\mathbf{T}(\mathbf{V})$ be a gödelian term which represents ω (w.r.t. x, z in **TC**). We say that $\omega(x, z)$ *well describes a symmetric* **g**-*filter* (or that $\mathbf{T}(\mathbf{V})$ *well defines a class of symmetric* **g**-*filters*) in a theory **T** stronger than **TSS**, U3 if the following is provable in **T**:

(1) $\mathbf{T}(\mathbf{V}) \subseteq \{\langle x, z\rangle; z$ is a symmetric **g**-filter on $x \,\&\, (\forall y \in x)(\mathbf{Inv}_x(y) \in z)\}$,

(2) $\mathbf{T}(\mathbf{V}) \neq 0$,

(3) $x \approx y \,\&\, x \in \mathbf{W}(\mathbf{T}(\mathbf{V}))\,.\rightarrow y \in \mathbf{W}(\mathbf{T}(\mathbf{V}))$,

(4) $\mathbf{T}(\mathbf{V}) \cap \mathbf{Ker} = \mathbf{T}(\mathbf{Ker})$.

3343. Remark. (1) is equivalent to $\omega(x, z) \rightarrow z$ is a well- symmetric **g**-filter on x; (2) is equivalent to $(\exists x, z)\, \omega(x, z)$; (3) is equivalent to $x \approx y \,\&$ $\&\, (\exists z)\, \omega(x, z)\,.\rightarrow (\exists z)\, \omega(y, z)$.

3344. METADEFINITION. We say that an $\boldsymbol{\alpha}$-boundable formula $\varphi(x)$ *has a permutation model* in a theory **T** stronger than **TSS**, U3 if there exists a formula $\omega(x, z)$ which well describes a symmetric **g**-filter and such that $\mathbf{T} \vdash \mathrm{Ind}(x) \,\&\, \omega(x, z)\,.\rightarrow \varphi^{\bar{p}_\alpha^x(z)}(x)$.
(We also say that ω *determines a permutation model of* φ.)

3345. In the following two examples we shall suppose that a formula $\varphi(x)$ is $\boldsymbol{\alpha}$-boundable (for some $\boldsymbol{\alpha}$ defined by a good definition) and $\varphi(x)$ has a permutation model determined by $\omega(x, z)$ (everything w.r.t. **TSS** + U3). We shall show that then the formula $(\exists x)\, \varphi(x)$ is consistent with **TSS** + + U3. (In Chapter VI we shall show that it is consistent even with **TSS** + D3.)

3346. Example 1. The axiom $(\exists z)\, \omega(\mathbf{Ur}, z)$ is consistent with **TSS** + + U3 + E2.

We proceed in the theory **TSS** + D3 + E2. The formula $(\exists x, z)\, \omega(x, z)$ is provable; choose a fixed $\mathbf{x}, \mathbf{z}$ such that $\omega(\mathbf{x}, \mathbf{z})$ and define the relation $\mathbf{r} = \{\langle u, u\rangle;\, u \in \mathbf{x}\}$. Consider the relation $\mathbf{R} = \mathbf{RUpw}(\mathbf{r})$ and add ∂**TSS** / / $\partial\mathfrak{N}(\mathbf{R})$. By Theorem 3221 there exists a unique real isomorphism F of **Ker** and **Ker*** w.r.t. **E**, **R**; therefore we have $F``\langle \mathbf{x}, \mathbf{z}\rangle \in^* \mathbf{T}^*(\mathbf{Ker}^*)$, hence, by (4) in 3342, $F``\langle \mathbf{x}, \mathbf{z}\rangle \in^* \mathbf{T}^*(\mathbf{V}^*)$ and therefore $\omega^*(F``\mathbf{x}, F``\mathbf{z})$.

Since $\mathbf{x} = \mathbf{Ur}^*$ and $\mathbf{x} \approx F``\mathbf{x}$, we have $\mathbf{x} \approx^* F``\mathbf{x}$ by Theorem 3221 and, by (3) in 3342, there exists some z^* which is a set in the sense of $\mathfrak{N}(\mathbf{R})$ and which satisfies $\omega^*(\mathbf{Ur}^*, z^*)$. This completes the proof and hence the specification of $\mathfrak{N}(\mathbf{R})$ by $\omega(\mathbf{x}, \mathbf{z})$ & $\mathbf{R} = \mathbf{RUpw}(\mathbf{I} \upharpoonright \mathbf{x})$ determines a model of the theory **TSS** + U3 + E2 + $(\exists z)\, \omega(\mathbf{Ur}, z)$ in **TSS** + D3 + E2.

3347. COROLLARY. The axiom $\varphi(\mathbf{Ur})$ is consistent with **TSS** + U3 + E2.

This follows from the fact that $\varphi(\mathbf{Ur})$ holds in $\mathfrak{St}(\mathbf{HS}(\mathbf{z}))$ as a model in **TSS** + U3 + E2 + $\omega(\mathbf{Ur}, \mathbf{z})$; $\varphi^{\bar{p}_{\boldsymbol{\alpha}}^{\mathbf{Ur}}}(\mathbf{Ur})$ holds in the sense of $\mathbf{HS}(\mathbf{z})$ since $\bar{p}_{\boldsymbol{\alpha}}^{\mathbf{Ur}}(\mathbf{z})$ is $\bar{p}_{\boldsymbol{\alpha}}^{\mathbf{Ur}}$ in the sense of $\mathbf{HS}(\mathbf{z})$.

3348. Example 2. We say that a is a *set of $\boldsymbol{\alpha}$-collections of urelements* (col $(a, \boldsymbol{\alpha})$) if a is a disjointed family of sets of urelements, $\bigcup a = \mathbf{Ur}$ and every element of a is of power $\aleph_\beta$, where $\aleph_\beta$ is the least regular cardinal greater than the cardinality of $\bar{p}_{\boldsymbol{\alpha}}^{a}$.

The axiom $(\exists z, a)\, (\mathrm{col}\,(a, \boldsymbol{\alpha})\ \&\ \omega(a, z))$ is consistent with **TSS** + + U3 + E2.

Proceeding in the theory **TSS** + D3 + E2 and fixing $\omega(\mathbf{a}, \mathbf{z})$ & Ind $(\mathbf{a})$ we define

$$\boldsymbol{\omega}_\beta = \mathbf{Min}\,\{\boldsymbol{\omega}_\gamma;\, \boldsymbol{\omega}_\gamma \text{ regular } \&\ \bar{p}_{\boldsymbol{\alpha}}^{\mathbf{a}} \prec \boldsymbol{\omega}_\gamma\}\,,$$

$$\mathbf{r} = \{\langle\langle x, u\rangle, \langle x, u\rangle\rangle;\, x \in \mathbf{a}\ \&\ u \in \boldsymbol{\omega}_\beta\}\,,$$

$$\mathbf{R} = \mathbf{RUpw}(\mathbf{r})\,.$$

Further we add the definitions ∂**TSS**/$\partial\mathfrak{N}(\mathbf{R})$. We have $\mathbf{Ur}^* = \mathbf{a} \times \boldsymbol{\omega}_\beta$ and $(\mathbf{V} = \mathbf{Ker}\,(\mathbf{Ur}))^*$. If F is a real isomorphic embedding of **Ker** into $\mathbf{C}\,(\mathbf{R})$

then we have $F''\langle \mathbf{a}, \mathbf{z}\rangle \in^* \mathbf{T}^*(\mathbf{Ker}^*) \subseteq \mathbf{T}^*(\mathbf{V}^*)$ and hence $\omega^*(F''\mathbf{a}, F''\mathbf{z})$. If $x \in \mathbf{a}$ let $h'x$ be the unique element $v \in \mathbf{C}(\mathbf{R})$ such that $\mathbf{Ext}_{\mathbf{R}}(v) = \{\langle x, \mu\rangle; \mu \in \boldsymbol{\omega}_\beta\}$ and let $\mathbf{a}^* = h''\mathbf{a}$. Then $\mathbf{a}^* \approx \mathbf{a}$ so that $\mathbf{a}^* \approx^* F''\mathbf{a}$ and therefore there exists z^* such that $\omega^*(\mathbf{a}^*, z^*)$. Furthermore, if $\boldsymbol{\alpha}^*$ is defined by the same definition as $\boldsymbol{\alpha}$ but in the sense of $\mathfrak{N}(\mathbf{R})$ then we have $\boldsymbol{\alpha}^* = F''\boldsymbol{\alpha}$ and $F''\bar{p}^{\mathbf{a}}_{\boldsymbol{\alpha}}$ is equivalent to $\bar{p}^{\mathbf{a}^*}_{\boldsymbol{\alpha}^*}$ in the sense of $\mathfrak{N}(\mathbf{R})$; hence $F''\boldsymbol{\omega}_\beta$, is, in the sense of $\mathfrak{N}(\mathbf{R})$, the least regular cardinal $\succ \bar{p}^{\mathbf{a}^*}_{\boldsymbol{\alpha}^*}$ and in addition $F''\boldsymbol{\omega}_\beta \approx^* v^*$ for $v^* \in^* \mathbf{a}^*$. Thus we have $\mathrm{col}^*(\mathbf{a}^*, \boldsymbol{\alpha}^*)$.

In the following we continue to assume that $\boldsymbol{\alpha}$ has a good definition, that ω well describes a symmetric **g**-filter and that φ is $\boldsymbol{\alpha}$-boundable and has a permutation model determined by ω.

3349. (1) DEFINITION (**TSS**, U3, E2, col $(\mathbf{a}, \boldsymbol{\alpha})$). If $\hat{q}$ denotes the extension of $q \in \mathbf{g}(\mathbf{Ur})$ to $\mathbf{a}$ (i.e., $\hat{q} = \check{q} \restriction \mathbf{a}$) then we define $h\!\downarrow\, = \{q \in \mathbf{g}(\mathbf{Ur}); \hat{q} \in h\}$ for each group h of permutations of $\mathbf{a}$. If $\omega(\mathbf{a}, z)$ let $z\!\downarrow\, = \{k;\ k$ is a group of permutations of $\mathbf{Ur}\ \&\ (\exists h \in z)(\exists e \in \mathbf{P}_{\mathbf{fin}}(\mathbf{Ur}))(k \supseteq h\!\downarrow \cap\, \mathbf{PInv}_{\mathbf{Ur}}(e))\}$.

(2) THEOREM (**TSS**, U3, E2, col $(\mathbf{a}, \boldsymbol{\alpha})$). If $\omega(\mathbf{a}, z)$ then $z\!\downarrow$ is a well-symmetric **g**-filter on **Ur** such that $\mathbf{a} \in \mathbf{HS}(z\!\downarrow)$ and $\bar{p}^{\mathbf{a}}_{\boldsymbol{\alpha}} \cap \mathbf{HS}(z\!\downarrow) = \bar{p}^{\mathbf{a}}_{\boldsymbol{\alpha}}(z)$.

(3) COROLLARY (**TSS**, U3, E2, col $(\mathbf{a}, \boldsymbol{\alpha})\ \&\ \omega(\mathbf{a}, \mathbf{z})$, $\partial\mathbf{TSS}/\partial\mathfrak{S}\mathrm{t}(\mathbf{HS}(\mathbf{z}\!\downarrow))$). Put $\mathbf{a}^* = \mathbf{a}$; then we have $\varphi^*(\mathbf{a}^*)$.

Proof of (2). $h\!\downarrow$ is the set of all permutations of **Ur** whose extensions to **a** belong to h. Clearly, $z\!\downarrow$ is the least **g**-filter containing all the groups $h\!\downarrow$ where $h \in z$ and in addition all the groups $\mathbf{PInv}_{\mathbf{Ur}}(e)$ where $e \in \mathbf{P}_{\mathbf{fin}}(\mathbf{Ur})$. It is not difficult to show that $p \,.\, (h\!\downarrow) \,.\, p^{-1} = (\hat{p} \,.\, h \,.\, \hat{p}^{-1})\!\downarrow$ for each $p \in \mathbf{Inv}_{\mathbf{Ur}}(\mathbf{a})$ and that $\mathbf{Inv}_{\mathbf{a}}(z)\!\downarrow\, \subseteq \mathbf{Inv}_{\mathbf{Ur}}(z\!\downarrow)$; hence from $\mathbf{Inv}_{\mathbf{a}}(z) \in z$ we obtain $\mathbf{Inv}_{\mathbf{Ur}}(z\!\downarrow) \in z\!\downarrow$ so that $z\!\downarrow$ is a well-symmetric **g**-filter. Further we have $h\!\downarrow\, \subseteq k\!\downarrow\, \equiv h \subseteq k$; in fact, the implication $\leftarrow$ is trivial and conversely, if $q \in h - k$ then any permutation of **Ur** which extends to q is in $h\!\downarrow$ but not in $k\!\downarrow$. It follows that

$$(*) \qquad h \in z \equiv h\!\downarrow\, \in z\!\downarrow \text{ for any group } h \text{ of permutations of } \mathbf{a}.$$

Indeed, if $h \in z$ then $h\!\downarrow\, \in z\!\downarrow$; on the other hand, if $h\!\downarrow\, \in z\!\downarrow$ then for some $k \in z$ and $e \in \mathbf{P}_{\mathbf{fin}}(\mathbf{Ur})$ we have $h\!\downarrow\, \supseteq k\!\downarrow \cap\, \mathbf{PInv}_{\mathbf{Ur}}(e)$. Put $\hat{e} = \{u \in \mathbf{a};\ u \cap e \neq 0\}$. Then $\hat{e}$ is a finite subset of **a**, whence $\mathbf{PInv}_{\mathbf{a}}(\hat{e}) \in z$ (z being well-symmetric). Consequently, we may suppose without loss of generality that k is included in $\mathbf{PInv}_{\mathbf{a}}(\hat{e})$ (if not, replace k by $k \cap \mathbf{PInv}_{\mathbf{a}}(\hat{e})$). We prove $h\!\downarrow\, \supseteq k\!\downarrow$. If $p \in k\!\downarrow$ then $\hat{p} \in k$ and $\hat{p}$ is identical on $\hat{e}$. We can change p to p_1

such that $\hat{p} = \hat{p}_1$ and p_1 is identical on e. Then $p_1 \in k\downarrow \cap \mathbf{PInv}_{\mathbf{Ur}}(e)$, hence $p_1 \in h\downarrow$ and $\hat{p}_1 \in h$. We have $h\downarrow \supseteq k\downarrow$ and so $h \supseteq k$. This proves $h \in z$.

Now observe that the following holds true:

$$(**) \quad x \in \mathbf{Ker}'(\mathbf{a}) \to [\mathbf{Inv}_{\mathbf{a}}(x)\downarrow = \mathbf{Inv}_{\mathbf{Ur}}(x) \cap \mathbf{Inv}_{\mathbf{Ur}}(\mathbf{a})] .$$

Indeed, if $p \in \mathbf{Inv}_{\mathbf{Ur}}(\mathbf{a})$ then p extends to $\mathbf{Ker}'(\mathbf{a})$ in the same manner as $\hat{p}$ does. Consequently, we have the following:

$$(***) \quad x \in \mathbf{Ker}'(\mathbf{a}) \to . \mathbf{Inv}_{\mathbf{a}}(x) \in z \equiv \mathbf{Inv}_{\mathbf{Ur}}(x) \in z\downarrow .$$

In fact, if $\mathbf{Inv}_{\mathbf{a}}(x) \in z$ then $\mathbf{Inv}_{\mathbf{a}}(x)\downarrow \in z\downarrow$ but, by $(**)$, $\mathbf{Inv}_{\mathbf{a}}(x)\downarrow \subseteq \mathbf{Inv}_{\mathbf{Ur}}(x)$ and so $\mathbf{Inv}_{\mathbf{Ur}}(x) \in z\downarrow$. On the other hand, suppose $\mathbf{Inv}_{\mathbf{Ur}}(x) \in z\downarrow$. Obviously, $\mathbf{Inv}_{\mathbf{Ur}}(\mathbf{a}) = \mathbf{g}(\mathbf{a})\downarrow \in z\downarrow$; hence $\mathbf{Inv}_{\mathbf{a}}(x)\downarrow = \mathbf{Inv}_{\mathbf{Ur}}(x) \cap \mathbf{Inv}_{\mathbf{Ur}}(\mathbf{a}) \in z\downarrow$ and, by $(*)$, $\mathbf{Inv}_{\mathbf{a}}(x) \in z$.

For every $x \in \mathbf{Ur}$ we have $\mathbf{Inv}_{\mathbf{Ur}}(x) \in z\downarrow$. The same holds for every $u \in \mathbf{a}$ (since z is well symmetric) and $\mathbf{Inv}_{\mathbf{Ur}}(\mathbf{a}) \in z\downarrow$ so that $\mathbf{a} \in \mathbf{HS}(z\downarrow)$. Finally, we obtain from $(***)$ by an easy induction that, for every $\beta \in \mathbf{On}$, $\bar{p}^{\mathbf{a}}_{\beta} \cap \mathbf{HS}(z\downarrow) = \bar{p}^{\mathbf{a}}_{\beta}(z)$. This completes the proof.

The proof of (3) is analogous to the proof of 3347.

3350. LEMMA. Under the assumptions of the preceding Theorem let x be a set of urelements such that $x \prec \aleph_{\beta}$, and let $k = h\downarrow \cap \mathbf{PInv}(e)$ for some $h \in z$ and $e \in \mathbf{P}_{\mathbf{fin}}(\mathbf{Ur})$. If $p \in k$ then there exists $q \in k$ such that $\hat{p} = \hat{q}$ and for every $\iota \in x$, $q'\iota = \iota \vee q'\iota \notin x$.

Proof. Let $u \in \mathbf{a}$ and let $\hat{p}'u = v$. Consider $(x \cap u) - e$ and let $y \subseteq v$ be a set of the same cardinality as $(x \cap u) - e$ and disjoint from $x \cup e \cup p''((x \cap u) - e)$. Finally, let f be a one-to-one mapping of $p''((x \cap u) - e)$ onto y. We define

$$\begin{aligned} q'\iota &= f'(p'\iota) \quad \text{for } \iota \in (x \cap u) - e , \\ &= f^{-1}(p'\iota) \text{ for } \iota \in p^{-1}{}''(y) , \\ &= p'\iota \text{ otherwise in } u . \end{aligned}$$

The mapping q has the required properties.

We shall now investigate certain relationships between complete Boolean algebras and complete Boolean algebras in the sense of a permutation model-class $\mathbf{HS}(\mathbf{z})$ where $\mathbf{z}$ is a symmetric $\mathbf{g}$-filter on $\mathbf{Ur}$; similarly for separative orderings. We first observe that if p is a permutation of **Ur** and $\boldsymbol{b}$ is a complete Boolean algebra then $p \in \mathbf{Inv}_{\mathbf{Ur}}(\boldsymbol{b})$ iff $\check{p} \upharpoonright b$ is an automorphism of $\boldsymbol{b}$ as a complete Boolean algebra; a similar remark holds for separatively ordered sets. For a given symmetric **g**-filter **z** let us denote by P the model-class **HS** (**z**) of all sets hereditarily symmetric w.r.t. **z**. Let $\boldsymbol{a}$ be an ordered set such that $\mathbf{Inv}(\boldsymbol{a}) \in \mathbf{z}$ (i.e. $\boldsymbol{a} \in \mathbf{S}(\mathbf{z})$). If we set $\boldsymbol{a}_1 = \langle a \cap P, \leqq \cap P \rangle$ then $\mathbf{Inv}(\boldsymbol{a}_1) \supseteq \mathbf{Inv}(\boldsymbol{a}) \cap \mathbf{Inv}(\mathbf{z})$; hence $\boldsymbol{a}_1 \in \mathbf{S}(\mathbf{z})$ and, since $a_1 \subseteq P$, we have $\boldsymbol{a}_1 \in P$. Clearly $\boldsymbol{a}_1$ is an ordered set (in the sense of $\mathfrak{St}(\mathbf{HS}(\mathbf{z}))$). Suppose $\boldsymbol{a}$ is separatively ordered. The ordered set $\boldsymbol{a}_1$ need not be separatively ordered (in the sense of $\mathfrak{St}(\mathbf{HS}(\mathbf{z}))$), but, if it is, it determines a complete Boolean algebra in the sense of $\mathfrak{St}(\mathbf{HS}(\mathbf{z}))$ and it is important to determine the relationship between this algebra and the algebra determined by $\boldsymbol{a}$.

We therefore make the following definition:

3351. Definition (**TSS** + U3). Let $\boldsymbol{a}$ be a separatively ordered set and let z be a symmetric **g**-filter on **Ur**. As in the preceding paragraph we let $\boldsymbol{a}_1 = \langle a \cap P, \leqq \cap P \rangle$. We say that $\boldsymbol{a}$ is *fruitful* w.r.t. z if it satisfies the following condition:

(1) $\mathbf{Inv}(\boldsymbol{a}) \in z$;

(2) $\boldsymbol{a}_1$ is separatively ordered;

(3) For every $u \in a$ and every $h \in z$ such that $h \subseteq \mathbf{Inv}(a)$ there exists $v \in a_1$ such that $v \leqq \bigvee_{p \in h} \check{p}'u$, the join being taken in an arbitrary complete Boolean algebra with $\boldsymbol{a}$ as base.

3352. Theorem (**TSS**, U3, WSym (**z**), ∂**TSS**$/\partial$(**HS** (**z**))). Let $\boldsymbol{a}$ be a separatively ordered set which is fruitful w.r.t. **z**. If $\boldsymbol{b}$ is a complete Boolean algebra with the base $\boldsymbol{a}$ and if $\boldsymbol{b}_1$ is a complete Boolean algebra in the sense of $\mathfrak{St}(\mathbf{HS}(\mathbf{z}))$ with the base $\boldsymbol{a}_1$ then there exists a one-to-one operation-preserving mapping of b_1 into b, i.e. a one-to-one mapping g such that $g'(-_1 u) = -g'u$ whenever $u \in b_1$ and $g'(\bigwedge_1 q) = \bigwedge g''q$ whenever $q \in \mathbf{P}^*(b_1)$ (where $-_1$ and $\bigwedge_1$ are the operations in $\boldsymbol{b}_1$ and $-$, $\bigwedge$ are the operations in $\boldsymbol{b}$).

Proof. Without loss of generality we may assume that the algebras $\boldsymbol{b}$ and $\boldsymbol{b}_1$ are constructed in the same way as in the proof of Theorem 2442; thus, in particular, the elements of b and b_1 are saturated subsets of the corresponding bases and the meet coincides with intersection. Every permutation in $\mathbf{Inv}(\boldsymbol{a})$ induces an automorphism of $\boldsymbol{b}$ as a complete Boolean algebra; a similar remark holds for $\boldsymbol{b}_1$. We recall that $u \in b$ is said to be symmetric if $\mathbf{Inv}(u) \in \mathbf{z}$. If u is a symmetric element of b we set $f'u = u \cap P$. We shall show that $g = \mathbf{Cnv}(f)$ is the required mapping.

1) We prove that $\mathbf{W}(f) \subseteq b_1$, i.e. $u \cap P \in b_1$ whenever u is symmetric. Since $\mathbf{Inv}(u \cap P) \supseteq \mathbf{Inv}(u) \cap \mathbf{Inv}(\mathbf{z})$, we have $(u \cap P) \in P$. It is clear that if $y^* \leqq x^* \in u \cap P$ then $y^* \in u \cap P$. Suppose that $(\forall y^* \leqq x^*)(\exists t^* \leqq y^*)(t^* \in u \cap P)$; we prove that $x^* \in u \cap P$. If we suppose the contrary then $x^* \notin u$ and there exists $y \leqq x^*$ such that the segment determined by y is disjoint from u. If we set $h = \mathbf{Inv}(x^*) \cap \mathbf{Inv}(u) \cap \mathbf{Inv}(\boldsymbol{a})$ then for each $p \in h$ the segment determined by $\check{p}'y$ is disjoint from u and $\check{p}'y \leqq x^*$. Since a is fruitful there exists $y^* \leqq \bigvee_{p \in h} \check{p}'y$ and, since the segment determined by y^* is disjoint from u, we arrive at a contradiction.

2) We now prove that $b_1 \subseteq \mathbf{W}(f)$. If $v \in b_1$ then we define

$$x \in u \equiv . x \in a \,\&\, (\forall y \leqq x)(\exists t \leqq y)(\exists w \geqq t)(w \in v) .$$

Clearly u is a saturated subset of a and hence $u \in b$. We shall prove that $u \cap P = v$. Let $x^* \in u$ and let y^* be an arbitrary element such that $y^* \leqq x^*$. There exist $t \leqq y^*$ and $w \geqq t$ such that $w \in v$. If we let $h = \mathbf{Inv}(y^*) \cap \mathbf{Inv}(w) \cap \mathbf{Inv}(\boldsymbol{a})$ and $t^* \leqq \bigvee_{p \in h} \check{p}'t$ then $t^* \leqq y$ and $t^* \leqq w$ so that $t^* \in v$. This proves that $u \cap P \subseteq v$; the converse inclusion is trivial. Evidently, u is symmetric since $\mathbf{Inv}(u) \supseteq \mathbf{Inv}(\boldsymbol{a}) \cap \mathbf{Inv}(v)$. Hence $u \in b$ and $v = f'u$.

3) We now prove that f is one-to-one. Suppose that $u, v \in \mathbf{D}(f)$ and that $u \neq v$. Then there exists x such that $x \in u - v$, say. If we let $h = \mathbf{Inv}(u) \cap \mathbf{Inv}(v) \cap \mathbf{Inv}(\boldsymbol{a})$ and $x^* \leqq \bigvee_{p \in h} \check{p}'x$ then $x^* \in u$ and $x^* \notin v$. (Exercise.)

4) Finally we prove that f preserves the operations. If $k \in \mathbf{P}^*(b_1)$ and $k' = \mathbf{Cnv}(f)'' k$ then $\bigwedge_1 k = \bigcap_{u \in k} u = \bigcap_{v \in k'} (v \cap P) = (\bigcap_{v \in k'} v) \cap P = \bigwedge k' \cap P = f'(\bigwedge k')$, since $\mathbf{Inv}(\bigcap k') \supseteq \mathbf{Inv}(k') \supseteq \mathbf{Inv}(k) \cap \mathbf{Inv}(\boldsymbol{a}) \cap \mathbf{Inv}(\mathbf{z})$. Similarly, if u is symmetric then $-u$ is also symmetric; we show that
$\{x \in a;\ \neg\,(\exists t \leqq x)(t \in u)\} \cap P = \{x^* \in a_1;\ \neg\,(\exists t^* \leqq x^*)(t^* \in u \cap P)\}$.
The inclusion $\subseteq$ is obvious. Let x^* be an element of a_1 such that $\neg\,(\exists t^* \leqq x^*)(t^* \in u)$. Suppose there exists an element t of u such that $t \leqq x^*$. There exists $t^* \leqq \bigvee_{p \in h} \check{p}'t$, where $h = \mathbf{Inv}(u) \cap \mathbf{Inv}(x^*) \cap \mathbf{Inv}(\boldsymbol{a})$; hence $t^* \leqq x^*$ and $t^* \in u$ which is a contradiction. Hence $-_1(u \cap P) = (-u) \cap P$ and the theorem is proved.

3353. *Examples.* Let $\boldsymbol{b}_0 \in \mathbf{Ker}$ be a complete Boolean algebra having more than two elements and let l be an ideal on $\mathbf{Ur}$. We shall investigate the product $\prod_{\mathbf{Ur}}^{l} \dot{\boldsymbol{b}}_0$ of $\dot{\boldsymbol{b}}_0$ over l (as usual $\dot{\boldsymbol{b}}_0$ is the set of all elements of $\boldsymbol{b}_0$ except for $0_{\boldsymbol{b}_0}$ and $1_{\boldsymbol{b}_0}$, with the canonical ordering). Let z be a well-symmetric **g**-filter on **Ur**; we denote $\prod_{\mathbf{Ur}}^{l} \dot{\boldsymbol{b}}_0$ by $\boldsymbol{a} = \langle a, \leqq \rangle$ and **HS** (z) by P.

3354. LEMMA (**TSS** + U3). $\mathrm{Inv}(l) \subseteq \mathrm{Inv}(\boldsymbol{a})$; if $\mathbf{Inv}(l) \in z$ then $\boldsymbol{a}_1 = \langle a \cap P, \leqq \cap P \rangle$ is a symmetric separatively ordered set.

Proof. (We proceed in **TSS**, U3, WSym (**z**), ∂**TSS**/$\partial\mathfrak{S}$t (**HS** (**z**)).) Put $P = \mathbf{HS}(\mathbf{z})$. If $p \in \mathbf{Inv}(l)$ and $f \in a$ then $\mathbf{D}(\breve{p}'f) \in l$ and $\langle u, \iota \rangle \in f \equiv \langle u, p'\iota \rangle \in \breve{p}'f$; hence $\mathbf{Inv}(l) \subseteq \mathbf{Inv}(\boldsymbol{a})$. If $\mathbf{Inv}(l) \in \mathbf{z}$ then $\boldsymbol{a}$ is symmetric and $\boldsymbol{a}_1$ is also symmetric. It remains to prove that $\boldsymbol{a}_1$ is a separatively ordered set. Let f^*, g^*, h^* and k^* denote elements of a_1. Suppose that $(\forall g^* \leqq f^*)(\exists h^* \leqq g^*)(h^* \leqq k^*)$; we prove that $f^* \leqq k^*$. To do this we first prove that $\mathbf{D}(k^*) \subseteq \mathbf{D}(f^*)$. Suppose on the contrary that there exists $\iota \in \mathbf{D}(k^*) - \mathbf{D}(f^*)$, and let $-_0 k^{*}{}'\iota = x \in \dot{b}_0$. If we let $g^* = f^* \cup \{\langle x, \iota \rangle\}$ then $g^* \in P$, since $\mathbf{Inv}(\iota) \in \mathbf{z}$. The inequality $h^* \leqq k^*$ does not hold for any $h^* \leqq g^*$ and so we have a contradiction. We shall now prove that $f^{*}{}'\iota \preccurlyeq_0 k^{*}{}'\iota$ for any $\iota \in \mathbf{D}(k^*)$, where $\preccurlyeq_0$ is the canonical ordering of $\boldsymbol{b}_0$. If we let $x \preccurlyeq_0 f^{*}{}'\iota$ and $g^* = (f^* - \{\langle f^{*}{}'\iota, \iota \rangle\}) \cup \{\langle x, \iota \rangle\}$ then $g^* \in P$ and there exists $h^* \leqq g^*$ such that $h^* \leqq k^*$; hence $h^{*}{}'\iota \leqq x$ and $h^{*}{}'\iota \leqq k^{*}{}'\iota$. Thus we have $(\forall x \preccurlyeq_0 f^{*}{}'\iota)(\exists y \preccurlyeq_0 x)(y \preccurlyeq_0 k^{*}{}'\iota)$ and, since $\preccurlyeq_0$ is separative, $f^{*}{}'\iota \preccurlyeq_0 k^{*}{}'\iota$.

3355. THEOREM (**TSS**, U3). Let $l = l_{\aleph_\beta}$ be the ideal of all subsets of **Ur** of power less than $\aleph_\beta$, where $\aleph_\beta$ is the power of **Ur** and let z be the **g**-filter determined by this ideal. If $\boldsymbol{b}_0 \in \mathbf{Ker}$ is a complete Boolean algebra and if $\boldsymbol{a} = \prod_{\mathbf{Ur}}^{l} \dot{\boldsymbol{b}}_0$ then $\boldsymbol{a}$ is fruitful w.r.t. z.

Proof. By Lemma 3354 $\boldsymbol{a}$ is symmetric and $\boldsymbol{a}_1$ is symmetric and separative. Since every element of $\boldsymbol{a}$ is hereditary symmetric we have $\boldsymbol{a}_1 = \boldsymbol{a}$ and the last condition for fruitfulness is satisfied vacuously.

3356. METATHEOREM. Let $\boldsymbol{\alpha}$ be a well defined constant for an ordinal number, let $\omega(x, z)$ well describe a symmetric **g**-filter and suppose that $\varphi(x)$ is $\boldsymbol{\alpha}$-boundable and has a permutation model determined by ω. Then the following assertion is provable in **TSS**, U3, E2, col $(\mathbf{q}, \boldsymbol{\alpha})$ & $\omega(\mathbf{q}, \mathbf{z})$:

Let $\boldsymbol{b}_0 \in \mathbf{Ker}$ be a complete Boolean algebra, and let $\mathbf{z}\!\downarrow$ be the filter on **Ur** defined in 3349. If l is the ideal of all subsets of **Ur** of cardinality less than $\aleph_\beta$, where **Ur** is of power $\aleph_\beta$, then $\boldsymbol{a} = \langle a, \leqq \rangle = \prod_{\mathbf{Ur}}^{l} \dot{\boldsymbol{b}}_0$ is fruitful w.r.t. $\mathbf{z}\!\downarrow$.

Demonstration. We proceed in the above-mentioned theory. Let $\boldsymbol{b}$ be an algebra with base $\boldsymbol{a}$. Let $k = h\!\downarrow \cap \mathbf{PInv}(e)$ where $h \in \mathbf{z}$ and $e \in \mathbf{P}_{\text{fin}}(\mathbf{Ur})$. Let $u \in a$ and put $v = u \upharpoonright e$; we prove $v \in a_1$ and $v \leqq \bigvee_{p \in h} \check{p}'u$. Evidently, v is symmetric. If $w \in a$ and $w \leqq v$ then we shall find a $p \in k$ such that $\check{p}'u \wedge \wedge\, w \neq 0_{\boldsymbol{b}}$; this will prove $v \leqq \bigvee_{p \in k} \check{p}'u$. Let p be a permutation such that (1) $\hat{p}$ is identical on $\mathbf{q}$, (2) $p \in \mathbf{PInv}(e)$ and (3) $\mathbf{D}(w) \cap p''(\mathbf{D}(u) - e) = 0$. (On each $x \in \mathbf{q}$, p maps $\mathbf{D}(u) - e$ onto a subset of x disjoint from $\mathbf{D}(w) \cup e$; recall $x \approx \aleph_\beta$!) Then $p \in k$ and $w \wedge \check{p}'u \neq 0_{\boldsymbol{b}}$.

Finally we prove a lemma which will be useful later on:

3357. LEMMA (**TSS**, U3). Let $l = l_{\aleph_\beta}$ be the ideal of all subsets of $\mathbf{Ur}$ of power less than $\aleph_\beta$, where $\aleph_\beta$ is the power of $\mathbf{Ur}$. Let $\boldsymbol{b}_0 \in \mathbf{Ker}$ be a complete Boolean algebra and let $\boldsymbol{b}$ be an algebra with base $\boldsymbol{a} = \prod^{l}_{\mathbf{Ur}} \dot{\boldsymbol{b}}_0$. For $e \in l$ let $\boldsymbol{b}/e$ be the subalgebra of $\boldsymbol{b}$ with base $\boldsymbol{a}/e = \prod^{l/e}_{\mathbf{Ur}} \boldsymbol{b}_0$. (We suppose that the algebra $\boldsymbol{b}$ is in "canonical form", i.e. its elements are the saturated subsets of its base.)

(1) If $u \in b$ and $\mathbf{Inv}(u) \supseteq \mathbf{PInv}(e)$ then $u \in b/e$.

(2) If $q \subseteq b$ is a disjointed system such that $\mathbf{Inv}(q) \supseteq \mathbf{PInv}(e)$, then $q \subseteq b/e$.

Proof. (1) We prove that $v \upharpoonright e \leqq u$ whenever $v \in a$, $v \leqq u$, $u \in b$. Given $v \in a$ such that $v \leqq u$, let w be an arbitrary element of a such that $w \leqq v \upharpoonright e$. Set $e_1 = \mathbf{D}(v)$ and $e_2 = \mathbf{D}(w)$; let p be a permutation identical on e and such that the image of $(e_1 \cup e_2) - e$ by p is disjoint from $(e_1 \cup e_2) - e$. Then $p \in \mathbf{PInv}(e)$ so that $\check{p}'v \leqq u$, and $w \wedge \check{p}'v \neq 0_{\boldsymbol{b}}$. Hence setting $\bar{w} = = w \wedge \check{p}'v$ we have: for each $w \leqq v \upharpoonright e$, there is a $\bar{w} \leqq w$ such that $\bar{w} \leqq u$. If follows that $v \upharpoonright e \leqq u$ and hence u is the sum of all $v \upharpoonright e$ such that $v \leqq u$. Consequently $u \in b/e$.

(2) Again it suffices to prove $v \upharpoonright e \leqq u$ for $u \in q$, $v \leqq u$, $v \in a$. We set $e_1 = \mathbf{D}(v)$; let p be a permutation which is identical on e and maps $e_1 - e$ onto a set disjoint from $e_1 - e$. We have $\check{p}'v \wedge v \neq 0_{\boldsymbol{b}}$ and $\check{p}'v \leqq \check{p}'u \in q$. Since $\check{p}'u \wedge u \neq 0_{\boldsymbol{b}}$ and q is disjointed, we have $\check{p}'u = u$ and $\check{p}'v \leqq u$. We prove that $v \upharpoonright e \leqq \bar{v}$ where $\bar{v}$ is the join of all $\check{p}'v$ for which p has the property just described. Let w be an element of a such that $w \leqq v \upharpoonright e$; we shall find some p having the property and such that $w \wedge \check{p}'v \neq 0_{\boldsymbol{b}}$. To do this it suffices to take a permutation which is identical on e and which maps the set $(e_1 \cup \mathbf{D}(w)) - e$ onto a disjoint set.

SECTION 4

Definable sets and the consistency of the axiom of choice (E2); the third axiom of choice

In the present Section we shall prove the consistency of the axiom of choice. Most of our work will be done in the theory **TSS**, (D3); we shall define a certain model-class **HDf** such that the axiom of choice holds in the sense of $\mathfrak{St}$ (**HDf**), i.e. there is a relation R such that $\mathrm{Real}^{\mathbf{HDf}}(R)$ and R is a well-ordering of **HDf**.

We start with the class $\mathscr{P}_{\mathbf{On}}$ of all sets p_λ such that λ is a limit; all of these sets are **V**-like.

The closure of the class $\mathscr{P}_{\mathbf{On}}$ under the gödelian operations will be denoted by **Df**; the elements of **Df** will be called definable sets*).

The class $\mathscr{P}_{\mathbf{On}}$ can be well-ordered and so the same is true of **Df**. In fact we can define a well-ordering of **Df**.

To speak imprecisely, **Df** is the maximal definable class with a definable well-ordering; a precise version of this statement is given below. The class **Df** itself need not be a model-class; however, the class **HDf** of all hereditarily definable sets (i.e. definable sets of definable sets of definable sets ...) is a model-class. The construction of this class can be generalized in a natural way and we can define the class of all (hereditarily) definable sets relative to a given real class X.

We now carry out in detail the program sketched above.

3401. Definition (**TSS**′).

$$\mathscr{P}_{\mathbf{On}} = \{p_\lambda;\ \lambda \in \mathbf{On}_{\mathrm{II}}\}\,,$$

$$\mathbf{Cg}_0(x) = x\,,$$

*) Although the term "definable" is not unnatural in this context, the notion **Df** is not intended as an explication of the intuitive notion of definability.

$$\mathbf{Cg}_{n+1}(x) = \mathbf{Cg}_n(x) \cup \{\mathbf{F}_i(u, v);\, u, v \in \mathbf{Cg}_n(x),\, i = 1, \ldots, 7\},$$
$$\mathbf{Cg}(x) \quad = \bigcup_{n \in \omega} \mathbf{Cg}_n(x)$$

(*closure under the gödelian operations*).

3402. LEMMA (**TSS′**). If $x \in \mathbf{Cg}(y)$ and $y \subseteq \mathbf{Cg}(z)$ then $x \in \mathbf{Cg}(z)$.

Proof. By induction.

3403. DEFINITION (**TSS′**). $\mathbf{Df} = \bigcup_{\lambda \in \mathbf{On}_{II}} \mathbf{Cg}(\{p_\lambda\})$ (the *class of all definable sets*).

3404. LEMMA (**TSS′**). If X is real and closed and if $\mathscr{P}_{\mathbf{On}} \subseteq X$ then $\mathbf{Df} \subseteq X$.

Proof. Obvious.

3405. LEMMA (**TSS′**). The class **Df** is closed.

Proof. Let x and y be in **Df**; suppose that $x \in \mathbf{Cg}(\{p_\alpha\})$ and $y \in \mathbf{Cg}(\{p_\beta\})$; (throughout the present Section, $\alpha, \beta, \gamma \ldots$, will denote limit ordinals). By Lemma 3402 it suffices to prove that p_α and p_β are both in $\mathbf{Cg}(\{p_\gamma\})$ for some γ. We prove this by the Reflection Principle. Suppose that there exist p_α and p_β for which there is no γ such that both p_α and p_β are in $\mathbf{Cg}(\{p_\gamma\})$. There is a gödelian term $\mathbf{T}(\mathbf{V})$ such that $\mathbf{T}(\mathbf{V}) = \{\langle p_\alpha, p_\beta \rangle\}$ where $\alpha < \beta$ and $\{\alpha, \beta\}$ is the **Sd**-least pair of ordinals with the property $\neg (\exists \gamma)(p_\alpha, p_\beta \in \in \mathbf{Cg}(\{p_\gamma\}))$. By the Reflection Principle there exists $\delta > \alpha, \beta$ such that $\mathbf{T}(p_\delta) = \mathbf{T}(\mathbf{V}) \cap p_\delta = \{\langle p_\alpha, p_\beta \rangle\}$; this is a contradiction, since then $p_\alpha, p_\beta \in \in \mathbf{Cg}(\{p_\delta\})$ *).

3406. METATHEOREM. There is a gödelian term $\mathbf{T}(X)$ without constants such that the following is provable in **TSS′** : $\mathbf{T}(\mathbf{V})$ is a one-to-one mapping of **Df** onto **On**.

Demonstration. It suffices to define in **TSS′** a 1-1 mapping of $\mathbf{Cg}(\{p_\alpha\})$ into $\boldsymbol{\omega}$. It is then easy to define a mapping $S = \mathbf{S}(\mathbf{V})$ which enumerates all elements of **Df** by means of pairs $\{\alpha, n\}$ as follows: $S'x = = \{\alpha, n\} \equiv x$ is the n-th element in $\mathbf{Cg}(\{p_\alpha\}) - \bigcup_{\substack{\gamma < \alpha \\ \gamma \in \mathbf{On}_{II}}} \mathbf{Cg}(\{p_\gamma\})$. Using the ordering **Sd** we can define a 1-1 mapping $\mathbf{T}(\mathbf{V})$ of **Df** onto **On**. We shall now define in **TSS′** a 1-1 mapping of $\mathbf{Cg}(\{p_\alpha\})$ into $\boldsymbol{\omega}$. First we define a set $t = \mathbf{t}(\mathbf{V})$ together with a well-ordering of t. Let $t_0 = \{0\}$, $t_{n+1} = \{\langle i, u, v \rangle;$

*) Cf. Metatheorem 3128.

$i = 1, \ldots, 7,\ u, v \in \bigcup_{i \leq n} t_i\}$, $\bar{t}_n = t_n - \bigcup_{i<n} t_i$ and $t = \bigcup_{i \in \omega} t_i$. The sets $\bar{t}_n$ are mutually disjoint. The set t_0 is well-ordered by $r_0 = 0$. Suppose that $\bigcup_{i \leq n} t_i$ is well ordered by r_n; we define a well-ordering r_{n+1} of $\bigcup_{i \leq n+1} t_i$ as follows:

$$\langle u, v \rangle \in r_{n+1} \equiv \langle u, v \rangle \in r_n \vee (u \in \bigcup_{i \leq n} t_i \,\&\, v \in \bar{t}_{n+1}) \vee$$
$$\vee (u, v \in \bar{t}_{n+1} \,\&\, (\exists j, k, u_1, u_2, v_1, v_2)$$
$$(u = \langle j, u_1, u_2 \rangle \,\&\, v = \langle k, v_1, v_2 \rangle \,\&\, [j < k \vee (j = k \,\&\, \langle u_1, v_1 \rangle \in r_n) \vee$$
$$\vee (j = k \,\&\, u_1 = v_1 \,\&\, \langle u_2, v_2 \rangle \in r_n)])) .$$

Thus the elements of t are well ordered first according to the indices of the sets $\bar{t}_n$ to which they belong; if two elements are in the same set $\bar{t}_n$ then we order them according to their first members (number 1, ..., 7); if the first members are the same then we order them according to their second members which are already ordered by hypothesis; finally, if the second members are the same then we order them according to their third members.

The relation $r = \bigcup_{n \in \omega} r_n$ well orders t and since each t_n is finite, r is isomorphic to $\mathbf{E} \upharpoonright \omega_0$; hence there is a gödelian term $\mathbf{i}(\mathbf{V})$ which defines a 1-1 mapping i of ω_0 onto t. For a limit number α we define a mapping f of t onto $\mathbf{Cg}(\{p_\alpha\})$ as follows: $f`0 = p_\alpha$ and $f`\langle i, u, v \rangle = \mathbf{F}_i(f`u, f`v)$. Finally, there is a gödelian term $\mathbf{R}(\mathbf{V}, \alpha)$ which defines a one-to-one mapping R of $\mathbf{Cg}(\{p_\alpha\})$ into ω_0 as follows:

$R`x = n \equiv n$ is the least number such that $f`i`n = x$. (The reader should verify in detail that $\mathbf{R}$ is a gödelian term and that we can prove in **TSS′** that $\mathbf{R}(\mathbf{V}, \alpha)$ is a one-to-one mapping of $\mathbf{Cg}(\{p_\alpha\})$ into ω_0.) This completes the demonstration.

3407. Definition (**TSS′**). A set x is in **HDf** if $x \in$ **Df** and **Unv** $(x) \subseteq$ **Df**.

3408. Theorem (**TSS**, D3, ∂**TSS**$/\partial\mathfrak{S}$**t** (**HDf**)). **HDf** is a model-class and Axiom (E2) holds in $\mathfrak{S}$**t** (**HDf**).

Proof. 1) **HDf** is closed, since **Df** is closed.

2) **HDf** is complete by definition.

3) **HDf** is almost universal. If $x \subseteq$ **HDf** then $x \subseteq$ **HDf** $\cap\, p_\alpha$ for some α. **HDf** is defined by a normal definition; i.e. **HDf** $=$ **T**(**V**) where **T**(X) is a certain gödelian term. By the Reflection Principle there exists $\lambda > \alpha$

such that $z = \mathbf{HDf} \cap p_\lambda = \mathbf{T}(p_\lambda)$. Hence $z \in \mathbf{Cg}(\{p_\lambda\})$ and $z \subseteq \mathbf{HDf}$; hence $x \subseteq z$ and $z \in \mathbf{HDf}$.

4) If $\mathbf{T}(\mathbf{V})$ is the one-to-one mapping of **Df** onto **On** constructed by Metatheorem 3406, then $\mathbf{T}(\mathbf{V}) \upharpoonright \mathbf{HDf}$ is a one-to-one mapping of **HDf** into **On**; hence there is a gödelian term $\hat{\mathbf{T}}(\mathbf{V})$ which defines a one-to-one mapping of **HDf** onto **On** since **HDf** is a proper real class. By the following Metatheorem we get $\mathrm{Real}^{\mathfrak{St}(\mathbf{HDf})}(\hat{\mathbf{T}}(\mathbf{V}))$:

3409. METATHEOREM. The following is provable in **TSS** + D3 for any gödelian term $\mathbf{T}(X)$ without constants:

$$(\forall x \in \mathbf{HDf})(\mathbf{T}(\mathbf{V}) \cap x \in \mathbf{HDf}).$$

Demonstration. Let $\mathbf{T}(X)$ be such a gödelian term; we prove the above formula in **TSS** + D3. If $x \in \mathbf{HDf}$ then $\mathbf{T}(\mathbf{V}) \cap x \subseteq \mathbf{HDf}$; hence it suffices to prove $\mathbf{T}(\mathbf{V}) \cap x \in \mathbf{Df}$. By the Reflection Principle there exists p_α such that $x \subseteq p_\alpha$ and $\mathbf{T}(\mathbf{V}) \cap p_\alpha = \mathbf{T}(p_\alpha)$; hence $\mathbf{T}(\mathbf{V}) \cap x = \mathbf{T}(\mathbf{V}) \cap p_\alpha \cap x = \mathbf{T}(p_\alpha) \cap x \in \mathbf{Df}$.

3410. COROLLARY. Axiom (E2) is consistent with **TSS**, D3; the model $\mathfrak{Dir}(\partial\mathbf{TSS}/\partial\mathfrak{St}(\mathbf{HDf}))$ is a model of **TSS**, D3, E2 in **TSS**, D3, $\partial\mathbf{TSS}/\partial\mathfrak{St}(\mathbf{HDf})$.

The above construction may be generalized. We start now with elements of some fixed real class X (in addition to the p_λ).

3411. DEFINITION (**TSS**′).

$$\mathbf{Df}(X) = \bigcup_{\substack{\lambda \in \mathbf{On}_{II} \\ e \in \mathbf{P}_{\mathrm{fin}}(X)}} \mathbf{Cg}(e \cup \{p_\lambda, X \cap p_\lambda\}),$$

$$x \in \mathbf{HDf}(X) \equiv x \in \mathbf{Df}(X) \,\&\, \mathbf{Unv}(x) \subseteq \mathbf{Df}(X).$$

3412. *Remark*. (**TSS**′). $\mathbf{Df} = \mathbf{Df}(0) = \mathbf{Df}(\mathscr{P}_{\mathbf{On}})$.

3413. LEMMA (**TSS**′). If $\mathbf{Df}(X) \subseteq \mathbf{Df}(Y)$ then $\mathbf{HDf}(X) \subseteq \mathbf{HDf}(Y)$.
Proof. Obvious.

3414. THEOREM (**TSS**, D3). $\mathbf{HDf}(X)$ is a model-class.
Proof. 1) $\mathbf{Df}(X)$ is closed and hence $\mathbf{HDf}(X)$ is closed. If

$$x \in \mathbf{Cg}(e_1 \cup \{p_{\lambda_1}, X \cap p_{\lambda_1}\}) \text{ and } y \in \mathbf{Cg}(e_2 \cup \{p_{\lambda_2}, X \cap p_{\lambda_2}\})$$

then p_{λ_1} and p_{λ_2} are both in some $\mathbf{Cg}(\{p_\lambda\})$ and $\mathbf{F}_i(x, y) \in \mathbf{Cg}(e_1 \cup e_2 \cup \cup \{p_\lambda, X \cap p_\lambda\})$ since $X \cap p_{\lambda_1} = (X \cap p_\lambda) \cap p_{\lambda_1}$.

2) $\mathbf{HDf}(X)$ is complete by definition.

3) $\mathbf{HDf}(X)$ is almost universal by the Reflection Principle. Indeed, if $x \subseteq \subseteq \mathbf{HDf}(X) \cap p_\alpha$ and $\mathbf{HDf}(X) = \mathbf{T}(\mathbf{V}, X)$ then there exists $p_\lambda \supseteq p_\alpha$ such that $p_\lambda \cap \mathbf{T}(\mathbf{V}, X) = \mathbf{T}(p_\lambda, X \cap p_\lambda)$ and $p_\lambda \cap \mathbf{T}(\mathbf{V}, X) \supseteq x$; hence $p_\lambda \cap \cap \mathbf{T}(\mathbf{V}, X) \in \mathbf{Df}(X)$. Since $p_\lambda \cap \mathbf{T}(\mathbf{V}, X) \subseteq \mathbf{HDf}(X)$ we have $p_\lambda \cap \cap \mathbf{T}(\mathbf{V}, X) \in \mathbf{HDf}(X)$.

3415. THEOREM (**TSS′**). Let X, Y be real classes and suppose $Y \subseteq X$. Then $\mathbf{Df}(Y) \subseteq \mathbf{Df}(X)$ iff $p_\alpha \cap Y \in \mathbf{Df}(X)$ for every limit number α.

Proof. If $\mathbf{Df}(Y) \subseteq \mathbf{Df}(X)$ then the condition obviously holds. If the condition holds then for any $e \in \mathbf{P}_{\mathbf{fin}}(Y)$ and any $\alpha \in \mathbf{On}_{\mathbf{II}}$ we have $e \cup \cup \{p_\alpha, Y \cap p_\alpha\} \subseteq \mathbf{Df}(X)$ and hence $\mathbf{Cg}(e \cup \{p_\alpha, Y \cap p_\alpha\}) \subseteq \mathbf{Df}(X)$; thus $\mathbf{Df}(Y) \subseteq \mathbf{Df}(X)$.

3416. COROLLARY (**TSS′**). 1) $\mathbf{Df}(0) \subseteq \mathbf{Df}(X)$.

2) If $z \in \mathbf{Df}(X)$ then $\mathbf{Df}(X \cap z) \subseteq \mathbf{Df}(X)$.

3417. METATHEOREM. Let $\mathbf{O}(Y, X, Z_1, \bullet)$ be a gödelian term (containing no constants). The following is provable in **TSS**, D3, Real $(\mathbf{X})$, $\partial\mathbf{TSS}/\partial\mathfrak{St}(\mathbf{HDf}(\mathbf{X}))$:

Let $\mathbf{Z}_1, \bullet \subseteq \mathbf{HDf}(\mathbf{X})$ and $\mathbf{O}(\mathbf{V}, \mathbf{X}, \mathbf{Z}_1, \bullet) \subseteq \mathbf{HDf}(\mathbf{X})$. If the classes $\mathbf{Z}_1, \bullet$ are real in the sense of $\mathfrak{St}(\mathbf{HDf}(\mathbf{X}))$ then $\mathbf{O}(\mathbf{V}, \mathbf{X}, \mathbf{Z}_1, \bullet)$ is also real in the sense of $\mathfrak{St}(\mathbf{HDf}(\mathbf{X}))$. In particular, if $\mathbf{X} \subseteq \mathbf{HDf}(\mathbf{X})$ then $\mathrm{Real}^{\mathfrak{St}(\mathbf{HDf}(\mathbf{X}))}(\mathbf{X})$.

Demonstration. We proceed in **TSS**, D3, Real $(\mathbf{X})$, $\partial\mathbf{TSS}/\partial\mathfrak{St}(\mathbf{HDf}(\mathbf{X}))$. If $x \in \mathbf{HDf}(\mathbf{X})$ then we prove $x \cap \mathbf{O}(\mathbf{V}, \mathbf{X}, \mathbf{Z}_1, \bullet) \in \mathbf{HDf}(\mathbf{X})$. By the Reflection Principle there exists p_λ such that $x \cap \mathbf{O}(\mathbf{V}, \mathbf{X}, \mathbf{Z}_1, \bullet) \subseteq p_\lambda$ and $\mathbf{O}(\mathbf{V}, \mathbf{X}, \mathbf{Z}_1, \bullet) \cap p_\lambda = \mathbf{O}(p_\lambda, \mathbf{X} \cap p_\lambda, \mathbf{Z}_1 \cap p_\lambda, \bullet)$. Since $x, p_\lambda, \mathbf{X} \cap p_\lambda, \mathbf{Z}_1 \cap p_\lambda, \bullet \in \mathbf{Df}(\mathbf{X})$ we have $x \cap \mathbf{O}(\mathbf{V}, \mathbf{X}, \mathbf{Z}_1, \bullet) = x \cap \mathbf{O}(p_\lambda, \mathbf{X} \cap p_\lambda, \mathbf{Z}_1 \cap p_\lambda, \bullet) \in \mathbf{Df}(\mathbf{X})$; moreover, $x \cap \mathbf{O}(\mathbf{V}, \mathbf{X}, \mathbf{Z}_1, \bullet) \subseteq \mathbf{HDf}(\mathbf{X})$ since $x \subseteq \subseteq \mathbf{HDf}(\mathbf{X})$ and therefore $x \cap \mathbf{O}(\mathbf{V}, \mathbf{X}, \mathbf{Z}_1, \bullet) \in \mathbf{HDf}(\mathbf{X})$.

3418. THEOREM (**TSS**, D3, Real $(\mathbf{X})$ & $\mathbf{X} \subseteq \mathbf{On}$, $\partial\mathbf{TSS}/\partial\mathfrak{St}(\mathbf{HDf}(\mathbf{X}))$). Axiom (E2) holds in the sense of $\mathfrak{St}(\mathbf{HDf}(\mathbf{X}))$.

Proof. We construct a real mapping of $\mathbf{HDf}(\mathbf{X})$ into $\mathbf{On}$. We have a regular real well-ordering $\mathbf{Sd}$ of $\mathbf{P}_{\mathbf{fin}}(\mathbf{On})$ and hence a real regular well-ordering of $\mathbf{P}_{\mathbf{fin}}(\mathbf{X})$. For every $e \in \mathbf{P}_{\mathbf{fin}}(\mathbf{X})$ and every $\lambda \in \mathbf{On}_{\mathbf{II}}$ we can define a one-one set-mapping $\mathbf{R}(e, \lambda)$ of $\mathbf{Cg}(e \cup \{p_\lambda, \mathbf{X} \cap p_\lambda\})$ onto $\boldsymbol{\omega}$. The theorem follows in the same way as Theorem 3408.

We shall now return to the class **HDf** = **HDf** (0). We know that **HDf** is defined by a gödelian term (in **TSS** + D3) and we have defined a constant **S** such that **TSS** + D3 ⊢ (**S** is a 1-1 mapping of **Df** onto **On**). We shall show that the class **Df** is, in a sense, maximal among classes having this property.

3419. METATHEOREM. If $\mathbf{T}(X)$ and $\mathbf{S}(X)$ are gödelian terms without constants then the following can be proved in **TSS** + D3:
If $\mathbf{S}(\mathbf{V})$ is a well-ordering of $\mathbf{T}(\mathbf{V})$ then $\mathbf{T}(\mathbf{V}) \subseteq \mathbf{Df}$.

Demonstration. We proceed in **TSS** + D3. Suppose that $\mathbf{T}(\mathbf{V}) - \mathbf{Df} \neq 0$ and let $\mathbf{A}(\mathbf{T}(\mathbf{V}) - \mathbf{Df})$ be the set of elements of least rank in $\mathbf{T}(\mathbf{V}) - \mathbf{Df}$. From $\mathbf{S}(\mathbf{V})$ we may define a selector, i.e. a mapping $\hat{\mathbf{S}}(\mathbf{V})$ such that $(\forall x \neq 0)$ $(x \subseteq \mathbf{T}(\mathbf{V}) \rightarrow \hat{\mathbf{S}}(\mathbf{V})\text{'}\, x \in x)$ & $\mathrm{Un}(\hat{\mathbf{S}}(\mathbf{V}))$. Set $z = \hat{\mathbf{S}}(\mathbf{A}(\mathbf{T}(\mathbf{V}) - \mathbf{Df}))$. Clearly $z = \mathbf{o}(\mathbf{V})$ where $\mathbf{o}$ is some gödelian term; by the Reflection Principle there exists $\lambda \in \mathbf{On}_{\mathrm{II}}$ such that $z \in p_\lambda$ and $p_\lambda \cap \mathbf{o}(\mathbf{V}) = \mathbf{o}(p_\lambda) = z$. Hence $z \in \mathbf{Df}$, a contradiction.

3420. AXIOM OF DEFINABILITY (E3). $\mathbf{V} = \mathbf{Df}$.

3421. COROLLARY (**TSS** + D3). (a) (E3) implies $\mathbf{V} = \mathbf{HDf}$.
(b) (E3) implies (E2).

3422. Remark. The consistency of Axiom (E3) will be established in the next Section.

The above Metatheorem shows that Axiom (E3) is the weakest of the so-called strong axioms of choice. By a strong axiom of choice we mean an axiom of the form "$\mathbf{T}(\mathbf{V})$ is a one-to-one mapping of **V** onto **On**" where $\mathbf{T}(X)$ is a gödelian term without constants. Every strong axiom of choice implies $\mathbf{V} \subseteq \mathbf{Df}$ and hence Axiom (E3). The authors feel that the notion of a strong axiom of choice corresponds exactly to the original intuitive conception of the axiom of choice: a rule can be presented which assigns to each nonempty set one of its elements.

To conclude this Section we show that there is a gödelian term $\mathbf{T}(X)$ without constants and such that the following is provable in **TSS** + D3:

$$(\forall x \in \mathbf{Df})(\exists \alpha \in \mathbf{On}_{\mathrm{II}})(x = \mathbf{T}(p_\alpha)).$$

We may call such a term *universal for definable sets.* (This result will not be used anywhere in the sequel.)

3423. METALEMMA. There is a gödelian term $\mathbf{S}(X, Y)$ such that

$$\mathbf{TSS} + \mathrm{D3} \vdash (\forall x \in \mathbf{Df})(\exists \alpha, \beta \in \mathbf{On}_{\mathrm{II}})(\alpha > \beta \;\&\; x = \mathbf{S}(p_\alpha, p_\beta)).$$

Demonstration. Clearly, we can define a gödelian term $\mathbf{G}(X)$ such that

$$\mathbf{TSS} + \mathrm{D3} \vdash \mathrm{Un}_2(\mathbf{G}(\mathbf{V}), \mathscr{P}_{\mathbf{On}}, \mathbf{Df}).$$

Hence

$$\mathbf{TSS} + \mathrm{D3} \vdash (\forall x \in \mathbf{Df})(\exists \beta \in \mathbf{On}_{\mathrm{II}})(x = \mathbf{G}(\mathbf{V})\text{'}\, p_\beta).$$

By the Reflection Principle we have in **TSS** + D3: There exists $\alpha > \beta$ such that $x \in p_\alpha \,\&\, \mathbf{G}(\mathbf{V}) \cap p_\alpha = \mathbf{G}(p_\alpha)$ and hence

$$(\forall x \in \mathbf{Df})(\exists \beta \in \mathbf{On}_{\mathrm{II}})(\exists \alpha > \beta)(\alpha \in \mathbf{On}_{\mathrm{II}} \,\&\, x = [\mathbf{G}(p_\alpha)]\text{'}\, p_\beta).$$

3424. METALEMMA. There are gödelian terms $\mathbf{R}(X)$ and $\mathbf{Q}(X)$ such that

$$\mathbf{TSS} + \mathrm{D3} \vdash (\forall \alpha, \beta \in \mathbf{On}_{\mathrm{II}})(\alpha > \beta \rightarrow (\exists \gamma \in \mathbf{On}_{\mathrm{II}})(p_\alpha = \mathbf{R}(p_\gamma) \,\&\, p_\beta = \\ = \mathbf{Q}(p_\gamma)).$$

Sketch of the demonstration. (1) We proceed in **TSS** + D3. Let $\mathbf{Pa} = \{\langle \alpha, \beta \rangle; \alpha, \beta \in \mathbf{On} \,\&\, \alpha > \beta\}$,
$\mathbf{Sdp} = \{\langle \langle \alpha_1, \beta_1 \rangle, \langle \alpha_2, \beta_2 \rangle \rangle; \langle \alpha_1, \beta_1 \rangle, \langle \alpha_2, \beta_2 \rangle \in \mathbf{Pa} \,\&\, \{\alpha_1, \beta_1\}\, \mathbf{Sd}\, \{\alpha_2, \beta_2\}\}$ (see 2226 for the definition of **Sd**).

Sdp is a real regular well-ordering of **Pa**; hence we can define a unique isomorphism **F** between **Pa** and **On** w.r.t. **Sdp** and **E**. The isomorphism **F** has the following properties:

(a) $\mathbf{F}\text{'}\langle \alpha, \beta \rangle \geqq \alpha$,

(b) $\beta \neq 0 \rightarrow \mathbf{F}\text{'}\langle \alpha, \beta \rangle > \alpha$.

Further, let $\mathscr{P}$ be as in 3131 i.e., $\mathscr{P}\text{'}\alpha = p_\alpha$.

(2) There are gödelian terms $\mathbf{T}_i(X)$ $(i = 1, \dots, 4)$ such that the following is provable in **TSS** + D3:

(c) $\mathbf{T}_1(\mathbf{V}) = \mathbf{Pa}$, $\mathbf{T}_2(\mathbf{V}) = \mathbf{Sdp}$, $\mathbf{T}_3(\mathbf{V}) = \mathbf{F}$, $\mathbf{T}_4(\mathbf{V}) = \mathscr{P}$,

(d) for every $\lambda \in \mathbf{On}_{\mathrm{II}}$, $\mathbf{T}_i(p_\lambda) = p_\lambda \cap \mathbf{T}_i(\mathbf{V})$.

(Hint: for $i = 1, 2$ there are RF's $\varphi_i(x)$ such that

$$\mathbf{TSS} + \mathrm{D3} \vdash x \in \mathbf{T}_i(\mathbf{V}) \equiv \varphi_i(x);$$

for $i = 3, 4$ there are RF's $\varphi_i(x, f, z)$ such that

$$\mathbf{TSS} + \text{D3} \vdash \lambda \in \mathbf{On}_{\text{II}} \,\&\, x \in p_\lambda \,.\rightarrow.\, x \in \mathbf{T}_i(\mathbf{V}) \equiv (\exists f \in p_\lambda)\, \varphi(x, f, p_\lambda)\,.$$

Hence the result follows by 3126.

(3) Let $\mathbf{T}_0(p_\lambda) = \mathbf{D}(\mathbf{T}_3(p_\lambda))$. Then the following is provable in $\mathbf{TSS}$ + + D3: Let $\langle \alpha, \beta \rangle \in \mathbf{Pa}$, $\alpha, \beta \in \mathbf{On}_{\text{II}}$ and let $\mathbf{F}\text{'}\langle \alpha, \beta \rangle = \lambda$. Then $\lambda \in \mathbf{On}_{\text{II}}$, $\alpha = \bigcup(\mathbf{W}(\mathbf{T}_0(p_\lambda)))$ and $\beta = \bigcup[(\{\alpha\} \times \lambda) \cap \mathbf{D}(\mathbf{T}_0(p_\lambda))]$. Further, if $x \subseteq \subseteq p_\lambda$ then $\bigcup x = \mathbf{W}(\mathbf{E}(p_\lambda) \upharpoonright x)$. Hence we can find gödelian terms $\mathbf{R}_0(X)$, $\mathbf{S}_0(X)$ such that

$$\mathbf{TSS} + \text{D3} \vdash \langle \alpha, \beta \rangle \in \mathbf{Pa} \,\&\, \alpha, \beta \in \mathbf{On}_{\text{II}} \,.\rightarrow$$
$$(\exists \lambda \in \mathbf{On}_{\text{II}})\,(\alpha = \mathbf{R}_0(p_\lambda) \,\&\, \beta = \mathbf{S}_0(p_\lambda))\,.$$

(4) The following is provable in $\mathbf{TSS}$ + D3: Let $\langle \alpha, \beta \rangle \in \mathbf{Pa}$, $\alpha, \beta \in \mathbf{On}_{\text{II}}$ and let $\lambda = \mathbf{F}\text{'}\langle \alpha, \beta \rangle$; then $\mathscr{P} \cap p_\lambda = \mathbf{T}_4(p_\lambda)$,

$$\{p_\alpha\} = \{\mathscr{P}\text{'}\alpha\} = (\mathscr{P} \cap p_\lambda) \cap (p_\lambda \times \{\mathbf{R}_0(p_\lambda)\})\,,$$
$$\{p_\beta\} = \{\mathscr{P}\text{'}\beta\} = (\mathscr{P} \cap p_\lambda) \cap (p_\lambda \times \{\mathbf{S}_0(p_\lambda)\})\,,$$
$$p_\alpha = \bigcup\{p_\alpha\}\,, \quad p_\beta = \bigcup\{p_\beta\}\,.$$

Hence we can find gödelian terms $\mathbf{R}(X)$, $\mathbf{S}(X)$ such that

$$\mathbf{TSS} + \text{D3} \vdash \langle \alpha, \beta \rangle \in \mathbf{Pa} \rightarrow (\exists \lambda \in \mathbf{On}_{\text{II}})\,(p_\alpha = \mathbf{R}(p_\lambda) \,\&\, p_\beta = \mathbf{S}(p_\lambda))\,.$$

This completes the demonstration.

If $\mathbf{S}$, $\mathbf{R}$ and $\mathbf{Q}$ are as in the two preceding Metalemmas and if we let $\mathbf{T}(X) = \mathbf{S}(\mathbf{R}(X), \mathbf{Q}(X))$ then we have the following

3425. METATHEOREM. There is a gödelian term $\mathbf{T}(X)$ such that

$$\mathbf{TSS} + \text{D3} \vdash (\forall x \in \mathbf{Df})\,(\exists \alpha \in \mathbf{On}_{\text{II}})\,(x = \mathbf{T}(p_\alpha))\,.$$

3426. COROLLARY. There is a natural number p such that the following can be proved in $\mathbf{TSS}$ + D3:

$$(\forall x \in \mathbf{Df})\,(\exists \alpha \in \mathbf{On}_{\text{II}})\,(x \in \bigcup_{i \leq p} \mathbf{Cg}_i(\{p_\alpha\}))\,.$$

The number p is determined by the complexity of the term $\mathbf{T}(X)$ and can actually be calculated.

SECTION 5

Constructible sets and the axiom of constructibility. The consistency of Axioms (E3) and (Cont)

In the present Section we shall demonstrate the consistency of the axiom of Continuum and of the axiom of Definability. For this purpose (and others) we shall prove in **TSS**, D3 that for any set there is a least model-class containing this set; in particular there exists a model-class which is included in all other model-classes. We shall show that a certain axiom of constructibility (Constr) holds in the sense of this least model-class; from this axiom we shall derive the axiom of definability and also the axiom of continuum. This is the principal aim of our work in the theory **TSS**, D3. We shall conclude this Section with a more general construction; we shall define for any real class X a model class which in a certain sense is least w.r.t. X.

3501. DEFINITION (**TSS'**). In accordance with Section 4 we define

$$\mathbf{Cg}_1(x) = x \cup \{z; (\exists u, v \in x)(z = \{u, v\} \vee z = \mathbf{E}(u) \vee z =$$
$$= u - v \vee z = \mathbf{D}(u) \vee z = u \upharpoonright v \vee z = \mathbf{Cnv}(u) \vee z = \mathbf{Cnv}_3(u))\}\,.$$

3502. METALEMMA. The formula $y = \mathbf{Cg}_1(x)$ is restricted in **TSS'**.

Demonstration. The formula $y \in \mathbf{Cg}_1(x)$ is clearly restricted; hence the formula $y \subseteq \mathbf{Cg}_1(x)$ is restricted. To show that $\mathbf{Cg}_1(x) \subseteq y$ is restricted notice that this formula is equivalent to the formula $x \subseteq y \,\&\, (\forall u, v \in x)$ $(\exists z \in y)(z = \{u, v\} \vee \ldots \vee z = \mathbf{Cnv}_3(u))$.

In order to find the least model-class containing a set x we ask: given a model-class M such that x is an element of M, what else must belong to M? The notion of absoluteness helps us to show that some sets must necessarily belong to M. First note that since we are interested in **TSS**, D3, we may use Theorem 3225 which tells us that all ordinal numbers are in M. Furthermore we have the following

3503. LEMMA ($\mathbf{TSS}$, D3, $\mathrm{Mcl}(\mathbf{M})$, $\partial\mathbf{TSS}/\partial\mathfrak{S}\mathrm{t}(\mathbf{M})$). If x^* is an element of $\mathbf{M}$ then

(a) $\bigcup^*(x^*) = \bigcup(x^*)$,
(b) $\mathbf{Unv}^*(x^*) = \mathbf{Unv}(x^*)$,
(c) $\mathbf{Cg}_1^*(x^*) = \mathbf{Cg}_1(x^*)$,
(d) $\mathbf{Cg}^*(x^*) = \mathbf{Cg}(x^*)$.

Proof. (a) The formula $y = \bigcup(x)$ is restricted in $\mathbf{TSS}$.

(b) $\mathbf{Unv}(x^*) = \bigcup \mathbf{W}(H)$ where $H'0 = x^*$, $H'(n+1) = \bigcup(H'n)$ and $\mathbf{D}(H) = \boldsymbol{\omega}$. By induction we get $H^{*\prime}n = H'n$ for $n \in \boldsymbol{\omega}$.

(c) $y = \mathbf{Cg}_1(x)$ is restricted in $\mathbf{TSS}'$ by 3502.

(d) Is proved analogously to (b).

This lemma leads us to the following

3504. DEFINITION ($\mathbf{TSS}'$). The *constructing function* $\mathscr{G}_x$ is defined inductively as follows: $\mathscr{G}_x'0 = \mathbf{Unv}(\{x\})$,

$$\mathscr{G}_x'(\alpha + 1) = \mathbf{Cg}_1(\mathscr{G}_x'\alpha) \cup \{\mathscr{G}_x'\alpha\};$$

$$\mathscr{G}_x'\lambda = \bigcup_{\alpha<\lambda} \mathscr{G}_x'\alpha \text{ for } \lambda \text{ limit;}$$

$$\mathbf{Cstr}(x) = \bigcup_{\alpha\in\mathbf{On}} \mathscr{G}_x'\alpha.$$

$\mathbf{Cstr}(x)$ is the class of all *sets constructible from* x.

3505. LEMMA ($\mathbf{TSS}$, D3, $\mathrm{Mcl}(\mathbf{M})$, $\partial\mathbf{TSS}/\partial\mathfrak{S}\mathrm{t}(\mathbf{M})$). If x^* is an arbitrary element of $\mathbf{M}$, then

$$\mathbf{Cstr}^*(x^*) = \mathbf{Cstr}(x^*).$$

Proof. It follows by induction (using Lemma 3503) that for every α, $\mathscr{G}_{x*}^{*\prime}\alpha = \mathscr{G}_{x*}'\alpha$; hence

$$\mathbf{Cstr}^*(x^*) = \bigcup_{\mathbf{On}}^* \mathscr{G}_{x*}^{*\prime}\alpha = \bigcup_{\mathbf{On}} \mathscr{G}_{x*}'\alpha = \mathbf{Cstr}(x^*).$$

3506. THEOREM. (1) ($\mathbf{TSS}'$). For any x, $\mathbf{Cstr}(x)$ is a model-class containing x.

(2) ($\mathbf{TSS}$, D3). If X is an arbitrary model-class such that $x \in X$ then $\mathbf{Cstr}(x) \subseteq X$.

Proof. (1) Since $x \in \mathbf{Unv}(\{x\})$, we have $x \in \mathbf{Cstr}(x)$. The class $\mathbf{Cstr}(x)$ is complete, since, by induction, $\mathcal{G}'_x\lambda$ is complete for every limit λ. Also $\mathbf{Cstr}(x)$ is closed, since $\mathbf{Cg}_1(\mathcal{G}'_x\alpha) \subseteq \mathcal{G}'_x(\alpha + 1)$. Finally, $\mathbf{Cstr}(x)$ is almost universal, since if $z \subseteq \mathbf{Cstr}(x)$ then for some α we have $z \subseteq \mathcal{G}'_x\alpha \in \mathbf{Cstr}(x)$.

(2) By Lemma 3505 $x \in \mathbf{M} \rightarrow \mathbf{Cstr}(x) \subseteq \mathbf{M}$ is provable in **TSS**, D3, Mcl $(\mathbf{M})$, ∂**TSS**/$\partial\mathfrak{St}(\mathbf{M})$. By 1257 the last formula is provable in **TSS**, D3, Mcl $(\mathbf{M})$ and therefore, by 1245, (b) is provable in **TSS**, D3.

3507. COROLLARY (**TSS**, D3). For every model-class X, $\mathbf{Cstr}(0) \subseteq X$.

3508. Remark. $\mathbf{Cstr}(x)$ is usually denoted by $\mathbf{L}_x$ and $\mathbf{Cstr}(0)$ by $\mathbf{L}$.

3509. THEOREM (**TSS**, D3, Mcl $(\mathbf{M})$, ∂**TSS**/$\partial\mathfrak{St}(\mathbf{M})$). If $\mathbf{M} = \mathbf{Cstr}(x^*)$ then $\mathbf{V}^* = \mathbf{Cstr}^*(x^*)$; in particular if $\mathbf{M} = \mathbf{Cstr}(0)$ then $\mathbf{V}^* = \mathbf{Cstr}^*(0^*)$.

Proof. $\mathbf{V}^*$ is $\mathbf{M}$, hence $\mathbf{V}^* = \mathbf{Cstr}(x^*)$; by Lemma 3505 $\mathbf{Cstr}(x^*) = \mathbf{Cstr}^*(x^*)$.

We are led to the following axioms:

3510. (Constr) $\mathbf{V} = \mathbf{Cstr}(0)$,
(SConstr) $(\exists x \subseteq \mathbf{On})(\mathbf{V} = \mathbf{Cstr}(x))$.

(Constr) is called the *axiom of constructibility* and (SConstr) is called the *axiom of constructibility relative to a set of ordinals.*

3511. METATHEOREM. (Constr) is consistent with **TSS**, D3; in fact, $\mathfrak{Dir}$ (∂**TSS**/$\partial\mathfrak{St}(\mathbf{Cstr}(0))$) is a faithful model of **TSS**, D3, (Constr) in **TSS**, D3, ∂**TSS**/$\partial\mathfrak{St}(\mathbf{Cstr}(0))$.

Demonstration. By 1445, 3147, 3228 and 3509 $\mathfrak{Dir}$ (∂**TSS**/$\partial\mathfrak{St}(\mathbf{Cstr}(0))$) is a model of **TSS**, D3, (Constr); to prove the faithfulness consider the following diagram:

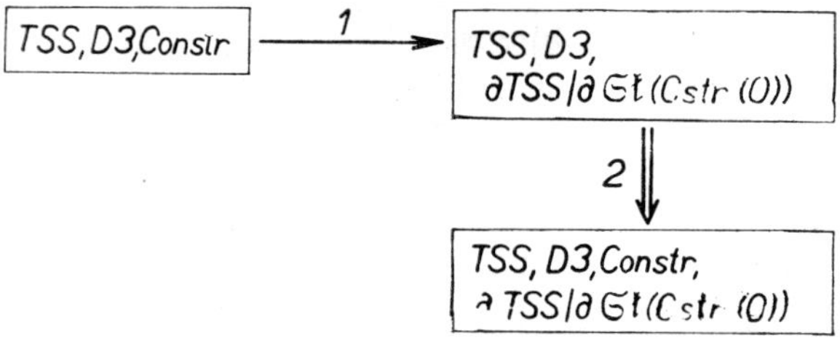

By Theorem 3509 we obtain $\mathbf{V}^* = \mathbf{V}$ in the third theory; i.e. the composition $1 * 2$ is equivalent to the identity model of the first theory in the third one and obviously the third theory is a conservative extension of the first. Hence 1 ∗ 2 is faithful and consequently 1 is also faithful.

3512. Remark. It is not difficult to show that **TSS′** ⊢ (SConstr) → (D3) and that $\mathfrak{Dir}$ (∂**TSS**/∂$\mathfrak{St}$ (**Cstr** (0))) is a (faithful) model of **TSS′**, (Constr) in **TSS′**, ∂**TSS**/∂$\mathfrak{St}$ (**Cstr** (0)). Moreover, one can prove in **TSS′** that **Cstr** (0) is a subclass of every model-class containing all ordinal numbers. But we shall not need these facts; their easy proofs are left to the reader.

Let us consider the axiom (SConstr). It is weaker than (Constr); instead of saying that the universe equals **Cstr** (0) it only says that **V** equals **Cstr** (y) for some set y of ordinal numbers. The question why y is assumed to be a set of ordinal numbers is answered by the following

3513. Theorem (**TSS**, D3). (a) If **V** = **Cstr** (y) and $y \subseteq$ **On** then **V** = = **HDf** (y) and hence (E2) holds.

(b) If **V** = **Cstr** (x) and (E1) holds then there is an $y \subseteq$ **On** such that **V** = **Cstr** (y).

Proof. (a) If $y \subseteq$ **On** then **HDf** (y) is a model-class containing y (evidently, $y \in$ **Df** (y); if $y \subseteq$ **On** then $y \subseteq$ **HDf** (y) and hence $y \in$ **HDf** (y)).

Consequently, **Cstr** (y) ⊆ **HDf** (y) and so **V** = **HDf** (y). (E2) follows by 3418.

(b) Suppose **V** = **Cstr** (**x**). By (E1), **Unv** ({**x**}) can be mapped one-one onto an ordinal number (by a function which is a set). Therefore there is a relation $r \subseteq$ **On**^{2} isomorphic to **E** ↾ **Unv** ({**x**}) (there is a set-isomorphism). Let **r** be a fixed relation with these properties and let **Is** be the isomorphism of **On**^{2}, **On** w.r.t. **Maxlex** and **E** (see 2241; note that **Is** can be defined by a formula restricted in **TSS′**). Finally denote **Is**"**r** by **y**; we prove that **Cstr** (**y**) = **V**. It suffices to prove **x** ∈ **Cstr** (**y**). By absoluteness, **Is** is a class of $\mathfrak{Dir}$ (∂**TSS**/∂$\mathfrak{St}$ (**Cstr** (**y**))) and **Is** = **Is***. (∗ denotes the notions of the direct model in question.) Moreover, Real* (**Is***) holds since **Is** is defined by a RSF and consequently, by a NSF. Hence **r** = **Cnv** (**Is**)" **y** is a set in the sense of $\mathfrak{St}$ (**Cstr** (**y**)). By 3229, **x** ∈ **Cstr** (**y**).

3514. Remark. Let $\varphi(x, r)$ be a formula saying "$r \subseteq$ **On**^{2} and r is isomorphic to **E** ↾ **Unv** ({x})". We proceeded in **TSS**, D3, E1, **V** = **Cstr** (**x**), φ(**x**, **r**), **y** = **Is**"**r**, ∂**TSS**/∂$\mathfrak{St}$ (**Cstr** (**y**)) and proved **V** = **Cstr** (**y**). Hence by 1245, 1257 the following is provable in **TSS**, D3, E1 (x, r, y now being variables): For any x, r, y, if **V** = **Cstr** (x), $\varphi(x, r)$ and y = **Is**" r then **V** = **Cstr** (y) holds. Consequently (b) is provable in **TSS**, D3.

3515. COROLLARY (**TSS**, D3). (1) (SConstr) implies (E2); (Constr) implies (E3).

(2) $(\exists x)(\mathbf{V} = \mathbf{Cstr}(x))$ implies (E1) $\equiv$ (E2).

3516. METACOROLLARY. (E3) is consistent with **TSS**, D3. (By Metatheorem 3511 and Corollary 3515.)

We now study powers of cardinals in **TSS**, D3, SConstr.

3517. LEMMA (**TSS**, D3, E1). If $\mathbf{Unv}(x)$ is subvalent to $\aleph_\alpha$ then (a) for any $\beta < \omega_\alpha$, $\mathcal{G}_x'\beta$ is subvalent to $\aleph_\alpha$; (b) for any β and γ, if $\beta \geqq \aleph_\alpha$ and $\beta \approx \aleph_\gamma$ then $\mathcal{G}_x'\beta \approx \aleph_\gamma$.

3518. THEOREM (**TSS**, D3). If $\mathbf{V} = \mathbf{Cstr}(x)$ and $x \subseteq \omega_\alpha$ then $2^{\aleph_\beta} = \aleph_{\beta+1}$ for any $\beta \geqq \alpha$.

Proof. To prove the theorem it suffices to show that $\mathbf{P}(\mathcal{G}_x'\omega_\alpha) \approx \omega_{\alpha+1}$. We prove this by finding for each $y \subseteq \mathcal{G}_x'\omega_\alpha$ some $\gamma < \omega_{\alpha+1}$ such that $y \in \mathcal{G}_x'\gamma$. First some metamathematical considerations. Consider the formula $x \subseteq \mathbf{On} \,\&\, u = \mathcal{G}_x \upharpoonright (\xi + 1)$ (with free variables u, x, ξ). We claim that it is restricted in **TSS**, D3. Indeed, it is equivalent to

$$x \subseteq \mathbf{On} \,\&\, \mathrm{Un}(u) \,\&\, u'0 = \{x\} \cup x \cup \bigcup x \,\&\, \mathbf{D}(u) =$$
$$= \xi + 1 \,\&\, (\forall \beta \in \mathbf{D}(u)) \left[(\beta + 1) \in \mathbf{D}(u) \rightarrow u'(\beta + 1) =\right.$$
$$\left.= \mathbf{Cg}_1(u'\beta) \cup \{u'\beta\} \,.\&.\, \beta \in \mathbf{On}_{\mathrm{II}} \rightarrow u'\beta = \bigcup u''\beta\right];$$

cf. the Metalemmas at the end of Chapt. I Sect 3 and Metalemma 3502. Hence let $\varphi(u, x, y)$ be a restricted **TSS**-formula such that **TSS**, D3 $\vdash \varphi(u, x, y) \equiv x \subseteq \mathbf{On} \,\&\, y \in \mathbf{On} \,\&\, u = \mathcal{G}_x \upharpoonright (y + 1)$. Further, let $\mathbf{H}(X)$ be a gödelian term without constants such that $\mathbf{H}(\mathbf{V})$ represents $\varphi(u, x, y)$ in **TC** w.r.t. u, x, y. Hence **TSS**, D3 $\vdash \langle u, x, \xi\rangle \in \mathbf{H}(\mathbf{V}) \equiv u = \mathcal{G}_x \upharpoonright (\xi + 1) \,\&\, x \subseteq \mathbf{On}$.

We return to our proof. Let $y \subseteq \mathcal{G}_x'\omega_\alpha$ and let ξ be such that $y \in \mathcal{G}_x'\xi$. Set $c_0 = \mathcal{G}_x'\omega_\alpha \cup \{y, \mathcal{G}_x \upharpoonright (\xi + 1), \xi\}$. Hence c_0 contains all elements of $\mathcal{G}_x'\omega_\alpha$ and, in addition, the three elements y, $\mathcal{G}_x \upharpoonright (\xi + 1)$, ξ. We have $c_0 \approx \aleph_\alpha$. By the Reflection Principle there is a set c of power $\aleph_\alpha$ such that $c_0 \subseteq c$ and $\mathbf{HRefl_H}(c, \mathbf{V})$ (in particular, $\mathrm{Vlk}(c)$ and $\mathbf{H}(\mathbf{V}) \cap c = \mathbf{H}(c)$). By Theorem 3215 there exists an isomorphism f w.r.t. $\mathbf{E}$, $\mathbf{E}$ which maps c onto a complete set d, again of power $\aleph_\alpha$. By Lemma 1321 we have $f''(\mathbf{H}(c)) = \mathbf{H}(d)$ and by 3126 we have $\mathbf{H}(d) = \mathbf{H}(\mathbf{V}) \cap d$. Since $\mathbf{Unv}(x) \subseteq c$ and $x \in c$ we have $f'x = x$; similarly we have $f'y = y$. Furthermore, $f'\langle\mathcal{G}_x \upharpoonright (\xi + 1), x, \xi\rangle = \langle f'(\mathcal{G}_x \upharpoonright (\xi + 1), x, f'\xi\rangle \in \mathbf{H}(d) = \mathbf{H}(\mathbf{V}) \cap$ $\cap\, d$, hence $f'(\mathcal{G}_x \upharpoonright (\xi + 1)) = \mathcal{G}_x \upharpoonright (f'\xi + 1)$ and since $\langle y, \xi\rangle \in \mathcal{G}_x \upharpoonright$ $\upharpoonright (\xi + 1)$ we have $\langle y, f'\xi\rangle \in \mathcal{G}_x \upharpoonright (f'\xi + 1)$. Since $d \approx \aleph_\alpha$ and $\mathcal{G}_x \upharpoonright f'\xi +$ $+ 1 \subseteq d$ we have $f'\xi < \boldsymbol{\omega}_{\alpha+1}$, q.e.d.

3519. Remark. 1) The same assertion can be proved under the assumption that $\mathbf{V} = \mathbf{Cstr}(x)$ where x is a subset of some ordinal ξ of power $\aleph_\alpha$.

2) The preceding theorem describes properties of the operation $\beth(2, \aleph_\beta)$ for $\beta \geq \alpha$. Properties of this operation for $\beta < \alpha$ can be derived from the following

3520. Theorem (**TSS** + D3). Let $x \subseteq \boldsymbol{\omega}_\alpha$ where $\aleph_\alpha$ is regular and let $\mathbf{V} = \mathbf{Cstr}(x)$. If $y \subseteq \boldsymbol{\omega}_\beta$ and $\beta < \alpha$ then there exists δ such that $\boldsymbol{\omega}_\beta \leqq \beta < \boldsymbol{\omega}_\alpha$ and $y \in \mathbf{Cstr}(x \cap \delta)$.

Proof. Set $c_0 = \boldsymbol{\omega}_\beta \cup \{y, \mathcal{G}_x \upharpoonright (\xi + 1), \xi, x\}$ where $y \in \mathcal{G}_x'\xi$. Since $c_0 \approx$ $\approx \aleph_\beta < \aleph_\alpha$ there exists a $\mathbf{V}$-like set $c \supseteq c_0$ such that $c \prec \boldsymbol{\omega}_\alpha$, $c \cap \boldsymbol{\omega}_\alpha$ is complete and c is $\mathbf{H}$-reflecting ($\mathbf{H}$ having the same meaning as in the proof of 3518; cf. Metatheorem 3130). We let f be an isomorphism of c onto a complete set d (w.r.t. $\mathbf{E}$, $\mathbf{E}$). Since $c \cap \boldsymbol{\omega}_\alpha$ is complete, we have $f'x =$ $= x \cap \delta$ where $\delta = c \cap \boldsymbol{\omega}_\alpha$. We also have $\langle f'(\mathcal{G}_x \upharpoonright (\xi + 1)), f'x, f'\xi\rangle \in$ $\in \mathbf{H}(d) = \mathbf{H}(\mathbf{V}) \cap d$. It follows that $f'(\mathcal{G}_x \upharpoonright (\xi + 1)) = \mathcal{G}_{x\cap\delta} \upharpoonright (f'\xi + 1)$. Furthermore $y \in d$, $f'y = y$ and $f'\xi \in d$ so that $y \in \mathcal{G}_{x\cap\delta}'(f'\xi)$.

3521. Theorem (**TSS** + D3). If $x \subseteq \boldsymbol{\omega}_\alpha$ where $\aleph_\alpha$ is regular and $\mathbf{V} =$ $= \mathbf{Cstr}(x)$ then for every $\beta < \alpha$, $2^{\aleph_\beta} \preccurlyeq \aleph_\alpha$.

Proof. For any $\delta < \boldsymbol{\omega}_\alpha$ we let $k_\delta = \{y \subseteq \boldsymbol{\omega}_\beta;\ y \in \mathbf{Cstr}(x \cap \delta)\}$; we have $\mathbf{P}(\boldsymbol{\omega}_\beta) = \bigcup_{\boldsymbol{\omega}_\beta \leqq \delta < \boldsymbol{\omega}_\alpha} k_\delta$. For any δ let $\aleph_\nu^\delta$ mean the cardinality of δ in the sense of $\mathbf{Cstr}(x \cap \delta)$. Since $\aleph_\nu^\delta \prec \aleph_\alpha$, we have $\aleph_{\nu+1}^\delta \preccurlyeq \aleph_\alpha$ and by the remark 3519 $2^{\aleph_\nu^\delta} = \aleph_{\nu+1}^\delta$ holds in the sense of $\mathbf{Cstr}(x \cap \delta)$. Hence $k_\delta \preccurlyeq \aleph_{\nu+1}^\delta \preccurlyeq \aleph_\alpha$ and so $\mathbf{P}(\boldsymbol{\omega}_\beta) \preccurlyeq \aleph_\alpha$, q.e.d.

We shall now present a more general notion of constructibility relative to a class.

3522. Definition (**TSS** + D3). For any real X,

$$\mathbf{Cstr}_1(X) = \bigcup_{\alpha \in \mathbf{On}} \mathbf{Cstr}(X \cap p_\alpha)\,.$$

3523. Theorem (**TSS**, D3). For each real class X, $\mathbf{Cstr}_1(X)$ is the least model-class Y such that $(\forall y \in Y)(X \cap y \in Y)$ and $X \subseteq Y$.

Proof. (a) We prove that $\mathbf{Cstr}_1(X)$ is a model-class. If $\alpha < \beta$ then $X \cap \cap p_\alpha \subseteq X \cap p_\beta \in \mathbf{Cstr}(X \cap p_\beta)$ and $p_\alpha \cap \mathbf{Cstr}(X \cap p_\beta) \in \mathbf{Cstr}(X \cap p_\beta)$ (cf. 3225 (3)) so that $X \cap p_\alpha = p_\alpha \cap \mathbf{Cstr}(X \cap p_\beta) \cap X \cap p_\beta \in \mathbf{Cstr}(X \cap p_\beta)$. Hence $\mathbf{Cstr}(X \cap p_\alpha) \subseteq \mathbf{Cstr}(X \cap p_\beta)$. This implies that $\mathbf{Cstr}_1(X)$ is closed. To prove that $\mathbf{Cstr}_1(X)$ is an almost universal class we suppose $x \subseteq \subseteq \mathbf{Cstr}_1(X)$. Then we have $x \subseteq \mathbf{Cstr}(X \cap p_\alpha)$ for some α and by what we proved there exists y such that $x \subseteq y \in \mathbf{Cstr}(X \cap p_\alpha)$. Thus $\mathbf{Cstr}_1(X)$ is a model-class. Further we have $X \subseteq \mathbf{Cstr}_1(X)$ and for all α, $X \cap p_\alpha \in \in \mathbf{Cstr}_1(X)$; hence, for any $y \in \mathbf{Cstr}_1(X)$, $X \cap y = X \cap p_\alpha \cap y$ for some α and so $X \cap y \in \mathbf{Cstr}_1(X)$.

(b) If Y is a model-class and $X \subseteq Y \,\&\, (\forall y \in Y)(X \cap y \in Y)$ then, for all α, $X \cap p_\alpha = X \cap (Y \cap p_\alpha) \in Y$ (since $Y \cap p_\alpha \in Y$). Hence $\mathbf{Cstr}(X \cap p_\alpha) \subseteq Y$ and so $\mathbf{Cstr}_1(X) \subseteq Y$.

3524. Lemma (**TSS**, D3). If x is a set then $\mathbf{Cstr}_1(x) = \mathbf{Cstr}(x)$.

Proof. There is an α such that $x = x \cap p_\alpha$; hence $x \in \mathbf{Cstr}_1(x)$ and so $\mathbf{Cstr}(x) \subseteq \mathbf{Cstr}_1(x)$. Conversely, since $x \in \mathbf{Cstr}(x)$, we have $(\forall y \in \mathbf{Cstr}(x))(x \cap y \in \mathbf{Cstr}(x))$ and so $\mathbf{Cstr}_1(x) \subseteq \mathbf{Cstr}(x)$.

3525. Corollary (**TSS**, D3, Mcl ($\mathbf{M}$), $\partial\mathbf{TSS}/\partial\mathfrak{S}\mathrm{t}(\mathbf{M})$). For any X^* such that Real* (X^*) we have $\mathbf{Cstr}_1^*(X^*) = \mathbf{Cstr}_1(X^*)$.

Proof. By 3225 (3) and 3505, $\mathbf{Cstr}(X^* \cap p_\alpha) = \mathbf{Cstr}^*(X^* \cap p_\alpha^*)$ for all α.

3526. Corollary (**TSS**, D3, Real ($\mathbf{X}$), $\partial\mathbf{TSS}/\partial\mathfrak{S}\mathrm{t}(\mathbf{Cstr}_1(\mathbf{X}))$).

$$\mathbf{V}^* = \mathbf{Cstr}_1^*(\mathbf{X}).$$

3527. Theorem (**TSS**, D3). (E2) implies that there exists a real class $Y \subseteq \subseteq \mathbf{On}$ such that $\mathbf{V} = \mathbf{Cstr}_1(Y)$.

Proof is analogous to the proof of (b) in Theorem 3513 (note that $\mathbf{V} = = \mathbf{Cstr}_1(\mathbf{V})$).

3528. The axiom of constructibility relative to a class:

(CConstr) $$(\exists Y \subseteq \mathbf{On})(\mathrm{Real}(Y) \,\&\, \mathbf{V} = \mathbf{Cstr}_1(Y)).$$

3529. Theorem (**TSS** + D3). Let X be a real class. If $\mathbf{V} = \mathbf{Cstr}_1(X)$ and $X \subseteq \mathbf{On}$ then $\mathbf{V} = \mathbf{Df}(X)$.

Proof. $\mathbf{HDf}(X)$ is a model-class and if $X \subseteq \mathbf{On}$ then by Theorem 3417, X is a real class in the sense of $\mathfrak{S}\mathrm{t}(\mathbf{HDf}(X))$; hence $\mathbf{Cstr}_1(X) \subseteq \mathbf{HDf}(X)$ and so $\mathbf{Df}(X) = \mathbf{V}$.

3530. COROLLARY (**TSS** + D3). Axiom (CConstr) is equivalent to Axiom (E2).

Proof. (E2) → (CConstr) by Theorem 3527; (CConstr) → (E2) by the preceding theorem and Theorem 3418.

3531. Remark. $\mathfrak{Dir}(\partial\mathbf{TSS}/\partial\mathfrak{St}(\mathbf{Cstr}_1(\mathbf{X})))$ is a faithful model of **TSS**, D3, (CConstr) in **TSS**, D3, Real$(\mathbf{X})$ & $\mathbf{X} \subseteq \mathbf{On}$, $\partial\mathbf{TSS}/\partial\mathfrak{St}(\mathbf{Cstr}_1(\mathbf{X}))$.

Finally we give some results which will be used in demonstrating the independence of the axiom of constructibility from the axiom of definability (E3), i.e. the consistency of $\mathbf{V} = \mathbf{HDf} + \mathbf{L}$. We show that $\mathbf{V} = \mathbf{HDf} + \mathbf{L}$ is a consequence of certain statements whose consistency will be established in Chapter VI.

3532. METADEFINITION. Let $\mathbf{T}(X)$ be a gödelian term without constants. We make the following definition in **TSS**, D3:

A relation r is called $\mathbf{T}$-*determining* $(\mathrm{Det}_{\mathbf{T}}(r))$ if

$$\left.\begin{array}{ll}(1) & \alpha \in \mathbf{T}(\mathbf{Cstr}(r \upharpoonright \{\alpha\}))\\ (2) & \alpha \notin \mathbf{T}(\mathbf{Cstr}(r \upharpoonright (\mathbf{D}(r) - \{\alpha\})))\end{array}\right\}\text{ for each } \alpha \in \mathbf{D}(r).$$

(3) $\quad \mathrm{Mcl}(X) \,\&\, \mathrm{Mcl}(Y) \,\&\, X \subseteq Y . \to \mathbf{T}(X) \subseteq \mathbf{T}(Y)$ for each X, Y.

3533. DEFINITION. A relation $\bar{r}$ is called an *estimate* of a relation r $(\mathrm{Est}(r, \bar{r}))$ if

1) $\mathbf{D}(r) = \mathbf{D}(\bar{r})$,

2) $r \subseteq \bar{r}$,

3) $(\forall \alpha \in \mathbf{D}(r))(\exists \beta \in \mathbf{D}(r))(\exists f \in \mathbf{Cstr}(0))(\mathrm{Un}_2(f, (\bar{r} \upharpoonright \alpha), \beta))$,

4) $\bar{r} \in \mathbf{Cstr}(0)$.

3534. METATHEOREM. For every gödelian term $\mathbf{T}(X)$ without constants there is a gödelian term $\mathbf{It}(r, \bar{r})$ with set images property such that (E3) & & $\neg$ (Constr) holds in $\mathfrak{Dir}(\partial\mathbf{TSS}/\partial\mathfrak{St}(\mathbf{Cstr}(\mathbf{r} \upharpoonright \mathbf{It}(\mathbf{r}, \bar{\mathbf{r}}))))$ as a model of (**TSS**, D3) in the following theory:

$$\mathbf{TSS},\ \mathrm{D3},\ \mathrm{Det}_{\mathbf{T}}(\mathbf{r}),\ \mathrm{Est}(\mathbf{r}, \bar{\mathbf{r}}),\ \mathbf{r} \upharpoonright \{0\} \notin \mathbf{Cstr}(0),$$

$$\partial\mathbf{TSS}/\partial\mathfrak{St}(\mathbf{Cstr}(\mathbf{r} \upharpoonright \mathbf{It}(\mathbf{r}, \bar{\mathbf{r}}))).$$

Demonstration. We proceed in the latter theory. Let $\xi = \mathbf{D}(\mathbf{r}) = \mathbf{D}(\bar{\mathbf{r}})$, $\beta_0 = 1$, $x_0 = \{0\}$. We define constructible sequences $\{\beta_n; n \in \boldsymbol{\omega}\}$ and $\{g_n; n \in \boldsymbol{\omega}\}$ by the following conditions: $\beta_n < \beta_{n+1}$ for each n; each g_n is the first constructible one-one mapping of $\bar{\mathbf{r}} \upharpoonright \beta_n$ onto $\beta_{n+1} - \beta_n$. We have $\beta_n < \xi$ for each n. We let $x_{n+1} = g_n''(r \upharpoonright x_n)$ and $x_\omega = \bigcup_{n \in \boldsymbol{\omega}} x_n$. It is obvious that x_ω is defined from $\bar{\mathbf{r}}$ and $\mathbf{r}$ by a gödelian term $\mathbf{It}(\mathbf{r}, \bar{\mathbf{r}})$. We let $\mathbf{M} = \mathbf{Cstr}\,(\mathbf{It}\,(\mathbf{r}, \bar{\mathbf{r}}))$.

We first prove $x_\omega = \mathbf{T}(\mathbf{M})$. If $\alpha \in x_\omega$ then $\mathbf{r} \upharpoonright \{\alpha\} \in \mathbf{M}$, $\mathbf{Cstr}\,(\mathbf{r} \upharpoonright \{\alpha\}) \subseteq \mathbf{M}$ and hence $\alpha \in \mathbf{T}(\mathbf{M})$ follows from $\alpha \in \mathbf{T}(\mathbf{Cstr}\,(\mathbf{r} \upharpoonright \{\alpha\}))$. On the other hand, suppose $\alpha \notin x_\omega$ and let $\mathbf{M}_{(\alpha)} = \mathbf{Cstr}\,(\mathbf{r} \upharpoonright (\mathbf{D}(\mathbf{r}) - \{\alpha\}))$. Then $\mathbf{r} \upharpoonright x_\omega \subseteq \mathbf{M}_{(\alpha)}$. Indeed, $\alpha = 0$ is impossible and we have $\mathbf{r} \upharpoonright x_0 \in \mathbf{M}_{(\alpha)}$. Moreover, if $\mathbf{r} \upharpoonright x_n \in \mathbf{M}_{(\alpha)}$ then $x_{n+1} = g_n''(\mathbf{r} \upharpoonright x_n) \in \mathbf{M}_{(\alpha)}$ since $\mathbf{r} \upharpoonright x_n = (\mathbf{r} \upharpoonright (\mathbf{D}(\mathbf{r}) - \{\alpha\})) \upharpoonright x_n$. Evidently, the whole sequence $\{x_n; n \in \boldsymbol{\omega}\}$ is an element of $\mathbf{M}_{(\alpha)}$ and hence $x_\omega \in \mathbf{M}_{(\alpha)}$. This implies $\mathbf{Cstr}\,(\mathbf{r} \upharpoonright x_\omega) \subseteq \mathbf{Cstr}\,(\mathbf{r} \upharpoonright (\mathbf{D}(\mathbf{r}) - \{\alpha\}))$ and $\alpha \notin \mathbf{T}(\mathbf{M})$. We have proved $x_\omega = \mathbf{T}(\mathbf{M})$ and consequently $x_\omega \in \mathbf{Df}^{\mathfrak{St}(\mathbf{M})}$. But $x_\omega \subseteq \mathbf{On}$ so that we have $x_\omega \in \mathbf{HDf}^{\mathfrak{St}(\mathbf{M})}$.

It follows that the sequence $\{x_n; n \in \boldsymbol{\omega}\}$ is an element of $\mathbf{HDf}^{\mathfrak{St}(\mathbf{M})}$. Since $\mathbf{r} \upharpoonright x_n = \mathbf{Cnv}\,(g_{n+1})''\, x_{n+1}$ we have $\{\mathbf{r} \upharpoonright x_n; n \in \boldsymbol{\omega}\} \in \mathbf{HDf}^{\mathfrak{St}(\mathbf{M})}$. Consequently, $\mathbf{r} \upharpoonright x_\omega \in \mathbf{HDf}^{\mathfrak{St}(\mathbf{M})}$ and hence $\mathbf{HDf}^{\mathfrak{St}(\mathbf{M})} \supseteq \mathbf{Cstr}\,(\mathbf{r} \upharpoonright x_\omega) = \mathbf{V}^{\mathfrak{St}(\mathbf{M})}$. We have proved $(\mathrm{E3})^{\mathfrak{St}(\mathbf{M})}$. Evidently, $\mathbf{r} \upharpoonright \{0\} \in \mathbf{M}$. This implies $\mathbf{M} \neq \mathbf{Cstr}\,(0)$ or, equivalently, $\neg\,(\mathrm{Constr})^{\mathfrak{St}(\mathbf{M})}$. This completes the proof.

To establish the consistency of $\mathbf{V} = \mathbf{HDf} \neq \mathbf{L}$ it remains to show that for a certain $\mathbf{T}$ it is consistent to assume the existence of an estimated $\mathbf{T}$-determining relation. This will be done in Chapter VI.

CHAPTER IV

SECTION 1

Axioms of support

We consider the theory **TSS′**. The basic concept throughout the chapter is the dependence of one class on another, defined in Chapter I, Section 4. We recall that X is said to be dependent on Z ($\mathrm{Dep}(X, Z)$) if $X = r``Z$ for some r; if X is dependent on Z then X is a semiset. We shall consider another more special concept of dependence; we first define

4101. Definition (**TSS′**).

$$\mathrm{Dr}(R) \equiv \mathrm{Rel}(R) \,\&\, (\forall x, y \in \mathbf{D}(R))\,(x \neq y \rightarrow \mathbf{Ext}_R(x) \cap \mathbf{Ext}_R(y) = 0)$$

(R is a *disjointed relation*).

4102. Lemma (**TSS′**). R is a disjointed relation iff $\mathbf{Cnv}(R)$ is a function.

4103. Definition (**TSS′**).

$$\mathrm{Dep}_{\mathrm{d}}(X, Z) \equiv (\exists r)\,(\mathrm{Dr}(r) \,\&\, X = r``Z)$$

(X is *disjointedly dependent* on Z).

4104. Lemma (**TSS′**). $\mathrm{Dep}(\sigma, Z) \equiv (\exists a)\, \mathrm{Dep}(\sigma, Z \cap a)$.

Proof. If $\sigma = r``Z$ then $\sigma = r``(Z \cap a)$ where $a = \mathbf{D}(r)$. Conversely, $\mathrm{Dep}(Z \cap a, Z)$ because $Z \cap a = (\mathbf{I} \upharpoonright a)``\, Z$, hence $\mathrm{Dep}(\sigma, Z \cap a)$ implies $\mathrm{Dep}(\sigma, Z)$.

4105. Lemma (**TSS′**). For every a, Z, $\mathrm{Dep}(\mathbf{P}(a) - \mathbf{P}(a - Z), Z \cap a)$.

Proof. We let $\langle u, x\rangle \in r \equiv x \in u \,\&\, u \subseteq a$. We have $r``(Z \cap a) = \mathbf{P}(a) - \mathbf{P}(a - Z)$ since $\mathbf{P}(a) - \mathbf{P}(a - Z)$ is the class of all subsets of a which are not disjoint from Z.

4106. THEOREM (**TSS'**). If A is a real class then $\mathrm{Dep}(\sigma, \mathrm{Z} \cap A) \equiv \equiv \mathrm{Dep_d}(\sigma, \mathbf{P}(A) - \mathbf{P}(A - \mathrm{Z}))$.

Proof. (1) Suppose $\mathrm{Dep_d}(\sigma, \mathbf{P}(A) - \mathbf{P}(A - \mathrm{Z}))$. Then there is a disjointed relation r such that $\sigma = r``(\mathbf{P}(A) - \mathbf{P}(A - \mathrm{Z})) = r``(x \cap (\mathbf{P}(A) - - \mathbf{P}(A - \mathrm{Z})))$ for every $x \supseteq \mathbf{D}(r)$. We may suppose $\mathbf{D}(r) \subseteq \mathbf{P}(A)$. Let $a = \bigcup(\mathbf{D}(r))$; then $\mathbf{P}(a) \supseteq \mathbf{D}(r)$ and $\mathbf{P}(a) \cap (\mathbf{P}(A) - \mathbf{P}(A - \mathrm{Z})) = = \mathbf{P}(a) - \mathbf{P}(a - (\mathrm{Z} \cap a))$. Hence $\mathrm{Dep}(\sigma, \mathbf{P}(a) - \mathbf{P}(a - (\mathrm{Z} \cap a)))$, furthermore $\mathrm{Dep}(\mathbf{P}(a) - \mathbf{P}(a - (\mathrm{Z} \cap a)), \mathrm{Z} \cap a)$ and $\mathrm{Dep}(\mathrm{Z} \cap a, \mathrm{Z} \cap A)$, and so we obtain $\mathrm{Dep}(\sigma, \mathrm{Z} \cap A)$.

(2) Conversely, if $\sigma = r``(\mathrm{Z} \cap A)$ then we let $\langle x, u\rangle \in S \equiv x \in \mathbf{W}(r) \,\&\, \& \, u = \{y \in \mathbf{D}(r) \cap A; x \in \mathbf{Ext}_r(y)\}$. Since the converse of S is a function, S is a disjointed relation; since $S \subseteq \mathbf{W}(r) \times \mathbf{P}(\mathbf{D}(r))$ and S is real, S is a set. We shall prove that $\sigma = S``W$ where $W = \mathbf{P}(A) - \mathbf{P}(A - \mathrm{Z})$ is the class of all subsets of $A$ which are not disjoint from Z. If $x \in \sigma$ then there exists $u$ such that $\langle x, u\rangle \in S$ and $u \cap \mathrm{Z} \neq 0$; clearly $u \in W$ and so $x \in S``W$. Conversely, if $u \in W$ and $\langle x, u\rangle \in S$ then there exists $y \in \mathrm{Z} \cap u$ such that $\langle x, y\rangle \in r$; hence $x \in r``\mathrm{Z} \cap A = \sigma$.

4107. COROLLARY and DEFINITION (**TSS'**).

1) $\mathrm{Dep}(\sigma, \mathrm{Z} \cap a) \equiv \mathrm{Dep_d}(\sigma, \mathbf{P}(a) - \mathbf{P}(a - \mathrm{Z}))$,

2) $\varrho \subseteq a \rightarrow (\mathrm{Dep}(\sigma, \varrho) \equiv \mathrm{Dep_d}(\sigma, \mathbf{P}(a) - \mathbf{P}(a - \varrho)))$.

We recall that in Chapter I, Section 4, we defined a support to be a class Z such that $\mathrm{Dep}(X, \mathrm{Z})$ and $\mathrm{Dep}(Y, \mathrm{Z})$ implies $\mathrm{Dep}(X - Y, \mathrm{Z})$ for all X and Y. Let us further define: $\mathrm{Supp_d}(\mathrm{Z}) \equiv \mathrm{Supp}(\mathrm{Z}) \,\&\, (\forall\sigma)(\mathrm{Dep}(\sigma, \mathrm{Z}) \rightarrow \rightarrow \mathrm{Dep_d}(\sigma, \mathrm{Z}))$.

We also defined the model $\mathfrak{Supp}(\mathrm{Z})$ determined by the F-definition $\mathfrak{Supp}(\mathrm{Z})$ with the specification $\mathrm{Supp}(\mathrm{Z})$ (see 1463). We shall sometimes write $\mathrm{Supp_1}(\mathrm{Z})$ instead of $\mathrm{Supp}(\mathrm{Z})$. We have shown in **TSS'**, $\mathrm{Supp_1}(\mathbf{Z})$, ∂**TSS**$/\partial\mathfrak{Supp}(\mathbf{Z})$ that every x is a set of the model; hence every normal formula is absolute. In particular, the formula $\mathrm{Dep}(X, \mathrm{Z})$ is absolute and the formula $\mathrm{Supp}(\mathrm{Z})$ is absolute from above. Clearly $(\forall x)\,\mathrm{Dep}(x \cap \mathbf{Z}, \mathbf{Z})$; hence $\mathbf{Z}$ is a class of the model and so it is a support in the sense of the model. Since $(\forall\sigma)\,\mathrm{Dep}(\sigma, \mathbf{Z})$ holds in the model, we say that $\mathbf{Z}$ is a total support in the sense of the model. This leads us to the following definition:

4108. DEFINITION (**TSS'**). Z is called a *total support* ($\mathrm{TSupp}(\mathrm{Z})$) if $(\forall\sigma)\,\mathrm{Dep}(\sigma, \mathrm{Z})$.

4109. The first axiom of support (S1)

$$(\exists Z)(\forall\sigma)\,\mathrm{Dep}(\sigma, Z)\,.$$

Note that every total support is a support. Axiom (S1) states that a total support exists.

4110. Metatheorem. The model $\mathfrak{Dir}(\partial\mathbf{TSS}/\partial\mathfrak{Supp}(\mathbf{Z}))$ is a faithful model of $\mathbf{TSS}'$, S1 in $\mathbf{TSS}'$, $\mathrm{Supp}_1(\mathbf{Z})$, $\partial\mathbf{TSS}/\partial\mathfrak{Supp}(\mathbf{Z})$.

Demonstration. In the following diagram the composition $1 * 2$ is equivalent to the identity model of the first theory in the third and the third theory is a conservative extension of the first.

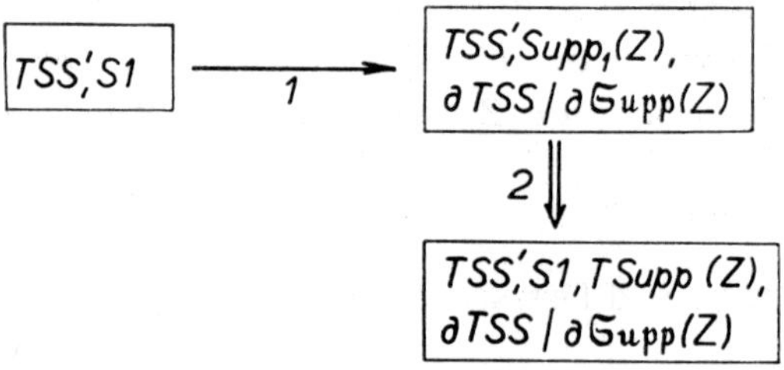

Remark. By Theorem 4106 Axiom (S1) is equivalent (in $\mathbf{TSS}'$) to the following formula: $(\exists Z)(\forall\sigma)\,\mathrm{Dep}_{\mathrm{d}}(\sigma, Z)$.

At the end of the present Section and in the next Section we shall study consequences of the axiom of support and of some stronger forms of this axiom. Before doing this, we shall devote our attention to supports in $\mathbf{TSS}'$.

Supports which are semisets are of special importance. We shall find simple necessary and sufficient conditions for a semiset to be a support.

4111. Lemma ($\mathbf{TSS}'$). If σ is dependent on $Z \cap \boldsymbol{a}$ then, for each c such that $\sigma \subseteq c$, $c - \sigma$ is disjointedly dependent on $\mathbf{P}(a - Z)$.

Proof. By 4107 there exists a disjointed relation r such that $\sigma = r''\varrho$ where $\varrho = \mathbf{P}(a) - \mathbf{P}(a - Z)$. If we let $b = r''\,\mathbf{P}(a)$ then $\sigma \subseteq b$, and, since r is disjointed, $b - \sigma = r''\,\mathbf{P}(a - Z)$. Put $s = ((c - b) \times \{0\}) \cup \cup (r \cap ((b \cap c) \times \mathbf{V}))$; s is disjointed and $c - \sigma = s''\,\mathbf{P}(a - Z)$.

4112. Theorem ($\mathbf{TSS}'$).

(a) $\quad \mathrm{Supp}(Z) \equiv (\forall a)\,\mathrm{Dep}\,(\mathbf{P}(a - Z), Z)\,,$

(b) $\quad \mathrm{Supp}_{\mathrm{d}}(Z) \equiv (\forall a)\,\mathrm{Dep}_{\mathrm{d}}(\mathbf{P}(a - Z), Z)\,.$

Proof. (a) Let Z be a support. By 4105 and 4104 we have Dep $(\mathbf{P}(a) - \mathbf{P}(a - Z), Z)$ and since $\mathbf{P}(a)$ is a set we have also Dep $(\mathbf{P}(a), Z)$. Hence Dep $(\mathbf{P}(a - Z), Z)$. Conversely, suppose $(\forall a)$ Dep $(\mathbf{P}(a - Z), Z)$. It suffices to prove

$$(\forall \sigma)(\text{Dep}(\sigma, Z) \to (\exists b \supseteq \sigma) \text{Dep}(b - \sigma, Z)\,.$$

If Dep (σ, Z) then by 4104 and 4111 there are sets $b \supseteq \sigma$ and a such that $\text{Dep}_{\mathbf{d}}(b - \sigma, \mathbf{P}(a - Z))$. Using Dep $(\mathbf{P}(a - Z), Z)$ we obtain Dep $(b - \sigma, Z)$. Hence Z is a support by 1466.

(b) follows from (a) by 4111.

By definition, Z is a support if every pair of semisets satisfies a certain condition, while by the preceding Theorem it is sufficient to consider a particular semiset. Indeed, the theorem implies the following

4113. COROLLARY (**TSS′**). If $\sigma \subseteq a$ then

(a) $\text{Supp}(\sigma) \equiv \text{Dep}(\mathbf{P}(a - \sigma), \sigma)$,

(b) $\text{Supp}_{\mathbf{d}}(\sigma) \equiv \text{Dep}_{\mathbf{d}}(\mathbf{P}(a - \sigma), \sigma)$.

Proof. The implication $\to$ follows immediately by the preceding theorem. And it is easy to modify the above proof in order to prove the converse implications of the present corollary.

4114. DEFINITION (**TSS′**).

$$\text{Sim}(Z_1, Z_2) \equiv .\ \text{Dep}(Z_1, Z_2) \,\&\, \text{Dep}(Z_2, Z_1)$$

(Z_1, Z_2 are *similar* if they depend on each other).

4115. LEMMA (**TSS′**). If Z_1, Z_2 are similar then they are semisets.

4116. LEMMA (**TSS′**).

(a) $\text{Sim}(\sigma, \sigma)$,
$\text{Sim}(\sigma, \varrho) \to \text{Sim}(\varrho, \sigma)$,
$\text{Sim}(\sigma_1, \sigma_2) \,\&\, \text{Sim}(\sigma_2, \sigma_3) .\to \text{Sim}(\sigma_1, \sigma_3)$.

(b) If σ_1, σ_2 are similar then $\text{Dep}(\varrho, \sigma_1) \equiv \text{Dep}(\varrho, \sigma_2)$ for every ϱ.

(c) If σ_1, σ_2 are similar and σ_1 is a total support then σ_2 is a total support.

4117. THEOREM (**TSS′**). If Z is a disjointed support and σ is dependent on Z then there is a set b such that σ and $Z \cap b$ are similar.

Proof. Let r be a disjointed relation such that $\sigma = r``Z$ and set $b = \mathbf{D}(r)$. Then $\sigma = r``(Z \cap b)$ and $Z \cap b = (\mathbf{Cnv}(r))``\sigma$.

4118. DEFINITION (**TSS′**). A relation R is *totally universal* if each semiset is the extension of some element.

$$\mathrm{Totunvr}(R) \equiv (\forall \sigma)(\exists x)(\sigma = \mathbf{Ext}_R(x)).$$

4119. THEOREM (**TSS′**). Axiom (S1) is equivalent to the existence of a totally universal relation.

Proof. Suppose Axiom (S1) holds and let Z be a total support. The relation defined by $\langle x, r\rangle \in R \equiv x \in r``Z$ is totally universal, since each semiset X is an extension of some r.

Conversely, if R is a totally universal relation then we show that R is a total support. If $X = \mathbf{Ext}_R(u)$ and $X \subseteq a$ we let

$$\langle x, \langle y, z\rangle\rangle \in r \equiv .\ x = y \,\&\, z = u \,\&\, x \in a\,.$$

Clearly $r``R = \mathbf{Ext}_R(u) = X$.

We shall now establish an extremely important Metatheorem concerning the theory **TSS′**, S1.

4120. METADEFINITION. (a) The formulas $x \in y$, $x \in \sigma$ and $x \in X$ are *seminormal* (SNF).

(b) If φ and ψ are SNF then $\varphi \,\&\, \psi$ and $\neg\varphi$ are SNF.

(c) If φ is a SNF then $(\exists x)\,\varphi$ and $(\exists \sigma)\,\varphi$ are SNF.

(d) Every SNF is obtained by a finite number of applications of the rules (b) and (c) to SNF sub (a).

4121. METATHEOREM. Every SNF is normal in **TSS′**, S1.

Demonstration. Every atomic SNF is a NF. Let φ and ψ be SNF and let R be a variable for regular totally universal relations; if $\hat{\varphi}(R)$ and $\hat{\psi}(R)$ are normal formulas equivalent to φ and ψ in **TSS′**, S1 then $\hat{\varphi} \,\&\, \hat{\psi}$ and $\neg\hat{\varphi}$ are normal formulas equivalent to $\varphi \,\&\, \psi$ and $\neg\varphi$ in **TSS′**, S1 while $(\exists x)\,\hat{\varphi}$ is a normal formula equivalent to $(\exists x)\,\varphi$ in **TSS′**, S1. If z is a set-variable which does not occur in φ then the formula $(\exists z)\,(\varphi(R''\{z\}))$ is equivalent to $(\exists\sigma)\varphi\,(\sigma)$ in **TSS′**, S1.

4122. Corollary. Every SNF is representable by a Gödelian term in **TSS′**, S1.

We have already mentioned that those supports which are semisets are of special importance. Before considering such supports more closely, we shall study so-called locally semiset supports.

4123. Definition (**TSS′**). Z is a *locally semiset support* ($\mathrm{Supp}_2\,(\mathrm{Z})$) if

$$\mathrm{Supp}_1\,(\mathrm{Z})\,\&\,(\forall a)\,(\exists b)\,(\forall\sigma \subseteq a)\,(\mathrm{Dep}\,(\sigma, \mathrm{Z}) \rightarrow \mathrm{Dep}\,(\sigma, \mathrm{Z} \cap b))\,;$$

Z is a *locally semiset total suppport* if

$$(\forall a)\,(\exists b)\,(\forall\sigma \subseteq a)\,(\mathrm{Dep}\,(\sigma, \mathrm{Z} \cap b))$$

4124. The second axiom of support (S2)

$$(\exists \mathrm{Z})\,(\forall a)\,(\exists b)\,(\forall\sigma \subseteq a)\,(\mathrm{Dep}\,(\sigma, \mathrm{Z} \cap b))\,.$$

4125. Lemma (**TSS′**). If Z is a locally semiset total support then

1) Z is a locally semiset support,

2) Z is a total support.

Proof. Obvious.

4126. Remark. A support Z is "locally semiset" iff for every set a there exists a semiset σ (of the form $b \cap \mathrm{Z}$) included in Z such that all semisets dependent on Z and included in a are dependent on σ.

4127. Theorem (**TSS′**, S2).

$$(\forall a)\,(\exists\sigma)\,(\forall\varrho \subseteq a)\,(\varrho \neq 0 \rightarrow (\exists x \in \mathbf{D}\,(\sigma))\,(\varrho = \mathbf{Ext}_\sigma\,(x)))\,.$$

This formula will be denoted by (Pot).

Proof. It suffices to put

$$\langle x, r\rangle \in \sigma \equiv r \subseteq (a \cdot\!\times\, b)\,\&\, x \in r''\mathrm{Z}$$

where b has the property given in Axiom (S2); σ is a semiset, since it is included in $a \times (\mathbf{P}(a \times b))$.

4128. Theorem (**TSS′**). (S2) ≡ . (S1) & (Pot).

Proof. It suffices to prove the implication from right to left. Let Z be a total support; we prove that Z is locally semiset. Let a be a set and let σ be a semiset relation which codes all subclasses of a, i.e. $(\forall \varrho \subseteq a)(\exists x)(\varrho = \sigma``\{x\})$. Suppose that $\sigma = r``Z$ and $b = \mathbf{D}(r)$. We have $\sigma = r``(Z \cap b)$; we show that $(\forall \varrho \subseteq a)(\mathrm{Dep}(\varrho, Z \cap b))$.

There exists x_0 such that $\varrho = \mathrm{Ext}_\sigma(x_0)$; setting $\varrho = S``\sigma$ where $\langle y, \langle y, x_0\rangle\rangle \in S \equiv y \in a$, we have $\mathrm{Dep}(\varrho, \sigma)$. This completes the proof.

4129. Metatheorem. $\mathfrak{Dir}(\partial\mathbf{TSS}/\partial\mathfrak{Supp}(\mathbf{Z}))$ is a faithful model of **TSS′**, S2 in **TSS′**, $\mathrm{Supp}_2(\mathbf{Z})$, $\partial\mathbf{TSS}/\partial\mathfrak{Supp}(\mathbf{Z})$.

Demonstration. (1) We prove that (S2) holds in the model. We have $\mathrm{Sm}^*(X^*) \equiv \mathrm{Dep}(X^*, \mathbf{Z})$; since the predicate Dep is absolute it follows by the definition of a locally semiset support that the formula

$$(\forall a)(\exists b)(\forall \sigma^* \subseteq a)\,\mathrm{Dep}(\sigma^*, \mathbf{Z} \cap b)$$

is provable in **TSS′**, $\mathrm{Supp}_2(\mathbf{Z})$, $\partial\mathbf{TSS}/\partial\mathfrak{Supp}(\mathbf{Z})$.

(2) Consider the following diagram (where $\mathrm{TSupp}_2(\mathbf{Z})$ means $\mathrm{TSupp}(\mathbf{Z})$ & $\mathrm{Supp}_2(\mathbf{Z})$):

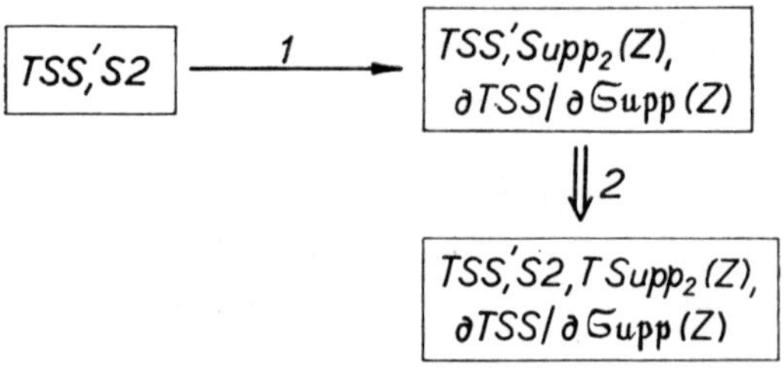

The demonstration is analogous to the demonstration of 4110.

4130. Definition. (**TSS′**) Z is a *semiset support* ($\mathrm{Supp}_3(\mathrm{Z})$) if $\mathrm{Supp}_1(\mathrm{Z})$ & & Sm (Z).

4131. The third axiom of support (S3)

$$(\exists \sigma)(\forall \varrho)(\mathrm{Dep}(\varrho, \sigma)).$$

4132. LEMMA (**TSS′**). If $(\forall \varrho)(\mathrm{Dep}(\varrho, \sigma))$ then

1) σ is a semiset support,

2) σ is a total support.

Proof. Obvious.

4133. DEFINITION (**TSS′**). A set a is called *bounding* $(\mathrm{Bd}(a))$ if for every semiset σ there exist a set $b \prec a$ and a semiset $\varrho \subseteq b$ such that $\sigma = \bigcup(\varrho)$; in other words, σ is the union of a collection of sets which is subvalent to a.

Axiom of boundedness (Bd)

$$(\exists a)(\mathrm{Bd}(a)).$$

4134. THEOREM (**TSS′**). $(S3) \equiv (\mathrm{Pot}) \& (\mathrm{Bd})$.

Proof. Since a semiset total support is locally semiset we have $(S3) \rightarrow \rightarrow (\mathrm{Pot})$. We prove that $(S3) \rightarrow (\mathrm{Bd})$.

Let $Z \subseteq a$ be a total support, suppose Z to be a disjointed total support. We shall show that a is bounding. Let $\sigma = r''Z$; we set $u \in b \equiv (\exists x \in a)(u = \mathbf{Ext}_r(x))$. Clearly $b \preccurlyeq a$; if we set $u \in \varrho \equiv (\exists x \in \mathbf{D}(r) \cap Z)(u = \mathbf{Ext}_r(x))$ then $\varrho \subseteq b$ and $\sigma = \bigcup(\varrho)$.

Conversely, suppose that $(\mathrm{Pot}) \& (\mathrm{Bd})$ holds; we shall prove (S3). Let a be bounding and let σ code all subclasses of a, i.e. $(\forall \varrho \subseteq a)(\exists x)(\varrho = \mathbf{Ext}_\sigma(x))$. We claim that σ is a total support. Let ϱ be a semiset and suppose $\varrho = \bigcup(\varrho_0)$ where $\varrho_0 \subseteq b$ and $b \preccurlyeq a$; let f be a 1-1 mapping of b into a and let $\varrho_1 = f''\varrho_0 = \mathbf{Ext}_\sigma(x)$. It follows that $\mathrm{Dep}(\varrho_1, \sigma)$, $\mathrm{Dep}(\varrho_0, \varrho_1)$ and $\mathrm{Dep}(\varrho, \varrho_0)$; hence $\mathrm{Dep}(\varrho, \sigma)$, q.e.d.

4135. METATHEOREM. $\mathfrak{Dir}(\partial\mathbf{TSS}/\partial\mathfrak{Supp}(\mathbf{Z}))$ is a faithful model of **TSS′**, S3 in **TSS′**, $\mathrm{Supp}_3(\mathbf{Z})$, $\partial\mathbf{TSS}/\partial\mathfrak{Supp}(\mathbf{Z})$.

The demonstration is analogous to the demonstration of Metatheorem 4110 and may be left to the reader.

4136. THEOREM (**TSS′**, $\mathrm{Supp}_1(\mathbf{Z}_1)$, $\partial\mathbf{TSS}/\partial\mathfrak{Supp}(\mathbf{Z}_1)$).

(S2) implies that $\mathbf{Z}_1$ is a locally semiset total support in the sense of $\mathfrak{Supp}(\mathbf{Z}_1)$. (In other words, if we assume that there is a locally semiset total support and $\mathbf{Z}_1$ is an arbitrary fixed support then it follows that $\mathbf{Z}_1$ is a locally semiset total support in the sense of $\mathfrak{Supp}(\mathbf{Z}_1)$.)

Proof. Let $\mathbf{Z}$ be a locally semiset total support and let $\mathbf{Z}_1$ be an arbitrary support. To prove that $\mathbf{Z}_1$ is a locally semiset total support in the sense of

$\mathfrak{Supp}(\mathbf{Z}_1)$ it suffices to prove $(\mathrm{Pot})^{\mathfrak{Supp}(\mathbf{Z}_1)}$. Let a be a set and let $T = \{r;\ \mathrm{Rel}(r) \& \mathbf{D}(r) \subseteq b\}$ (b as in 4123). If we let $\langle y, r\rangle \in R \equiv r \in T \& y \in r''\mathbf{Z}_1$, then R is a class of the model. Let σ code all subclasses of a, i.e. $(\forall \varrho \subseteq a)(\exists x)(\varrho = \mathbf{Ext}_\sigma(x))$. We define

$$\langle r, z\rangle \in S \equiv z \in \mathbf{D}(\sigma) \& r \in T \& \mathbf{Ext}_\sigma(z) = \mathbf{Ext}_R(r)\,.$$

$\mathbf{D}(S)$ is a semiset and hence by D1 there exists $\varrho_0 \subseteq S$ such that $\mathbf{D}(\varrho_0) = \mathbf{D}(S)$. Thus $\mathbf{W}(\varrho_0)$ is a semiset and $\mathbf{W}(\varrho_0) \subseteq T$; hence $\mathbf{W}(\varrho_0) \subseteq q$ for some q. Let $R_0 = R \upharpoonright q$; R_0 is a semiset of the model. If $r''\mathbf{Z}_1 \subseteq a$ then $r''\mathbf{Z}_1 = \mathbf{Ext}_R(z)$; hence there exists $s \in \mathbf{D}(\varrho_0)$ such that $\mathbf{Ext}_\sigma(s) = r''\mathbf{Z}_1$; if $\langle r_1, s\rangle \in \varrho_0$ then $\mathbf{Ext}_{R_0}(r_1) = r''\mathbf{Z}_1$. Hence, in the sense of $\mathfrak{Supp}(\mathbf{Z}_1)$, R_0 codes the subclasses of a.

4137. THEOREM (**TSS**′, $\mathrm{Supp}_1(\mathbf{Z}_1)$, $\partial\mathbf{TSS}/\partial\mathfrak{Supp}(\mathbf{Z}_1)$).

$$(\mathrm{Bd}) \rightarrow (\mathrm{Bd})^{\mathfrak{Supp}(\mathbf{Z}_1)}\,.$$

Proof. Let a be a bounding set; we shall prove that a is bounding in the sense of $\mathfrak{Supp}(\mathbf{Z}_1)$. If σ is a semiset in the sense of $\mathfrak{Supp}(\mathbf{Z}_1)$ then there exist $b \preccurlyeq a$ and $\varrho \subseteq b$ such that $\sigma = \bigcup(\varrho)$. We let $\varrho_0 = \{x \in b;\ x \subseteq \sigma\}$; we have $\mathrm{Sm}^{\mathfrak{Supp}(\mathbf{Z}_1)}(\varrho_0)$. Since $\sigma = \bigcup(\varrho_0)$ the set a is bounding in the sense of $\mathfrak{Supp}(\mathbf{Z}_1)$.

4138. COROLLARY (**TSS**′, $\mathrm{Supp}(\mathbf{Z}_1)$, $\partial\mathbf{TSS}/\partial\mathfrak{Supp}(\mathbf{Z}_1)$).

(1) $\quad (\mathrm{S2}) \rightarrow (\mathrm{S2})^{\mathfrak{Supp}(\mathbf{Z}_1)}$;

(2) $\quad (\mathrm{S3}) \rightarrow (\mathrm{S3})^{\mathfrak{Supp}(\mathbf{Z}_1)}$.

We shall prove a stronger result:

4139. THEOREM (**TSS**′, $\mathrm{Supp}_1(\mathbf{Z}_1)$, $\partial\mathbf{TSS}/\partial\mathfrak{Supp}(\mathbf{Z}_1)$). If Z is an arbitrary disjointed semiset total support then there is a set b such that $Z \cap b$ is a semiset total support in the sense of $\mathfrak{Supp}(\mathbf{Z}_1)$.

Proof. We know that (Pot) and (Bd) hold in the sense of $\mathfrak{Supp}(\mathbf{Z}_1)$; hence there exists σ such that σ is a semiset total support in the sense of $\mathfrak{Supp}(\mathbf{Z}_1)$. By 4117 there is a set b such that σ and $Z \cap b$ are similar and therefore $Z \cap b$ is a semiset total support in the sense of $\mathfrak{Supp}(\mathbf{Z}_1)$.

4140. METATHEOREM. $\mathfrak{Dir}(\partial\mathbf{TSS}/\partial\mathfrak{Supp}(\mathbf{Z}))$ is a faithful model of **TSS**′, (Si) in **TSS**′, (Si), $\mathrm{Supp}_1(\mathbf{Z})$, $\partial\mathbf{TSS}/\partial\mathfrak{Supp}(\mathbf{Z})$ ($i = 1, 2, 3$).

Demonstration. Denote the theory **TSS′**, Supp_1 (**Z**), ∂**TSS**/∂𝔖𝔲𝔭𝔭 (**Z**) by $\mathbf{TSS}'_{\mathrm{Supp}}$ and consider the following diagram:

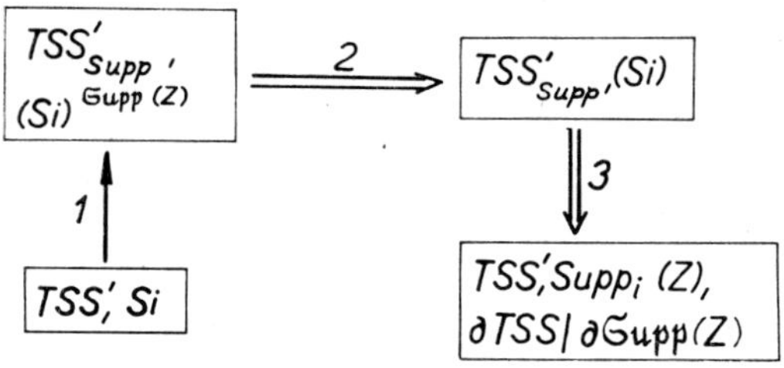

Arrow 1 is 𝔇𝔦𝔯 (∂**TSS**/∂𝔖𝔲𝔭𝔭 (**Z**)). By 4109 and 1223 it is a model. Arrow 2 is an identity; by 4138 and 1230 it is a model. The model we are interested in is the composition 1 ∗ 2. Now, arrow 3 is also a model because evidently **TSS′**, TSupp_i (**Z**) ⊢ (Supp_1 (**Z**) & (S_i)). By 4109, 4129 and 4135 1 ∗ 2 ∗ 3 is faithful; hence 1 ∗ 2 is also faithful (cf. 1233).

SECTION 2

Boolean supports

In the present section we shall consider a special type of supports, namely complete ultrafilters on generalized complete Boolean algebras. Generalized complete Boolean algebras are like complete Boolean algebras except for the fact that the equality relation need not be the identity; such algebras will be called Boolean embryos. We shall find necessary and sufficient conditions for the existence of a total support which is a complete ultrafilter on a Boolean embryo. In particular, we shall give necessary and sufficient conditions for the existence of a total support which is a complete ultrafilter on a complete Boolean algebra $\boldsymbol{b}$. Furthermore we shall show that in the theory **TSS** + D2 the existence of a total support which is a complete ultrafilter on a Boolean embryo is equivalent to the existence of a total support which is a complete ultrafilter on a complete Boolean algebra. We also prove various properties of Boolean embryos which are of importance for set theory; in particular, we shall prove in **TSS** + D2 the existence of a universal complete Boolean algebra, i.e. complete Boolean algebra $\boldsymbol{B}$ such that every complete Boolean algebra is a homomorphic image of $\boldsymbol{B}$. We recall the definition of a complete Boolean algebra: Let B, C and F be nonempty real classes; suppose that C is a mapping of B into B and that F is a mapping of $\mathbf{P}(B)$ into B. Then $\boldsymbol{B} = [B, C, F]$ is a complete Boolean algebra if the following conditions are satisfied, where $F\text{'}0 = 1_B$, $C\text{'}x = -x$, $-1_B = 0_B$, $F\text{'}\{x, y\} = x \wedge y$:

(1) $x \wedge -x = 0_B$ for all $x \in B$,

(2) $x \wedge -y = 0_B \equiv x \wedge y = x$ for all $x, y \in B$,

(3) $F\text{'}z = F\text{'}(F\text{''}q)$ for all $z \subseteq B$ and $q \subseteq \mathbf{P}(B)$ such that $z = \bigcup(q)$.

We shall now give a more general

4201. DEFINITION (**TSS'**). A real class $\boldsymbol{B} = [B, C, F]$ is a *Boolean embryo* (Bemb ($\boldsymbol{B}$)) if B is nonempty, C is a 1-1 mapping of B onto B such that $C'C'x = x$ for each $x \in B$ and if F is a mapping of $\mathbf{P}(B)$ into B.

We again write $C'x = -x$, $F'z = \bigwedge z$ and $F'\{x, y\} = x \wedge y$.

Remark. If $\boldsymbol{B}$ is a complete Boolean algebra then $\boldsymbol{B}$ is a Boolean embryo.

4202. DEFINITION (**TSS'**). If $\boldsymbol{B}$ and $\overline{\boldsymbol{B}}$ are Boolean embryos then H is called a *homomorphism* of $\boldsymbol{B}$ onto $\overline{\boldsymbol{B}}$ if H is a real mapping of B onto $\bar{B}$ such that $H'C'x = \bar{C}'H'x$ for all $x \in B$ and $H'(F'z) = \bar{F}'(H''z)$ for all $z \subseteq B$.

Remark. If $\boldsymbol{B}$ and $\overline{\boldsymbol{B}}$ are complete Boolean algebras then the concept of homomorphism coincides with that defined previously. The same remark applies also to other concepts to be defined below.

4203. DEFINITION (**TSS'**). Let $\boldsymbol{B}$ be a Boolean embryo and let R be a real equivalence on $\boldsymbol{B}$; we write $x \sim y$ if $\langle x, y\rangle \in R$ and $z \sim\sim w$ if

$$(\forall x \in z)(\exists y \in w)(x \sim y) \,\&\, (\forall y \in w)(\exists x \in z)(x \sim y)\,.$$

R is called a *congruence* if the following conditions are satisfied:

(i) $x \sim y \rightarrow C'x \sim C'y$ for any $x, y \in B$,

(ii) $z \sim\sim w \rightarrow F'z \sim F'w$ for any $z, w \subseteq B$.

A congruence R on $\boldsymbol{B}$ is called *canonical* if it is the weakest congruence (the least according to the ordering by inclusion) having the following properties:

(1) $x \wedge -x \sim 0_B$ if $x \in B$,

(2) $x \wedge -y \sim 0_B \equiv x \wedge y \sim x$ if $x, y \in B$,

(3) $F'z \sim F'(F''q)$ if $z \subseteq B$ and $z = \bigcup(q)$.

4204. THEOREM (**TSS'**). The canonical congruence exists on every Boolean embryo $\boldsymbol{B}$.

Proof. We use Metatheorem 2145 and define a certain real relation H by induction; in what follows we use the notation $x \sim_\alpha y$ rather than $\langle x, y\rangle \in H''\{\alpha\}$: We let $x \sim_0 y$ if one of the following conditions is satisfied:

1) $x = y$,

2) $x = 0_B$ and $y = u \wedge -u$ for some $u \in B$,

3) $x = F'z$ and $y = F'F''q$ for some $q \subseteq \mathbf{P}(B)$ and $z = \bigcup(q)$.

If $\alpha > 0$ then we let $x \sim_\alpha y$ if one of the following conditions is satisfied (where $x \sim_{<\alpha} y$ denotes $(\exists\beta < \alpha)(x \sim_\beta y)$ and $z \sim\sim_{<\alpha} w$ denotes $(\forall x \in z)(\exists y \in w)(\exists\beta < \alpha)(x \sim_\beta y) \& (\forall x \in w)(\exists y \in z)(\exists\beta < \alpha)(x \sim_\beta y))$:

1) $x \sim_{<\alpha} y \vee y \sim_{<\alpha} x \vee -x \sim_{<\alpha} -y$,

2) $(\exists u)(x \sim_{<\alpha} u \& u \sim_{<\alpha} y)$,

3) $(\exists z, w \subseteq B)(z \sim\sim_{<\alpha} w \& x = F\text{'}z \& y = F\text{'}w)$,

4) $(\exists u, v)(u \wedge -v \sim_{<\alpha} 0_B \& x = u \wedge v \& y = u)$,

5) $(\exists u, v)(u \wedge v \sim_{<\alpha} u \;\& x = u \wedge -v \& y = 0_B)$.

We now define

$$x \sim y \equiv (\exists\alpha)(x \sim_\alpha y).$$

The relation $\sim$ is obviously an equivalence and we have $x \wedge -x \sim 0_B$ and $x \sim y \to -x \sim -y$. Suppose $z \sim\sim w$; we shall prove $(\exists\alpha)(z \sim\sim_{<\alpha} w)$. We define a real function G on $(\{0\} \times z) \cup (\{1\} \times w)$ such that $G\text{'}\langle 0, x\rangle$ is the least α such that $(\exists y \in w)\,(x \sim_{<\alpha} y)$ and similarly for $\langle 1, y\rangle$. The domain of the function G is a set and so the range of G is also a set; hence there is some α which is greater than all elements of $\mathbf{W}(G)$. Thus we have $z \sim\sim_{<\alpha} w$ and $F\text{'}z \sim_\alpha F\text{'}w$, so that $F\text{'}z \sim F\text{'}w$. If $x \wedge -y \sim 0_B$ then $x \wedge$ $\wedge -y \sim_{<\alpha} 0_B$ for some α; hence $x \wedge y \sim_\alpha x$ and $x \wedge y \sim x$; analogously, $x \wedge y \sim x$ implies $x \wedge -y \sim 0_B$. Finally, if $z = \bigcup(q)$ then $F\text{'}z \sim_0 F\text{'}F\text{''}q$ and hence $F\text{'}z \sim F\text{'}F\text{''}q$.

By Metatheorem 2146, $\sim$ is the least congruence satisfying the above conditions.

Remark. If $\boldsymbol{B}$ is a complete Boolean algebra then the canonical congruence coincides with the identity. (Obvious.)

4205. DEFINITION (**TSS'**). If $\boldsymbol{B}$ is a Boolean embryo then we define

$$x \lesssim y \equiv x \wedge -y \sim 0_B \text{ for } x, y \in B.$$

The relation $\lesssim$ is called the *canonical quasi-ordering* of $\boldsymbol{B}$.

4206. LEMMA (**TSS'**). The quasi-ordering defined above has the following properties: $x \lesssim x$; if $x \lesssim y$ and $y \lesssim z$ then $x \lesssim z$.

4207. Remark. As in the preceding lemma, we may prove results corresponding to each of the theorems of Chapter II, Section 4 (replacing =

and $\leqq$ by $\sim$ and $\lesssim$ respectively); we shall leave this to the reader. The following is of special interest: if, for $x \in B$ and $z \subseteq B$ we define

$$\inf(x, z) \equiv (\forall y \in z)(x \lesssim y) \& (\forall \bar{x}) [(\forall y \in z)(\bar{x} \lesssim y) \rightarrow \bar{x} \lesssim x],$$

then we have $\inf(\bigwedge z, z)$.

4208. DEFINITION (**TSS'**). If $\mathrm{Bemb}(\boldsymbol{B})$ then for all $z \subseteq \boldsymbol{B}$ we define

$$\bigvee z = -\bigwedge C\text{''}z .$$

In particular, $x \vee y = -(-x \wedge -y)$.

4209. LEMMA (**TSS'**). If $z \subseteq B$ then $\bigwedge z \sim -\bigvee C\text{''}z$.

4210. Similarly if, for $x \in B$ and $z \subseteq B$, we define $\sup(x, z) \equiv (\forall y \in z)(y \lesssim x) \& (\forall \bar{x}) [(\forall y \in z)(y \lesssim \bar{x}) \rightarrow x \lesssim \bar{x}]$ for all $x \in B$ and $z \subseteq B$, then we have $\sup(\bigvee z, z)$.

4211. THEOREM (**TSS'**). Let $\boldsymbol{B}$ and $\boldsymbol{B}_1$ be embryos, and let $\sim$ and $\sim_{(1)}$ be the corresponding canonical congruences. If H is a homomorphism of $\boldsymbol{B}$ onto $\boldsymbol{B}_1$ then

$$x \sim y \rightarrow H\text{'}x \sim_{(1)} H\text{'}y \text{ for all } x, y \in B .$$

Proof. Let α be the least ordinal for which there exist x and y such that $x \sim_\alpha y$ and $\neg(H\text{'}x \sim_{(1)} H\text{'}y)$. We first prove that $\alpha > 0$. If $\alpha = 0$ then there are two possible cases: a) $x = u \wedge -u$ and $y = 0_B = -F\text{'}0$, so that $H\text{'}x = H\text{'}u \wedge_{(1)} -_{(1)} H\text{'}u$ and $H\text{'}y = -_{(1)} \bigwedge_{(1)} 0 = 0_{B_1}$, whence $H\text{'}x \sim_{(1)} \sim_{(1)} H\text{'}y$; b) $x = \bigwedge z$ and $y = \bigwedge_{t \in q} (\bigwedge t)$ where $z = \bigcup(q)$, so that $H\text{'} \bigwedge z = = \bigwedge_{(1)} H\text{''}z \sim_{(1)} \bigwedge_{(1)} (\bigcup_{t \in q} H\text{''}t) \sim_{(1)} \bigwedge_{t \in q}{}_{(1)} (\bigwedge_{(1)} H\text{''}t) \sim_{(1)} \bigwedge_{t \in q}{}_{(1)} (H\text{'} \bigwedge t) \sim_{(1)} \sim_{(1)} H\text{'}(\bigwedge_{t \in q} (\bigwedge t))$. Thus we have $\alpha > 0$, we shall investigate Cases 1 to 5 from the proof of 4204. If one of the conditions 1 to 5 is satisfied then $H\text{'}x \sim_{(1)} H\text{'}y$; we establish this for Case 3). If $x = \bigwedge z$, $y = \bigwedge w$ and $z \sim\sim \sim\sim_{<\alpha} w$, then we have $H\text{''}z \sim\sim_{(1)} H\text{''}w$ by induction hypothesis; hence $\bigwedge_{(1)} H\text{''}z \sim_{(1)} \bigwedge_{(1)} H\text{''}w$ and $H\text{'}x \sim_{(1)} H\text{'}y$. This completes the proof.

4212. METATHEOREM. There is a gödelian term $\mathbf{B}(\mathbf{V})$ such that the following is provable in **TSS'**:

$\mathbf{B}(\mathbf{V})$ is a triple of real classes; if we denote $\mathbf{B}(\mathbf{V})$ by $\boldsymbol{B}_0$ (or $[B_0, C_0, F_0]$) then (a) $\boldsymbol{B}_0$ is a Boolean embryo and (b) Every Boolean embryo is a homomorphic image of $\boldsymbol{B}_0$. ($\boldsymbol{B}_0$ is called the *free Boolean embryo.*)

Demonstration. Define a real relation H by the following condition:

$$H``\{0\} = \{\langle x, \varepsilon, 0\rangle; x \in \mathbf{V} - \{0\} \,\&\, (\varepsilon = 0 \vee \varepsilon = 1)\},$$
$$\alpha > 0 \rightarrow H``\{\alpha\} = \{\langle z, \varepsilon, \mathbf{s}(z)\rangle; z \subseteq H``\alpha \,\&\, (\varepsilon = 0 \vee \varepsilon = 1)\}$$

where $\mathbf{s}(z) = \mathbf{Min}\{\alpha; \neg (\exists \beta \geqq \alpha)(\exists u, v)(\langle u, v, \beta\rangle \in z)\}$.
The definition is correct in **TSS′** by Metatheorem 2145. Indeed, if X° is a variable for real classes, it suffices to show that the term

$$\mathbf{T}(X^\circ) = \{\langle x, \varepsilon, \mathbf{s}(x)\rangle; x \subseteq X^\circ\} \text{ is iterable .}$$

We further define

$$B_0 = \mathbf{W}(H),$$
$$C_0`\langle z, 0, \alpha\rangle = \langle z, 1, \alpha\rangle,$$
$$C_0`\langle z, 1, \alpha\rangle = \langle z, 0, \alpha\rangle,$$
$$F_0`q = \langle q, 0, \mathbf{s}(q)\rangle .$$

B_0, C_0 and F_0 are real classes. Let $\boldsymbol{B}_0 = [B_0, C_0, F_0]$. We call $\boldsymbol{B}_0$ the free Boolean embryo. Evidently, $\boldsymbol{B}_0$ is a Boolean embryo. We now prove that every Boolean embryo is a homomorphic image of $\boldsymbol{B}_0$. Thus, if $\boldsymbol{B}$ is a Boolean embryo then using Metatheorem 2145 we suppose $0 \notin B$ and let

$$\langle x, y\rangle \in K``\{0\} \equiv .\, (x \in B \,\&\, (y = \langle x, 0, 0\rangle \vee y = \langle -x, 1, 0\rangle)) \vee$$
$$\vee\, (x = 0_{\boldsymbol{B}} \,\&\, (\exists z \notin B)(z \neq 0 \,\&\, y = \langle z, 0, 0\rangle)) \vee$$
$$\vee\, (x = 1_{\boldsymbol{B}} \,\&\, (\exists z \notin B)(z \neq 0 \,\&\, y = \langle z, 1, 0\rangle))$$
$$\alpha > 0 \rightarrow .\, \langle x, y\rangle \in K``\{\alpha\} \equiv$$
$$\equiv (\exists z \subseteq \mathbf{D}(K``\alpha))\, [(y = \langle z, 0, \mathbf{s}(z)\rangle \,\&\, x = \bigwedge(K``\alpha)`` z) \vee$$
$$\vee\, (y = \langle z, 1, \mathbf{s}(z)\rangle \,\&\, x = -\bigwedge(K``\alpha)`` z)] .$$

We claim that $Hm = \mathbf{W}(K)$ is a homomorphism of $\boldsymbol{B}_0$ onto $\boldsymbol{B}$. The class Hm is real and is a function, since every $K``\alpha$ is a function; this can be proved by induction. The function Hm obviously preserves intersections and complements.

4213. In the theory **TSS** + (D2) we shall prove the existence of a free complete Boolean algebra, i.e. a complete Boolean algebra $\bar{\boldsymbol{B}}$ such that every complete Boolean algebra is a homomorphic image of $\bar{\boldsymbol{B}}$. Let $\sim_0$ be the canonical congruence on $\boldsymbol{B}_0$ and let $R = \mathbf{Rg}(\sim_0)$ (cf. Definition 3113). For any $x \in B_0$, we let $\bar{x}$ be the set $\{y; \langle y, x\rangle \in R\}$. We define $y \in \bar{B} \equiv$ $\equiv (\exists x \in B_0)(y = \bar{x})$, $\bar{C}'\bar{x} = \overline{C_0'x}$ for $x \in B_0$ and $\bar{F}'q = \overline{F_0' \bigcup(q)}$ for $q \subseteq \bar{B}$, $\bar{\boldsymbol{B}} = [\bar{B}, \bar{C}, \bar{F}]$. It follows from the definition of the canonical congruence that $\bar{\boldsymbol{B}}$ is a complete Boolean algebra; we call it the *free complete Boolean algebra.* If $\boldsymbol{B}$ is a complete Boolean algebra then there is a homomorphism H of $\boldsymbol{B}_0$ onto $\boldsymbol{B}$; we define $\bar{H}'\bar{x} = H'x$ for $x \in B_0$. To prove that $\bar{H}$ is a homomorphism of $\bar{\boldsymbol{B}}$ onto $\boldsymbol{B}$, it suffices to show that $\bar{H}$ is a mapping of $\bar{B}$ onto B, i.e. that $\bar{x} = \bar{y} \rightarrow H'x = H'y$; this follows from Theorem 4211.

Thus we have

4214. Theorem (**TSS**, D2). Every complete Boolean algebra is a homomorphic image of $\bar{\boldsymbol{B}}$.

4215. Definition (**TSS'**). If $\boldsymbol{B}$ is a Boolean embryo then we say that a class $Z \subseteq \boldsymbol{B}$ is a *complete ultrafilter* of $\boldsymbol{B}$ if it satisfies the following conditions:

$$x \in Z \equiv -x \notin Z \text{ for any } x \in B\,,$$

$$a \subseteq Z \rightarrow \bigwedge a \in Z \text{ for any set } a \subseteq Z\,,$$

$$x \in Z \,\&\, x \lesssim y \rightarrow y \in Z \text{ for any } x, y \in B\,.$$

4216. Lemma (**TSS'**). If $\boldsymbol{B}$ is a Boolean embryo and Z is a complete ultrafilter on $\boldsymbol{B}$, then

a) $0_B \notin Z$,

b) $\bigwedge u \in Z \rightarrow u \subseteq Z$ for all $u \subseteq B$.

Proof. a) Since $0 \subseteq Z$ we have $1_B = \bigwedge 0 \in Z$ and $0_B = -1_B \notin Z$.

b) Suppose $\bigwedge u \in Z$ and let $y \in u$; it suffices to prove $\bigwedge u \lesssim y$. If we let $x = \bigwedge u$ then we have $x \wedge y = F'\{F'u, y\} \sim F'\{F'u, F'\{y\}\} = F'F''\{u, \{y\}\} =$ $= F'(u \cup \{y\}) = F'u = x$; this follows from the fact that $y \sim F'\{y\} =$ $= y \wedge y$, since $y \wedge -y \sim 0_B$.

4217. Lemma (**TSS'**). If $\boldsymbol{B}$ is a Boollean embryo and Z is a complete ultrafilter on $\boldsymbol{B}$ then for all $u \subseteq B$ we have $\bigvee u \in Z$ iff $u \cap Z \neq 0$.

Proof.

$$\bigvee u \notin Z \equiv -\bigwedge C''u \notin Z \equiv \bigwedge C''u \in Z \equiv C''u \subseteq Z \equiv u \cap Z = 0\,.$$

4218. LEMMA (**TSS′**). Let Z be a complete ultrafilter on $\boldsymbol{B}$. If $x \in Z$ and $y \sim x$ then $y \in Z$.

Proof. Since $x \sim y$, we have $-x \sim -y$ and hence $x \wedge -y \sim x \wedge -x \sim \sim 0_B$, so that $x \precsim y$.

4219. THEOREM (**TSS′**). If Z is a complete ultrafilter on a Boolean embryo $\boldsymbol{B}$ then Z is a disjointed support.

Proof. We shall prove $\mathrm{Dep_d}(\mathbf{P}(a - Z), Z)$ for every a. Evidently, it suffices to suppose $a \subseteq B$. If $\langle y, x\rangle \in r \equiv (y \subseteq a \;\&\; x = -\bigvee y)$, then $\mathbf{P}(a - Z) = r\text{''}Z$ and r is a disjointed relation. Hence Z is a disjointed support by 4112.

4220. DEFINITION (**TSS′**). (1) Z is a *Boolean support* on $\boldsymbol{B}$ iff $\boldsymbol{B}$ is a Boolean embryo and Z is a complete ultrafilter on $\boldsymbol{B}$.

(2) $\mathrm{Supp}_4(Z) \equiv (\exists B)$ (Z is a Boolean support on B).

(3) Z is a *total Boolean support* iff there is a $\boldsymbol{B}$ such that Z is a Boolean support on $\boldsymbol{B}$ and Z is a total support.

4221. AXIOM (S4). $(\exists Z)$ (Z is a total Boolean support.)

4222. METATHEOREM. $\mathfrak{Dir}\,(\partial\mathbf{TSS}/\partial\mathfrak{Supp}\,(\mathbf{Z}))$ is a faithful model of **TSS′**, S4 in **TSS′**, $\mathrm{Supp}_4(\mathbf{Z})$, $\partial\mathbf{TSS}/\partial\mathfrak{Supp}\,(\mathbf{Z})$.

Demonstration. The following is provable in **TSS′**, $\mathrm{Supp}_4(\mathbf{Z})$, $\partial\mathbf{TSS}/\partial\mathfrak{Supp}\,(\mathbf{Z})$: **Z** is a class of the model and, in the sense of $\mathfrak{Supp}\,(\mathbf{Z})$, **Z** is a complete ultrafilter on a Boolean embryo. Hence (S4) holds in the model. A standard argument shows that the model is faithful.

4223. THEOREM (**TSS′**, S4). Every nonempty class of ordinals has a least element.

Proof. It suffices to prove that every nonempty semiset of ordinals has a least element. Suppose that Z is a total support and that Z is a complete ultrafilter on some $\boldsymbol{B}$. Let X be a nonempty subclass of some ordinal number ξ. We have $X = r\text{''}Z = r\text{''}(Z \cap a)$ where $a = \mathbf{D}(r)$; we may suppose that $\mathbf{W}(r) \subseteq \xi$. Obviously $a \cap Z \neq 0$; for any $x \in a$ we let $f\text{'}x$ be the least element of the set $\mathbf{Ext}_r(x)$. If $\alpha \leqq \xi$ then we define $u_\alpha = \bigvee\{x; f\text{'}x = \alpha\} - \bigvee\{u_\beta; \beta < \alpha\}$. If $\beta \neq \alpha$ then $u_\beta \wedge u_\alpha \sim 0_B$ and if $f\text{'}x = \alpha$ then $x \precsim \precsim \bigvee\{u_\beta; \beta \leqq \alpha\}$; hence $\bigvee a \precsim \bigvee\{u_\alpha, \alpha \leqq \xi\}$. Since $\bigvee a \in Z$, we have $\bigvee\{u_\alpha; \alpha \leq \xi\} \in Z$ and $(\exists\alpha)(u_\alpha \in Z)$. This α is unique and there is some $x \in Z$ such that $\alpha \in \mathbf{Ext}_r(x)$. The ordinal α is the least element of X; for, if $\beta < \alpha$ and $\beta \in \mathbf{Ext}_r(y)$ then $y \precsim \bigvee\{u_\gamma; \gamma \leqq \beta\} \notin Z$ since $u_\gamma \notin Z$ whenever $\gamma < \alpha$.

4224. The axiom of standardness (St)

$$(\forall X)(X \subseteq \mathbf{On} \,\&\, X \neq 0 \rightarrow (\exists \alpha)(\alpha \in X \,\&\, (\forall \beta < \alpha)(\beta \notin X))) .$$

We want to show that, in **TSS'**, Axiom (S4) is equivalent to (S1) & (St). We first prove a number of preliminary assertions.

We shall consider the theory **TSS'** + (S1) + (St). (Note that since both (S1) and (St) can be proved in **TS** the following considerations also hold for the theory **TS** + (D1).) To begin with we demonstrate a certain Metatheorem on induction which is analogous to those proved at the end of Section 1, Chapter II. We use the letters σ and ϱ as variables for semisets and the letter Z for total supports.

4225. METADEFINITION. Let **T** be a theory stronger than **TSS'**.

(a) A gödelian term $\mathbf{T}(X)$ is Sm-*local* in **T** if

$$\mathsf{T} \vdash x \in \mathbf{T}(X) \equiv (\exists \sigma \subseteq X)(x \in \mathbf{T}(\sigma)) .$$

(b) A gödelian term $\mathbf{T}(X)$ is Sm-iterable in **T** if it is Sm-local in **T** and has the *semiset-images property* in **T**, i.e. $\mathsf{T} \vdash \mathrm{Sm}(\mathbf{T}(\sigma))$.

We now consider an arbitrary theory **T** stronger than (**TSS'**, S1, St); **R** is a constant for a totally universal relation. We write $\sigma \vartriangle F$ for $(\exists x \in F)$ $(\sigma = \mathbf{Ext_R}(x))$.

4226. DEFINITION (**TSS'**, Sl, St). Let $F \subseteq \mathbf{C}(\mathbf{R})$ and let X be a class. F is said to *code an* $\subseteq$*-cofinal system of subsemisets of* X ($F[\subseteq \vartriangle] X$) if (1) $(\forall \sigma \vartriangle F)(\sigma \subseteq X)$ and (2) $(\forall \varrho \subseteq X)(\exists \sigma \vartriangle F)(\varrho \subseteq \sigma)$.

4227. METALEMMA. Let $\mathbf{T}(X)$ be a gödelian term. $\mathbf{T}(X)$ is Sm-local in **T** iff

(∗) $$\mathsf{T} \vdash (\forall F)(F[\subseteq \vartriangle] X \rightarrow \mathbf{T}(X) = \bigcup_{\sigma \vartriangle F} \mathbf{T}(\sigma))\ ^{*)}$$

Demonstration. (a) If (∗) holds then $\mathbf{T}(X)$ is obviously Sm-local (take for F a coding of all subsemisets of X). (b) Let $\mathbf{T}(X)$ be Sm-local. Proceeding in **T**, we suppose $F[\subseteq \vartriangle] X$. Since $\mathbf{T}(X)$ is Sm-local we have $\mathbf{T}(\sigma) \subseteq$ $\subseteq \mathbf{T}(X)$ whenever $\sigma \vartriangle F$; hence $\bigcup_{\sigma \vartriangle F} \mathbf{T}(\sigma) \subseteq \mathbf{T}(X)$. On the other hand, if $x \in \mathbf{T}(X)$ then there is a $\varrho \subseteq X$ such that $x \in \mathbf{T}(\varrho)$; further, there is

*) The second part of this implication may be written more exactly as $(\forall u)$ $(u \in \mathbf{T}(X) \equiv$ $\equiv (\exists \sigma \vartriangle F)(u \in \mathbf{T}(\sigma)))$. Note that the last formula is normal in **T**.

a $\sigma \vartriangle F$ such that $\varrho \subseteq \sigma$ which implies $x \in \mathbf{T}(\sigma)$. Thus we have $\mathbf{T}(X) \subseteq \subseteq \bigcup_{\sigma \vartriangle F} \mathbf{T}(\sigma)$.

We are now ready to state our Metatheorem on induction (cf. 2145).

4228. Metatheorem. Let $\mathbf{T}$ be a theory stronger than ($\mathbf{TSS'}$, S1, St) and let $\mathbf{R}$ be a constant of $\mathbf{T}$ for a totally universal relation. For every Sm-iterable gödelian term $\mathbf{T}(X)$ there is a gödelian term $\mathbf{S}(X, \mathbf{R})$ such that the following is provable in $\mathbf{T}$: For each $X \neq 0$, $S = \mathbf{S}(X, \mathbf{R})$ is a relation, $S``\{0\} = X$, $\mathbf{D}(S)$ is an ordinal and $S``\{\alpha\} = \mathbf{T}(S``\alpha)$ for every $\alpha > 0$.

Demonstration. Define

$$\Delta(\varrho, \alpha) \equiv (\forall \iota)(0 < \iota < \alpha \rightarrow \varrho``\{\iota\} = \mathbf{T}(\varrho``\iota)),$$

$$\varrho \vartriangle \mathbf{C}(X, \mathbf{R}) \equiv . \operatorname{Rel}(\varrho) \,\&\, \mathbf{D}(\varrho) \in \mathbf{On} \,\&\, \varrho``\{0\} \subseteq X \,\&\, \Delta(\varrho, \mathbf{D}(\varrho)) \text{*)}$$

Write C for $\mathbf{C}(X, \mathbf{R})$ and define $\mathbf{S}(X, \mathbf{R}) = \bigcup_{\varrho \vartriangle C} \varrho = S$. We proceed in $\mathbf{T}$.

(1) $\mathbf{D}(S)$ is an ordinal because it is a complete subclass of $\mathbf{On}$.

(2) $S``\{0\} = X$ for, if $x \in X$ then defining $\varrho = \{\langle x, 0 \rangle\}$, we have $\varrho \vartriangle \mathbf{C}(X, \mathbf{R})$.

(3) $S``\{\alpha\} = \bigcup_{\varrho \vartriangle C} \varrho``\{\alpha\}$ and $S``\alpha = \bigcup_{\varrho \vartriangle C} \varrho``\alpha$; hence $S``\{\alpha\} = \bigcup_{\varrho \vartriangle C} \mathbf{T}(\varrho``\alpha)$ and $\mathbf{T}(S``\alpha) = \mathbf{T}(\bigcup_{\varrho \vartriangle C} \varrho``\alpha)$. It remains to prove that $\mathbf{T}(\bigcup_{\varrho \vartriangle C} \varrho``\alpha) = \bigcup_{\varrho \vartriangle C} \mathbf{T}(\varrho``\alpha)$. Let $\sigma \vartriangle D \equiv (\exists \varrho \vartriangle C)(\sigma = \varrho``\alpha)$; we prove that $D$ codes an $\subseteq$-cofinal system of subsemisets of $\bigcup_{\varrho \vartriangle C} \varrho``\alpha$; in other words, for every $\sigma \subseteq \bigcup_{\varrho \vartriangle C} \varrho``\alpha$ there is a $\bar{\varrho} \vartriangle C$ such that $\sigma \subseteq \bar{\varrho}``\alpha$. This will complete the proof by Metalemma 4227.

(4) Hence let $\sigma \subseteq \bigcup_{\varrho \vartriangle C} \varrho``\alpha$. Define $\langle x, u \rangle \in Q \equiv . u \in \sigma \,\&\, (\exists \varrho \vartriangle C)(\varrho = = \mathbf{Ext}_{\mathbf{R}}(x) \,\&\, u \in \varrho``\alpha)$. Q is a relation and $\mathbf{D}(Q)$ is a semiset (in fact, $\mathbf{D}(Q) = \sigma$). By Axiom (D1) let $Q_0 \subseteq Q$ be a semiset relation such that $\mathbf{D}(Q) = \mathbf{D}(Q_0)$; then $\mathbf{W}(Q_0)$ is a semiset and $\bigcup_{\varrho \vartriangle \mathbf{W}(Q_0)} \varrho``\{0\}$ is also a semiset (by (C2)); denote this semiset by $\varrho_0$. Evidently $\varrho_0 \subseteq X$. We claim that there is a $\bar{\varrho} \vartriangle C$ such that $\bar{\varrho}``\{0\} = \varrho_0$ and $\Delta(\bar{\varrho}, \alpha)$. Indeed, set $\beta \in A \equiv$

*) This is a slightly imprecise but short way of expressing the following idea: $\mathbf{C}(X, \mathbf{R})$ contains all codes of all semisets which satisfy the right hand side of the defining equivalence and contains no code of any semiset which does not satisfy it. Obviously this definition is normal and $\mathbf{C}(X, \mathbf{R})$ can be written as a gödelian term.

$\equiv \neg (\exists \varrho \vartriangle C)(\varrho''\{0\} = \varrho_0 \,\&\, \Delta(\varrho, \beta))$ (the condition defining A is seminormal; recall Metatheorem 4121). If $A \neq 0$ then it has a first element β. A standard proof (using St) shows that for every $\gamma < \beta$ there is exactly one $\varrho_\gamma \vartriangle C$ such that $\varrho_\gamma''\{0\} = \varrho_0$ and $\Delta(\varrho_\gamma, \gamma)$ and that $\gamma_1 < \gamma_2 < \beta$ implies $\varrho_{\gamma_1} \subseteq \varrho_{\gamma_2}$. Hence if β is limit then setting $\varrho_\beta = \bigcup_{\gamma<\beta} \varrho_\gamma$ we obtain $\beta \notin A$, a contradiction (note that **R** is used in the definition of ϱ_β; the reader may easily write down a seminormal formula equivalent to $x \in \varrho_\beta$). If β is isolated, say $\beta = \gamma + 1$, then defining $\varrho_\beta = \varrho_\gamma \cup \mathbf{T}(\mathbf{W}(\varrho_\gamma)) \times \{\gamma\}$ we obtain $\beta \notin A$, also a contradiction. We denote $\varrho_{\alpha+1}$ by $\bar{\varrho}$; thus our claim is proved. It follows by induction (using St) that for every $\gamma \leqq \alpha$ and every $\varrho \vartriangle \mathbf{W}(Q_0)$ we have $\varrho''\gamma \subseteq \bar{\varrho}''\gamma$. Consequently $\bigcup_{\varrho \vartriangle \mathbf{W}(Q_0)} \varrho''\alpha \subseteq \bar{\varrho}''\alpha$ and so $\sigma \subseteq \bar{\varrho}''\alpha$. This completes the proof that $\mathbf{T}(\bigcup_{\varrho \vartriangle C} \varrho''\alpha) = \bigcup_{\varrho \vartriangle C} \mathbf{T}(\varrho''\alpha)$; the Metatheorem follows (cf. (3)).

We shall now consider the notion of well-ordering in **TSS′** with the axiom of standardness and S1.

4229. LEMMA (**TSS′**, St). If a set r is a well-ordering then every nonempty semiset $\sigma \subseteq \mathbf{C}(r)$ has a least element (w.r.t. the ordering r).

Proof. There is an isomorphism f of $\mathbf{C}(r)$ onto an ordinal α (w.r.t. r, **E**) (see 2141). Set $\varrho = f''\sigma$; then $\varrho \subseteq \alpha$ and $\varrho \neq 0$, hence ϱ has a least element γ, and the $u \in \sigma$ such that $f'u = \gamma$ is the least element of σ.

4230. DEFINITION (**TSS′**). A semiset ϱ is a Sm-*well-ordering* if it is an ordering such that every nonempty subsemiset of $\mathbf{C}(\varrho)$ has a least element.

Remark. Lemma 4229 states that every well-ordering which is a set is a Sm-well-ordering.

4231. THEOREM (**TSS′**, St, S1). If ϱ is a Sm-well-ordering then there exist a uniquely determined ordinal number γ and a uniquely determined one-one mapping σ of $\mathbf{C}(\varrho)$ onto γ such that σ is an isomorphism of ϱ, $(\mathbf{E} \upharpoonright \gamma)$.

Proof 1) By the metatheorem on induction 4228 a relation H can be defined by the following conditions: $H''\{0\} = \{$the least element of $\mathbf{C}(\varrho)\}$; for $\alpha > 0$, if $\mathbf{C}(\varrho) - H''\alpha$ is nonempty then $H''\{\alpha\} = \{$the least element of $\mathbf{C}(\varrho) - H''\alpha\}$; otherwise $H''\{\alpha\} = 0$. H is an isomorphism and $\mathbf{W}(H) = \mathbf{C}(\varrho)$; hence $\mathbf{W}(H)$ is a semiset. H is one-one and thus an exact functor, hence $\mathbf{D}(H)$ and H are semisets. Furthermore, $\mathbf{D}(H)$ is a complete subsemiset of **On** and is therefore an ordinal. ($\mathbf{D}(H) = \gamma$ where γ is the first ordinal not in $\mathbf{D}(H)$.)

2) To prove the uniqueness it suffices to show that if σ is an isomorphism of $\mathbf{E} \upharpoonright \gamma$, $\mathbf{E} \upharpoonright \delta$ (where γ and δ are ordinals) then $\gamma = \delta$ and σ is the identity. Suppose σ is not the identity; then (by St) there is a least ξ such that $\sigma'\xi \neq \xi$; hence $\sigma'\xi > \xi$. It follows that ξ has no pre-image, a contradiction.

In **TSS**′, (S1), (St) the definition of a complete ultrafilter can be simplified.

4232. THEOREM (**TSS**′, S1, St). If $\boldsymbol{B}$ is a Boolean embryo then $Z \subseteq B$ is a complete ultrafilter if and only if it has the following properties:

1) $x \in Z \equiv -x \notin Z$ for all $x \in B$,

2) $a \subseteq Z \equiv \bigwedge a \in Z$ for all $a \subseteq B$.

Proof. Every complete ultrafilter obviously has both properties. If Z has properties 1) and 2) then to show that Z is a complete ultrafilter we need only prove that

$$y \gtrsim x \in Z \to y \in Z ,$$

where $\sim$ is the canonical congruence. We prove that for all $x, y \in B$; if $x \sim y$ then $x \in Z \equiv y \in Z$. We first prove this in case $x \sim_0 y$. a) We have $0_B \notin Z$; for $0 \subseteq Z$ so that $\bigwedge 0 \in Z$ whence $0_B = -\bigwedge 0 \notin Z$. Since $\{u, -u\} \nsubseteq Z$, we have $u \wedge -u \notin Z$. b) If $z = \bigcup(q)$ then

$$\bigwedge z \in Z \equiv z \subseteq Z \equiv (\forall t \in q)(t \subseteq Z) \equiv$$
$$\equiv (\forall t \in q)(\bigwedge t \in Z) \equiv \bigwedge_{t \in q}(\bigwedge t) \in Z .$$

We now suppose that $\alpha > 0$ is the least ordinal for which there exist x and y such that $x \sim_\alpha y$ and $\neg (x \in Z \equiv y \in Z)$ (note that (St) is used!):

a) If $-x \sim_{<\alpha} -y$ then $-x \in Z \equiv -y \in Z$ and hence $x \in Z \equiv y \in Z$.

b) If $x \sim_{<\alpha} u \sim_{<\alpha} y$ then $x \in Z \equiv u \in Z$ and $u \in Z \equiv y \in Z$; hence $x \in Z \equiv y \in Z$.

c) If $x = \bigwedge z$, $y = \bigwedge w$ and $z \sim\sim_{<\alpha} w$ then $z \subseteq Z \equiv w \subseteq Z$; hence $x \in Z \equiv y \in Z$.

d) If $x = u \wedge v$, $y = u$ and $u \wedge -v \sim_{<\alpha} 0_B$ then $x \in Z \equiv \{u, v\} \subseteq Z$; since $u \wedge -v \sim_{<\alpha} 0_B \notin Z$ we have $\{u, v\} \subseteq Z \equiv u \in Z$.

e) If $x = u \wedge -v$, $y = 0_B$ and $u \wedge v \sim_{<\alpha} u$ then $y \notin Z$. We also have $x \notin Z$; for, if $x \in Z$ then $u, -v \in Z$ whereas $u \wedge v \in Z$ since $u \wedge v \sim_{<\alpha} u$, so that $v, -v \in Z$, a contradiction.

This completes the proof of the assertion $x \sim y \rightarrow (x \in Z \equiv y \in Z)$. We can now easily prove $y \gtrsim x \in Z \rightarrow y \in Z$; if $x \in Z$ and $x \wedge y \sim x$ then $x \wedge y \in Z$ and hence $\{x, y\} \subseteq Z$, i.e. $y \in Z$. Hence Z is a complete ultrafilter.

Finally we show that Axiom (S4) can be proved in **TSS'**, (S1), (St):

4233. THEOREM (**TSS'**, S1, St). There exists a total support which is a complete ultrafilter on $\boldsymbol{B}_0$.

Proof. We let

$$K''\{0\} = \{\langle x, \varepsilon, 0\rangle; (x \in Z \,\&\, \varepsilon = 0) \vee (x \notin Z \,\&\, \varepsilon = 1 \,\&\, x \neq 0)\},$$

$$K''\{\alpha\} = \{\langle z, \varepsilon, \alpha\rangle; \langle z, \varepsilon, \alpha\rangle \in B_0 \,\&$$

$$\& [(z \subseteq K''\alpha \,\&\, \varepsilon = 0) \vee (z \nsubseteq K''\alpha \,\&\, \varepsilon = 1)]\}$$

(cf. 4212; Z is a total support).

Let $Z_0 = K''\,\mathbf{On}$. We shall prove: 1) that Z_0 is a complete ultrafilter on $\boldsymbol{B}_0$, and 2) that Z_0 is a total support.

ad 1): We obviously have $x \in Z_0 \equiv -x \notin Z_0$ for all $x \in B_0$ and $a \subseteq Z_0 \equiv \equiv \bigwedge a \in Z_0$ for all $a \subseteq B_0$. By Theorem 4232, Z_0 is a complete ultrafilter on $\boldsymbol{B}_0$.

ad 2): If $\sigma = r''Z$ we define $\langle u, \langle x, 0, 0\rangle\rangle \in \bar{r} \equiv \langle u, x\rangle \in r$ so that $\sigma = \bar{r}''Z_0$. Hence Z_0 is a total support.

Thus we have proved

4234. THEOREM (**TSS'**). Axiom (S4) is equivalent to (S1) & (St).

We shall now consider (locally) semiset total supports which are complete ultrafilters.

4235. DEFINITION (**TSS'**). $\mathrm{Supp}_5(Z) \equiv \mathrm{Supp}_2(Z) \,\&\, \mathrm{Supp}_4(Z)$ (*locally semiset Boolean support*).

4236. AXIOM (S5) (The axiom of locally semiset total Boolean support) There exists a locally semiset total support which is a complete ultrafilter on a Boolean embryo.

4237. THEOREM (**TSS'**). Axiom (S5) is equivalent to (S4) & (Pot).

Proof. (S5) implies (S4) and (S2); (S2) implies (Pot). Conversely, (S4) & & (Pot) implies (S5); for, if Z is a support satisfying the conditions of (S4), then by (Pot) and by the proof of 4128 Z is a locally semiset support.

4238. METATHEOREM. $\mathfrak{Dir}$ (∂**TSS**/$\partial\mathfrak{Supp}$ (**Z**)) is a faithful model of **TSS**′, S5 in **TSS**′, Supp_5 (**Z**), ∂**TSS**/$\partial\mathfrak{Supp}$ (**Z**). (Cf. the demonstration of 4222.)

4239. DEFINITION. $\text{Supp}_6(Z) \equiv \text{Supp}_3(Z) \,\&\, \text{Supp}_4(Z)$ (*semiset Boolean support*).

4240. AXIOM (S6). There exists a total support Z which is a complete ultrafilter on some complete Boolean algebra $\boldsymbol{b} = \langle b, c, f \rangle$. (The axiom of total semiset Boolean support.)

4241. THEOREM (**TSS**′). Axiom (S6) is equivalent to (S3) & (St).

Proof. It is obvious that Axiom (S6) implies (S3) and (S4) and hence (St), (Pot) and (Bd). Conversely, if we suppose that (S3) and (St) hold then we may use all results concerning Boolean embryos. We let $Z_0 \subseteq a$ be a total support and suppose, by 4107, that Z_0 is a disjointed support. Let $b = a \times \{0, 1\}$ and let $Z = (Z_0 \times \{0\}) \cup ((a - Z_0) \times \{1\})$; Z is a disjointed support. It follows that $\mathbf{P}(Z) = r_1'' Z$. Let $f_1 = \mathbf{Cnv}(r_1)$; f_1 is a mapping of a subset of $\mathbf{P}(b)$ into b and $f_1 \upharpoonright \mathbf{P}(Z)$ is a mapping of $\mathbf{P}(Z)$ into Z. Let y_0 be an arbitrary element of $b - Z$; we define $f'z = f_1'z$ for $z \in \mathbf{D}(f_1)$, $f'z = y_0$ for $z \in \mathbf{P}(b) - \mathbf{D}(f_1)$. Then f is a function which maps $\mathbf{P}(b)$ into b in such a way that $z \subseteq Z \equiv f'z \in Z$. For $x \in a$ we let $c'\langle x, 0\rangle = \langle x, 1\rangle$ and $c'\langle x, 1\rangle = \langle x, 0\rangle$, so that $x \in Z \equiv c'x \notin Z$; hence $\boldsymbol{b} = [b, c, f]$ is a Boolean embryo and Z is a complete ultrafilter. If r is the canonical congruence on $[b, c, f]$, then we let $\bar{x} = r''\{x\}$ for $x \in b$, $\bar{b} = \{\bar{x};\, x \in b\}$, $\bar{f}'z = \overline{f'\bigcup z}$, $\bar{c}'\bar{x} = \overline{c'x}$ and $\bar{Z} = \{\bar{x};\, x \in Z\}$; then $\bar{\boldsymbol{b}} = \langle \bar{b}, \bar{c}, \bar{f} \rangle$ is a complete Boolean algebra. We have $x \in Z \equiv \bar{x} \in \bar{Z}$ and hence $\bar{x} \in \bar{Z} \equiv \bar{c}'\bar{x} \notin \bar{Z}$; thus $z \subseteq \bar{Z} \equiv \bigcup(z) \subseteq Z \equiv \overline{\bigwedge\bigcup(z)} \in \bar{Z} \equiv \bar{f}'z \in \bar{Z}$ and $\bar{Z}$ is a complete ultrafilter on $\langle \bar{b}, \bar{c}, \bar{f} \rangle$. Since $Z = (\mathbf{Cnv}(r))''\,\bar{Z}$, we have $\text{Dep}(Z, \bar{Z})$ and $\bar{Z}$ is a total support.

4242. COROLLARY (**TSS**′) (S6) ≡ (Pot) & (Bd) & (St).

4243. METATHEOREM. $\mathfrak{Dir}$ (∂**TSS**/$\partial\mathfrak{Supp}$ (**Z**)) is a faithful model of **TSS**′, S6 in **TSS**′, Supp_6 (**Z**), ∂**TSS**/$\partial\mathfrak{Supp}$ (**Z**).

Demonstration. 1) We proceed in **TSS**′, Supp_6 (**Z**), ∂**TSS**/$\partial\mathfrak{Supp}$ (**Z**). Supp_6 (**Z**) implies that **Z** is a total semiset support on a real Boolean embryo in the sense of Supp (**Z**). It follows that the embryo in question is a set; we denote it by $[b, c, f]$. On factorizing by the canonical congruence we obtain a complete Boolean algebra $\langle \bar{b}, \bar{c}, \bar{f} \rangle$ and a complete ultrafilter $\bar{Z}$ on it; $\bar{Z}$ is a total support since Dep (Z, $\bar{Z}$). Thus we have proved $(\text{S6})^{\mathfrak{Supp}\mathbf{Z}}$.

2) One can show in the usual way that the model is faithful.

4244. THEOREM (**TSS'**). If $\bar{\boldsymbol{b}}$ is a complete subalgebra of $\boldsymbol{b}$ and Z is a complete ultrafilter on $\boldsymbol{b}$ then $Z \cap \bar{\boldsymbol{b}}$ is a complete ultrafilter on $\bar{\boldsymbol{b}}$.

Proof. Exercise.

We shall now consider semisets which correspond to a restriction of a complete ultrafilter on some complete Boolean algebra $\boldsymbol{b}$ to a base for $\boldsymbol{b}$.

4245. DEFINITION (**TSS'**). If $\boldsymbol{a} = \langle a, \leqq \rangle$ is a separatively ordered set then we say that $u \subseteq a$ is *dense* in $\leqq$ if

$$(\forall x \in a)(\exists y \in u)(\exists z \in a)(z \leqq x \,\&\, z \leqq y) .$$

4246. LEMMA (**TSS'**). Let $\boldsymbol{b}$ be a complete Boolean algebra. If $\leqq$ is the canonical ordering and b_0 is a base for $\boldsymbol{b}$, then $u \subseteq b_0$ is dense in $\leqq$ (as an ordering of b_0) iff $\bigvee u = 1_{\boldsymbol{b}}$.

Proof. Exercise.

4247. DEFINITION (**TSS'**). If $\leqq$ is a separative ordering of b then a semiset $\sigma \subseteq b$ is called a *complete ultrafilter* on $\langle b, \leqq \rangle$ if it has the following properties:

1) $(\forall x, y \in \sigma)(\exists z \in b)(z \leqq x \,\&\, z \leqq y)$,
2) $(\forall u)(u \text{ is dense} \rightarrow u \cap \sigma \neq 0)$.

4248. LEMMA (**TSS'**). Let $\leqq$ be a separative ordering and suppose that $\sigma \subseteq b$ is a complete ultrafilter on $\langle b, \leqq \rangle$. If $u \in \sigma$, $v \in b$ and $u \leqq v$ then $v \in \sigma$.

Proof. We let

$$w \in q \equiv w = v \vee \neg (\exists t \in b)(t \leqq v \,\&\, t \leqq w) .$$

The set q is a dense subset of b; for, if $x \in b$ then either $(\exists t \in b)(t \leqq x \,\&\, t \leqq v)$ or $x \in q$. Hence there is some $x \in q \cap \sigma$. We have $x = v$; for, if $x \neq v$ then $\neg (\exists t \in b)(t \leqq v \,\&\, t \leqq x)$ and $\neg (\exists t)(t \leqq u \,\&\, t \leqq x)$, a contradiction. Hence $v \in \sigma$.

4249. LEMMA (**TSS'**). Let $\boldsymbol{b}$ be a complete Boolean algebra, let $\leqq$ be the canonical ordering and suppose that b_0 is a base. If σ is a complete ultrafilter on $\boldsymbol{b}$ then $\sigma \cap b_0$ is a complete ultrafilter on $\langle b_0, \leqq \rangle$ (and obviously $\mathrm{Dep}(\sigma \cap b_0, \sigma)$).

Proof. If $x, y \in \sigma \cap b_0$, then $x \wedge y \in \sigma$ and there exists $z \leqq x \wedge y$ such that $z \in b_0$; this proves 1). If u is dense in $\langle b_0, \leqq \rangle$ then $\bigvee u = 1_b$ by Lemma 4246; hence there exists $z \in u$ such that $z \in \sigma$ and 2) is proved.

4250. THEOREM (**TSS′**). Let $\boldsymbol{b}$ be a complete Boolean algebra, let $\leqq$ be the canonical ordering and suppose that b_0 is a base. If Z_0 is a complete ultrafilter on $\langle b_0, \leqq \rangle$ then the semiset $Z = \{x \in b; (\exists y \in Z_0)(y \leqq x)\}$ is a complete ultrafilter on $\boldsymbol{b}$, $Z_0 = Z \cap b_0$ and $\mathrm{Dep}(Z, Z_0)$.

Proof. 1) We shall prove $x \in Z \equiv -x \notin Z$ for $x \in b$. Suppose that $x \in b$. If $x \in Z \mathbin{\&} -x \in Z$ then there exist $u \in Z_0$ and $v \in Z_0$ such that $u \leqq x$ and $v \leqq y$; hence there exists $z \in b_0$ such that $z \leqq u$ and $z \leqq v$; this is a contradiction since $0_b \notin b_0$. If $x \notin Z$ and $-x \notin Z$ then we let $v \in u \equiv v \in b_0 \mathbin{\&}$ $\mathbin{\&} (v \leqq x \vee v \leqq -x)$; u is obviously dense in b_0 since $\bigvee u = 1_b$. There exists $v \in u \cap Z_0$ such that either $v \leqq x$ or $v \leqq -x$. In the first case we obtain $x \in Z$, a contradiction; the second case is treated similarly.

We shall prove $u \subseteq Z \equiv \bigwedge u \in Z$ for $u \subseteq b$. It suffices to prove $u \cap Z \neq$ $\neq 0 \equiv \bigvee u \in Z$. The implication from left to right is obvious. To prove the converse we suppose $\bigvee u \in Z$. Let $v \in \bar{u} \equiv v \in b_0 \mathbin{\&} (v \leqq -\bigvee u \vee (\exists w \in u)$ $(v \leqq w))$. Since $\bar{u}$ is dense in $\langle b_0, \leqq \rangle$ there is some $v \in \bar{u} \cap Z_0$. It is obvious that $v \nleqq -\bigvee u$, since $\bigvee u \in Z$; hence there exists $w \in u$ such that $v \leqq w$, so that $u \cap Z \neq 0$.

3) We obviously have $0_b \notin Z$, since there is no element in the base which is $\leqq 0_b$; hence Z is a complete ultrafilter on $\boldsymbol{b}$. By Lemma 4248 we have $Z_0 = Z \cap b_0$. Set $r = \{\langle x, y \rangle; y \leqq x\}$. Then $Z = r''Z_0$, hence $\mathrm{Dep}(Z, Z_0)$.

4251. THEOREM (**TSS′**). If $\leqq$ is a separative ordering of b and Z_0 is a complete ultrafilter on $\langle b, \leqq \rangle$ then Z_0 is a support.

Proof. We may suppose that b is a base for a complete Boolean algebra $\langle \bar{b}, \bar{c}, \bar{f} \rangle$ and that $Z_0 = Z \cap b$ where Z is a complete ultrafilter on $\langle \bar{b}, \bar{c}, \bar{f} \rangle$. Hence Z is a (disjointed) support. Since $\mathrm{Dep}(Z, Z_0)$ and $\mathrm{Dep}(Z_0, Z)$, Z_0 is also a support (not necessarily disjointed).

4252. DEFINITION (**TSS′**). A relation r is *antimonotone* on an ordered set a if

$$x \leqq y \rightarrow r''\{x\} \supseteq r''\{y\}$$

for all $x, y \in a$.

4253. THEOREM (**TSS′**). Let $\boldsymbol{a}$ be a separatively ordered set and suppose that Z_0 is a complete ultrafilter on $\boldsymbol{a}$. If $\mathrm{Dep}(\sigma, Z_0)$ then there is a relation r antimonotone on a such that $\sigma = r''Z_0$.

Proof. There exists a complete Boolean algebra $\bar{\boldsymbol{b}} = \langle \bar{b}, \bar{c}, \bar{f} \rangle$ such that a is a base for $\bar{\boldsymbol{b}}$ and the canonical ordering of $\bar{\boldsymbol{b}}$ coincides on a with $\leqq$. Let Z be a complete ultrafilter on $\bar{\boldsymbol{b}}$ which is an extension of Z_0. We have $\sigma = \bar{r}\text{''}Z$ where $\bar{r}$ is a disjointed relation. If we let

$$\langle y, x \rangle \in r \equiv (\exists x_1)\,(x \in a \;\&\; x_1 \in \bar{b} \;\&\; x \leqq x_1 \;\&\; \langle y, x_1 \rangle \in \bar{r})$$

then r is the required relation.

4254. THEOREM (**TSS**′, $\text{Supp}_1(\mathbf{Z}_1)$, $\partial\mathbf{TSS}/\partial\mathfrak{Supp}(\mathbf{Z}_1)$).
$(\text{St}) \to (\text{St})^{\mathfrak{Supp}(\mathbf{Z}_1)}$.

Proof. If every nonempty semiset of ordinal numbers has a least element then in particular every nonempty semiset which is dependent on $\mathbf{Z}_1$ has a least element.

4255. COROLLARY (**TSS**′, $\text{Supp}_1(\mathbf{Z}_1)$, $\partial\mathbf{TSS}/\partial\mathfrak{Supp}(\mathbf{Z}_1)$).

(a) $(\text{S4}) \to (\text{S4})^{\mathfrak{Supp}(\mathbf{Z}_1)}$;
(b) $(\text{S5}) \to (\text{S5})^{\mathfrak{Supp}(\mathbf{Z}_1)}$;
(c) $(\text{S6}) \to (\text{S6})^{\mathfrak{Supp}(\mathbf{Z}_1)}$.

4256. METATHEOREM. The model $\mathfrak{Dir}(\partial\mathbf{TSS}/\partial\mathfrak{Supp}(\mathbf{Z}))$ is a faithful model of **TSS**′, (Si) in **TSS**′, (Si), $\text{Supp}_1(\mathbf{Z})$, $\partial\mathbf{TSS}/\partial\mathfrak{Supp}(\mathbf{Z})$ ($i = 4, 5, 6$).

We conclude with a theorem concerning (S6) which is analogous to the last theorem of the preceding Section.

4257. THEOREM (**TSS**′, $\text{Supp}(\mathbf{Z}_1)$, $\partial\mathbf{TSS}/\partial\mathfrak{Supp}(\mathbf{Z}_1)$). If Z is a total Boolean support on $\boldsymbol{b}$ then there exists a complete subalgebra $\bar{\boldsymbol{b}}$ of $\boldsymbol{b}$ such that $\text{Sim}(\mathbf{Z}_1, Z \cap \bar{b})$. Hence $Z \cap \bar{b}$ is a total support in the sense of $\mathfrak{Supp}(\mathbf{Z}_1)$.

Proof. By the last theorem of Section 1 there exists $a \subseteq b$ such that $a \cap Z$ is a total support in the sense of $\mathfrak{Supp}(\mathbf{Z}_1)$; if we let $\bar{\boldsymbol{b}}$ be the least subalgebra of $\boldsymbol{b}$ such that $a \subseteq \bar{b}$ then we have $\text{Dep}(a \cap Z, \bar{b} \cap Z)$. It suffices to prove $\text{Dep}(\bar{b} \cap Z, \mathbf{Z}_1)$.

To prove this we show that $\bar{b} \cap Z$ is a class in the sense of $\mathfrak{Supp}(\mathbf{Z}_1)$ or, to speak loosely, we construct $\bar{b} \cap Z$ inside $\mathfrak{Supp}(\mathbf{Z}_1)$. Since (St) implies $(\text{St})^{\mathfrak{Supp}(\mathbf{Z}_1)}$, we may use the Metatheorem on Induction in $\mathfrak{Supp}(\mathbf{Z}_1)$. Since $\bar{b}$ is a set, it is also a set in $\mathfrak{Supp}(\mathbf{Z}_1)$. We construct $\bar{b}$ as follows:

$$H\text{''}\{0\} = a\,,$$
$$H\text{''}\{\alpha\} = \{-x;\, x \in H\text{''}\alpha\} \cup \{\textstyle\bigwedge z;\, z \subseteq H\text{''}\alpha\} \quad (\alpha > 0)\,,$$
$$\bar{b} = H\text{''}\,\mathbf{On}\,.$$

Further we let

$$K``\{0\} = a \cap Z\,,$$
$$K``\{\alpha\} = \{-x;\, x \in H``\alpha - K``\alpha\} \cup \{\textstyle\bigwedge z;\, z \subseteq K``\alpha\} \quad (\alpha > 0)\,.$$

We can prove $\bar{b} \cap Z = K``\,\mathbf{On}$ by induction; we also have $\mathrm{Cls}^{\mathfrak{Supp}(Z_1)}(K``\,\mathbf{On})$. This completes the proof.

SECTION 3

Properties of complete ultrafilters

In the present section we shall investigate certain properties of complete ultrafilters; in particular we shall be interested in the relationship between similar complete ultrafilters. We also consider properties of semisets determined by the properties of a complete Boolean algebra with Boolean support; more details will be given in Chapt. V, Sect. 3. Throughout this section $\vartheta(\boldsymbol{b}, Z)$ means "Z is a Boolean support on $\boldsymbol{b}$", i.e. "$\boldsymbol{b}$ is a complete Boolean algebra and Z is a complete ultrafilter on $\boldsymbol{b}$".

4301. LEMMA (**TSS′**). Let a be some set, let $Z \subseteq a$ and suppose that $\vartheta(\boldsymbol{b}, Z)$. Then there exists $w \in Z$ such that $v \in a$ for every nonzero $v \leqq w$.

Proof. We have $\bigvee(b - a) \notin Z$ and $-\bigvee(b - a) = \bigwedge\{-v, v \notin a\} \in Z$. If we let $w = -\bigvee(b - a)$ then $(b \mid w)^{\bullet} \subseteq a$; for, if there exists $v \in b^{\bullet}$ such that $v \leqq w$ and $v \notin a$ then $v \leqq w \leqq -v$, so that $v = 0_b$, a contradiction.

4302. METATHEOREM. The following is provable in **TSS′** for any SF $\varphi(\boldsymbol{b}, x, y, \bullet)$:

If Z is a Boolean support on $\boldsymbol{b}$ and if $\varphi(\boldsymbol{b}, x, y, \bullet)$ holds for every $x \in Z$, then there exists $w \in Z$ such that $\varphi(\boldsymbol{b}, v, y, \bullet)$ holds whenever $v \leqq w$ and $v \neq 0_b$.

4303. DEFINITION (**TSS′**). Let $\boldsymbol{b}$ be a complete Boolean algebra. If $\sigma \subseteq b$, $u \in b$ and $u \neq 0_b$, we define $\sigma\,|_b\, u = \{v \wedge u, v \in \sigma\}$.

4304. DEFINITION (**TSS′**). Let $\boldsymbol{b}$ be a complete Boolean algebra and let $\boldsymbol{c}$ be a subalgebra of $\boldsymbol{b}$. We denote by $\boldsymbol{c}|_b\, u$ the algebra whose field is $c|_b\, u$ and whose operations are defined as follows: the meet of a nonempty subset of $c|_b\, u$ is the same as its meet in $\boldsymbol{b}$, while the meet of the empty set is u; if $v \in c$ then the complement of $v \wedge u$ is $-v \wedge u$ where $-$ and $\wedge$ are the operations of $\boldsymbol{b}$.

4305. *Remark.* In particular $\boldsymbol{b}|_b u$ is the same as $\boldsymbol{b}| u$ (see 2413), and $\boldsymbol{c}|_b 1_b = \boldsymbol{c}$.

4306. LEMMA (**TSS′**). The set $\boldsymbol{c}|_b u$, with the operations as defined above, is a complete Boolean algebra.

Proof. The operations are obviously uniquely defined. If we set $\boldsymbol{c}|_b u = \boldsymbol{d}$ then $0_d = 0_b$ and $1_b = u$. For $x \in c$ we have $(x \wedge u) \wedge (-x \wedge u) = 0_b = 0_d$. Similarly, $(x \wedge u) \wedge (y \wedge u) = x \wedge u$ iff $x \wedge u \wedge y = x \wedge u$ iff $x \wedge u \wedge \wedge -y = 0_b$ iff $(x \wedge u) \wedge (-y \wedge u) = 0_d$. The verification of associativity is left to the reader.

4307. LEMMA (**TSS′**). Let $\boldsymbol{c}$ be a subalgebra of $\boldsymbol{b}$, let $u \in Z$ and suppose that $\vartheta(\boldsymbol{b}, Z)$. If $Z_1 = Z \cap (\boldsymbol{c}|_b u)$ then

1) $Z_1 = (Z \cap c)|_b u$,
2) Z_1 is a complete ultrafilter on $\boldsymbol{c}|_b u$.

Proof. 1) If $x \in Z_1$ then $x \in Z$ and $x = y \wedge u$ for some $y \in c$. It follows that $y \in Z$ and so we have $y \in Z \cap c$ and $x = y \wedge u$.

Conversely, if $x \in (Z \cap c)|_b u$ then $x = y \wedge u$ for some $y \in Z \cap c$, i.e. $x \in Z$ and $y \in c$.

2) Suppose that $q \subseteq c|_b u$. We claim that $q \subseteq Z_1$ iff $\bigwedge q \in Z_1$. This is clear if q is empty; suppose it is nonempty. If $q \subseteq Z_1$ then $\bigwedge q \in Z$ and $\bigwedge q \in c|_b u$; hence $\bigwedge q \in Z_1$. If $\bigwedge q \in Z_1$ then $q \subseteq Z$; since $q \subseteq c|_b u$, it follows that $q \subseteq Z_1$. Finally it is clear that for any $v \in c$ we have $v \wedge u \in Z_1$ iff $-v \wedge u \notin Z_1$.

4308. COROLLARY (**TSS′**). Let Z be a Boolean support on $\boldsymbol{b}$.

1) If $\boldsymbol{c}$ is a subalgebra of $\boldsymbol{b}$ then $Z \cap c$ is a Boolean support on $\boldsymbol{c}$.
2) If $u \in Z$ then $Z|_b u$ is a Boolean support on $\boldsymbol{b}| u$.

4309. LEMMA (**TSS′**). Let Z be a Boolean support on $\boldsymbol{b}$ and let $\boldsymbol{c}$ be a subalgebra of $\boldsymbol{b}$. If $u \in Z$ then $Z \cap c$ and $(Z \cap c)|_b u$ are similar.

Proof. We have $(Z \cap c)|_b u = r``(Z \cap c)$ where, for $v_1, v_2 \in b$, $\langle v_1, v_2 \rangle \in \in r \equiv v_1 = v_2 \wedge u$. Hence $(Z \cap c)|_b u$ is dependent on $(Z \cap c)$. Conversely, we have $(Z \cap c) = s``((Z \cap c)|_b u)$, where, for $v_1, v_2 \in b$, $\langle v_1, v_2 \rangle \in s \equiv \equiv v_1 \in c \,\&\, v_2 = v_1 \wedge u$; hence $Z \cap c$ is dependent on $(Z \cap c)|_b u$.

4310. COROLLARY (**TSS′**). Let Z be a Boolean support on $\boldsymbol{b}$. If $u \in Z$ then Z and $Z|_b u$ are similar.

4311. DEFINITION (**TSS′**). 1) If $\boldsymbol{b}$ is a complete Boolean algebra then a disjointed relation r is said to be a *$\boldsymbol{b}$-set* if $\mathbf{D}(r) \subseteq b$.

2) If r is a $\boldsymbol{b}$-set then for $x \in \mathbf{W}(r)$ we denote by r_x the unique element of $\mathbf{D}(r)$ such that $\langle x, r_x \rangle \in r$. If $x \notin \mathbf{W}(r)$ then we let $r_x = 0_{\boldsymbol{b}}$.

3) If r is a $\boldsymbol{b}$-set and if $u \in b$ then we let $r \,|_{\boldsymbol{b}}\, u = \{\langle x, v \wedge u \rangle; \langle x, v \rangle \in r\}$.

4) If r, s are $\boldsymbol{b}$-sets then we write $r \doteq s$ for $(\forall x)(r_x = s_x)$.

4312. LEMMA (**TSS′**). Let Z be a Boolean support on $\boldsymbol{b}$. If r is a $\boldsymbol{b}$-set and $u \in Z$ then $r''Z = (r \,|_{\boldsymbol{b}}\, u)'' (Z \,|_{\boldsymbol{b}}\, u)$.

Proof. If $\langle x, v \rangle \in r$ and $v \in Z$ then $\langle x, v \wedge u \rangle \in r \,|_{\boldsymbol{b}}\, u$ and $v \wedge u \in$ $\in Z \,|_{\boldsymbol{b}}\, u$, so that $r''Z \subseteq (r \,|_{\boldsymbol{b}}\, u)'' (Z \,|_{\boldsymbol{b}}\, u)$. Conversely, let $\langle x, v \rangle \in r \,|_{\boldsymbol{b}}\, u$ and let $v \in Z \,|_{\boldsymbol{b}}\, u$. Then we have $v \in Z$ and there exists w such that $\langle x, w \rangle \in r$ and $v = w \wedge u$. Thus we have $w \in Z$ and $x \in r''Z$.

4313. THEOREM (**TSS′**), Let Z be a Boolean support on $\boldsymbol{b}$. If r and s are $\boldsymbol{b}$-sets then $r''Z = s''Z$ iff $r \,|_{\boldsymbol{b}}\, u \doteq s \,|_{\boldsymbol{b}}\, u$ for some $u \in Z$.

Proof. The implication from right to left follows from the preceding lemma. Suppose that $r''Z = s''Z$, i.e. that $r_x \in Z \equiv s_x \in Z$ for every $x \in \mathbf{W}(r)$. Let $u = \bigwedge_{x \in \mathbf{W}(r)} [(r_x \wedge s_x) \vee (-r_x \wedge -s_x)]$; clearly $u \in Z$. If $v = \bigwedge_{y \neq x} [(r_y \wedge$ $\wedge\, s_y) \vee (-r_y \wedge -s_y)]$ then

$$r_x \wedge u = v \wedge r_x \wedge [(r_x \wedge s_x) \vee (-r_x \wedge -s_x)] = v \wedge r_x \wedge s_x$$

and similarly $s_x \wedge u = v \wedge r_x \wedge s_x$; hence $r_x \wedge u = s_x \wedge u$ for every x and so we obtain $r \,|_{\boldsymbol{b}}\, u \doteq s \,|_{\boldsymbol{b}}\, u$.

4314. DEFINITION (**TSS′**). Let $\boldsymbol{b}$ be a complete Boolean algebra, let $u \in b$ and let $a \subseteq b$. We say that u is *layered by a* ($\text{Thru}_{\boldsymbol{b}}(u, a)$) if for each $x \in a$ either $u \leqq x$ or $u \leqq -x$. $\mathbf{Thru}_{\boldsymbol{b}}(a)$ denotes the join of all u which are layered by a. (This join itself need not be layered by a.)

4315. LEMMA (**TSS′**). Let $\boldsymbol{b}$ be a complete Boolean algebra and suppose that Z is a Boolean support on $\boldsymbol{b}$. If $a \subseteq b$ then $a \cap Z$ is a set iff $\mathbf{Thru}_{\boldsymbol{b}}(a) \in Z$.

Proof. Suppose that $a \cap Z = a_1$ is a set. Then $a_2 = a - Z$ is also a set and $a_3 = \{-x: x \in a_2\}$ is a subset of Z. Hence $u = \bigwedge(a_1 \cup a_3)$ belongs to Z and is layered by a; i.e. $u \leqq \mathbf{Thru}_{\boldsymbol{b}}(a)$ and so $\mathbf{Thru}_{\boldsymbol{b}}(a) \in Z$.

Conversely, if $\mathbf{Thru}_{\boldsymbol{b}}(a) \in Z$ then some $u \in Z$ is layered by a. Then $a \cap Z =$ $= \{x \in a; x \geq u\}$ so that $a \cap Z$ is a set.

4316. THEOREM (**TSS′**). Let Z be a Boolean support on $\boldsymbol{b}$ and let $\sigma = r''Z$ where r is a $\boldsymbol{b}$-set. Then σ is a set iff $\mathbf{Thru}_{\boldsymbol{b}}(\mathbf{D}(r)) \in Z$.

Proof. By 4315, $\mathbf{Thru}_{b}(\mathbf{D}(r)) \in Z$ iff $\mathbf{D}(r) \cap Z$ is a set. But since $r''Z$ is similar to $\mathbf{D}(r) \cap Z$, $\mathbf{D}(r) \cap Z$ is a set iff $r''Z$ is a set.

4317. DEFINITION (**TSS'**). Semisets σ and ϱ are called *strongly similar* (sSim (σ, ϱ)) if there exists a relation r such that $\sigma = r''\varrho$ and $\varrho = \mathbf{Cnv}(r)''\sigma$.

4318. DEFINITION (**TSS'**). Let $\boldsymbol{b}$ be a complete Boolean algebra and let $u \in b$. Subsets a_1, a_2 of b are said to be ***b**-strongly similar on u* ($\text{sSim}_{\boldsymbol{b}}(a_1, a_2, u)$) if

$$a_1 \,|_{\boldsymbol{b}}\, u - \{0_{\boldsymbol{b}}\} = a_2 \,|_{\boldsymbol{b}}\, u - \{0_{\boldsymbol{b}}\}$$

i.e. if the set of all nonzero meets of elements of a_1 with u coincides with the set of all nonzero meets of elements of a_2 with u. The join of all u such that $\text{sSim}_{\boldsymbol{b}}(a_1, a_2, u)$ is denoted by $\mathbf{sSim}_{\boldsymbol{b}}(a_1, a_2)$.

4319. LEMMA (**TSS'**). Let Z be a Boolean support on $\boldsymbol{b}$. If $a_1, a_2 \subseteq b$ then $a_1 \cap Z$ is strongly similar to $a_2 \cap Z$ iff $\mathbf{sSim}_{\boldsymbol{b}}(a_1, a_2) \in Z$.

Proof. If $\mathbf{sSim}_{\boldsymbol{b}}(a_1, a_2) \in Z$ then a_1 is $\boldsymbol{b}$-strongly similar to a_2 on some $u \in Z$. If, for $x \in a_1$ and $y \in a_2$, we let $\langle x, y\rangle \in r \equiv x \wedge u = y \wedge u$, then $a_1 \cap Z = r''(a_2 \cap Z)$ and $a_2 \cap Z = \mathbf{Cnv}(r)''(a_1 \cap Z)$; hence $a_1 \cap Z$ and $a_2 \cap Z$ are strongly similar.

Conversely, if $a_1 \cap Z$ is strongly similar to $a_2 \cap Z$ then we may suppose that the corresponding relation r is a subset of $a_1 \times a_2$. We have $(a_1 - \mathbf{W}(r)) \cap Z = (a_2 - \mathbf{D}(r)) \cap Z = 0$ and $v \notin Z$ where $v = \bigvee[(a_1 - \mathbf{W}(r)) \cup (a_2 - \mathbf{D}(r))]$, Set $w = \bigwedge_{\langle x,y\rangle \in r} [(x \wedge y) \vee (-x \wedge -y)]$; since $\langle x, y\rangle \in r$ implies $x \in Z \equiv y \in Z$, we have $w \in Z$. Consequently $u = w - v \in Z$. If $x \in a_1 - \mathbf{W}(r)$ then $x \wedge u = 0_{\boldsymbol{b}}$; similarly for $y \in a_2 - \mathbf{D}(r)$. If $x \in \mathbf{W}(r)$ and $\langle x, y\rangle \in r$ then $x \wedge u = y \wedge u$ and so $a_1 \,|_{\boldsymbol{b}}\, u - \{0_{\boldsymbol{b}}\} = a_2 \,|_{\boldsymbol{b}}\, u - \{0_{\boldsymbol{b}}\}$. Hence a_1 and a_2 are $\boldsymbol{b}$-strongly similar on u and u is in Z. Thus we have proved that $\mathbf{sSim}_{\boldsymbol{b}}(a_1, a_2) \in Z$.

4320. THEOREM (**TSS'**). Let Z be a Boolean support on $\boldsymbol{b}$ and let r and s be $\boldsymbol{b}$-sets. The semisets $r''Z$ and $s''Z$ are strongly similar iff $\mathbf{sSim}_{\boldsymbol{b}}(\mathbf{D}(r), \mathbf{D}(s)) \in Z$.

Proof. We have sSim $(r''Z, \mathbf{D}(r) \cap Z)$ and sSim $(s''Z, \mathbf{D}(s) \cap Z)$. Hence sSim $(r''Z, s''Z)$ iff sSim $(\mathbf{D}(r) \cap Z, \mathbf{D}(s) \cap Z)$; however, the latter is equivalent to $\mathbf{sSim}_{\boldsymbol{b}}(\mathbf{D}(r), \mathbf{D}(s)) \in Z$.

4321. LEMMA (**TSS'**). Let Z be a Boolean support on $\boldsymbol{b}$. If f is a mapping of b into itself such that $f''Z \subseteq Z$ then there exists $w \in Z$ such that $u \leqq f'u$ whenever $u \leqq w$.

Proof. For every $u \in b$ we have $(u - f'u) \notin Z$; hence $\bigvee_{u \in b} (u - f'u) \notin Z$. We set $w = -\bigvee_{u \in b} (u - f'u)$; if $u \leqq w$ then $(u - f'u) \leqq w$ and $(u - f'u) \leqq$ $\leqq -w$; hence $u - f'u = 0_b$, so that $u \leqq f'u$.

4322. LEMMA (**TSS'**). Let Z be a Boolean support on $\boldsymbol{b}$ and let f be a mapping of b into itself such that $f''Z \subseteq Z$. If $f'u \leqq u$ whenever $u \in Z$, then there exists $w \in Z$ such that $f'u = u$ whenever $u \leqq w$, $u \neq 0_b$.

Proof. By Metatheorem 4302 there exists $w_1 \in Z$ such that $f'u \leqq u$ for every $u \leqq w_1$, $u \neq 0_b$. By the preceding lemma there exists $w_2 \in Z$ such that $u \leqq f'u$ for every $u \leqq w_2$, $u \neq 0_b$. If we let $w = w_1 \wedge w_2$ then we have $w \in Z$ and $f'u = u$ holds for every $u \leqq w$, $u \neq 0_b$.

We shall now investigate under what conditions two Boolean supports are similar.

4323. THEOREM (**TSS'**). A Boolean support Z_1 on $\boldsymbol{b}_1$ and a Boolean support Z_2 on $\boldsymbol{b}_2$ are similar iff there exist $u_1 \in Z_1$, $u_2 \in Z_2$ and an isomorphism f between $\boldsymbol{b}_1 \mid u_1$ and $\boldsymbol{b}_2 \mid u_2$ such that $Z_2 |_{\boldsymbol{b}_2} u_2 = f''(Z_1 |_{\boldsymbol{b}_1} u_1)$.

Proof. If the condition holds then $Z_1 |_{\boldsymbol{b}_1} u_1$ and $Z_2 |_{\boldsymbol{b}_2} u_2$ are similar; by Lemma 4309, $Z_1 |_{\boldsymbol{b}_1} u_1$ is similar to Z_1 and $Z_2 |_{\boldsymbol{b}_2} u_2$ is similar to Z_2; hence Z_1 is similar to Z_2.

To prove the converse, suppose that Z_1 and Z_2 are similar; let r_i $(i = 1, 2)$ be disjointed relations and suppose that $\mathbf{D}(r_i) \subseteq b_i$, $\mathbf{W}(r_1) = b_2$, $\mathbf{W}(r_2) =$ $= b_1$, $Z_2 = r_1''Z_1$ and $Z_1 = r_2''Z_2$. We use indices 1 and 2 to distinguish the operations in $\boldsymbol{b}_1$ and $\boldsymbol{b}_2$. For $u \in b_i$ we set $(b_i)^u = \{v \in b_i, v \geqq_i u\}$; for $x \in b_1$ we define

$$h_1'x = \bigwedge\nolimits_2(r_1''(b_1)^x) \wedge_2 \bigwedge\nolimits_2(\mathbf{Cnv}(r_2)''(b_1)^x),$$

and for $x \in b_2$ we define

$$h_2'x = \bigwedge\nolimits_1(r_2''(b_2)^x) \wedge_1 \bigwedge\nolimits_1(\mathbf{Cnv}(r_1)''(b_2)^x).$$

If $u \leqq_1 v$ then obviously $h_1'u \leqq_2 h_1'v$ and, since r_2 is disjointed, $u \in Z_1$ implies $h_1'u \in Z_2$; similarly for h_2. We let $g_1'x = h_2'(h_1'x)$ for $x \in b_1$ and $g_2'x =$ $= h_1'(h_2'x)$ for $x \in b_2$. If $x \in Z_i$ then $g_i'x \in Z_i$ $(i = 1, 2)$; we shall prove that $g_i'x \leqq_i x$. If $x \in b_1$ and $\langle x, y\rangle \in r_2$ then $h_1'x \leqq_2 y$, so that $h_2'h_1'x = g_1'x \leqq_1$ $\leqq_1 h_2'y \leqq_1 \bigwedge\nolimits_1 r_2''\{y\} \leqq_1 x$; similarly for $x \in b_2$ and g_2. By Lemma 4322 there exists $w_i \in Z_i$ such that $g_i'u = u$ whenever $u \leqq_i w_i$. If we let $u_1 =$ $= w_1 \wedge_1 h_2'w_2$ and $u_2 = h_1'u_1$ then $u_1 \in Z_1$ and $u_2 \in Z_2$; we shall prove that

$f = h_1 \restriction (b_1 \mid u_1)$ is an isomorphism between $\boldsymbol{b}_1 \mid u_1$ and $\boldsymbol{b}_2 \mid u_2$. We have $u_2 = h_1'u_1 \leqq_2 h_1'(h_2'w_2) = g_2'w_2 = w_2$, so that $u_2 \leqq_2 w_2$ and therefore $g_2'u = u$ for any $u \leqq_2 u_2$. If $x \leqq_1 y \leqq_1 u_1$ then $h_1'x \leqq_2 h_1'y \leqq_2 u_2$ and hence f is an order-preserving mapping of $b_1 \mid u_1$ into $b_2 \mid u_2$. The function f is one-to-one; for, if $x, y \in (b_1 \mid u_1)$ and $h_1'x = h_1'y$ then $x = g_1'x = h_2'h_1'x = g_1'y = y$. The range of f is $b_2 \mid u_2$; for, if $y \leqq_2 u_2 = h_1'u_1$ then $h_2'y \leqq_1 g_1'u_1 = u_1$ and, setting $x = h_2'y$ we have $x \in b_1 \mid u_1$ and $h_1'x = h_1'(h_2'y) = g_2'y = y$.

Since f is an order-preserving one-one mapping of $b_1 \mid u_1$ onto $b_2 \mid u_2$, we can easily check that f preserves meets and complements and hence f is an isomorphism between $\boldsymbol{b}_1 \mid u_1$ and $\boldsymbol{b}_2 \mid u_2$. Since f is a restriction of h_1, we have $Z_2 \mid_2 u_2 \supseteq f''(Z_1 \mid_1 u_1)$ and since g_1 is the identity on $b_1 \mid u_1$ we finally obtain the equality $Z_2 \mid_2 u_2 = f''(Z_1 \mid_1 u_1)$.

4324. DEFINITION (**TSS**′). Let p be an isomorphism between complete Boolean algebras $\boldsymbol{b}_1$ and $\boldsymbol{b}_2$. If r is a $\boldsymbol{b}_1$-set then we let $\hat{p}'r = \{\langle x, p'v\rangle; \langle x, v\rangle \in r\}$.

4325. LEMMA (**TSS**′). Let p be an isomorphism between complete Boolean algebras $\boldsymbol{b}_1$ and $\boldsymbol{b}_2$. If Z_1 is a complete ultrafilter on $\boldsymbol{b}_1$ then $Z_2 = p''Z_1$ is a complete ultrafilter on $\boldsymbol{b}_2$ and we have

$$r''Z_1 = (\hat{p}'r)''\, Z_2$$

for any $\boldsymbol{b}_1$-set r.

Proof. Exercise.

4326. THEOREM (**TSS**′). Let Z_1 and Z_2 be complete ultrafilters on $\boldsymbol{b}_1$ and $\boldsymbol{b}_2$ respectively and suppose that Z_1 and Z_2 are similar. If r_i is a $\boldsymbol{b}_i$-set $(i = 1, 2)$ then $r_1''Z_1 = r_2''Z_2$ iff there exist $u_1 \in Z_1$, $u_2 \in Z_2$ and an isomorphism p between $\boldsymbol{b}_1 \mid u_1$ and $\boldsymbol{b}_2 \mid u_2$ such that $r_2 \mid_{\boldsymbol{b}_2} u_2 \doteq \hat{p}''(r_1 \mid_{\boldsymbol{b}_1} u_1)$ and $p''(Z_1 \mid_{\boldsymbol{b}_1} u_1) = Z_2 \mid_{\boldsymbol{b}_2} u_2$.

Proof. If the condition holds then

$$r_2''Z_2 = (r_2 \mid u_2)''\,(Z_2 \mid u_2) = [\hat{p}'(r_1 \mid u_1)]''\,[p''(Z_1 \mid u_1)] = \\ = (r_1 \mid u_1)''\,(Z_1 \mid u_1) = r_1''Z_1\,.$$

Conversely, suppose that $r_1''Z_1 = r_2''Z_2$; since Z_1 is similar to Z_2, there exist $u_{10} \in Z_1$, $u_{20} \in Z_2$ and an isomorphism p_0 between $\boldsymbol{b}_1 \mid u_{10}$ and $\boldsymbol{b}_2 \mid u_{20}$ such that $p_0''(Z_1 \mid u_{10}) = Z_2 \mid u_{20}$. If we let $\bar{r}_1 = r_1 \mid u_{10}$, $\bar{r}_2 = r_2 \mid u_{20}$, $\bar{Z} = Z_1 \mid u_{10}$, $\bar{Z}_2 = Z_2 \mid u_{20}$ and $s_2 = \hat{p}_0'\bar{r}_1$ then we have $\bar{r}_2''\bar{Z}_2 = s_2''\bar{Z}_2$

and there exists $u_2 \in \bar{Z}_2$ such that $s_2 \mid u_2 \doteq \bar{r}_2 \mid u_2 (= r_2 \mid u_2)$. Finally, choose u_1 such that $p_0' u_1 = u_2$ and let $p = p_0 \upharpoonright (b_1 \mid u_1)$. Clearly $u_1 \in Z_1$ and p is an isomorphism between $\boldsymbol{b}_1 \mid u_1$ and $\boldsymbol{b}_2 \mid u_2$. We also have

$$\hat{p}'(r_1 \mid u_1) = \hat{p}_0'((r_1 \mid u_{10}) \mid u_1) = [\hat{p}_0'(r_1 \mid u_{10})] \mid u_2 = s_2 \mid u_2 \doteq r_2 \mid u_2 .$$

4327. THEOREM (**TSS'**). Two Boolean supports Z_1 and Z_2 on $\boldsymbol{b}$ are similar iff $Z_2 = f''Z_1$ for some automorphism f of $\boldsymbol{b}$.

Proof. If $Z_2 = f''Z_1$ and if f is an automorphism then Z_1 is (strongly) similar to Z_2. Conversely, suppose that Z_1 is similar to Z_2 and suppose that $Z_1 \neq Z_2$. Choose $v_1 \in Z_1 - Z_2$, and set $v_2 = -v_1$; thus $v_2 \in Z_2$ and $Z_i \mid v_i$ is similar to Z_i $(i = 1, 2)$. By Theorem 4323 there exist $u_i \in Z_i$ $(u_i \leqq v_i)$ and an isomorphism g between $\boldsymbol{b} \mid u_1$ and $\boldsymbol{b} \mid u_2$ such that $Z_2 \mid u_2 = g''(Z_1 \mid u_1)$; clearly, $u_1 \wedge u_2 = 0_{\boldsymbol{b}}$. To prove the theorem, it is enough to extend the mapping g to an automorphism of $\boldsymbol{b}$. For $x \in b$ we let $f'x = [x - (u_1 \vee u_2)] \vee g'(x \wedge u_1) \vee \mathbf{Cnv}(g)'(x \wedge u_2)$. The function f is one-to-one and maps b onto itself; for any $x, y \in b$ we have $x \leqq y \equiv f'x \leqq f'y$; hence f is an automosphism. Clearly $x \leqq u_1$ implies $f'x = g'x$. We shall prove $Z_2 = f''Z_1$. If $x \in Z_1$ then $x \wedge u_1 \in Z_1$ so that $g'(x \wedge u_1) \in Z_2$; since $f'x \geqq g'(x \wedge u_1)$ we have $f'x \in Z_2$. If $y \in Z_2$ then $y \wedge u_2 \in Z_2$ so that $\mathbf{Cnv}(g)'(y \wedge u_2) \in Z_1$; if $x \in b$ and $y = f'x$ then $x \wedge u_1 = \mathbf{Cnv}(g)'(y \wedge u_2)$ and hence $x \wedge u_1 \in Z_1$. Since $x \geqq x \wedge u_1$ we have $x \in Z_1$ and we are done.

4328. THEOREM (**TSS'**). Let Z be a Boolean support on $\boldsymbol{b}$ and suppose that $\boldsymbol{b}_1$ and $\boldsymbol{b}_2$ are subalgebras of $\boldsymbol{b}$. $Z \cap b_1$ is similar to $Z \cap b_2$ iff $b_1 |_{\boldsymbol{b}} u = b_2 |_{\boldsymbol{b}} u$ for some $u \in Z$.

Proof. Let $u \in Z$ be such that $b_1 |_{\boldsymbol{b}} u = b_2 |_{\boldsymbol{b}} u$. By Lemmas 4307 and 4309, $Z \cap b_i$ is similar to $Z \cap (b_i |_{\boldsymbol{b}} u)$ $(i = 1, 2)$ and hence $Z \cap b_1$ is similar to $Z \cap b_2$. Conversely, suppose that $Z \cap b_2$ and $Z \cap b_1$ are similar. There are disjointed relations r_1, r_2 such that $Z \cap b_1 = r_1''(Z \cap b_2)$ and $Z \cap b_2 = r_2''(Z \cap b_1)$. It follows that $Z \cap b_1$ and $Z \cap b_2$ are strongly similar (consider $r_1 \cup \mathbf{Cnv}(r_2)$!). By Lemma 4319 we obtain a $u \in Z$ such that $b_1 |_{\boldsymbol{b}} u = b_2 |_{\boldsymbol{b}} u$.

We shall now investigate properties of complete ultrafilters on products of separatively ordered sets (in particular, on products of two separatively ordered sets).

We first prove a theorem which concerns complete ultrafilters on products of any number of factors; the products are taken over a cut l having the singleton property.

4329. THEOREM (**TSS**′). Let $\boldsymbol{a} = \prod^l_{x \in s} \boldsymbol{a}_x$ be a product of nonempty separatively ordered sets (without greatest element) and let Z be a complete ultrafilter on $\boldsymbol{a}$. If we set $Z_x = \{u \in a_x; \langle u, x\rangle \in Z\}$ for each $x \in s$ then

1) Z_x is a complete ultrafilter on $\boldsymbol{a}_x$,
2) $f \in Z \equiv (\forall x \in \mathbf{D}(f))(f\text{'}x \in Z_x)$ for any $f \in a$.

Proof. 1) If $u, v \in Z_x$ then $\{\langle u, x\rangle\}, \{\langle v, x\rangle\} \in Z$, and so there exists $f \in a$ such that $f \leqq \{\langle u, x\rangle\}$ and $f \leqq \{\langle v, x\rangle\}$. It follows that $f\text{'}x \in a_x$, $f\text{'}x \leqq_x u$ and $f\text{'}x \leqq_x v$; hence we have $(\forall u, v \in Z_x)(\exists w \in a_x)(w \leqq_x u \,\&\, w \leqq_x v)$. If q is dense in $\boldsymbol{a}_x$ we shall prove that $q \times \{x\}$ is dense in $\boldsymbol{a}$. If $f \in a$ and $x \in \mathbf{D}(f)$ then $f\text{'}x \in a_x$ and there exist $u \in q$ and $v \in a_x$ such that $v \leqq_x u$ and $v \leqq_x f\text{'}x$. Similarly for $x \notin \mathbf{D}(f)$. For $y \in \mathbf{D}(f)$ $(y \neq x)$ we define $g\text{'}x = v$ and $g\text{'}y = f\text{'}y$; we have $g \leqq f$ and $g \leqq \{\langle u, x\rangle\}$, so that $q \times \{x\}$ is dense in $\boldsymbol{a}$. Consequently, if q is dense in $\boldsymbol{a}$ then there exists an element $\{\langle u, x\rangle\}$ of $q \times \{x\}$ which is in Z and hence $u \in Z_x$. Hence Z_x intersects every dense subset of $\boldsymbol{a}_x$, so that Z_x is a complete ultrafilter on $\boldsymbol{a}_x$ as asserted.

2) The implication from left to right is obvious. If the right-hand side holds and if $\boldsymbol{b}$ is the complete Boolean algebra with base $\boldsymbol{a}$ then $f = \bigwedge_{x \in \mathbf{D}(f)} \{\langle f\text{'}x, x\rangle\}$ (where the meet is taken in $\boldsymbol{b}$); since $f\text{'}x \in Z_x$ we have $\{\langle f\text{'}x, x\rangle\} \in Z$ and hence $f \in Z$.

4330. COROLLARY. We have $f \in Z \equiv f \subseteq \bigcup(Z)$ for every $f \in a$.

Analogous considerations can be carried out for the operation $\odot$ (cf. 2521): If $\boldsymbol{a}_i$ is separatively ordered with greatest element $1_{\boldsymbol{a}_i}$ $(i = 0, 1)$, and Z is a complete ultrafilter on $\boldsymbol{a}_0 \odot \boldsymbol{a}_1$, then we let $Z_0 = \{u \in a_0; \langle u, 1_{\boldsymbol{a}_1}\rangle \in Z\}$ and $Z_1 = \{v \in a_1; \langle 1_{\boldsymbol{a}_0}, v\rangle \in Z\}$. It follows that $Z = Z_0 \times Z_1$ (the ordinary cartesian product) and that Z_i is a complete ultrafilter on $\boldsymbol{a}_i$. (Prove !)

CHAPTER V

SECTION 1

The full upward extension of the universe

a) Parallels between **TSS″** *and* **TSS″**, S5

Our first aim in the present section is to generalize certain constructions already made in **TSS″** or in weaker theories. We consider **TSS″** and **TSS″**, S5 simultaneously, observing certain analogies between the results in these two theories. The second theory is stronger than the first, and in it we can present the promised "upward" method for constructing models. By this method we obtain a model of a theory of sets (**TS″**) in a theory of semisets (**TSS″**, S5). The results obtained for **TSS″** will be used in the next section.

We already have certain parallels between **TSS′** and **TSS″**, S5. We recall Metatheorem 2145 on recursion for iterable terms and Metatheorem 4228 on recursion for Sm-iterable terms. The first result concerns sets and real classes while the second concerns semisets and arbitrary classes. Actually the first result refers to **TSS′** and the second to **TSS′**, S4; however **TSS″** is stronger than **TSS′** and **TSS″**, S5 is stronger than **TSS′**, S4. The reason for dealing with these stronger theories is that we shall need the notion of rank and, for stronger results, not only the existence of a total support and the Standardness Axiom, but also the axiom (Pot) which gives for each set a "semiset coding of all subsemisets".

There are other analogies. In **TSS″** we have Theorem 3109 which states the existence of a (real) regulator for every real equivalence relation. In **TSS″**, S5 we have the following result:

5101. Theorem (**TSS″**, S5). For every equivalence relation S there is a class $P \subseteq \mathbf{C}(S)$ such that for each $x \in \mathbf{C}(\mathbf{S})$ the class $\{y \in P; \langle y, x\rangle \in S\}$ is a nonempty semiset.

Proof. Define $x \in P \equiv [x \in \mathbf{C}(S) \,\&\, (\forall y \in \mathbf{C}(S))(\langle x, y\rangle \in S \rightarrow \tau(x) \leq$

$\leqq \tau(y))]$; thus for each equivalence class P contains all elements of smallest rank. If x is an arbitrary element of $\mathbf{C}(S)$ then $\{y \in P; \langle y, x \rangle \in S\}$ is a semiset (since all elements have the same rank); by (St), this semiset is non-empty.

5102. DEFINITION (**TSS**). R is an *economical functor* (or an *economical relation*, Econ (R)) if, for each $x \in \mathbf{D}(R)$, $\{y \in \mathbf{D}(R); \mathbf{Ext}_R(x) = \mathbf{Ext}_R(y)\}$ is a semiset (i.e. each R-extension has only a semiset of R-codes).

Note that **TSS** $\vdash$ Ncon $(R) \rightarrow$ Econ (R), i.e. every nowhere constant functor is economical. We can prove a certain generalization of the axiom (C2) concerning economical functors.

5103. THEOREM. (1) (**TSS″**) If R is a real economical functor and if $\mathbf{W}(R)$ is a set then $\mathbf{D}(R)$ is a set.

(2) (**TSS″**, S5) If R is an economical functor and $\mathbf{W}(R)$ is a semiset then $\mathbf{D}(R)$ is a semiset.

Proof. (1) Each $\mathbf{Ext}_R(x)$ is a set; if we define $F = \{\langle x, q \rangle; x \in \mathbf{D}(R) \& q = \mathbf{Ext}_R(x)\}$ then F is a real relation. $\mathbf{W}(F) = \mathbf{D}(R)$ and $\mathbf{D}(F) \subseteq \mathbf{P}(\mathbf{W}(R))$ is a set; since R is economical, F is regular; hence $\mathbf{W}(F)$ is a set by 1414, 1454, 1403(e).

(2) Set $\sigma = \mathbf{W}(R)$ and let a be a set which includes σ. By (Pot) there is a coding ϱ of all subsemisets of a; let

$$S = \{\langle y, x \rangle; x \in \mathbf{D}(\varrho) \& y \in \mathbf{D}(R) \& \mathbf{Ext}_\varrho(x) = \mathbf{Ext}_R(y)\}.$$

We have $\mathbf{D}(S) \subseteq \mathbf{D}(\varrho)$ and so $\mathbf{D}(S)$ is a semiset; for each $x \in \mathbf{D}(S)$, $\mathbf{Ext}_S(x)$ is the semiset of all R-codes of $\mathbf{Ext}_\varrho(x)$. (The fact that R is economical is used here.) Hence, by (C2), $\mathbf{W}(S)$ is a semiset. Obviously $\mathbf{D}(R) = \mathbf{W}(S)$.

5104. LEMMA. (1) (**TSS″**) Every real extensional relation is morphic to a real strongly extensional (a fortiori, economical) relation.

(2) (**TSS″**, S5) Every extensional relation is morphic to an extensional economical relation.

Proof. (for both (1) and (2)). Let R be a relation with the assumed properties. Set $S = \{\langle u, v \rangle; u, v \in \mathbf{C}(R) \& \mathbf{Ext}_R(u) = \mathbf{Ext}_R(v)\}$. S is an equivalence; take a regulator P of S (in case (1) P is chosen to be real). Set $\bar{R} = R \cap P^2$, $H = S \restriction P$. Evidently H is a morphism of R, $\bar{R}$ and therefore $\bar{R}$ is extensional. For $x \in P$ we have $\mathbf{Ext}_{\bar{R}}(x) = \mathbf{Ext}_R(x) \cap P$; $\{y \in \mathbf{C}(\bar{R}); \mathbf{Ext}_{\bar{R}}(y) = \mathbf{Ext}_{\bar{R}}(x)\} = \{y \in P; \langle y, x \rangle \in S\}$ is a semiset and so $\bar{R}$ is economical.

The proof of (2) is ready; to prove (1), we continue as follows: Put $\hat{S} = S \cap P^2$; $\hat{S}$ is a real regular equivalence. Define

$$\hat{R} = \{\langle \mathbf{Ext}_{\hat{S}}(u), \mathbf{Ext}_{\hat{S}}(v)\rangle; \langle u, v\rangle \in R \,\&\, u, v \in P\},$$

$$\hat{H} = \{\langle u, \mathbf{Ext}_{\hat{S}}(v)\rangle; \langle u, v\rangle \in S \,\&\, v \in P\}.$$

Evidently, $\hat{R}$ is the required relation and $\hat{H}$ is the required morphism.

5105. DEFINITION (**TSS**). R is an almost regular relation (AReg (**R**)) if for every $x \in \mathbf{C}(R)$ there is a semiset $\sigma \subseteq \mathbf{Ext}_R(x)$ such that $\mathbf{Ext}_R(x) = \mathbf{SAT}_R(\sigma)$.

Evidently every regular extensional relation is almost regular.

5106. THEOREM. (1) (**TSS**″, Elk (**R**), ∂**TSS**/$\partial\mathfrak{N}$ (**R**)). If the relation **R** is real, almost regular and almost universal then Axiom (C2) holds in the sense of $\mathfrak{N}$ (**R**).

(2) (**TSS**″, S5, Elk (**R**), ∂**TSS**/$\partial\mathfrak{N}$ (**R**)). If the relation **R** is almost regular and almost universal then Axiom (C2) holds in the sense of $\mathfrak{N}$ (**R**).

Proof. (1) By 5104 (1), we can define a relation $\hat{\mathbf{R}}$ and a morphism $\hat{\mathbf{H}}$ of **R**, $\hat{\mathbf{R}}$ such that $\hat{\mathbf{R}}$ is strongly extensional. Evidently, $\hat{\mathbf{R}}$ is pairing, regular and almost universal provided **R** is pairing, almost regular and almost universal. Hence the theorem follows by 1313 and 1416.

(2) Since **R** is E-like, the axioms of **TC** hold in the sense of $\mathfrak{N}$ (**R**). By 5104 (2), we can construct a relation $\overline{\mathbf{R}}$ and a morphism **H** of **R**, $\overline{\mathbf{R}}$ such that $\overline{\mathbf{R}}$ is extensional and economical. Moreover, $\overline{\mathbf{R}}$ is pairing, regular and almost universal provided **R** is pairing, almost regular and almost universal. Hence using 1313 we can suppose without loss of generality that **R** is regular and economical. Under this assumption we prove (C2). First we have $\mathrm{Sm}^*(X^*) \equiv \mathrm{Sm}(X^*)$. Indeed, if $\mathrm{Sm}^*(X^*)$ then $X^* \subseteq \mathbf{Ext}_{\mathbf{R}}(u)$ for some u; consequently, $\mathrm{Sm}(X)$ by the regularity of **R**, while conversely if X^* is a semiset then $X^* \subseteq \mathbf{Ext}_{\mathbf{R}}(u)$ for some u by the almost-universality of **R**.

Now suppose $X^* \subseteq \mathbf{C}(\mathbf{R})$ and $\mathrm{Exct}^*(X^*)$ and set $X = \mathbf{Dec}_{\mathbf{R}}(X^*)$. We have $\mathbf{D}^*(X^*) = \mathbf{D}(X)$, $\mathbf{Ext}_X(u) = \mathbf{Ext}^*_{X^*}(\mathbf{Ext}_{\mathbf{R}}(u))$ etc., hence we obtain $\mathrm{Reg}(X)$ from $\mathrm{Reg}^*(X^*)$ and $\mathrm{Econ}(X)$ from $\mathrm{Ncon}^*(X^*)$. Thus $\mathrm{Sm}(\mathbf{D}(X)) \equiv \mathrm{Sm}(\mathbf{W}(X))$ ($\rightarrow$ by 1414 and $\leftarrow$ by 5103 (2)) and therefore

$$\mathrm{Sm}^*(\mathbf{D}^*(X^*)) \equiv \mathrm{Sm}(\mathbf{D}(X)) \equiv \mathrm{Sm}(\mathbf{W}(X)) \equiv \mathrm{Sm}^*(\mathbf{W}^*(X^*)).$$

5107. DEFINITION (**TSS**). A relation R is saturated-universal (denotation: Satunvr (R)) if for every $\sigma \subseteq \mathbf{C}(R)$ there is a $x \in \mathbf{C}(R)$ such that $\mathbf{Ext}_R(x) = \mathbf{SAT}_R(\sigma)$.

Note that every saturated-universal relation is almost universal. We shall prove in **TSS**″, S5 that there are extensional almost regular saturated-universal relations. This is important because of the following

5108. METATHEOREM. The F-definition $\mathfrak{N}(\mathbf{R})$ with the specification Extl $(\mathbf{R})$ & AReg $(\mathbf{R})$ & Satunvr $(\mathbf{R})$ determines a model of **TS**″ in **TSS**″, S5.

Demonstration. We proceed in the theory **TSS**″, S5, (Extl $(\mathbf{R})$ & & AReg $(\mathbf{R})$ & Satunvr $(\mathbf{R})$), ∂**TS**$/\partial\mathfrak{N}(\mathbf{R})$. It follows immediately from Satunvr $(\mathbf{R})$ that $\mathbf{R}$ is pairing. Hence by Theorem 5106 all axioms of **TC** and also (C2) hold in the sense of $\mathfrak{N}(\mathbf{R})$. We prove (C1) and (C3) in the sense of $\mathfrak{N}(\mathbf{R})$. First $\mathbf{R}$ can again be supposed to be regular and economical (cf. the proof of Theorem 5106); if $\mathbf{R}$ is saturated-universal then $\overline{\mathbf{R}}$ is also saturated-universal). Thus every saturated subsemiset σ of $\mathbf{R}$ is the extension of some element of $\mathbf{C}(\mathbf{R})$, and so we have (C3)*. Define

$$H\text{''}\{0\} = \{x \in \mathbf{C}(\mathbf{R}); \mathbf{Ext}_{\mathbf{R}}(x) = 0\},$$

$$H\text{''}\{n+1\} = \{x \in \mathbf{C}(\mathbf{R}); \mathbf{Ext}_{\mathbf{R}}(x) = H\text{''}\{n\}\}.$$

(Using Metatheorem 4228.) Since H is regular, $H\text{''}\boldsymbol{\omega}$ is a semiset; obviously $H\text{''}\boldsymbol{\omega}$ is saturated. Hence there is an x^* such that $x^* = H\text{''}\boldsymbol{\omega}$ and we have

$$0^* \in^* x^* \mathbin{\&} (\forall n^* \in^* x^*)(\{n^*\}^* \in^* x^*).$$

Indeed, if $u \in x^*$ then $u \in H\text{''}\{n\}$ for some n. We have $(\forall u' \in H\text{''}\{n\})(\mathbf{Ext}_R(u') = \mathbf{Ext}_{\mathbf{R}}(u))$ and for each $v \in H\text{''}\{n+1\}$ we have $\mathbf{Ext}_{\mathbf{R}}(v) = H\text{''}\{n\}$; hence setting $u^* = \mathbf{Ext}_{\mathbf{R}}(u)$, $v^* = \mathbf{Ext}_{\mathbf{R}}(v)$ we have $v^* = \{u^*\}^*$. Thus Axiom (C1)* is proved. The proof of (D2)* is easy.

It remains to prove the existence of extensional almost regular saturated-universal relations in **TSS**″, S5. This will be done in a quite general way in an analogy with the construction of a real upward extension (of a real regular strongly extensional relation) described in Chapt. III Sect. 2. (Cf. Metatheorem 3223.)

5109. METATHEOREM. There are gödelian terms **Fupw** (Q, S) and **Imb** (Q) such that (1) the following is provable in **TSS**″, S5: If Q is a regular extensional economical relation and if S is an arbitrary totally universal relation then **Fupw** (Q, S) is a regular extensional economical and saturated-universal relation. Moreover,

(2) in **TSS**″, S5 we can fix constants **Q** for a regular extensional economical relation, **S** for a totally universal relation and **R** for **Fupw** (**Q**, **S**). Adding ∂**TS**$/\partial\mathfrak{N}$ (**R**), we obtain a theory in which the following is provable: There is a complete subrelation $\bar{Q}$ of **R** such that **Imb** (**Q**) is an isomorphism of **Q** and $\bar{Q}$; $\mathbf{C}(\bar{Q})$ is a class in the sense of $\mathfrak{N}$ (**R**) and $\mathbf{V}^* = \mathbf{Ker}^*(\mathbf{C}(\bar{Q}))$ holds.

Fupw (Q, S) is called the *full upward extension* of Q (constructed with the aid of S).

Demonstration. We proceed in **TSS**″, S5. Given Q define $F\text{'}x = \langle 0, x\rangle$ for $x \in \mathbf{C}(Q)$ and set $\bar{Q} = \{\langle\langle 0, x\rangle, \langle 0, y\rangle\rangle; \langle x, y\rangle \in Q\}$. Evidently F can be defined by a gödelian term **Imb** (Q). Given S define another totally universal relation $\bar{S}$ such that $\bar{Q} \subseteq\, \subseteq \bar{S}$ and with the following additional properties: (a) semisets which are $\bar{Q}$-extensions have the same codes in $\bar{S}$ and in $\bar{Q}$, (b) $\bar{S}$ is economical and regular. Indeed, define

$$\begin{aligned} \bar{S} = \bar{Q} \cup \{\langle u, \langle 1, x\rangle\rangle;\ & u \in \mathbf{Ext}_S(x) \,\&\, \mathrm{Sm}(\mathbf{Ext}_S(x))\, \& \\ & \& \neg (\exists y \in \mathbf{C}(\bar{Q}))(\mathbf{Ext}_S(x) = \mathbf{Ext}_{\bar{Q}}(y))\, \& \\ & \& \neg (\exists y \in \mathbf{D}(S))(\mathbf{Ext}_S(x) = \mathbf{Ext}_S(y) \,\&\, \tau(y) < \tau(x)\}\,. \end{aligned}$$

Thus $\bar{Q}$-extensions are coded by their $\bar{Q}$-codes and each other semiset is coded by all elements of the form $\langle 1, x\rangle$ where x is an S-code for this semiset of last rank. By (St) $\bar{S}$ is a totally universal relation; obviously $\bar{S}$ is economical.

We now define H by recursion as follows: $H\text{''}\{0\} = \mathbf{C}(\bar{Q})$, $\alpha > 0 \rightarrow H\text{''}\{\alpha\} = \{y \in \mathbf{D}(\bar{S}); \mathbf{Ext}_{\bar{S}}(y) \subseteq H\text{''}\alpha \,\&\, \mathrm{Sat}_{\bar{S}}(\mathbf{Ext}_{\bar{S}}(y))\}$, The definition is correct by Metatheorem 4228 (note that we use (Pot) to have the Sm-iterability). Further set $R = \bar{S} \upharpoonright (H\text{''}\,\mathbf{On})$. Evidently R can be defined by a gödelian term **Fupw** (Q, S). It follows from the condition defining $H\text{''}\{\alpha\}$ that $x \in \mathbf{C}(R)$ implies $\mathbf{Ext}_R(x) = \mathbf{Ext}_{\bar{S}}(x)$ and therefore $\mathbf{Ext}_R(x)$ is R-saturated, hence R is extensional, economical and regular and $\bar{Q} \subseteq\, \subseteq R \subseteq\, \subseteq \bar{S}$. We prove that R is saturated-universal. Take a subsemiset σ of $\mathbf{C}(R)$; since R is economical $\mathbf{SAT}_R(\sigma) = \mathbf{SAT}_{\bar{S}}(\sigma)$ is a semiset and therefore there is an α and an $x \in H\text{''}\{\alpha\}$ such that $\mathbf{SAT}_R(\sigma) = \mathbf{Ext}_R(x)$. Thus R is saturated-universal.

Let us proceed now in the theory indicated in (2). Note that this theory is a conservative extension of **TSS**″, S5 and that all axioms of **TS**″ hold in the sense of $\mathfrak{N}(\mathbf{R})$. Denote $\mathbf{Imb}(\mathbf{Q})$ by F. F is evidently an isomorphism of $\mathbf{Q}$ and $\overline{\mathbf{Q}}$, and $\overline{\mathbf{Q}}$ is a complete subrelation of $\mathbf{R}$. Moreover, by our construction, $\mathbf{C}(\overline{\mathbf{Q}})$ is a saturated subclass of $\mathbf{C}(\mathbf{R})$ and therefore $\mathrm{Cls}^*(\mathbf{C}(\overline{\mathbf{Q}}))$ and $\mathrm{Comp}^*(\mathbf{C}(\overline{\mathbf{Q}}))$. We prove $\mathbf{V}^* = \mathbf{Ker}^*(\mathbf{C}(\overline{\mathbf{Q}}))$. Let Z^* be a class in the sense of $\mathfrak{N}(\mathbf{R})$ such that $\mathbf{C}(\overline{\mathbf{Q}}) \subseteq Z^*$ and $\mathbf{P}^*(Z^*) \subseteq Z^*$. Then it follows by induction (using (St)) that $H''\alpha \subseteq Z^*$ for every α; hence $\mathbf{V}^* = H''\,\mathbf{On} \subseteq$ $\subseteq Z^*$. This proves $\mathbf{V}^* = \mathbf{Ker}^*(\mathbf{C}(\overline{\mathbf{Q}}))$.

5110. Remark. Note that if S_1, S_2 are two totally universal relations and Q is a regular extensional economical relation then $\mathbf{Fupw}(Q, S_1)$ and $\mathbf{Fupw}(Q, S_2)$ are morphic. The proof of this fact is left to the reader as an exercise.

b) *Full upward extension of* E; *the model* $\mathfrak{Up}$

An extremally important case of the full upward extension is $\mathbf{Fupw}(\mathbf{E}, S)$. This contrasts with the real upward extensions because $\mathbf{Rupw}(\mathbf{E})$ is evidently isomorphic to $\mathbf{E}$ and therefore of no interest.

5111. DEFINITION ($\mathbf{TSS}''$, S5). A relation R is a *full upward extension* ($\mathrm{Fupw}(R)$) if there is a totally universal relation S such that $R = \mathbf{Fupw}(\mathbf{E}, S)$. We denote $\mathbf{Imb}(\mathbf{E})$ by $\mathbf{Imb}$ and $\mathbf{Imb}''\,\mathbf{V}$ by $\mathbf{Tor}$ (the *torso* of the full upward extension).

5112. METADEFINITION AND DIAGRAM. $\mathbf{TSS}^{\mathrm{e}}$ is the theory

$$\mathbf{TSS}'',\ \mathrm{S5},\ \mathrm{Fupw}(\mathbf{Fup}),\ \partial\mathbf{TS}/\partial\mathfrak{N}(\mathbf{Fup})$$

(the theory of semisets in which the upward extending model will be studied).

Thus $\mathbf{Fup}$ is a constant fixed for an arbitrary full upward extension. It follows by Metatheorem 5109 that $\mathfrak{Dir}(\partial\mathbf{TS}/\partial\mathfrak{N}(\mathbf{Fup}))$ is a direct model of $\mathbf{TS}''$ in $\mathbf{TSS}^{\mathrm{e}}$; hence $\mathbf{TSS}^{\mathrm{e}}$ is a conservative extension of $\mathbf{TSS}''$, S5.

Consider the following diagram:

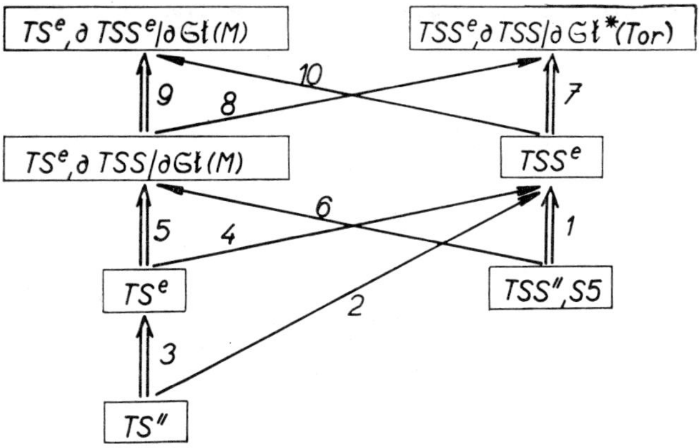

Arrow 1 is the identity and arrow 2 is $\mathfrak{Dir}\,(\partial\mathbf{TS}/\partial\mathfrak{N}(\mathbf{Fup}))$. (Translations of formulas in the last model will be denoted by asterisk.) All arrows in this diagram denote some direct translations (and we shall show that they are models); double arrows denote identities. The reader is recommended to draw this diagram step by step as it is being studied.

We defined the constant **Tor** in **TSS**″, S5 by $\mathbf{Tor} = \mathbf{Imb}''\,\mathbf{V}$ and we know that $\mathbf{TSS}^{e} \vdash \mathrm{Cls}^{*}(\mathbf{Tor})\,\&\,\mathrm{Comp}^{*}(\mathbf{Tor})\,\&\,\mathbf{V}^{*} = \mathbf{Ker}^{*}(\mathbf{Tor})$. We shall establish other properties of the class **Tor**. First, the following lemma follows immediately from the properties of the isomorphism **Imb**:

5113. LEMMA ($\mathbf{TSS}^{e}$).

(a) $(\forall x)\,(\exists!\, x^{*} \in^{*} \mathbf{Tor})\,(x^{*} = \mathbf{Imb}''\,x)$,

$(\forall x^{*} \in^{*} \mathbf{Tor})\,(\exists!\, x)\,(x^{*} = \mathbf{Imb}''\,x)$;

(b) $(\forall \sigma)\,(\exists!\, x^{*} \subseteq \mathbf{Tor})\,(x^{*} = \mathbf{Imb}''\,\sigma)$,

$(\forall x^{*} \subseteq \mathbf{Tor})\,(\exists!\, \sigma)\,(x^{*} = \mathbf{Imb}''\,\sigma)$;

(c) $(\forall X)\,(\exists!\, X^{*} \subseteq \mathbf{Tor})\,(X^{*} = \mathbf{Imb}''\,X)$,

$(\forall X^{*} \subseteq \mathbf{Tor})\,(\exists!\, X)\,(X^{*} = \mathbf{Imb}''\,X)$;

(d) $(x^{*} = \mathbf{Imb}''\,x\,\&\,Y^{*} = \mathbf{Imb}''\,Y) \rightarrow [x \in Y \equiv x^{*} \in^{*} Y^{*}]$.

This has the following

5114. METACOROLLARY. If $\varphi(x, \bullet, \sigma, \bullet, X, \bullet)$ is a RF then

$$\mathbf{TSS}^e \vdash (x^* = \mathbf{Imb}\text{''}\, x \;\&\; \bullet \;\&\; y^* = \mathbf{Imb}\text{''}\, \sigma \;\&\; \bullet \;\&\; X^* = \mathbf{Imb}\text{''}\, X \;\&\; \bullet) \rightarrow$$
$$\rightarrow \left[\varphi(x, \bullet, \sigma, \bullet, X, \bullet) \equiv \varphi^*(x^*, \bullet, y^*, \bullet, X^*, \bullet)\right].$$

In particular, let $\varphi_i(x, y, z)$ be RF's equivalent to $z = \mathbf{F}_i(x, y)$ $(i = 1, \ldots, 7)$ where the $\mathbf{F}_i$ are gödelian operations. Then

$$\mathbf{TSS}^e \vdash (x^* = \mathbf{Imb}\text{''}\, x \;\&\; \bullet) \rightarrow \left[\varphi_i(x, y, z) \equiv \varphi_i^*(x^*, y^*, z^*)\right].$$

Hence we can proceed as follows in $\mathbf{TSS}^e$. Let $x^*, y^* \in^* \mathbf{Tor}$ and let $x^* = \mathbf{Imb}\text{''}\, x$, $y^* = \mathbf{Imb}\text{''}\, y$. Set $z = \mathbf{F}_i(x, y)$ and $z^* = \mathbf{Imb}\text{''}\, z$. Then $z^* \in^* \mathbf{Tor}$ and $\varphi_i^*(x^*, y^*, z^*)$. Hence $(\forall x^*, y^*)(\exists z^*)(z^* = \mathbf{F}_i^*(x^*, y^*))$, i.e. we have proved $\mathrm{Clos}^*(\mathbf{Tor})$ in $\mathbf{TSS}^e$. Further we prove $\mathrm{AUncl}^*(\mathbf{Tor})$. Recall that all axioms of **TS** hold in the sense of $\mathfrak{N}(\mathbf{Fup})$. Hence we must prove that in the sense of $\mathfrak{N}(\mathbf{Fup})$ every subset of **Tor** is a subset of an element of **Tor**. Let $x^* \subseteq \mathbf{Tor}$. There is a σ such that $x^* = \mathbf{Imb}\text{''}\, \sigma$ and there is an a such that $\sigma \subseteq a$. For $a^* = \mathbf{Imb}\text{''}\, a$ we obtain $x^* \subseteq a^*$. Hence we have proved

5115. LEMMA ($\mathbf{TSS}^e$). $\mathrm{Mcl}^*(\mathbf{Tor})$.

5116. DEFINITION ($\mathbf{TSS}'$). (a) $\mathrm{Mcl}_{\mathrm{S1}}(X) \equiv .\ \mathrm{Mcl}(X) \;\&\; (\exists Z \subseteq X)(\forall \sigma \subseteq X)(\exists r \in X)(\sigma = r\text{''}Z)$;

(b) $\mathrm{Mcl}_{\mathrm{D2}}(X) \equiv .\ \mathrm{Mcl}(X) \;\&\; (\exists F)(\mathrm{Un}(F, \mathbf{On}, X) \;\&\; \bigcup(\mathbf{W}(F)) = X \;\&\; (\forall u \in X)(F \cap u \in X))$.

$\mathrm{Mcl}_{\mathrm{S1}}(X)$ is read "X is a *model-class with* S1",
$\mathrm{Mcl}_{\mathrm{D2}}(X)$ is read "X is a *model-class with* D2".
These names are justified by the following obvious

5117. METALEMMA. (a) $\mathfrak{St}(\mathbf{M})$ with the specification $\mathrm{Mcl}_{\mathrm{S1}}(\mathbf{M})$ determines a faithful model of $\mathbf{TSS}'$, S1 in $\mathbf{TSS}'$. (b) $\mathfrak{St}(\mathbf{M})$ with the specification $\mathrm{Mcl}_{\mathrm{D2}}(\mathbf{M})$ determines a faithful model of **TSS**, D2 in $\mathbf{TSS}'$.

5118. LEMMA ($\mathbf{TSS}^e$) In the sense of $\mathfrak{N}(\mathbf{Fup})$ **Tor** is a model-class with S1 and D2 and, moreover, $\mathbf{On}^* \subseteq \mathbf{Tor}$.

Proof. There is a total support Z (which is a local semiset Boolean support, although we do not use this fact). Denote $\mathbf{Imb}\text{''}\, Z$ by Z^*. We have $Z^* \subseteq \mathbf{Tor}$.

If $u^* \subseteq \mathbf{Tor}$ then there is a σ such that $u^* = \mathbf{Imb}''\,\sigma$ and there is an r such that $\sigma = r''Z$. Denoting $\mathbf{Imb}''\,r$ by r^* we have $r^* \in^* \mathbf{Tor}$. Evidently the formula $\sigma = r''Z$ is a RF; thus we obtain $u^* = r^{*}{}''{}^{*}Z^*$ by Metacorollary 5114. We have thus proved $\mathrm{Mcl}^*_{s1}(\mathbf{Tor})$; we now prove $x^* = \mathbf{Imb}''\,x \to \to [\mathrm{Ord}(x) \equiv \mathrm{Ord}^*(x^*)]$. Suppose $x^* = \mathbf{Imb}''\,x$; then $\mathrm{Comp}(x) \equiv \equiv \mathrm{Comp}^*(x^*)$ since $\mathrm{Comp}(x)$ is a RF. Similarly if $\mathrm{Trich}(x)$ means $(\forall u, v \in x)(u \in v \vee u = v \vee v \in u)$ then we have $\mathrm{Trich}(x) \equiv \mathrm{Trich}^*(x^*)$. We prove $\mathrm{Ord}^*(x^*) \equiv \mathrm{Ord}(x)$. Let $\mathrm{First}(x, y)$ be the formula $y \subseteq x \,\& \,\& [y \neq 0 \to (\exists u \in y)(u \cap y = 0)]$. $\mathrm{First}(x, y)$ is a RF (in **TSS**). Suppose $\mathrm{Ord}(x)$. Let $y^* \subseteq x^*$ and $y^* = \mathbf{Imb}''\,\sigma$. Then $\sigma \subseteq x$ and if $\sigma \neq 0$ then by (St) there is a $u \in \sigma$ such that $u \cap \sigma = 0$, and so we obtain $\mathrm{First}^*(x^*, y^*)$. Conversely, suppose $\mathrm{Ord}^*(x^*)$. Then $\mathrm{Comp}(x)$ and $\mathrm{Trich}(x)$. If $y \subseteq x$ then for $y^* = \mathbf{Imb}''\,y$ we have $y^* \subseteq x^*$ so that $\mathrm{First}^*(x^*, y^*)$. $\mathrm{First}(x, y)$ follows and we proved $\mathrm{Ord}(x)$.

We now prove $\mathbf{On}^* \subseteq \mathbf{Tor}$. Suppose not; then there is a least α^* in $\mathbf{On}^* - \mathbf{Tor}$ and there is a $\sigma \subseteq \mathbf{On}$ such that $\alpha^* = \mathbf{Imb}''\,\sigma$. By (St) $\mathbf{On} - \sigma$ has a minimal element α; since σ is complete it follows that $\sigma = \alpha$. Then $\alpha^* = \mathbf{Imb}''\,\alpha$, hence $\alpha^* \in^* \mathbf{Tor}$, a contradiction.

We now prove $\mathrm{Mcl}_{D2}{}^*(\mathbf{Tor})$. By D2 there is a real function F such that $\mathbf{D}(F) = \mathbf{On}$ and $\bigcup(\mathbf{W}(F)) = \mathbf{V}$. Denote $\mathbf{Imb}''\,F$ by F^*. Evidently $\langle x, \alpha\rangle \in \in F \equiv \langle x^*, \alpha^*\rangle \in^* F^*$ for $x^* = \mathbf{Imb}''\,x$, $\alpha^* = \mathbf{Imb}''\,\alpha$. It follows that $\mathbf{D}^*(F^*) = \mathbf{On}^*$ and $\mathbf{W}^*(F^*) \subseteq \mathbf{Tor}$. Moreover, if $x^* \in^* \mathbf{Tor}$, $x^* = \mathbf{Imb}''\,x$ then, for some α, $x \in F'\alpha$ and hence $x^* \in^* F^{*}{}'{}^{*}\alpha^*$. Thus we obtain $\bigcup^*(\mathbf{W}^*(F^*)) = \mathbf{Tor}$ and $\mathrm{Mcl}_{D2}(\mathbf{Tor})$ follows.

This leads us to the following definition:

5119. Definition (**TS**″). $\mathrm{FMcl}(X) \equiv .\ \mathrm{Mcl}_{S1}(X) \,\&\, \mathrm{Mcl}_{D2}(X) \,\&\, \mathbf{On} \subseteq \subseteq X \,\&\, \mathbf{V} = \mathbf{Ker}(X)$. ($X$ is a *full model-class* if it is a model-class with S1 and D2 such that $\mathbf{On} \subseteq X$ and $\mathbf{V} = \mathbf{Ker}(X)$.)

5120. Theorem (**TSS**^{e}). **Tor** is a full model-class in the sense of $\mathfrak{N}$ (**Fup**).

In theories with stronger regularity axioms the notion of a full model-class can be simplified.

5121. Lemma. (1) (**TS**, U3) M is a full model-class iff it is a model-class with S1 which contains all urelements.

(2) (**TS**, D3) M is a full model-class iff it is a model-class with S1.

Proof. (1) Every full model-class must contain all urelements since $\mathbf{Ur} \nsubseteq M$ implies $\mathbf{Ker}(M) \neq \mathbf{V}$. Conversely if $\mathbf{Ur} \subseteq M$ then $\mathbf{V} = \mathbf{Ker}(M)$.

We proved in **TSS**, U3 that every model-class contains all ordinal numbers. Further it follows easily in **TSS**, U3 that every model-class is a model-class with D2 (moreover, in **TSS**, U3, ∂**TSS**/$\partial\mathfrak{St}$ (**M**) we have (U3) $^{\mathfrak{St}(\mathbf{M})}$).

(2) is an immediate consequence of (1).

5122. METADEFINITION. (1) **TS**^{e} is the theory **TS**$''$, FMcl (**M**). (2) $\mathfrak{Up}$ is the extension of $\mathfrak{Dir}$ (∂**TS**/$\partial\mathfrak{N}$(**Fup**)) interpreting **M** as **Tor**.

FMcl (**M**) is a fixing in **TS**$''$ (cf. arrow 3 of Diagram 5112); by 5120, the direct translation $\mathfrak{Up}$ is a model of **TS**^{e} in **TSS**^{e} (arrow 4). The model $\mathfrak{Up}$ is of great importance; we shall prove in the course of the subsequent discussion that $\mathfrak{Up}$ is faithful.

5123. METALEMMA. $\mathfrak{Dir}$ (∂**TSS**/$\partial\mathfrak{St}$ (**M**)) is a model of **TSS**$''$, S5 in **TS**^{e}, ∂**TSS**/$\partial\mathfrak{St}$ (**M**).

Demonstration. (See arrows 5 and 6.) $\mathfrak{Dir}$ (∂**TSS**/$\partial\mathfrak{St}$ (**M**)) is a model of **TSS**; (S1) and (D2) hold in this model. Hence it suffices to show that (St) and (Pot) hold. We proceed in **TS**^{e}, ∂**TSS**/$\partial\mathfrak{St}$ (**M**). Denote the notions in the sense of $\mathfrak{St}$ (**M**) by means of the superscript $\square$. Recall that $\mathbf{On} \subseteq \mathbf{M}$. It is obvious from the definition of Ord (x) that Ord $(x^\square) \rightarrow$ Ord$^\square$ $(x^\square)$; hence $\mathbf{On} \subseteq \subseteq \mathbf{On}^\square$. Suppose $\mathbf{On}^\square - \mathbf{On} \neq 0$ and let $\beta^\square \in \mathbf{On}^\square - \mathbf{On}$. Then $\beta^\square \in \mathbf{M}$ and so $\beta^\square$ is a set. On the other hand, for every $\alpha \in \mathbf{On}$ we have $\alpha \in \beta^\square$ because $\alpha \in \beta^\square \vee \beta^\square \in \alpha$ holds and $\beta^\square \in \alpha$ implies $\beta^\square \in \mathbf{On}$. But this means that $\mathbf{On} \subseteq \beta^\square$ which is a contradiction. Thus we have proved that $\mathbf{On} = = \mathbf{On}^\square$. Now let $\sigma^\square \subseteq \mathbf{On}^\square$. Then $\sigma^\square$ is a set of ordinal numbers (we are working in a set theory!) and so if it is non-empty then it has a first element. This proves (St)$^\square$. We now prove (Pot)$^\square$. If $S^\square$ is a totally universal relation in the sense of $\mathfrak{St}$ (**M**) and a is an element of **M** then, for each $u \subseteq a$, let $F'u$ be the set of all $S^\square$-codes of x of the lest rank. (All semisets are sets!). Then F is a function, $\mathbf{D}(F)$ is a set and hence $\bigcup(\mathbf{W}(F))$ is a set. Hence if we define $S_0 = S^\square \upharpoonright \bigcup(\mathbf{W}(F))$ then S_0 is a set (since S_0 is regular and $\mathbf{D}(S_0)$ is a set) and $S_0 \subseteq \mathbf{M}$, hence S_0 is a semiset in the sense of $\mathfrak{St}$ (**M**); let $\sigma^\square = S_0$. Evidently $\sigma^\square$ is a "semiset coding" of all subsemisets of a in the sense of $\mathfrak{St}$ (**M**). This proves (Pot)$^\square$ and hence (S5)$^\square$ follows.

5124. The model $\mathfrak{Up}$ (arrow 4) can be extended to a direct model of **TS**^{e}, ∂**TSS**/$\partial\mathfrak{St}$ (**M**) in **TSS**^{e}, ∂**TSS**/$\partial\mathfrak{St}^*$ (**Tor**) (arrows 5, 7, 8) where $\mathfrak{St}^*$ (**Tor**) is the following F-definition:

$$(\forall X^*)\left([\exists X^\square)(X^\square = X^*) \equiv X^* \subseteq \mathbf{Tor}\right] \,\&\, sec$$

$$(\forall X^\square, Y^\square)(X^\square \in^\square Y^\square \equiv X^\square \in^* Y^\square) \,\&\, sec\,.$$

(This definition is induced by the model $\mathfrak{Up}$ and by the F-definition $\mathfrak{St}(\mathbf{M})$ added to **TS**^{e}.) We prove that the composition of arrows 6 and 8 is equivalent to the identity 1 $*$ 7. Lemma 5113 which was proved in **TSS**^{e}, has the following

5125. COROLLARY (**TSS**^{e}, ∂**TSS**$/\partial\mathfrak{St}^*$ (**Tor**))

$$(\forall X)(\exists!\, X^{\square})(\mathbf{Imb}\text{''}\, X = X^{\square}) \,\&\, (\forall X^{\square})(\exists!\, X)(\mathbf{Imb}\text{''}\, X = X^{\square})\,;$$
$$X^{\square} = \mathbf{Imb}\text{''}\, X \,\&\, Y^{\square} = \mathbf{Imb}\text{''}\, Y. \rightarrow [X \in Y \equiv X^{\square} \in^{\square} Y^{\square}]\,.$$

Our result follows by 1269. Consequently 6 $*$ 8 is faithful and therefore 6 is faithful.

We have also the following useful

5126. METACOROLLARY. Let $\varphi(X, \bullet)$ be an arbitrary **TSS**-formula. Then

$$\mathbf{TSS}^e, \partial\mathbf{TSS}/\partial\mathfrak{St}^*(\mathbf{Tor}) \vdash (X^{\square} = \mathbf{Imb}\text{''}\, X \,\&\, \bullet) \rightarrow \varphi(X, \bullet) \equiv \varphi^{\square}(X^{\square}, \bullet)\,.$$

Similarly, the model described by arrow 6 can be extended to a direct model of **TSS**^{e} in **TS**^{e}, ∂**TSS**$^e/\partial\mathfrak{St}(\mathbf{M})$ (arrows 1, 9, 10). In this latter theory $\square$-variables, and all $\square$-notions corresponding to **TSS**-notions, are first defined and we know that all axioms of **TSS**$''$, S5 in the $\square$-sense are provable in the terminal theory of arrow 6. In particular, we have the constants $\mathbf{Fup}^{\square}$, $\mathbf{Tor}^{\square}$ and $\mathbf{Imb}^{\square}$ and we can give the definition of $*$-variables as saturated parts of $\mathbf{C}(\mathbf{Fup}^{\square})$ (note that it does not matter if we say "in the sense of $\mathfrak{St}(\mathbf{M})$" or not because the notions of a relation and a saturated part are absolute). Furthermore, we can derive the definitions of all $*$-notions corresponding to **TS**-notions. We know that FMcl* ($\mathbf{Tor}^{\square}$) is provable in **TS**^{e}, ∂**TSS**$^e/\partial\mathfrak{St}(\mathbf{M})$. We have the following

5127. LEMMA (**TS**^{e}, ∂**TSS**$^e/\partial\mathfrak{St}(\mathbf{M})$). There is a morphism K of $\mathbf{E}$ and $\mathbf{Fup}^{\square}$ such that $K\text{''}\mathbf{M} = \mathbf{Tor}^{\square}$.

Proof. We start with $\mathbf{Imb}^{\square}$ which is an isomorphism of $\mathbf{E} \restriction \mathbf{M}$ and $\mathbf{Fup}^{\square} \restriction \mathbf{Tor}^{\square}$. Define $H\text{''}\{0\} = \mathbf{Imb}^{\square}$;

$$\alpha > 0 \rightarrow H\text{''}\{\alpha\} = \{\langle y, x\rangle;\, x \subseteq \mathbf{D}(H\text{''}\alpha) \,\&\, \mathbf{Ext}_{\mathbf{Fup}^{\square}}(y) = (H\text{''}\alpha)\text{''}\, x\}\,.$$

It follows from $\mathbf{Ker}(\mathbf{M}) = \mathbf{V}$, by the construction of $\mathbf{Fup}^{\square}$ and by the absoluteness of ordinal numbers, that if we set $\mathbf{K} = H\text{''}\mathbf{On}$ then $\mathbf{K}$ is a morphism with the required properties.

5128. COROLLARY ($\mathbf{TS}^e$, $\partial\mathbf{TSS}^e/\partial\mathfrak{St}(\mathbf{M})$)

$$(\forall X)(\exists! X^*)(X^* = \mathbf{K}\text{''}X) \,\&\, (\forall X^*)(\exists! X)(X^* = \mathbf{K}\text{''}X)\,;$$
$$X^* = \mathbf{K}\text{''}X \,\&\, Y^* = \mathbf{K}\text{''}Y. \rightarrow [X \in Y \equiv X^* \in^* Y^*]\,;$$
$$\mathbf{K}\text{''}\mathbf{M} = \mathbf{Tor}^{\square}\,.$$

By Lemma 1269 and the remark following it, the composition of arrows 4, 10 is a faithful model. Hence the model $\mathfrak{Up}$ (arrow 4) is also faithful.

We summarize our discussion into the folowing

5129. METATHEOREM. All arrows in the diagram 5112 are direct models and the diagram commutes. Hence (1) the model $\mathfrak{Up}$ (arrow 4), i.e. the extension of $\mathfrak{Dir}(\partial\mathbf{TS}/\partial\mathfrak{R}(\mathbf{Fup}))$ interpreting $\mathbf{M}$ as $\mathbf{Tor}$, is a faithful model of $\mathbf{TS}^e$ in $\mathbf{TSS}^e$; (2) the model $\mathfrak{Dir}(\partial\mathbf{TSS}^e/\partial\mathfrak{St}(\mathbf{M}))$ (arrow 6) is a faithful model of $\mathbf{TSS}''$, S5 in ($\mathbf{TS}^e$, $\partial\mathbf{TSS}/\partial\mathfrak{St}(\mathbf{M})$).

c) *Relations, cardinalities, regularity and choice in the sense of the model* $\mathfrak{Up}$

As we have shown, $\mathfrak{Up}$ is a model of $\mathbf{TS}''$, FMcl ($\mathbf{M}$) in $\mathbf{TSS}^e$. Recall that $*$-classes (classes in the sense of $\mathfrak{Up}$) were defined in $\mathbf{TSS}^e$ as saturated subclasses of the relation **Fup** (full upward extension of the relation **E**) and $\in^*$ was defined as usual by $\mathfrak{R}(\mathbf{Fup})$. We also defined the class **Tor** and proved FMcl* (**Tor**) and defined the isomorphism **Imb** of **V** and **Tor**. We proved Lemma 5113 which can be interpreted as follows: **Imb** makes sets into $*$-elements of **Tor** and semisets into $*$-subsets of **Tor**. Using this lemma and Metatheorem 5114 it is easy to characterize various notions of the model ($*$-notions) in terms containing no asterisked symbols; in particular, we can characterize some important $*$-notions concerning $*$-elements of **Tor** in terms of simple notions concerning sets and semisets. The following lemmas should be understood as the most important examples of such characterizations.

5130. LEMMA ($\mathbf{TSS}^e$). If $x^* = \mathbf{Imb}\text{''}\,\sigma$ then

(a) $\mathrm{Rel}^*(x^*) \equiv \mathrm{Rel}(\sigma)\,;$

(b) $\mathrm{Un}^*(x^*) \equiv \mathrm{Un}(\sigma)\,;$

(c) $\mathbf{D}^*(x^*) = \mathbf{D}(\sigma)\,.$

Proof. (a) and (b) follow by Metatheorem 5114. By the same Metatheorem, we have $y \in \mathbf{D}(\sigma) \equiv \mathbf{Imb}'' \, y \in^* \mathbf{D}^*(x^*)$ and (c) follows.

Furthermore, in the course of the discussion 5118 we proved the following

5131. LEMMA ($\mathbf{TSS}^e$). $\mathbf{On}^* = \mathbf{Imb}'' \, \mathbf{On}$; i.e. $x \in \mathbf{On}$ iff $(\mathbf{Imb}'' \, x) \in^* \mathbf{On}^*$ and if $x^* \in^* \mathbf{On}^*$ then there is an $x \in \mathbf{On}$ such that $x^* = \mathbf{Imb}'' \, x$.

We now consider equivalence of sets. Suppose $x^* = \mathbf{Imb}'' \, x$ and $y^* = \mathbf{Imb}'' \, y$. If $x^* \approx^* y^*$ then there is an $f^* \subseteq x^* \times^* y^*$ such that $\mathrm{Un}_2^*(f^*, x^*, y^*)$. From $x^*, y^* \in^* \mathbf{Tor}$ we deduce $f^* \subseteq \mathbf{Tor}$ and hence by Lemma 5113 and Metatheorem 5114 we obtain $(\exists \sigma) \, \mathrm{Un}_2(\sigma, x, y)$ i.e. $x \mathrel{\hat{\approx}} y$ (x is absolutely equivalent to y, cf. the definition 2201). Conversely, if $x \mathrel{\hat{\approx}} y$ then obviously $x^* \approx^* y^*$. This proves the following.

5132. LEMMA ($\mathbf{TSS}^e$). If $x^* = \mathbf{Imb}'' \, x$ and $y^* = \mathbf{Imb}'' \, y$ then $x^* \approx^* y^* \equiv x \mathrel{\hat{\approx}} y$; similarly, $x^* \preccurlyeq^* y^* \equiv x \mathrel{\hat{\preccurlyeq}} y$.

We are thus led to definitions which differ from the customary definitions in that certain bound set variables are replaced by bound semiset variables.

5133. DEFINITION ($\mathbf{TSS}'$).

$$\hat{\mathrm{Card}}(\alpha) \equiv \neg (\exists \beta < \alpha)(\beta \mathrel{\hat{\approx}} \alpha) \quad (\alpha \text{ is an } \textit{absolute cardinal}),$$

$$\hat{\mathrm{Conf}}(\alpha, \beta) \equiv (\alpha \leqq \beta \,\&\, (\exists \sigma)(\mathrm{Un}(\sigma, \alpha, \beta) \,\&\, \bigcup(\mathbf{W}(\sigma)) = \beta)$$

(α is *absolutely cofinal* with β).

5134. LEMMA. (1) ($\mathbf{TSS}'$) $\hat{\mathrm{Card}}(\alpha) \to \mathrm{Card}(\alpha)$ and $\mathrm{Conf}(\alpha, \beta) \to \hat{\mathrm{Conf}}(\alpha, \beta)$.

(2) ($\mathbf{TSS}^e$) Let $\alpha^* = \mathbf{Imb}'' \, \alpha$, $\beta^* = \mathbf{Imb}'' \, \beta$. Then

(a) $\mathrm{Card}^*(\alpha^*) \equiv \hat{\mathrm{Card}}(\alpha)$,

(b) $\mathrm{Conf}^*(\alpha^*, \beta^*) \equiv \hat{\mathrm{Conf}}(\alpha, \beta)$.

5135. LEMMA ($\mathbf{TSS}'$, S4). Every natural number is an absolute cardinal; $\boldsymbol{\omega}$ is an absolute cardinal.

Proof. We prove that if n is a natural number and σ is a mapping with domain n then σ is a set. We denote this statement by $\varphi(n)$ and let $A = \{n; \varphi(n)\}$ (this class exists by Metatheorem 4121). It follows by induction, using (St), that $A = \boldsymbol{\omega}$. The lemma follows.

5136. LEMMA ($\mathbf{TSS}^e$). The following are equivalent: (1) Cardinals are absolute (i.e. every cardinal is an absolute cardinal), (2) for every α, if $\alpha^* = \mathbf{Imb}'' \, \alpha$ then $\aleph^*_{\alpha^*} = \mathbf{Imb}''(\aleph_\alpha)$.

Proof. Obviously **Imb**" $\aleph_0$ is the first limit ordinal number in the sense of $\mathfrak{Up}$. (Note that $\beta = \alpha + 1$ is a restricted formula.) Hence $\aleph^*_{0*} =$ **Imb**" $(\aleph_0)$. Suppose that all cardinals are absolute. If $\aleph^*_{\alpha*} =$ **Imb**" $(\aleph_\alpha)$ holds for all $\alpha < \lambda$ and if $\lambda^* =$ **Imb**" λ then $\aleph^*_{\lambda*}$ is some **Imb**" β; certainly $\beta \geq \aleph_\lambda$ (because $\widehat{\mathrm{C}}\mathrm{ard}\,(\beta)$ implies $\mathrm{Card}\,(\beta)$); the converse inequality holds by the absoluteness; **Imb**" $\aleph_\lambda$ is the $*$-least cardinal number $*$-greater than all $\aleph^*_{\alpha*}\,(\alpha^* <^* \lambda^*)$ and (2) is proved. The implication (2) → (1) is obvious.

5137. LEMMA (**TSS**^{e}). Suppose $x^* =$ **Imb**" x and $y^* =$ **Imb**" y. The following are equivalent:

(1) $x^* \preccurlyeq^* \mathbf{P}^*(y^*) - \{0^*\}^*$,

(2) $(\exists \varrho)(\mathrm{Exct}\,(\varrho)\ \&\ \mathbf{D}(\varrho) = x\ \&\ \mathbf{W}(\varrho) \subseteq y)$.

Proof. In **TSS**′, $x \preccurlyeq \mathbf{P}(y) - \{0\} \equiv (\exists f)(\mathrm{Un}\,(f, x, \mathbf{P}(y))\ \&\ \mathrm{Un}_2(f)) \equiv$ $\equiv (\exists r)(\mathrm{Exct}\,(r)\ \&\ \mathbf{D}(r) = x\ \&\ \mathbf{W}(r) \subseteq y)$. Hence (1) is equivalent to

$$(\exists r^* \subseteq y^* \times^* x^*)(\mathrm{Exct}^*(r^*)\ \&\ \mathbf{D}^*(r^*) = x^*\ \&\ \mathbf{W}^*(r^*) \subseteq y^*)$$

which is equivalent to (2) by Metatheorem 5114.

We wish now to characterize some notions concerning cardinalities of sets. To do this we need (E1)*, so we ask what must be added to **TSS**^{e} in order that (E1)* be provable. In the following metatheorem we answer this question and we also give conditions for certain other axioms of regularity and choice to hold in the sense of $\mathfrak{Up}$.

5138. METATHEOREM. $\mathfrak{Up}$ is a faithful model of

(a) **TS**, U3, FMcl (**M**) in **TSS**^{e}, U3;

(b) **TS**, D3, FMcl (**M**) in **TSS**^{e}, D3;

(c) **TS**″, E1, FMcl (**M**) in **TSS**^{e}, E10, where (E10) is

$$(\forall x)(\exists \sigma)(\exists \alpha)(\mathrm{Un}_2(\sigma, x, \alpha))$$

(consequently, $\mathfrak{Up}$ is a model of **TS**″, E1, FMcl (**M**) in **TSS**^{e}, E1);

(d) **TS**, E2, FMcl (**M**) in **TSS**^{e}, E20, where (E20) is $(\exists F)(\mathrm{Un}_2(F, \mathbf{V}, \mathbf{On}))$ (consequently, $\mathfrak{Up}$ is a model of **TS**, E2, FMcl (**M**) in **TSS**^{e}, E2);

(e) **TS**, D3, (**M** = **Cstr**) in **TSS**^{e}, D3, (Constr);

(f) **TS**, D3, FMcl (**M**) & **M** ≠ **V** in **TSS**^{e}, D3, ¬ (C3)

(consequently, $\mathfrak{Up}$ is a model of **TS**, D3, ¬ (Constr) in **TSS**^{e}, D3, ¬ (C3)).

Demonstration. (a) It suffices to prove $(U3) \equiv (U3)^{\mathfrak{N}(\mathbf{Fup})}$ in $\mathbf{TSS^e}$. By Metacorollary 5126 we have $(U3) \equiv [(U3)^{\mathfrak{St}(\mathbf{M})}]^{\mathfrak{N}(\mathbf{Fup})}$ in $\mathbf{TSS^e}$, $\partial\mathbf{TSS}/\partial\mathfrak{St}^*(\mathbf{Tor}))$ (arrows, 6, 8); hence if we prove $(U3) \equiv (U3)^{\mathfrak{St}(\mathbf{M})}$ in $(\mathbf{TS^e}, \partial\mathbf{TSS}/\partial\mathfrak{St}(\mathbf{M}))$ (the terminal theory of arrow 6) we obtain $(U3)^{\mathfrak{N}(\mathbf{Fup})} \equiv \equiv [(U3)^{\mathfrak{St}(\mathbf{M})}]^{\mathfrak{N}(\mathbf{Fup})}$ in $(\mathbf{TSS^e}, \partial\mathbf{TSS}/\partial\mathfrak{St}^*(\mathbf{Tor}))$ (by arrow 8) and hence $(U3) \equiv (U3)^{\mathfrak{N}(\mathbf{Fup})}$ in the last mentioned theory. Since the formula $(U3) \equiv \equiv (U3)^{\mathfrak{N}(\mathbf{Fup})}$ is a $\mathbf{TSS^e}$-formula, and the identity arrow 7 is conservative, we obtain $\mathbf{TSS^e} \vdash (U3) \equiv (U3)^{\mathfrak{N}(\mathbf{Fup})}$. We proceed therefore in $(\mathbf{TS^e}, \partial\mathbf{TSS}/\partial\mathfrak{St}(\mathbf{M}))$. Obviously $(U3) \rightarrow (U3)^{\mathfrak{St}(\mathbf{M})}$. Conversely, suppose $(U3)^{\square}$ holds. Since $\mathbf{M}$ is full, $\mathbf{V} = \mathbf{Ker}(\mathbf{M})$ from which it follows that $\mathbf{Ur} \subseteq \mathbf{M}$ and $\mathbf{Ur}^{\square} = \mathbf{Ur}$; thus $\mathrm{M}^{\square}(\mathbf{Ur}^{\square})$ implies $\mathrm{M}(\mathbf{Ur})$. Furthermore $\mathbf{M} = = \mathbf{Ker}^{\square}(\mathbf{Ur}^{\square}) \subseteq \mathbf{Ker}(\mathbf{Ur}^{\square}) = \mathbf{Ker}(\mathbf{Ur})$; hence if $Z \supseteq \mathbf{Ur}$ and $\mathbf{P}(Z) \subseteq Z$ then $Z \supseteq \mathbf{Ker}(\mathbf{Ur}) \supseteq \mathbf{Ker}^{\square}(\mathbf{Ur}^{\square}) = \mathbf{M}$, hence $Z \supseteq \mathbf{Ker}(\mathbf{M}) = \mathbf{V}$. Thus $\mathbf{V} = \mathbf{Ker}(\mathbf{Ur})$ and so (U3) holds.

(b) is demonstrated in the same way.

(c) We prove $(E1)^{\mathfrak{N}(\mathbf{Fup})} \equiv (E10)$ in $\mathbf{TSS^e}$. Suppose $(E1)^{\mathfrak{N}(\mathbf{Fup})}$ and let x be an arbitrary set. Set $x^* = \mathbf{Imb}''\, x$. There is a $y^* \subseteq \mathbf{Tor}$ and an α^* such that $\mathrm{Un}_2^*(y^*, x^*, \alpha^*)$. Setting $\mathbf{Imb}''\,\sigma = y^*$ and $\mathbf{Imb}''\,\alpha = \alpha^*$ we obtain $\mathrm{Un}_2(\sigma, x, \alpha)$ by 5130.

Conversely, suppose (E10) and let y^* be a set in the sense of $\mathfrak{N}(\mathbf{Fup})$. Since $\mathbf{Fup}$ is regular, y^* is a semiset, σ say, and it follows from (E10) by (St) that there is a semiset ϱ and an ordinal number α such that $\mathrm{Un}_2(\varrho, \sigma, \alpha)$. Hence σ can be Sm-well ordered and so by (St) there is a semiset $\sigma_0 \subseteq \sigma$ such that distinct elements of σ_0 have distinct extensions (in $\mathbf{Fup}$) and such that for each $u \in \sigma$ there is a $\bar{u} \in \sigma_0$ with the same extension. There exist ϱ_0, α_0 such that $\mathrm{Un}_2(\varrho_0, \sigma_0, \alpha_0)$. Now define

$$x \in \tau \equiv (\exists u \in \sigma)(\exists \gamma \in \alpha_0)(\mathbf{Ext}_{\mathbf{Fup}}(x) = \langle \mathbf{Imb}''\,\gamma, \mathbf{Ext}_{\mathbf{Fup}}(u)\rangle^* \;\& \\ \& \;\langle \gamma, \bar{u}\rangle \in \varrho_0).$$

τ is saturated and there is a $q \in \mathbf{C}(\mathbf{Fup})$ such that $\tau = \mathbf{Ext}_{\mathbf{Fup}}(q)$. Hence τ is a set in the sense of $\mathfrak{N}(\mathbf{Fup})$; let $q^* = \tau$; it is easy to see that $\mathrm{Un}_2^*(q^*, y^*, \alpha_0^*)$ where $\alpha_0^* = \mathbf{Imb}''\,\alpha_0$.

(d) is proved analogously.

(e) It suffices to prove $(\mathbf{M} = \mathbf{Cstr})^{\mathfrak{N}(\mathbf{Fup})} \equiv (\mathbf{V} = \mathbf{Cstr})$ in $(\mathbf{TSS^e}, \mathrm{D3})$. But $(\mathbf{V} = \mathbf{Cstr}) \equiv [(\mathbf{V} = \mathbf{Cstr})^{\mathfrak{St}(\mathbf{M})}]^{\mathfrak{N}(\mathbf{Fup})}$ is provable in $(\mathbf{TSS^e}, \mathrm{D3}, \partial\mathbf{TSS}/\partial\mathfrak{St}^*(\mathbf{Tor}))$ and $(\mathbf{V} = \mathbf{Cstr})^{\mathfrak{St}(\mathbf{M})} \equiv \mathbf{M} = \mathbf{Cstr}$ is provable in $(\mathbf{TS^e}, \mathrm{D3}, \partial\mathbf{TSS}/\partial\mathfrak{St}(\mathbf{M}))$. The assertion follows.

(f) We know that **TS**, D3 $\vdash (\mathbf{V} \neq \mathbf{Cstr}) \equiv (\exists X)(\mathrm{Mcl}(X) \& X \neq \mathbf{V})$. The identity model of **TS**, D3, $(\mathbf{V} \neq \mathbf{Cstr})$ is a model in **TS**, D3, $(\mathrm{FMcl}(\mathbf{M}) \& \& \mathbf{M} \neq \mathbf{V})$; hence it suffices to demonstrate that $\mathfrak{Up}$ is a faithful model of **TS**, D3, $(\mathrm{FMcl}(\mathbf{M}) \& \mathbf{M} \neq \mathbf{V})$ in **TSS**^{e}, D3, $\neg$ (C3). To do this we prove $(\mathbf{V} \neq \mathbf{M})^{\mathfrak{N}(\mathbf{Fup})} \equiv \neg$ (C3) in **TSS**^{e}. Suppose (C3), i.e. every semiset is a set. Then it follows from the construction of **Fup** that $\mathbf{C}(\mathbf{Fup}) = \mathbf{Tor}$; hence $(\mathbf{V} = \mathbf{M})^{\mathfrak{N}(\mathbf{Fup})}$. Conversely, if there is a semiset σ which is not a set, then $\sigma \times \{0\}$ is a subsemiset of **Tor** which is not a set (and which is obviously **Fup**-saturated), hence there is a $y \in \mathbf{C}(\mathbf{Fup})$ such that $\mathbf{Ext}_{\mathbf{Fup}}(y) = \sigma \times \{0\}$. Obviously $y \notin \mathbf{Tor}$ and hence $\mathbf{Tor} \neq \mathbf{C}(\mathbf{Fup})$, i.e. $(\mathbf{V} \neq \mathbf{M})^{\mathfrak{N}(\mathbf{Fup})}$.

We now consider **TSS**^{e}, E1. Suppose $x^* = \mathbf{Imb}'' x$ and $\alpha^* = \mathbf{Imb}'' \alpha$. Then $\alpha^* = \overline{\overline{x^*}}^*$ (i.e. α^* is the cardinality of x^* in the sense of $\mathfrak{Up}$) iff α is the smallest ordinal number γ such that $\gamma \approx x$. As a consequence it follows (in **TSS**^{e}, E1) that for every x there is the smallest γ such that $\gamma \approx x$. The reader can easily show that the last statement is provable in **TSS**$'$, S4, E1 (recall that S4 $\equiv_{\bullet}$ S1 & St), i.e. we need not use D2 and (Pot). Hence we introduce the following

5139. Definition (**TSS**$'$, S4, E1). For each x, $\hat{\hat{x}}$ is the smallest γ such that $\gamma \approx x$ ($\hat{\hat{x}}$ is called the *absolute cardinality* of x).

5140. Lemma (**TSS**^{e}, E1). If $x^* = \mathbf{Imb}'' x$ and $\alpha^* = \mathbf{Imb}'' \alpha$ then $\alpha^* = \overline{\overline{x^*}}^*$ iff $\alpha = \hat{\hat{x}}$.

In **TSS**$'$, S5, E1 we can prove that for any x there exists a smallest α such that all subsemisets of x can be one-one coded by elements of α, i.e. such that

$$(\exists \varrho)(\mathrm{Exct}(\varrho) \& \mathbf{D}(\varrho) = \alpha \& (\forall \sigma \subseteq x)(\exists \iota)(\sigma = \mathbf{Ext}_{\varrho}(\iota)) .$$

Indeed, by (Pot) there is a coding ϱ_0 of all subsemisets of x. By E1 and St, $\mathbf{D}(\varrho_0)$ can be Sm-well ordered and by St we can obtain a ϱ_1 which is a nowhere constant coding of all subsemisets of x (Note that ϱ_1 can be defined by a seminormal formula.) If the absolute cardinality of $\mathbf{D}(\varrho_1)$ is α we obtain from ϱ_1 a nowhere constant coding of all subsemisets of x by all elements of α. If there were a coding ϱ_2 with $\mathbf{D}(\varrho_2) \in \mathbf{On}$ and $\mathbf{D}(\varrho_2) < \alpha$ then we should have a one-one-mapping of α onto $\mathbf{D}(\varrho_2)$ which contradicts the fact that α is the absolute cardinality of some set and hence an absolute cardinal.

This justifies the following definition:

5141. DEFINITION (**TSS'**, S5, E1). (a) For every α, $\hat{\beth}(2, \alpha)$ is the smallest γ such that $(\exists\varrho)(\mathrm{Exct}(\varrho) \,\&\, \mathbf{D}(\varrho) = \gamma \,\&\, (\forall\sigma \subseteq \alpha)(\exists\iota)(\sigma = \mathbf{Ext}_\varrho(\iota)))$. (i.e. such that all subsemisets of α can be one-one coded by γ);

(b) For any α, β, $\hat{\beth}(\beta, \alpha)$ is the smallest γ such that

$$(\exists\varrho)(\mathrm{Exct}(\varrho) \,\&\, \mathbf{D}(\varrho) = \gamma \,\&\, (\forall\sigma)(\mathrm{Un}(\sigma, \alpha, \beta) \rightarrow (\exists\iota)(\sigma = \mathbf{Ext}_\varrho(\iota))) .$$

(i.e. such that all semiset mappings of α into β can be one-one coded by γ).

5142. LEMMA (**TSS**^{e}, E1). Let α, β be infinite absolute cardinals, set $\alpha^* =$ $= \mathbf{Imb}''\,\alpha$ and $\beta^* = \mathbf{Imb}''\,\beta$. Then

(a) $\beth^*(2^*, \alpha^*) = \mathbf{Imb}''\,\hat{\beth}(2, \alpha)$,

(b) $\beth^*(\beta^*, \alpha^*) = \mathbf{Imb}''\,\hat{\beth}(\beta, \alpha)$.

Proof. Recall that $\beth(\aleph_\mu, \aleph_\nu)$ is the same as $\aleph_\mu^{\aleph_\nu}$.

(*a*) If $\gamma^* = \mathbf{Imb}''\,\gamma$ then the following are equivalent: (1) $\gamma^* \approx^* \mathbf{P}^*(\alpha^*)$, (2) there is an exact functor ϱ with $\mathbf{D}(\varrho) = \gamma$ which codes all subsemisets of α. Therefore the smallest γ^* which is $*$-equivalent to $\mathbf{P}^*(\alpha^*)$ equals to $\mathbf{Imb}''\,\hat{\beth}(2, \alpha)$.

(b) is proved analogously.

In the same way the reader may show that the following definition is correct and prove the lemma following it.

5143. DEFINITION (1) (**TSS'**, S4).

$\hat{\mathbf{cf}}(\alpha) = \mathbf{Min}\,\{\beta;\, \hat{\mathrm{Conf}}(\beta, \alpha)\}$,

(2) (**TSS'**, S5, E1)

$\hat{\gimel}(\alpha) = \hat{\beth}(\alpha, \hat{\mathbf{cf}}(\alpha))$.

5144. LEMMA (**TSS**^{e}, E1). Let α be an absolute cardinal and set $\alpha^* =$ $= \mathbf{Imb}''\,\alpha$. Then

(a) $\mathbf{cf}^*(\alpha^*) = \mathbf{Imb}''\,(\hat{\mathbf{cf}}(\alpha))$,

(b) $\gimel^*(\alpha^*) = \mathbf{Imb}''\,\hat{\gimel}(\alpha)$.

5145. LEMMA (**TSS'**, S4). $\hat{\mathbf{cf}}(\hat{\mathbf{cf}}(\aleph_\alpha) = \hat{\mathbf{cf}}(\aleph_\alpha)$.

(The proof is analogous to the proof of 2236.)

SECTION 2

The ultraproduct model

In the preceding section we studied the extending model $\mathfrak{Up}$ which, roughly speaking, "extends" the universe of the theory of semisets. The theory **TSS**$^{\mathbf{e}}$ in which we studied this model has the general assumption of the existence of a locally semiset total Boolean support as an axiom. But, because of various relative consistency proofs, it is necessary to study the model $\mathfrak{Up}$ as a model in theories stronger than **TSS**$^{\mathbf{e}}$, in which it is assumed that there exists a total Boolean support on an algebra with some particular properties. Naturally, the question arises whether such theories are consistent (relative to **TSS**). The task of the present section is to obtain a positive answer to this question. (See Metatheorem 5215). The well-known ultraproduct method is used is used to construct appropriate models of **TSS**$^{\mathbf{e}}$.

5201. Definition (**TSS**$'$). If U is a real ultrafilter on a Boolean algebra $\boldsymbol{B}$ and P is a partition filter on $\boldsymbol{B}$ then we define

$$\mathbf{Ulc}(\boldsymbol{B}, P) = \{f; \mathrm{Un}(f) \,\&\, \mathbf{D}(f) \in P\},$$

(the *ultraproduct class*),

$$\mathbf{Ulr}(\boldsymbol{B}, P, U) \subseteq [\mathbf{Ulc}(\boldsymbol{B}, P)]^2 \,\&$$

$$\& (\forall f, g \in \mathbf{Ulc}(\boldsymbol{B}, P)) (\langle f, g\rangle \in \mathbf{Ulr}(\boldsymbol{B}, P, U) \equiv$$

$$\equiv \bigvee\{u \wedge v; u \in \mathbf{D}(f) \,\&\, v \in \mathbf{D}(g) \,\&\, f`u \in g`v\} \in U)$$

(the *ultraproduct relation*).

5202. Remark. If P is the partition filter containing all partitions of $\boldsymbol{B}$ then we write $\mathbf{Ulc}(\boldsymbol{B})$ and $\mathbf{Ulr}(\boldsymbol{B}, U)$ instead of $\mathbf{Ulc}(\boldsymbol{B}, P)$ and $\mathbf{Ulr}(\boldsymbol{B}, P, U)$;

if it is clear which $\boldsymbol{B}$, P and U are concerned we write **Ulc** and **Ulr**. In what follows Bool $(\boldsymbol{B}, P, U)$ means "$\boldsymbol{B}$ is a complete Boolean algebra, P is a partition filter on $\boldsymbol{B}$ and U is a real ultrafilter on $\boldsymbol{B}$". Note that **Ulc** and **Ulr** are real classes.

5203. DEFINITION (**TSS**, E2, Bool (**B**, **P**, **U**)). For any x we let k_x be the functions which assigns x to the unit element:

$$k_x = \{\langle x, 1_{\mathbf{B}}\rangle\} .$$

For any X we let

$$\bar{X} = \{f \in \mathbf{Ulc}; \bigvee\{u \in \mathbf{D}(f); f'u \in X\} \in \mathbf{U}\} .$$

Note that $\mathbf{Ext}_{\mathbf{Ulr}}(k_x) = \bar{x}$.

5204. LEMMA (**TSS**, E2, Bool (**B**, **P**, **U**)). **Ulr** is a nonempty extensional relation; for any f, $g \in \mathbf{Ulc}$ we have $\mathbf{Ext}_{\mathbf{Ulr}}(f) = \mathbf{Ext}_{\mathbf{Ulr}}(g)$ iff

$$\bigvee\{u \wedge v; u \in \mathbf{D}(f) \,\&\, v \in \mathbf{D}(g) \,\&\, f'u = g'v\} \in \mathbf{U} .$$

Proof. Clearly **Ulr** is nonempty; if $x \in y$ then $\langle k_x, k_y\rangle \in \mathbf{Ulr}$. We write $\hat{f}$ instead of $\mathbf{Ext}_{\mathbf{Ulr}}(f)$. We prove the equivalence

$$\hat{f} = \hat{g} \equiv \bigvee\{u \wedge v; u \in \mathbf{D}(f) \,\&\, v \in \mathbf{D}(g) \,\&\, f'u = g'v\} \in \mathbf{U} ,$$

the extensionality then follows immediately. We suppose first that $\bigvee\{u \wedge v; f'u = g'v\} \in \mathbf{U}$ (for brevity we omit the clause $u \in \mathbf{D}(f)$ etc.). If $h \in \hat{f}$ then $\bigvee\{w \wedge u; h'w \in f'u\} \in \mathbf{U}$; since

$$\bigvee\{u \wedge v; f'u = g'v\} \wedge \bigvee\{w \wedge u; h'w \in f'u\} =$$
$$= \bigvee\{u \wedge v \wedge w; f'u = g'v \,\&\, h'w \in f'u\} = \bigvee\{w \wedge v \wedge u;$$
$$h'w \in g'v \,\&\, f'u = g'v\} \leqq \bigvee\{w \wedge v; h'w \in g'v\} ,$$

it follows that the last join is in **U** and so $h \in \hat{g}$.

We now suppose that $\bigvee\{u \wedge v; f'u = g'v\} \notin \mathbf{U}$. It follows that $\bigvee\{u \wedge v; f'u \neq g'v\} \in \mathbf{U}$ so that e.g. $\bigvee\{u \wedge v; f'u - g'v \neq 0\} \in \mathbf{U}$. Using a selector we can define a function h on $\mathbf{D}(f) \mathbin{\triangle} \mathbf{D}(g)$ such that $h'w \in f'u - g'v$ whenever the conditions $w \leqq u \in \mathbf{D}(f)$, $w \leqq v \in \mathbf{D}(g)$ and $f'u - g'v \neq 0$ are fulfilled. Thus we have $h \in \hat{f}$ and $h \notin \hat{g}$.

5205. LEMMA (**TSS**, E2, Bool (**B**, **P**, **U**)). For any X, $\overline{X}$ is a saturated part of **Ulc**. (Exercise.)

We shall consider the model $\mathfrak{Dir}$ (∂**TSS**$/\partial\mathfrak{N}$ (**R**)) in **TSS**, E2, Bool (**B**, **P**, **U**) & **R** = **Ulr** (**B**, **P**, **U**), ∂**TSS**$/\partial\mathfrak{N}$ (**R**). We may introduce set-variables in the sense of $\mathfrak{N}$ (**R**). We have the following important metatheorem, in which $\hat{f}$ again denotes $\mathbf{Ext}_{\mathbf{Ulr}}(f)$.

5206. METATHEOREM. For any normal **TC**-formula $\varphi(x, \bullet, X, \bullet)$ without constants the following is provable in **TSS**, E2, Bool (**B**, **P**, **U**), ∂**TSS**$/\partial\mathfrak{N}$ (**Ulr** (**B**, **P**, **U**)):

$$(x_1^* = \hat{f}_1 \,\&\, \bullet) \,\&\, (X_1^* = \overline{X}_1 \,\&\, \bullet) \to \big[\varphi^*(x_1^*, \bullet, X_1^*, \bullet) \equiv$$
$$\equiv \bigvee\{u_1 \wedge \bullet;\, u_1 \in \mathbf{D}(f_1) \,\&\, \bullet \,\&\, \varphi(f_1{}^{\backprime}u_1, \bullet, X_1, \bullet)\} \in \mathbf{U}\big]\,.$$

Demonstration. By induction. If φ is atomic then φ is either $x \in y$ or $x \in X$; we have $\hat{f} \in^* \hat{g} \equiv \langle f, g\rangle \in \mathbf{Ulr} \equiv \bigvee\{u \wedge v;\, f^{\backprime}u \in g^{\backprime}v\} \in \mathbf{U}$ and

$$\hat{f} \in^* \overline{X} \equiv f \in \overline{X} \equiv \bigvee\{u;\, f^{\backprime}u \in X\} \in \mathbf{U}\,.$$

If φ is $\varphi_1 \,\&\, \varphi_2$ then we have

$$\varphi^*(\hat{f}_1, \bullet) \equiv \varphi_1^*(\hat{f}_1, \bullet) \,\&\, \varphi_2^*(\hat{f}_1, \bullet) \equiv$$
$$\equiv \bigvee\{u_1 \wedge \bullet;\, \varphi_1(f_1{}^{\backprime}u_1, \bullet)\} \in \mathbf{U} \,\&\, \bigvee\{u_1 \wedge \bullet;\, \varphi_2(f_1{}^{\backprime}u_1, \bullet)\} \in \mathbf{U} \equiv$$
$$\equiv \bigvee\{u_1 \wedge \bullet;\, \varphi_1(f_1{}^{\backprime}u_1, \bullet) \,\&\, \varphi_2(f_1{}^{\backprime}u_1, \bullet)\} \in \mathbf{U} \equiv$$
$$\equiv \bigvee\{u_1 \wedge \bullet;\, \varphi(f_1{}^{\backprime}u_1, \bullet)\} \in \mathbf{U}\,.$$

If φ is $\neg\psi$ then

$$\varphi^*(\hat{f}_1, \bullet) \equiv \neg\, \psi^*(\hat{f}_1, \bullet) \equiv \bigvee\{u_1 \wedge \bullet;\, \psi(f_1{}^{\backprime}u_1, \bullet)\} \notin \mathbf{U} \equiv$$
$$\equiv \bigvee\{u_1 \wedge \bullet;\, \varphi(f_1{}^{\backprime}u_1, \bullet)\} \in \mathbf{U}\,.$$

Suppose finally that φ is $(\exists y)\, \psi(y, x_1, \bullet)$. If $\varphi^*(\hat{f}_1, \bullet)$ then there exists g such that $\psi^*(\hat{g}, \hat{f}_1, \bullet)$; we have $\bigvee\{v \wedge u_1 \wedge \bullet;\, \psi(g^{\backprime}v, f_1{}^{\backprime}u_1, \bullet)\} \in \mathbf{U}$ and so $\bigvee\{u_1 \wedge \bullet;\, (\exists y)\, \psi(y_1, f_1{}^{\backprime}u_1, \bullet)\} \in \mathbf{U}$. Conversely, if $\bigvee\{u_1 \wedge \bullet;\, (\exists y)\, \psi(y, f_1{}^{\backprime}u_1, \bullet)\} \in \mathbf{U}$ then we let $j = \mathbf{D}(f_1) \,\triangle\, \bullet$; clearly $j \in \mathbf{P}$. We let

$$\langle y, u\rangle \in R \equiv u \in j \,\&\, (\exists u_1, \bullet)\,(u_1 \in \mathbf{D}(f_1) \,\&\, \bullet \,\&\, u \leqq$$
$$\leqq u_1 \,\&\, \bullet \,\&\, \psi(y, f_1{}^{\backprime}u_1, \bullet))\,.$$

Let $F_0 \subseteq R$ be a function such that $\mathbf{D}(F_0) = \mathbf{D}(R)$. Since $\bigvee \mathbf{D}(F_0) \in \mathbf{U}$ and F_0 is a set, we may let $f = F_0 \cup \{\langle 0_{\mathbf{B}}, -\bigvee \mathbf{D}(F_0)\rangle\}$. Clearly $f \in \mathbf{Ulc}$ and we have

$$\bigvee\{v \wedge u_1 \wedge \bullet;\ \psi(f\text{'}v, f_1\text{'}u_1, \bullet)\} \in \mathbf{U}\,,$$

hence $\psi^*(\hat{f}_1, \bullet)$.

Remark. In the course of the above proof we could use **TSS** + (D2 & E1) instead of **TSS** + E2.

5207. Corollary. For any closed SF formula φ the following is provable in **TSS**, E2, Bool (**B**, **P**, **U**), ∂**TSS**/$\partial\mathfrak{N}$ (**Ulr** (**B**, **P**, **U**)):

$$\varphi \equiv \varphi^*$$

5208. Metacorollary. Axioms (A1) to (A7), (B1) to (B7) and (C1) hold in $\mathfrak{Dir}$ (∂**TSS**/$\partial\mathfrak{N}$ (**Ulr** (**B**, **P**, **U**)) as a model in **TSS**, E2, Bool (**B**, **P**, **U**), ∂**TSS**/$\partial\mathfrak{N}$ (**Ulr** (**B**, **P**, **U**)).

Demonstration. Axioms A1 to A7 and C1 can be expressed equivalently (in **TE**, d1, d2) by normal set formulas. Therefore, by the preceding corollary, they hold in the model. In particular, A1 holds in the model and therefore **TSS**, E2, Bool (**B**, **P**, **U**) $\vdash$ (**Ulr** is a pairing relation). Consequently, **TSS**, E2, Bool (**B**, **P**, **U**) $\vdash$ (**Ulr** is an **E**-like relation) and so Axioms B1 to B7 hold in the model.

In the sequel we shall be interested in **TSS**, E2, Bool (**b**, **p**, **z**). The fixing of constants **b**, **p**, **z** by Bool (**b**, **p**, **z**) is permissible in **TSS**, E2 since E2 implies the existence of an ultrafilter which is a set on any complete Boolean algebra which is a set.

5209. Metatheorem. $\mathfrak{Dir}$ (∂**TSS**/$\partial\mathfrak{N}$ (**Ulr** (**b**, **p**, **z**)) is a model of **TSS**, E2 in **TSS**, E2, Bool (**b**, **p**, **z**), ∂**TSS**/$\partial\mathfrak{N}$ (**Ulr** (**b**, **p**, **z**)).

Demonstration. By Metacorollary 5208 it suffices to show that C2 and E2 hold in the model. To prove that C2 holds in the model it suffices (by 5106(1)) to prove in **TSS**, E2, Bool (**b**, **p**, **z**) that **Ulr** (**b**, **p**, **z**) is almost regular and almost universal. We proceed in **TSS**, E2, Bool (**b**, **p**, **z**). Let $f \in \mathbf{Ulc}$, let $q = \bigcup(\mathbf{W}(f))$ and let $g \in a \equiv .\ g \in \mathbf{Ulc}\ \&\ \mathbf{W}(g) \subseteq q\ \&\ g \in^* f$; we prove that $\mathbf{SAT}_{\mathbf{Ulr}}(a) = \mathbf{Ext}_{\mathbf{Ulr}}(f)$. Let $g \in \mathbf{Ext}_{\mathbf{Ulr}}(f)$ and suppose $x \in q$; set $u = -\bigvee\{v;\ g\text{'}v \subseteq q\}$. Clearly $u \notin z$. If $u \neq 0_{\mathbf{b}}$ let $h\text{'}u = x$ and let $h\text{'}v = g\text{'}v$ whenever $v \in \mathbf{D}(g)$ and $v \wedge u = 0_{\mathbf{b}}$. It follows that $h \in a$ and

$\hat{g} = \hat{h}$; hence $g \in \mathbf{SAT}_{\mathbf{Ulr}}(a)$. Conversely, if $g \in \mathbf{SAT}_{\mathbf{Ulr}}(a)$ then we choose $h \in a$ such that $\hat{h} = \hat{g}$; it follows that $h \in \hat{f}$ and so $g \in \hat{f}$.

We have proved that **Ulr** is almost regular. If $a \subseteq \mathbf{Ulc}$ then we let $x = \bigcup_{f \in a} \mathbf{W}(f)$; clearly $a \subseteq \bar{x}$ and so **Ulr** is almost universal.

We prove (E2)*. Let A be a mapping of **V** into **On**; i.e. suppose that $(\forall x)(\exists!\, y)(\mathrm{Ord}(y) \,\&\, \langle y, x\rangle \in A)$. It follows by Metatheorem 5206 that

$$[(\forall x)(\exists!\, y)(\mathrm{Ord}(y) \,\&\, \langle y, x\rangle \in \bar{A})]^* ;$$

hence $\bar{A}$ is a mapping of **V** into **On** in the sense of $\mathfrak{N}(\mathbf{Ulr})$. By the following lemma, $\bar{A}$ is a real class in the sense of the model.

LEMMA (**TSS**, E2, Bool $(\mathbf{b}, \mathbf{p}, \mathbf{z})$, $\partial\mathbf{TSS}/\partial\mathfrak{N}(\mathbf{Ulr}(\mathbf{b}, \mathbf{p}, \mathbf{z}))$. For any real class X we have Real* $(\bar{X})$.

Proof. If $f \in \mathbf{Ulr}$ then we let g be such that $\mathbf{D}(g) = \mathbf{D}(f)$ and $g'u = f'u \cap X$ for all $u \in \mathbf{D}(f)$. Clearly $g \in \mathbf{Ulc}$ and $\hat{g} = \hat{f} \cap \bar{X}$; hence Real* $(\bar{X})$.

5210. *Remark* (**TSS**, E2, Bool $(\mathbf{b}, \mathbf{p}, \mathbf{z})$, $\partial\mathbf{TSS}/\partial\mathfrak{N}(\mathbf{Ulr}(\mathbf{b}, \mathbf{p}, \mathbf{z}))$. In the sense of $\mathfrak{N}(\mathbf{Ulc}(\mathbf{b}, \mathbf{p}, \mathbf{z}))$ $\bar{\mathbf{b}}$ is a complete Boolean algebra (by Metatheorem 5206).

5211. We shall now restrict ourselves to the case where **p** is the partition filter consisting of all partitions. Bool $(\mathbf{b}, \mathbf{z})$ means "**b** is a complete Boolean algebra and **z** is an ultrafilter on **b**".

5212. THEOREM (**TSS**, E2, Bool $(\mathbf{b}, \mathbf{z})$, $\partial\mathbf{TSS}/\partial\mathfrak{N}(\mathbf{Ulr}(\mathbf{b}, \mathbf{z}))$. In the sense of $\mathfrak{N}(\mathbf{Ulr}(\mathbf{b}, \mathbf{z}))$ there exists a complete ultrafilter Z on $\bar{\mathbf{b}}$; moreover, for all $u \in \mathbf{b}$, $k_u \in Z$ iff $u \in \mathbf{z}$.

Proof. For $f \in \mathbf{Ulc}$ we let $f \in Z \equiv \bigvee\{u \wedge f'u;\, u \in \mathbf{D}(f) \,\&\, f'u \in \mathbf{b}\} \in \mathbf{z}$. We write $\beta(u, f)$ instead of $u \in \mathbf{D}(f) \,\&\, f'u \in \mathbf{b}$ and $\gamma(u, v, f, g)$ instead of $u \in \mathbf{D}(f) \,\&\, v \in \mathbf{D}(g) \,\&\, f'u = g'v$.

1) Z is a saturated part of **Ulc**. Suppose $f \in z$ and $\hat{g} = \hat{f}$, where $\hat{f} = \mathbf{Ext}_{\mathbf{Ulr}}(f)$. Since $\bigvee\{u \wedge f'u;\, \beta(u, f)\} \wedge \bigvee\{u \wedge v;\, \gamma(u, v, f, g)\} \leqq$
$= \bigvee\{u \wedge f'u \wedge v;\, \beta(u, f) \,\&\, \gamma(u, v, f, g)\} =$
$= \bigvee\{u \wedge v \wedge g'v;\, \beta(v, g) \,\&\, \gamma(u, v, f, g)\} \leqq$
$\leqq \bigvee\{v \wedge g'v;\, \beta(v, g)\}$,

it follows that $g \in Z$. (The first inequality follows from the fact that the elements of $\mathbf{D}(f)$ are mutually disjoint.) Hence Cls* (Z).

2) Clearly $Z \subseteq \bar{\mathbf{b}}$ and $k_{1_{\mathbf{b}}} \in Z$.

3) Let $f \in \bar{\mathbf{b}}$ and let $\mathbf{W}(f) \subseteq \mathbf{b}$. We let g be such that $\mathbf{D}(g) = \mathbf{D}(f)$ and $g'u = -f'u$ for all $u \in \mathbf{D}(f)$. Clearly, $\hat{g}$ is the complement of $\hat{f}$ in the algebra $\bar{\mathbf{b}}$ in the sense of **Ulr**. For any $u \in \mathbf{D}(f)$ we have $u = (u \wedge f'u) \vee \vee (u \wedge g'u)$ and so

$$1_{\mathbf{b}} = \bigvee\{u \wedge f'u; u \in \mathbf{D}(f)\} \vee \bigvee\{u \wedge g'u; u \in \mathbf{D}(g)\} ;$$

hence $f \in Z \equiv g \notin Z$.

4) Let $\hat{h} \subseteq \bar{\mathbf{b}}$ and let $\bigvee^* \hat{h} \in^* Z$; we prove $\hat{h} \cap Z \neq 0$. Since $\hat{h} \subseteq \bar{\mathbf{b}}$ we have $\bigvee\{u; h'u \subseteq \mathbf{b}\} \in \mathbf{z}$ and we may suppose $\bigvee\{u; h'u \subseteq \mathbf{b}\} = 1_{\mathbf{b}}$. Defining f by $\mathbf{D}(f) = \mathbf{D}(h)$ and $(\forall u \in \mathbf{D}(h))(f'u = \bigvee h'u)$ we have $\hat{f} = \bigvee^* \hat{h}$. Since $\hat{f} \in^* Z$ we have $\bigvee\{\bigvee h'u \wedge u; u \in \mathbf{D}(h)\} \in Z$. For every $u \in \mathbf{D}(h)$ let $\{v^u_\alpha; \alpha \in \zeta_u\}$ be a well-ordering of the elements of $h'u$. Let $\{w^u_\alpha; \alpha \in \zeta_u\}$ be a sequence of pairwise disjoint elements such that $w_\alpha \leqq v_\alpha$ and $\bigvee_{\xi_u} w^u_\alpha = \bigvee_{\xi_u} v^u_\alpha$ (i.e. $w^u_0 = v^u_0$ and $w^u_\alpha = v^u_\alpha - \bigvee_{\beta<\alpha} w^u_\beta$ for $\alpha > 0$). The set containing all elements $w^u_\alpha \wedge u$ $(u \in \mathbf{D}(h), \alpha \in \zeta_u)$ together with the element $-\bigvee(\bigvee h'u \wedge u)$ is a partition of $\mathbf{b}$. We define a function g whose domain is this partition by setting $g'(w^u_\alpha \wedge u) = v^u_\alpha$ for each α, u and $g'(-\bigvee(h'u \wedge u)) = 1_{\mathbf{b}}$. For this g we have the following: (a) $g \in Z$ because $\bigvee\{g'v \wedge v; v \in \mathbf{D}(g)\} \geq \geq \bigvee_u(\bigvee_\alpha(v^u_\alpha \wedge w^u_\alpha \wedge u)) = \bigvee_u(\bigvee_\alpha w^u_\alpha \wedge u) = \bigvee_u h'u \wedge u \in \mathbf{z}$;

(b) $g \in \hat{h}$ because $\bigvee\{v \wedge u; g'v \in h'u\} \geq \bigvee_{\alpha,u}(w^u_\alpha \wedge u) = \bigvee_u(\bigvee h'u \wedge u) \in \mathbf{z}$.

Hence $g \in Z \cap \hat{h}$.

In view of 1–4, Z is a complete ultrafilter on $\bar{\mathbf{b}}$ in the sense of $\mathfrak{N}(\mathbf{Ulr}(\mathbf{b},\mathbf{z}))$. As for the additional property of Z, we have

$$\bigvee\{v \wedge k'_u v; v \in \mathbf{D}(k_u)\} = \bigvee\{1_{\mathbf{b}} \wedge u\} = u ;$$

hence $u \in \mathbf{z} \equiv k_u \in Z$.

Using the above theorem we obtain some important consistency results.

5213. METADEFINITION. We say that a set-formula $\beta(x)$ *describes a complete Boolean algebra* in a theory **T** (stronger than **TSS**′, say) if $\mathbf{T} \vdash \vdash (\exists x)\, \beta(x) \,\&\, (\forall x)\, (\beta(x) \rightarrow x$ is a complete Boolean algebra with at least two elements).

5214. METALEMMA. Let Γ be a sequence of closed set formulas. If $\beta(x)$ describes a complete Boolean algebra in (**TSS**, E2, Γ), and if $\vartheta(\boldsymbol{b}, Z)$ is a formula which asserts in **TSS'** that b is a complete Boolean algebra and Z is a complete ultrafilter on b, then the axiom

$$(\exists b, Z)\,(\beta(b)\,\&\,\vartheta(b, Z))$$

is consistent with **TSS**, E2, Γ.

Indeed, by Metatheorems 5209 and 5206 and by the preceding theorem, $\mathfrak{Dir}\,(\partial\mathbf{TSS}/\partial\mathfrak{N}\,(\mathbf{Ulr}\,(\mathbf{b}, \mathbf{z}))$ is a model in the following diagram:

$$
\boxed{\begin{array}{l} TSS,E2,\Gamma \\ (\exists b,Z)\,(\beta(b)\,\&\,\vartheta(b,Z)) \end{array}}
\longrightarrow
\boxed{\begin{array}{l} TSS,E2,\Gamma,\beta\,(b), \\ Bool\,(b,z), \\ \partial TSS|\partial\,\mathfrak{N}(Ulr(b,z)) \end{array}}
$$

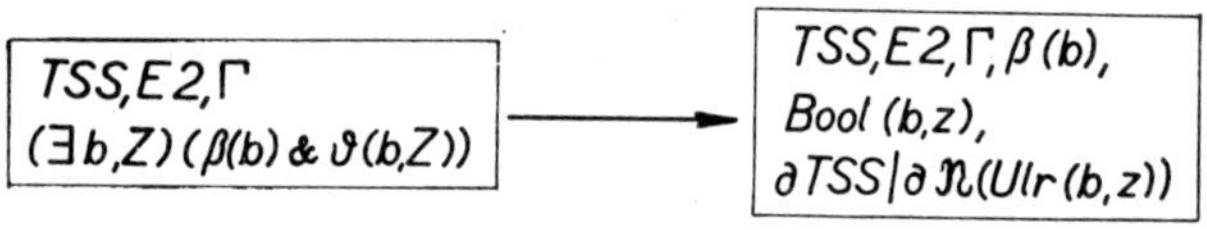

5215. METATHEOREM (Support Principle). Let Γ be a sequence of closed set formulas. If $\beta(x)$ describes a complete Boolean algebra in **TSS**, E2, Γ and if $\vartheta(b, Z)$ has the same meaning as in the preceding metalemma then the axiom

(Sβ) $\qquad\qquad (\exists b, Z)\,(\beta(b)\,\&\,\vartheta(b, Z)\,\&\,\mathrm{TSupp}\,(Z))$

is consistent with **TSS**, E2, Γ.

Indeed, $\mathfrak{Dir}\,(\partial\mathbf{TSS}/\partial\mathfrak{Supp}\,(Z))$ is a model in the following diagram:

$$
\boxed{\begin{array}{l} TSS,E2,\Gamma, \\ (S\,\beta) \end{array}}
\longrightarrow
\boxed{\begin{array}{l} TSS,E2,\Gamma, \\ \beta\,(b)\,\&\,\vartheta\,(b,Z) \\ \partial TSS|\partial\mathfrak{Supp}(Z) \end{array}}
$$

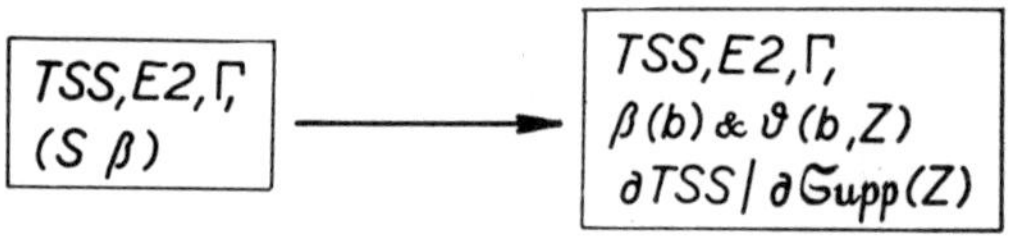

5216. Remarks. (1) Axiom (Sβ) is the strengthening of (S6) promised above; whereas by (S6) there exists a total support which is a complete ultrafilter on some complete Boolean algebra b, here we assert that the algebra in question satisfies $\beta(b)$.

(2) If **T** is a term such that $\mathbf{T} \vdash (\mathbf{T}$ is a complete Boolean algebra which is a set) then (S T) has the same meaning as (S($b = \mathbf{T}$)) i.e. $(\exists Z)\,(\vartheta(\mathbf{T}, Z)\,\&$ & Z is a total support).

(3) Consider the following diagram:

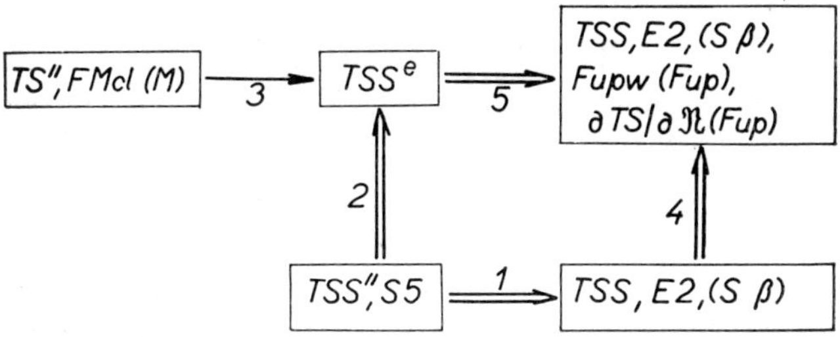

The theory **TSS**, E2, (Sβ) is stronger than **TSS**″, S5 (arrow 1); $\mathfrak{Up}$ is a model of **TS**″, FMcl (**M**) in **TSS**^{e} (arrow 3) and **TSS**^{e} is a conservative extension of **TSS**″, S5 (arrow 2). The theory **TSS**, E2, (Sβ), Fupw (**Fup**), ∂**TS**/$\partial\mathfrak{N}$ (**Fup**) is a conservative extension of **TSS**, E2, (Sβ) (arrow 4) and is stronger than **TSS**^{e} (arrow 5). Consequently $\mathfrak{Up}$ is a model of **TS**″, FMcl (**M**) in the terminal theory of 5 (composition of arrows 3 $*$ 5). As we know from Section 1, E2 holds in the last model. For the sake of brevity we denote the sequence Fupw (**Fup**), ∂**TS**/$\partial\mathfrak{N}$ (**Fup**) by (+ upw), so that the terminal theory of 5 is denoted by **TSS**, E2, (Sβ), (+upw).

The last theory is a conservative extension of **TSS**, E2, (Sβ). If Γ is a sequence of set formulas and β describes a complete Boolean algebra in **TSS**, E2, Γ, we know by the support principle the (Sβ) is consistent with the last theory and therefore **TSS**, E2, (Sβ) + upw is consistent relative to **TSS**, E2, Γ. This is a very important corollary of the support principle. Let us give an example to show how this principle is used.

5217. Example (Independence of the axiom of constructibility). Let Noat ($\boldsymbol{b}$) be a set formula which asserts in **TSS**′ that $\boldsymbol{b}$ is an algebra with no atoms. By 2536 ($\exists \boldsymbol{b}$) Noat ($\boldsymbol{b}$) is provable in **TSS**, E2 and so **TSS**, E2, (S Noat) + upw is consistent (relative to **TSS**). In this theory ($\exists\sigma$) (σ is not a set) (i.e. $\neg$ C3) is provable (the support on **b** cannot be a set) and is equivalent to ($\exists x^* \subseteq$ **Tor**) ($x^* \notin^*$ **Tor**). Hence $\mathfrak{Up}$ is a model in the following diagram.

But the former theory is stronger than **TS**, D3, E2, $\neg$ (Constr), so that $\neg$ (Constr) is consistent with **TS**, D3, E2.

For more details and other consistency results see Chapter VI.

The following metatheorem is a very important provability result.

5218. METATHEOREM. Let Γ be a sequence of closed set formulas and let φ be a closed set formula. Suppose that $\beta(x)$ describes a complete Boolean algebra in **TSS**, E2, Γ. The formula φ is provable in **TSS**, E2, Γ if and only if φ is provable in **TSS**, E2, Γ,(Sβ); in other words, (Sβ) extends **TSS**, E2, Γ conservatively w.r.t. set formulas.

Demonstration. **TSS**, E2, Γ, (Sβ) is an extension of **TSS**, E2, Γ; hence if φ is provable in the latter theory then it is provable in the former. Conversely, suppose that φ is provable in **TSS**, E2, Γ, (Sβ) and consider **TSS**, E2, Γ. The theory

$$\mathbf{TSS}_1 = (\mathbf{TSS}, \mathrm{E2}, \Gamma, \beta(\mathbf{b}), \mathrm{Bool}(\mathbf{b}, \mathbf{z}), \partial\mathbf{TSS}/\partial\mathfrak{N}(\mathbf{Ulr}(\mathbf{b}, \mathbf{z})))$$

is a conservative extension of **TSS**, E2, Γ; we denote the notions of $\mathfrak{N}(\mathbf{Ulr}(\mathbf{b}, \mathbf{z}))$ by the means of $*$. We construct a support model in $\mathfrak{N}(\mathbf{Ulr}(\mathbf{b}, \mathbf{z}))$ i.e. consider $\mathbf{TSS}_2 = (\mathbf{TSS}_1, \mathrm{Supp}^*(\mathbf{Z}^*), \partial\mathbf{TSS}/\partial\mathfrak{Supp}^*(\mathbf{Z}^*))$. By 5212, $\mathbf{TSS}_2$ is again a conservative extension of **TSS**, E2, Γ; we denote the notions of $\mathfrak{Supp}^*(\mathbf{Z}^*)$ by the means of $\square$. By 5214 and 5215 $\mathfrak{Supp}^*(\mathbf{Z}^*)$ is a model of **TSS**, E2, Γ, (Sβ) and so $\mathbf{TSS}_2 \vdash \varphi^\square$. By the absoluteness of set formulas in support models we have $\mathbf{TSS}_2 \vdash \varphi^*$ and so $\mathbf{TSS}_1 \vdash \varphi^*$. By 5207 we obtain $\mathbf{TSS}_1 \vdash \varphi$ and so **TSS**, E2, $\Gamma \vdash \varphi$.

SECTION 3

Characteristics of complete Boolean algebras and properties of relations and functions

We shall investigate the way in which the properties of semisets, and especially the properties of semiset relations and functions, depend on the properties of complete Boolean algebras with total Boolean support. Before doing this we shall consider the important notion of an approximable mapping. We shall investigate this notion first in the theory of semisets without axioms of supports. In the presence of axioms of support the notion of an approximable function can be characterized in terms of the properties of complete Boolean algebras with total Boolean support, and it closely related to the problem of cardinality in the model $\mathfrak{Up}$. We begin with some important notions concerning $\boldsymbol{b}$-sets.

5301. Remark. According to 4311, each $\boldsymbol{b}$-set r determines, for any x, a unique element r_x of b; we shall write r_{xy} for $r_{\langle x,y\rangle}$.

5302. DEFINITION (**TSS'**). A $\boldsymbol{b}$-set r is called a ***b**-relation* $(\mathrm{Rel}^{\boldsymbol{b}}(r))$ if $r \subseteq \mathbf{V}^2 \times b$; a $\boldsymbol{b}$-relation r is called a ***b**-relation on x and y* if $\mathbf{W}(r) \subseteq$ $\subseteq x \times y$; a $\boldsymbol{b}$-relation r is called a ***b**-mapping* or a $\boldsymbol{b}$-function $(\mathrm{Un}^{\boldsymbol{b}}(r))$ if

$$(\forall \iota, \kappa_1, \kappa_2)\left(\langle \kappa_1, \iota\rangle \in \mathbf{W}(r) \,\&\, \langle \kappa_2, \iota\rangle \in \mathbf{W}(r) \,\&\, \kappa_1 \neq \kappa_2 . \rightarrow \right.$$

$$\left. \rightarrow r_{\kappa_1\iota} \wedge r_{\kappa_2\iota} = 0_{\boldsymbol{b}}\right);$$

a $\boldsymbol{b}$-mapping r is called a ***b**-1-1-mapping* $(\mathrm{Un}_2^{\boldsymbol{b}}(r))$ if moreover

$$(\forall \iota_1, \iota_2, \kappa)\left(\langle \kappa, \iota_1\rangle \in \mathbf{W}(r) \,\&\, \langle \kappa, \iota_2\rangle \in \mathbf{W}(r) \,\&\, \iota_1 \neq \iota_2 . \rightarrow r_{\kappa\iota_1} \wedge r_{\kappa\iota_2} = 0_{\boldsymbol{b}}\right);$$

a $\boldsymbol{b}$-mapping r is called a ***b**-mapping of x into y* $(\mathrm{Un}^{\boldsymbol{b}}(r, x, y))$ if $r \subseteq (y \times x) \times b$ and if $\{r_{\kappa\iota}, \kappa \in y\}$ is a partition of $\boldsymbol{b}$ for every $\iota \in x$.

5303. LEMMA (**TSS'**). If Z is a total Boolean support on $\boldsymbol{b}$ then

(a) $\mathrm{Rel}(\sigma) \equiv (\exists r)(\mathrm{Rel}^{\boldsymbol{b}}(r) \,\&\, \sigma = r\text{''}Z)$;

(b) $\mathrm{Un}(\sigma) \equiv (\exists r)(\mathrm{Un}^{\boldsymbol{b}}(r) \,\&\, \sigma = r\text{''}Z)$;

(c) $\mathrm{Un}_2(\sigma) \equiv (\exists r)(\mathrm{Un}_2^{\boldsymbol{b}}(r) \,\&\, \sigma = r\text{''}Z)$;

(d) $\mathrm{Un}(\sigma, x, y) \equiv (\exists r)(\mathrm{Un}^{\boldsymbol{b}}(r, x, y) \,\&\, \sigma = r\text{''}Z)$.

Proof. The implications from right to left are obvious; we shall prove the converse. Suppose s is a disjointed relation such that $\sigma = s\text{''}Z$.

(a) If σ is a relation then $\sigma = r\text{''}Z$ where $r = s \cap (\mathbf{V}^2 \times b)$.

(b) If $\mathrm{Un}(\sigma)$ then we may assume that $\mathrm{Rel}^{\boldsymbol{b}}(s)$ and $s\text{''}Z = \sigma$. Let $a \subseteq \mathbf{V}^2$ be such that $\sigma \subseteq a$. Since $s_{\kappa_1\iota} \wedge s_{\kappa_2\iota} \notin Z$ for every $\langle\kappa_1,\iota\rangle, \langle\kappa_2,\iota\rangle \in a$ $(\kappa_1 \neq \kappa_2)$ we have

$$\bigvee_{\substack{\kappa_1 \neq \kappa_2 \\ \langle\kappa_1,\iota\rangle, \langle\kappa_2,\iota\rangle \in a}} (s_{\kappa_1\iota} \wedge s_{\kappa_2\iota}) \notin Z$$

and, denoting the above join by u, we have $-u \in Z$, so that

$$s\text{''}Z = r\text{''}Z \quad \text{where} \quad r = s \,|_{\boldsymbol{b}}\, (-u)\,;$$

we also have $\mathrm{Un}^{\boldsymbol{b}}(r)$. The proof of (c) is analogous.

(d) $\mathrm{Un}(\sigma, x, y)$ implies $\mathrm{Un}(\sigma)$ and hence there exists a $\boldsymbol{b}$-mapping r such that $r\text{''}Z = \sigma$. For each $\iota \in x$ the system $\{r_{\kappa\iota};\, \kappa \in y\}$ is disjointed in $\boldsymbol{b}$, its join belongs to Z and so that the complement of this join does not belong to Z. We choose a fixed element κ_0 in y independently of $\iota \in x$. Define $\bar{r}_{\kappa_0\iota} = r_{\kappa_0\iota} \vee -\bigvee_{\kappa\in y} r_{\kappa\iota}$, and $\bar{r}_{\kappa\iota} = r_{\kappa\iota}$ $(\kappa \neq \kappa_0)$. Clearly $\bar{r}\text{''}Z = r\text{''}Z$ and $\mathrm{Un}^{\boldsymbol{b}}(\bar{r}, x, y)$.

Remark. Thus $\boldsymbol{b}$-relations may be regarded as "codes" for those semisets which are relations. The elements of the range of a $\boldsymbol{b}$-relation can be considered as "indices" and the $\boldsymbol{b}$-relation itself can be considered as a "two parameter system of elements of $\boldsymbol{b}$"; i.e. a $\boldsymbol{b}$-relation r can be considered not as assigning ordered pairs to elements of a complete Boolean algebra but as assigning elements $r_{\kappa\iota}$ to ordered pairs $\langle\kappa, \iota\rangle$.

In the same sense, $\boldsymbol{b}$-functions may be considered as "parametric systems of disjointed systems in $\boldsymbol{b}$"; for, if $\iota \in x$ then the $r_{\kappa\iota}$ $(\kappa \in y)$ form a disjointed system in $\boldsymbol{b}$.

We shall now investigate semiset mappings.

5304. DEFINITION (**TSS**′). We say that a mapping σ of x into y is ***z-approximable*** if there exists a relation $q \subseteq (y \times x)$ such that

1) $\sigma \subseteq q$,
2) $(\forall u \in \mathbf{D}(q))(\mathbf{Ext}_q(u) \prec z)$

(σ can be embedded into a cylinder of gauge z). We say that ***mappings of x into y are z-approximable*** if every (semiset) mapping of x into y is z-approximable. (Notation: Appr (x, y, z).)

5305. LEMMA (**TSS**′). If Appr (x, y, z) and $x_1 \preccurlyeq x$, $y_1 \preccurlyeq y$, $z_1 \succcurlyeq z$ then Appr (x_1, y_1, z_1). (Obvious.)

In particular, we have

5306. LEMMA (**TSS**′). Appr $(x, y, 2)$ holds iff every mapping of x into y is a set; in particular, Appr $(x, 2, 2)$ holds iff every subclass of x is a set.

Proof. The fact that every mapping of x into y is 2-approximable means that every such mapping can be embedded in a set relation having only singleton extensions. Such a relation must coincide with the original function and so this function is a set. Every subclass σ of x is similar to its characteristic function; i.e. to the function ϱ such that, for any $u \in x$, $\varrho'u = 1 \equiv u \in \sigma$ and $\varrho'u = 0 \equiv u \notin \sigma$; ϱ is a mapping of x into 2.

5307. LEMMA (**TSS**′). If Appr $(x, 2, 2)$ and $(y \times x_1) \preccurlyeq x$ then Appr $(x_1, \mathbf{P}(y), 2)$.

Proof. Let σ be a mapping of x_1 into $\mathbf{P}(y)$; define $\langle v, u\rangle \in \varrho \equiv v \in \sigma'u$. Clearly $\varrho \subseteq y \times x_1$; hence ϱ is a set and so σ is a set.

5308. LEMMA (**TSS**′). If Appr $(\aleph_\beta, 2, 2)$ then Appr $(\aleph_\beta, \mathbf{P}(\boldsymbol{\omega}_\beta), 2)$.

Proof. The assertion follows from the preceding lemma since $\boldsymbol{\omega}_\beta \times \boldsymbol{\omega}_\beta \approx \boldsymbol{\omega}_\beta$.

Recall the definition 5133 of the predicate $\hat{\mathrm{C}}\mathrm{onf}(\alpha, \beta)$. In **TSS**′ we have $\hat{\mathrm{C}}\mathrm{onf}(\mathbf{cf}(\aleph_\beta), \aleph_\beta)$ for every β.

In addition we have

5309. LEMMA (**TSS**′). If $\aleph_\alpha \leqq \mathbf{cf}(\aleph_\beta)$ then

$$\hat{\mathrm{C}}\mathrm{onf}(\aleph_\alpha, \aleph_\beta) \equiv \hat{\mathrm{C}}\mathrm{onf}(\aleph_\alpha, \mathbf{cf}(\aleph_\beta)).$$

Proof. Let f be a (set) monotone mapping of $\mathbf{cf}(\aleph_\beta)$ into $\aleph_\beta$ such that $\mathsf{U}(\mathbf{W}(f)) = \boldsymbol{\omega}_\beta$.

1) If σ is a mapping of $\aleph_\alpha$ into $\mathbf{cf}(\aleph_\beta)$ such that $\bigcup(\mathbf{W}(\sigma)) = \mathbf{cf}(\aleph_\beta)$ then the composition of f and σ is a (semiset) mapping ϱ of $\aleph_\alpha$ into $\aleph_\beta$ such that $\bigcup(\mathbf{W}(\varrho)) = \aleph_\beta$.

2) Conversely, if ϱ is a mapping of $\aleph_\alpha$ into $\aleph_\beta$ such that $\bigcup(\mathbf{W}(\varrho)) = \aleph_\beta$, then we let σ be a mapping which assigns to each $\xi < \aleph_\alpha$ the least $\zeta < \mathbf{cf}(\aleph_\beta)$ such that $f'\zeta > \varrho'\xi$. We prove $\bigcup(\mathbf{W}(\sigma)) = \mathbf{cf}(\aleph_\beta)$. If $\eta < \mathbf{cf}(\aleph_\beta)$ then there exists $\xi < \boldsymbol{\omega}_\alpha$ such that $\sigma'\xi \geq \eta$; for, if we let $\eta_0 = \bigcup(f''\eta)$ then $\eta_0 < \boldsymbol{\omega}_\beta$ and there exists ξ such that $\varrho'\xi > \eta_0$ whence $\sigma'\xi \geq \eta$.

5310. Lemma (**TSS**′). If $\aleph_\alpha \approx \aleph_\beta$ and $\alpha \leqq \beta$ then $\widehat{\mathrm{Conf}}(\aleph_\alpha, \aleph_\beta)$.

Proof. If σ is a 1-1-mapping of $\boldsymbol{\omega}_\alpha$ onto $\boldsymbol{\omega}_\beta$ then $\bigcup(\mathbf{W}(\sigma)) = \boldsymbol{\omega}_\beta$ and hence we have $\widehat{\mathrm{Conf}}(\aleph_\alpha, \aleph_\beta)$.

5311. Lemma (**TSS**′). If $\aleph_\alpha < \aleph_\beta$ and $\neg\,\mathrm{Appr}(\aleph_\alpha, \aleph_\beta, \aleph_\beta)$ then $\widehat{\mathrm{Conf}}(\aleph_\alpha, \aleph_\beta)$.

Proof. Let $\mathrm{Un}(\sigma, \aleph_\alpha, \aleph_\beta)$ and suppose that σ is not $\aleph_\beta$- approximable. We shall prove that $\bigcup(\mathbf{W}(\sigma)) = \aleph_\beta$. If $\gamma < \boldsymbol{\omega}_\beta$ and $\gamma \notin \bigcup(\mathbf{W}(\sigma))$ then $\bigcup(\mathbf{W}(\sigma)) \subseteq \gamma$ and $\gamma < \aleph_\beta$, i.e. $\sigma \subseteq \gamma \times \boldsymbol{\omega}_\alpha$, which contradicts the fact that σ is not $\aleph_\beta$-approximable.

5312. Lemma (**TSS**′). Let $\aleph_\beta$ be regular and suppose that $\aleph_\alpha < \aleph_\beta$. If $\widehat{\mathrm{Conf}}(\aleph_\alpha, \aleph_\beta)$ then $\neg\,\mathrm{Appr}(\aleph_\alpha, \aleph_\beta, \aleph_\beta)$.

Proof. Let σ be such that $\mathrm{Un}(\sigma, \aleph_\alpha, \aleph_\beta)$ and $\bigcup(\mathbf{W}(\sigma)) = \boldsymbol{\omega}_\beta$; we shall show that σ is not $\aleph_\beta$-approximable. Let $\sigma \subseteq r$, let $r \subseteq \boldsymbol{\omega}_\beta \times \boldsymbol{\omega}_\alpha$ and suppose that $\mathbf{Ext}_r(\iota) \prec \boldsymbol{\omega}_\beta$ for every $\iota \in \boldsymbol{\omega}_\alpha$; then $\bigcup(\mathbf{W}(r)) = \boldsymbol{\omega}_\beta$ since $\sigma \subseteq r$; by regularity we have $\mathbf{Sup}(r''\{\iota\}) \prec \boldsymbol{\omega}_\beta$ for any $\iota \in \boldsymbol{\omega}_\alpha$. Hence $\mathbf{Sup}(r''\boldsymbol{\omega}_\alpha) = \bigcup(\mathbf{W}(r)) < \boldsymbol{\omega}_\beta$, a contradiction. Thus $\neg\,\mathrm{Appr}(\aleph_\alpha, \aleph_\beta, \aleph_\beta)$.

5313. Theorem (**TSS**′). If $\aleph_\alpha < \mathbf{cf}(\aleph_\beta)$ then

$$\widehat{\mathrm{Conf}}(\aleph_\alpha, \aleph_\beta) \equiv \neg\,\mathrm{Appr}(\aleph_\alpha, \mathbf{cf}(\aleph_\beta), \mathbf{cf}(\aleph_\beta)).$$

This follows from lemmas 5309, 5311 and 5312.

5314. Lemma (**TSS**′). If $\alpha < \beta$ and $\aleph_\alpha \not\approx \aleph_\beta$ then $\neg\,\mathrm{Appr}(\aleph_\alpha, 2, 2)$.

Proof. If $\mathrm{Un}_2(\sigma, \aleph_\alpha, \aleph_\beta)$ then there exists a relation $\varrho \subseteq \boldsymbol{\omega}_\alpha \times \boldsymbol{\omega}_\alpha$ which well-orders $\boldsymbol{\omega}_\alpha$ and which is isomorphic, by means of a semiset isomorphism, to $\mathbf{E} \upharpoonright \boldsymbol{\omega}_\beta$. This semiset is not a set. Since $\boldsymbol{\omega}_\alpha \times \boldsymbol{\omega}_\alpha \sim \boldsymbol{\omega}_\alpha$ there exists a subclass of $\boldsymbol{\omega}_\alpha$ which is not a set; hence $\neg\,\mathrm{Appr}(\aleph_\alpha, 2, 2)$.

5315. THEOREM (**TSS'**). If Appr $(\aleph_\alpha, 2, 2)$ holds for every $\alpha < \beta$ then every cardinal $\aleph_\alpha \leqq \aleph_\beta$ is an absolute cardinal.

5316. LEMMA (**TSS'**, S4). Appr $(n, 2, 2)$ holds for every $n \in \omega$.

This follows by 5135. We also know (in **TSS'**, S4) that ω is an absolute cardinal. Using the notion of approximability we shall now find a condition equivalent to

$$\alpha < \beta \;\&\; \aleph_\alpha \mathrel{\hat{\approx}} \aleph_\beta .$$

In what follows we use Axioms St, Pot and E1; thus the results hold in **TSS**^{e}, E1 for example.

5317. METATHEOREM. If $\varphi(x, Y)$ is a NF and if σ is a semiset variable then

$$\mathbf{TSS}', \text{Pot, St, E1} \vdash (\forall x \in a)(\exists \sigma \subseteq b)(\varphi(x, \sigma) \rightarrow$$
$$\rightarrow (\exists \varrho \subseteq b \times a)(\forall x \in a)\, \varphi(x, \varrho\text{''}\{x\}) .$$

Demonstration. We give a proof in the theory **TSS'**, Pot, St, E1 as follows. Let $\tau \subseteq b \times c$ be a coding of all subclasses of b; such a coding exists in view of Axiom Pot. Let f be a well-ordering of c; such a mapping exists in view of El. In view of (St) we may define ϱ_0 as follows: $\langle u, v\rangle \in \varrho_0 \equiv \equiv v \in a \;\&\; u \in c \;\&\; u$ is the least element in the ordering f of c such that $\varphi(v, \mathbf{Ext}_\tau(u))$.

Clearly ϱ_0 is a function. If we define

$$\langle u, v\rangle \in \varrho \equiv \;.\; u \in b \;\&\; v \in a \;\&\; u \in \mathbf{Ext}_\tau(\varrho_0\text{'}v) ,$$

then ϱ has the required properties.

5318. LEMMA (**TSS'**, St, Pot, E1). Let $\alpha < \beta$ and suppose that $\widehat{\text{Conf}}(\aleph_\alpha, \aleph_\beta)$. If $\aleph_\alpha \mathrel{\hat{\approx}} \aleph_\gamma$ for every γ such that $\alpha < \gamma < \beta$, then $\aleph_\alpha \mathrel{\hat{\approx}} \aleph_\beta$.

Proof. Let σ be a mapping of $\aleph_\alpha$ into $\aleph_\beta$ such that $\bigcup(\mathbf{W}(\sigma)) = \omega_\beta$. For every $\lambda < \omega_\beta$ there exists ϱ_λ such that $\text{Un}_2(\varrho_\lambda, \omega_\alpha, \lambda)$: hence, by Metatheorem 5317, there exists $\varrho \subseteq (\omega_\beta \times \omega_\alpha) \times \omega_\beta$ such that, for every $\lambda < \omega_\beta$, $\varrho\text{''}\{\lambda\}$ is a one-one mapping of ω_α onto λ. For $\iota, \kappa < \omega_\alpha$ we define $\tau\text{'}\langle \iota, \kappa\rangle = (\varrho\text{''}\{\sigma\text{'}\iota\})\text{'}\,\kappa$. Clearly $\tau\text{''}(\omega_\alpha \times \omega_\alpha) = \omega_\beta$; hence there exists a one-one mapping of ω_β into $\omega_\alpha \times \omega_\alpha$. Since $\omega_\alpha \approx \omega_\alpha \times \omega_\alpha$, there exists a one-one mapping of ω_β into ω_α; since $\omega_\alpha \preccurlyeq \omega_\beta$ this completes the proof.

5319. THEOREM (**TSS**′, S5, E1). If $\aleph_\alpha < \aleph_\beta$ then for $\aleph_\alpha \hat{\approx} \aleph_\beta$ to hold it is necessary and sufficient that $\neg$ Appr $(\aleph_\alpha, \aleph_\gamma, \aleph_\gamma)$ for every regular $\aleph_\gamma$ such that $\alpha < \gamma \leqq \beta$.

Proof. If $\aleph_\alpha \hat{\approx} \aleph_\beta$ then, by 2204, $\aleph_\alpha \hat{\approx} \aleph_\gamma$ for every γ such that $\alpha < \gamma \leqq \leqq \beta$; hence by Lemma 5310 $\widehat{\text{Conf}}(\aleph_\alpha, \aleph_\gamma)$. Conversely, suppose the condition holds. Let $\alpha < \lambda \leqq \beta$ and suppose that $\aleph_\alpha \hat{\approx} \aleph_\gamma$ for all γ such that $\alpha < \gamma < \lambda$. If $\aleph_\lambda$ is regular then $\neg$ Appr $(\aleph_\alpha, \aleph_\lambda, \aleph_\lambda)$ and by 5311 we have $\widehat{\text{Conf}}(\aleph_\alpha, \aleph_\lambda)$. If $\mathbf{cf}(\aleph_\beta) \leqq \aleph_\alpha$ then $\text{Conf}(\aleph_\alpha, \aleph_\lambda)$, hence $\widehat{\text{Conf}}(\aleph_\alpha, \aleph_\lambda)$; if $\aleph_\alpha < \mathbf{cf}(\aleph_\lambda)$ then using $\neg$ Appr $(\aleph_\alpha, \mathbf{cf}(\aleph_\beta), \mathbf{cf}(\aleph_\beta))$ we obtain $\widehat{\text{Conf}}(\aleph_\alpha, \aleph_\lambda)$. Thus we have proved $\widehat{\text{Conf}}(\aleph_\alpha, \aleph_\lambda)$ and by Lemma 5318 we have $\aleph_\alpha \hat{\approx} \aleph_\lambda$. Now, using S1 and St, we obtain $\aleph_\alpha \hat{\approx} \aleph_\beta$.

5320. THEOREM (**TSS**′, S5, E1).

(1) In order that $\aleph_\beta$ and all smaller cardinals be absolute, it is necessary and sufficient that

$$\text{Appr}(\aleph_\alpha, \aleph_{\alpha+1}, \aleph_{\alpha+1}) \text{ for every } \alpha < \beta .$$

(2) In order that $\aleph_\beta$ and all larger cardinals be absolute, it is necessary and sufficient that

$$\text{Appr}(\aleph_\alpha, \aleph_{\alpha+1}, \aleph_{\alpha+1}) \text{ for every } \alpha \geqq \beta .$$

Proof. (1) There is a non-absolute cardinal $\leqq \aleph_\beta$ iff $\aleph_\alpha \hat{\approx} \aleph_{\alpha+1}$ holds for some $\alpha < \beta$; but this latter condition holds iff $\neg$ Appr $(\aleph_\alpha, \aleph_{\alpha+1}, \aleph_{\alpha+1})$ for some $\alpha < \beta$. The proof of (2) is analogous. We observe that (1) also holds for $\aleph_0$ in view of 5135.

Under the assumption that $\aleph_\alpha \hat{\approx} \aleph_0$ for some $\alpha > 0$ we can prove the existence of complete ultrafilters on certain Boolean algebras.

5321. THEOREM (**TSS**′, S4, E1). Let $\aleph_\alpha \hat{\approx} \aleph_0$ and suppose that $\boldsymbol{b}$ is a complete Boolean algebra such that the cardinality of the set q of all partitions of $\boldsymbol{b}$ is $\aleph_\alpha$. Then there exists a complete ultrafilter on $\boldsymbol{b}$.

Proof. Let ϱ be a one-one mapping of $\boldsymbol{\omega}_0$ onto q and let A be a selector for b. We define H"$\{0\} = \{u\}$, where $u \neq 0_{\boldsymbol{b}}$ is some chosen element of ϱ'0, and H"$\{n + 1\} = \{v\}$, where $v \neq 0_{\boldsymbol{b}}$ is some chosen element of ϱ'$(n + 1)$ which has non-zero meet with $\bigwedge H$"$(n + 1)$. The meet $\bigwedge H$"$(n + 1)$ always exists, since H"$(n + 1)$ is a set. We let $Z_0 = H$"$\boldsymbol{\omega}$ and $Z = \{w; (\exists n)(w \geqq H\text{"}\{n\})\}$. Z is a complete ultrafilter on $\boldsymbol{b}$.

We shall now find out how the approximability of mappings depends on the properties of a Boolean algebra with total support.

5322. DEFINITION (**TSS′**). If $\boldsymbol{b}$ is a complete Boolean algebra and r is a $\boldsymbol{b}$--mapping of x into y then we say that an element u of $\boldsymbol{b}$ is *z-layered* by r (w.r.t. $\boldsymbol{b}$; $\text{Thru}_{\boldsymbol{b}}(u, r, z)$), if for every $p \in x$,

$$\{q \in y; (r_{qp} \wedge u) \neq 0_{\boldsymbol{b}}\} \prec z$$

(u intersects less than z elements in each "level").

We say that $\boldsymbol{b}$ is $x - y - z$*-distributive* ($\delta_{\boldsymbol{b}}(x, y, z)$) is, for every $\boldsymbol{b}$-mapping r of x into y, the join of all elements which are z-layered by r is $1_{\boldsymbol{b}}$.

5323. Remark. 1) In **TSS′** + E1 the notion of $\aleph_\alpha - \aleph_\beta - 2$-distributivity coincides with the usual notion of $\aleph_\alpha - \aleph_\beta$-distributivity for complete Boolean algebras.

2) In **TSS′** + E1, $\aleph_\alpha - \aleph_\beta - \aleph_\gamma$-distributivity is equivalent to the following property:

For every collection of $\aleph_\alpha$ partitions, each of power $\leqq \aleph_\beta$, there exists a partition consisting of $\aleph_\gamma$-layered elements.

5324. LEMMA (**TSS′**). If there is a total Boolean support Z on $\boldsymbol{b}$ and if $\boldsymbol{b}$ is $x - y - z$-distributive then $\text{Appr}(x, y, z)$.

Proof. If σ is a mapping of x into y then there exists a $\boldsymbol{b}$-mapping r of x into y such that $r\text{''}Z = \sigma$. Since the join of all z-layered elements is $1_{\boldsymbol{b}}$, there exists a z-layered element $u \in Z$. If we let $v \in q \equiv .\, v \in b \;\&\; v \wedge u \neq 0_{\boldsymbol{b}}$, then $s = r\text{''}q$ is a set and $\sigma \subseteq s$; for, if $p \in \sigma$ then $r_p \in Z$ and r_p intersects u. Clearly the extension of any element of $\mathbf{D}(s)$ is strictly subvalent to z.

5325. LEMMA (**TSS′**). If Z is a total Boolean support on $\boldsymbol{b}$ and $\text{Appr}(x, y, z)$ then there exists $u \in Z$ such that $\boldsymbol{b} \mid u$ is $x - y - z$-distributive.

Proof. For any $\boldsymbol{b}$-mapping r of x into y there exists an s such that $r\text{''}Z \subseteq \subseteq s \subseteq y \times x$ and every extension in s is strictly subvalent to z. Since $r\text{''}Z \subseteq s$ we have

$$\bigwedge_{p \in x} \bigvee_{q \in \text{Ext}_s(p)} r_{qp} \in Z$$

for any r and s. Thus, if we set $u = \bigwedge_r \bigvee_s \bigwedge_p \bigvee_q r_{pq}$ where r ranges over $\boldsymbol{b}$-mappings of x into y, s ranges over subrelations of $y \times x$ which satisfy the condition $(\forall p \in x)\,(\mathbf{Ext}_s(p) \prec z)$, p ranges over elements of x and q ranges over elements of $\mathbf{Ext}_s(p)$, then we have $u \in Z$. For every $(\boldsymbol{b} \mid u)$-mapping r of x into y there is a $\boldsymbol{b}$-mapping $\hat{r}$ of x into y such that $\hat{r}_{pq} \wedge u = r_{pq}$. We have $\bigvee_s \bigwedge_p \bigvee_q \hat{r}_{pq} \geq u$ and hence $\bigvee_s \bigwedge_p \bigvee_q r_{pq} = 1_{\boldsymbol{b}|u}$. Since $\bigwedge_p \bigvee_q r_{pq}$ is a z-layered set for any s, the algebra $\boldsymbol{b} \mid u$ is $x - y - z$-distributive.

5326. LEMMA (**TSS**′). If $\boldsymbol{b}$ is $x - y - z$-distributive then $\boldsymbol{b} \mid u$ is $x - y - z$-distributive for each $u > 0_{\boldsymbol{b}}$.

Proof. For every $\boldsymbol{b} \mid u$-mapping r of x into y there is a $\boldsymbol{b}$-mapping $\hat{r}$ of x into y such that $\hat{r}_{pq} \wedge u = r_{pq}$; if $v \in b$ is z-layered by r then $v \wedge u$ is z-layered by r.

5327. THEOREM (**TSS**′). If Z is a total Boolean support on $\boldsymbol{b}$, then $\mathrm{Appr}\,(x, y, z)$ iff $\boldsymbol{b} \mid u$ is $x - y - z$-distributive for some $u \in Z$.

We shall give two sufficient conditions for absoluteness of cardinals in the theory with Boolean total supports.

5328. THEOREM (**TSS**′). If $\boldsymbol{\mu}(\boldsymbol{b}) \leqq \aleph_\beta$ then $\boldsymbol{b}$ is $\aleph_\alpha - \aleph_\beta - \aleph_\beta$-distributive for all α.

Proof. Let r be a $\boldsymbol{b}$-mapping of $\aleph_\alpha$ into $\aleph_\beta$. Since $\{r_{\kappa\iota};\ \kappa \in \aleph_\beta\} \prec \aleph_\beta$ for every $\iota \in \aleph_\alpha$, it follows that every element of b is $\aleph_\beta$-layered by r.

5329. COROLLARY (1) (**TSS**′). If there is a total Boolean support on $\boldsymbol{b}$ and if $\aleph_\beta \geq \boldsymbol{\mu}(\boldsymbol{b})$ then $\mathrm{Appr}\,(\aleph_\alpha, \aleph_\beta, \aleph_\beta)$ for every α.

(2) (**TSS**′). Let $\boldsymbol{b}$ be a complete Boolean algebra with total Boolean support. If $\mathbf{cf}\,(\aleph_\beta) \geq \boldsymbol{\mu}(\boldsymbol{b})$ and $\alpha < \beta$ then

$$\widehat{\mathrm{Conf}}\,(\aleph_\alpha, \aleph_\beta) \equiv \mathrm{Conf}\,(\aleph_\alpha, \aleph_\beta)\,.$$

(3) (**TSS**′, E1). Let $\boldsymbol{b}$ be a complete Boolean algebra with total Boolean support. If $\aleph_\beta \geq \boldsymbol{\mu}(\boldsymbol{b})$ then $\aleph_\beta$ is an absolute cardinal.

Proof. Suppose $\aleph_\alpha \hat{\approx} \aleph_\beta$ for some $\alpha < \beta$; then we have $\widehat{\mathrm{Conf}}\,(\aleph_\alpha, \aleph_\beta)$.

(a) If $\aleph_\beta$ is regular then we have $\mathrm{Conf}\,(\aleph_\alpha, \aleph_\beta)$ by Corollary (2), a contradiction.

(b) If $\aleph_\beta$ is singular, then $\aleph_\beta > \boldsymbol{\mu}(\boldsymbol{b})$, since $\boldsymbol{\mu}(\boldsymbol{b})$ is regular (cf. 2510); hence there exists $\aleph_{\alpha_1} \geq \aleph_\alpha$ such that $\boldsymbol{\mu}(\boldsymbol{b}) \leqq \aleph_{\alpha_1} < \aleph_\beta$ and $\aleph_{\alpha_1} \hat{\approx} \aleph_{\alpha_1+1}$. We now apply (a) to the regular cardinal $\aleph_{\beta_1} = \aleph_{\alpha_1+1}$.

5330. THEOREM (**TSS**′, E1). If $\boldsymbol{b}$ is a complete Boolean algebra and if $\aleph_\alpha < \boldsymbol{\nu}(\boldsymbol{b})$ then $\boldsymbol{b}$ is $\aleph_\alpha - \aleph_\beta - 2$-distributive for every β.

Proof. Let r be a $\boldsymbol{b}$-mapping of $\boldsymbol{\omega}_\alpha$ into $\boldsymbol{\omega}_\beta$ and let a be an $\aleph_\alpha$-multiplicative base for $\boldsymbol{b}$. We have $\bigvee_{\delta\in\beta} r_{\delta\gamma} = 1_{\boldsymbol{b}}$ for every $\gamma \in \boldsymbol{\omega}_\alpha$. For each $v \in a$ we shall construct an element $u \in a$ such that $u \leqq v$ and u is 2-layered by r. We define a decreasing sequence $\{f'\xi\}_{\xi<\boldsymbol{\omega}_\alpha}$ of elements of a as follows. Let $f'0$ be an element of the base which is $\leqq v \wedge r_{\delta_0 0}$, where δ_0 is least δ such that $r_{\delta 0} \wedge v \neq 0_{\boldsymbol{b}}$. Suppose $f'\xi$ has been defined for all $\xi < \lambda$ $(\lambda < \boldsymbol{\omega}_\alpha)$ in such a way that for each ξ there exists some δ_ξ such that $f'\xi \leqq r_{\delta_\xi \xi}$; then we let $f'\lambda$ be an element of the base which is $\leqq (\bigwedge_{\xi<\lambda} f'\xi) \wedge r_{\delta_\lambda \lambda}$, where δ_λ is least δ such that $(\bigwedge_{\xi<\lambda} f'\zeta) \wedge r_{\delta\lambda} \neq 0_{\boldsymbol{b}}$. Such an element exists since $\bigwedge_{\lambda} f'\zeta \neq 0_{\boldsymbol{b}}$ by $\aleph_\alpha$-multiplicativity. Finally, we let u be some element of the base which is $\leqq \bigwedge_{\xi<\boldsymbol{\omega}_\alpha} f'\xi$. It follows that the join of all elements which are 2-layered by r is $1_{\boldsymbol{b}}$ and hence $\boldsymbol{b}$ is $\aleph_\alpha - \aleph_\beta - 2$-distributive.

5331. Corollary (1) (**TSS**$'$, E1). If $\boldsymbol{b}$ has a total Boolean support and if $\aleph_\alpha < \mathbf{v}(\boldsymbol{b})$ then

$$\mathrm{Appr}\,(\aleph_\alpha, \aleph_\beta, 2)$$

holds for every $\aleph_\beta$.

(2) (**TSS**$'$, E1). If $\boldsymbol{b}$ has a total Boolean support and if $\aleph_\alpha < \mathbf{v}(\boldsymbol{b})$ then every semiset $\sigma \mathrel{\hat{\approx}} \aleph_\alpha$ is a set.

(3) (**TSS**$'$, E1). If $\boldsymbol{b}$ has a total Boolean support and if $\aleph_\beta \leqq \mathbf{v}(\boldsymbol{b})$ then $\aleph_\beta$ is an absolute cardinal. (Otherwise, we would have $\aleph_\alpha \mathrel{\hat{\approx}} \aleph_\beta$ for some $\alpha < \beta$, contradicting Corollary (2).)

(4) (**TSS**$'$, E1). If $\boldsymbol{b}$ has a total Boolean support and $\mathbf{cf}\,(\aleph_\beta) \leqq \mathbf{v}(\boldsymbol{b})$ then

$$\mathrm{Conf}\,(\aleph_\alpha, \aleph_\beta) \equiv \widehat{\mathrm{Conf}}\,(\aleph_\alpha, \aleph_\beta)$$

for each $\alpha < \beta$. (By (1) we have $\mathrm{Appr}\,(\aleph_\alpha, \mathbf{cf}\,(\aleph_\beta), \mathbf{cf}\,(\aleph_\beta))$ for every $\aleph_\alpha < < \mathbf{cf}\,(\aleph_\beta)$.)

We now use the characteristics $\boldsymbol{\mu}$ and $\mathbf{v}$ to get more information about supports on the product of two algebras. Recall that if $\boldsymbol{b}_0$ and $\boldsymbol{b}_1$ are algebras then $\boldsymbol{b}_0 \odot \boldsymbol{b}_1$ denotes the algebra with base $\mathring{\boldsymbol{b}}_0 \odot \mathring{\boldsymbol{b}}_1$. Every complete ultrafilter on $\mathring{\boldsymbol{b}}_0 \odot \mathring{\boldsymbol{b}}_1$ is a product $Z_1 \times Z_2$ of complete ultrafilters on $\boldsymbol{b}_0$ and $\boldsymbol{b}_1$ respectively (cf. 2532).

5332. Theorem (**TSS**$'$, E1). Let $\boldsymbol{b}_0$ and $\boldsymbol{b}_1$ be complete Boolean algebras and suppose that $Z = Z_0 \times Z_1$ is a total Boolean support on $\mathring{\boldsymbol{b}}_0 \odot \mathring{\boldsymbol{b}}_1$.

If every exclusive system on $\mathring{\boldsymbol{b}}_0$ is of power $\leqq \aleph_\alpha$, i.e. $\boldsymbol{\mu}(\boldsymbol{b}_0) \leqq \aleph_{\alpha+1}$, and if $\mathring{\boldsymbol{b}}_1$ is $\aleph_\alpha$-multiplicative, i.e. $\boldsymbol{\nu}(\boldsymbol{b}_1) \geq \aleph_{\alpha+1}$, then every semiset $\sigma \mathrel{\hat{\preccurlyeq}} \aleph_\alpha$ depends on Z_0.

Proof. Let q be a partition of $\boldsymbol{b}_0 \odot \boldsymbol{b}_1$. We say that $v \in \mathring{\boldsymbol{b}}_1$ *stratifies* q if for every $w \in q$ such that $w \wedge \langle 1_{\boldsymbol{b}_0}, v\rangle \neq 0_{\boldsymbol{b}_0 \odot \boldsymbol{b}_1}$ there exists $u \in \mathring{\boldsymbol{b}}_0$ such that $w \wedge \langle 1_{\boldsymbol{b}_0}, v\rangle = \langle u, v\rangle$ ($\wedge$ is the meet in $\boldsymbol{b}_0 \odot \boldsymbol{b}_1$). Clearly if v stratifies q and $\bar{v} \leqq_1 v$, $\bar{v} \in \mathring{\boldsymbol{b}}_1$ then $\bar{v}$ also stratifies q.

1) We prove that if q is a partition of $\boldsymbol{b}_0 \odot \boldsymbol{b}_1$ then for every $v \in \mathring{\boldsymbol{b}}_1$ there exists $\bar{v} \in \mathring{\boldsymbol{b}}_1$, $\bar{v} \leqq_1 v$ such that $\bar{v}$ stratifies q. There exist $u_0 \in \mathring{\boldsymbol{b}}_0$ and $v_0 \in \mathring{\boldsymbol{b}}_1$ such that $\langle u_0, v_0\rangle \leqq w \wedge \langle 1_{\boldsymbol{b}_0}, v\rangle$ for some $w \in q$. Suppose that we already defined an exclusive system $\{u_\beta\}_{\beta<\lambda}$ in $\mathring{\boldsymbol{b}}_0$ and a decreasing system $\{v_\beta\}_{\beta<\lambda}$ in $\boldsymbol{b}_1$ such that for every $\beta < \lambda$ we have $\langle u_\beta, v_\beta\rangle \leqq w$ for some $w \in q$; it follows that $\lambda < \boldsymbol{\omega}_{\alpha+1}$. If $\bigvee_{0\,\lambda} u_\beta <_0 1_{\boldsymbol{b}_0}$ then we let $\langle u_\lambda, v_\lambda\rangle$ be such that $u_\lambda \leqq_0 1_{\boldsymbol{b}_0} -_0$ $-_0 \bigvee_{0\,\lambda} u_\beta$, $v_\lambda \leqq \bigwedge_{1\,\lambda} v_\beta$ and $\langle u_\lambda, v_\lambda\rangle \leqq w$ for some $w \in q$. There exists $\lambda <$ $< \boldsymbol{\omega}_{\alpha+1}$ such that $\bigvee_{0\,\lambda} u_\beta = 1_{\boldsymbol{b}_0}$, since $\boldsymbol{\mu}(\boldsymbol{b}_0) \leqq \aleph_{\alpha+1}$. Let $\bar{v} \in \mathring{\boldsymbol{b}}_1$ be such that $\bar{v} \leqq_1 \bigwedge_{1\,\lambda} v_\beta$; we prove that $\bar{v}$ stratifies q. If $w \in q$ and if $w \wedge \langle 1_{\boldsymbol{b}_0}, v\rangle \neq$ $\neq 0_{\boldsymbol{b}_0 \odot \boldsymbol{b}_1}$ then we let

$$\bar{u}_w = \bigvee_0 \{u_\beta, \beta < \lambda \,\&\, \langle u_\beta, \bar{v}\rangle \leqq w\}\,.$$

Clearly $\langle \bar{u}_w, \bar{v}\rangle = w \wedge \langle 1_{\boldsymbol{b}_0}, \bar{v}\rangle$.

2) We prove that if t is a set of partitions of $\boldsymbol{b}_0 \odot \boldsymbol{b}_1$ and if $t \preccurlyeq \boldsymbol{\omega}_\alpha$ then for every $v \in \mathring{\boldsymbol{b}}_1$ there exists $\bar{v} \in \mathring{\boldsymbol{b}}_1$, $\bar{v} \leqq_1 v$ such that $\bar{v}$ stratifies each $q \in t$; we say that $\bar{v}$ *uniformly stratifies* t. Consider an arbitrary well-ordering of t, $t = \{q_\beta, \beta < \boldsymbol{\omega}_\alpha\}$, and let $\bar{v}_0 \leqq_1 v$ be some element of b_1 which stratifies q_0. If $\{\bar{v}_\beta, \beta < \lambda\}$ is a decreasing sequence of elements of $\mathring{\boldsymbol{b}}_1$ such that every $\bar{v}_\beta$ stratifies q_β ($\beta < \lambda$) then we let $\bar{v}_\lambda$ be some element $\leqq \bigwedge_{1\,\beta<\lambda} \bar{v}_\beta$ which stratifies q_λ. Finally we let $\bar{v} \leqq_1 \bigwedge_{1\,\beta<\boldsymbol{\omega}_\alpha} \bar{v}_\beta$; $\bar{v}$ uniformly stratifies t.

3) It t is a set of partitions of $\boldsymbol{b}_0 \odot \boldsymbol{b}_1$ and $t \preccurlyeq \boldsymbol{\omega}_\alpha$, then there exists a partition q_1 of $\boldsymbol{b}_1$ such that every element of q_1 uniformly stratifies t. This partition q_1 can be defined by transfinite induction using 2).

4) We finally prove that if σ is a 1-1-mapping on $\boldsymbol{\omega}_\alpha$ then $\sigma = s\text{''}Z_0$ for some s. Let r be a $(\boldsymbol{b}_0 \odot \boldsymbol{b}_1)$-mapping such that $\sigma = r\text{''}Z$ and let

$$t = \{\{r_{x\xi}, r_{x\xi} \neq 0_{\boldsymbol{b}_0 \odot \boldsymbol{b}_1}\}, \xi \in \boldsymbol{\omega}_\alpha\}\,.$$

Since t is a system of partitions of $\boldsymbol{b}_0 \odot \boldsymbol{b}_1$ and has power $\leqq \aleph_\alpha$, there exists a partition q_1 of $\boldsymbol{b}_1$ whose elements uniformly stratify t. Let v be the unique element of $q_1 \cap Z_1$.

For every $w \in \bigcup(t)$ define $u_w \in b_0$ as follows: if $w \wedge \langle 1_{\boldsymbol{b}_0}, v\rangle \neq 0_{\boldsymbol{b}_0 \odot \boldsymbol{b}_1}$ then $\langle u_w, v\rangle = \langle 1_{\boldsymbol{b}_0}, v\rangle \wedge w$, otherwise $u_w = 0_{\boldsymbol{b}_0}$. If we let $s_{x\xi} = u_{r_{x\xi}}$ then s is a $\boldsymbol{b}_0$-mapping and $s_{x\xi} \in Z_0$ iff $r_{x\xi} \in Z$. Hence $\sigma = r''Z = s''Z_0$.

In an analogous way we can prove the following generalization:

5333. THEOREM (**TSS'**, E1). Let $\boldsymbol{a}_0$ and $\boldsymbol{a}_1$ be separatively ordered sets and suppose that $Z = Z_0 \times Z_1$ is a total Boolean support on $\boldsymbol{a}_0 \odot \boldsymbol{a}_1$. If every exclusive system on $\boldsymbol{a}_0$ has power at most $\aleph_\alpha$ and if $\boldsymbol{a}_1$ is $\aleph_\alpha$-multiplicative then every $\sigma \mathrel{\hat{\lessapprox}} \boldsymbol{\omega}_\alpha$ depends on Z_0. (Exercise.)

5334. COROLLARY (**TSS'**, E1). Let $\mathbf{Z} = \mathbf{Z}_0 \odot \mathbf{Z}_1$ be a total Boolean support on $\mathring{\boldsymbol{b}}_0 \odot \mathring{\boldsymbol{b}}_1$ and suppose that $\boldsymbol{\mu}(\boldsymbol{b}_0) \leqq \boldsymbol{\nu}(\boldsymbol{b}_1)$ and $\aleph_\beta \leqq \boldsymbol{\nu}(\boldsymbol{b}_1)$. If $\aleph_\beta$ is an absolute cardinal in the sense of $\mathfrak{Supp}(\mathbf{Z}_0)$ then $\aleph_\beta$ is an absolute cardinal.

5335. THEOREM (**TSS'**, St). If $\aleph_\alpha$ is the least cardinal such that $\neg\, \mathrm{Appr}(\aleph_\alpha, 2, 2)$, i.e. such that some part of $\aleph_\alpha$ is not a set, then

$$\neg\, \mathrm{Appr}(\mathbf{cf}(\aleph_\alpha), \mathbf{P}(\aleph_\alpha), 2)\,.$$

Proof. For regular $\aleph_\alpha$ the statement follows trivially from Lemma 5305. Hence we suppose that $\mathbf{cf}(\aleph_\alpha) = \aleph_\gamma < \aleph_\alpha$ and that $\sigma \subseteq \boldsymbol{\omega}_\alpha$ is not a set. Let ϱ be a mapping of $\boldsymbol{\omega}_\gamma$ into $\boldsymbol{\omega}_\alpha$ such that $\bigcup(\mathbf{W}(\varrho)) = \boldsymbol{\omega}_\alpha$. For each $\zeta < \boldsymbol{\omega}_\gamma$, $\sigma \cap (\varrho'\zeta)$ is a set, and so the function τ defined by $\tau'\zeta = \sigma \cap (\varrho'\zeta)$ is a semiset and maps $\boldsymbol{\omega}_\gamma$ into $\mathbf{P}(\boldsymbol{\omega}_\alpha)$; since $\bigcup(\mathbf{W}(\tau)) = \sigma$, τ is not a set. Hence we have

$$\neg\, \mathrm{Appr}(\mathbf{cf}(\aleph_\alpha), \mathbf{P}(\boldsymbol{\omega}_\alpha), 2)\,.$$

5336. THEOREM (**TSS'**, St). If $\aleph_\alpha$ is the least cardinal for which there is a y such that $\neg\, \mathrm{Appr}(\aleph_\alpha, y, 2)$ then $\aleph_\alpha$ is an absolutely regular cardinal number.

Proof. We have $\mathrm{Appr}(\aleph_\gamma, 2, 2)$ for every $\gamma < \alpha$; hence $\widehat{\mathrm{Card}}(\aleph_\alpha)$ by Lemma 5314. Suppose $\aleph_\alpha$ is absolutely singular. It follows that there exists a mapping σ from some $\aleph_\gamma$ into $\aleph_\alpha$ such that $\bigcup(\mathbf{W}(\sigma)) = \aleph_\alpha$. Since $\mathrm{Appr}(\aleph_\gamma, \aleph_\alpha, 2)$, the mapping σ is a set; hence $\aleph_\alpha$ is singular. If ϱ is a mapping on $\boldsymbol{\omega}_\alpha$ and if ϱ is not a set then the mapping τ on $\boldsymbol{\omega}_\gamma$ defined by

$\tau'\zeta = \varrho \upharpoonright \sigma'\zeta$ is not a set; every $\varrho \upharpoonright \sigma'\zeta$ is a set but $\bigcup(\mathbf{W}(\tau))$ is not. Hence for some y we have $\neg$ Appr $(\aleph_y, y, 2)$, a contradiction.

Relations were considered as functors, i.e. as "mappings" between semi-sets and sets. A similar approach can be adopted in connection with $\boldsymbol{b}$-relations:

5337. Definition (**TSS′**). Let $\boldsymbol{b}$ be a complete Boolean algebra. If r is a $\boldsymbol{b}$-relation such that $r \subseteq (y \times x) \times b$, then we define

$$r[\iota] = \{\langle\kappa, u\rangle, \langle\langle\kappa, \iota\rangle\, u\rangle \in r\},$$

for all $\iota \in x$.

5338. Lemma (**TSS′**). Let Z be a Boolean support on $\boldsymbol{b}$. If r is a $\boldsymbol{b}$-relation and $\sigma = r''Z$, then $\mathbf{Ext}_\sigma(\iota) = (r[\iota])''\, Z$ for all ι.

5339. Definition (**TSS′**). A $\boldsymbol{b}$-relation r is called a ***b**-exact functor on z* if r is a $\boldsymbol{b}$-mapping on z and if the following condition holds for all $\iota_1 \neq \iota_2$ in z:

$$\mathbf{sSim}_{\boldsymbol{b}}(\mathbf{D}(r[\iota_1]), \mathbf{D}(r[\iota_2]) = 0_{\boldsymbol{b}}\,.$$

5340. Lemma (**TSS′**). If Z is a Boolean support on $\boldsymbol{b}$ and r is a $\boldsymbol{b}$-exact functor on z, then $r''Z$ is an exact functor and $\mathbf{D}(r''Z) \subseteq z$.

Proof. Let ι_1, ι_2 be distinct elements of z and let $\sigma = r''Z$. We have $\mathbf{Ext}_\sigma(\iota_1) = (r[\iota_1])''\, Z$ and similarly for ι_2. By 4320, we have $\mathbf{Ext}_\sigma(\iota_1) \neq \mathbf{Ext}_\sigma(\iota_2)$; the rest is obvious.

5341. Definition (**TSS′**). We shall write $\mathrm{Beth}_{\boldsymbol{b}}(2, \aleph_\beta, \aleph_\gamma)$ if there exists a $\boldsymbol{b}$-exact functor r on $\boldsymbol{\omega}_\gamma$ such that $r \subseteq (\boldsymbol{\omega}_\beta \times \boldsymbol{\omega}_\gamma) \times b$. We shall write $\mathrm{Beth}_{\boldsymbol{b}}(\aleph_\alpha, \aleph_\beta, \aleph_\gamma)$ if there exists a $\boldsymbol{b}$-exact functor r on $\boldsymbol{\omega}_\gamma$ such that for each $\iota \in \boldsymbol{\omega}_\gamma$ $r[\iota]$ is a $\boldsymbol{b}$-mapping of $\boldsymbol{\omega}_\beta$ into $\boldsymbol{\omega}_\alpha$.

5342. Theorem (**TSS′**, S5, E1). If there is a total Boolean support on $\boldsymbol{b}$ then

1) $\mathrm{Beth}_{\boldsymbol{b}}(2, \aleph_\beta, \aleph_\gamma)$ implies $\aleph_\gamma \leqq \hat{\beth}(2, \aleph_\beta)$,
2) $\mathrm{Beth}_{\boldsymbol{b}}(\aleph_\alpha, \aleph_\beta, \aleph_\gamma)$ implies $\aleph_\gamma \leqq \hat{\beth}(\aleph_\alpha, \aleph_\beta)$.

Proof. 1) Suppose $\mathrm{Beth}_{\boldsymbol{b}}(2, \aleph_\beta, \aleph_\gamma)$. Then there exists an exact functor on $\boldsymbol{\omega}_\gamma$ such that every extension is included in $\boldsymbol{\omega}_\beta$. Hence $\aleph_\gamma \leqq \hat{\beth}(2, \aleph_\beta)$.
2) Similarly.

5343. Lemma (**TSS′**, S5, E1). If $\aleph_\alpha$ and $\aleph_\beta$ are absolute cardinals then $\aleph_\alpha^{\aleph_\beta} \leqq \hat{\beth}(\aleph_\alpha, \aleph_\beta)$ and $2^{\aleph_\beta} \leqq \hat{\beth}(2, \aleph_\beta)$. (Obvious.)

5344. LEMMA (**TSS**′, S5, E1). If there is a total Boolean support on $\boldsymbol{b}$ and if $b \approx \aleph_\gamma$ then

1) If $\aleph_\beta$ is an absolute cardinal and then $\hat{\beth}(2, \aleph_\beta) \leqq \aleph_\gamma^{\aleph_\beta}$.
2) If $\aleph_\alpha$ and $\aleph_\beta$ are absolute cardinals then

$$\hat{\beth}(\aleph_\alpha, \aleph_\beta) \leqq \aleph_\gamma^{\mathrm{Max}(\aleph_\alpha, \aleph_\beta)} .$$

Proof. 1) $\aleph_\gamma^{\aleph_\beta}$ is the number of all possible $\boldsymbol{b}$-sets $r \subseteq \boldsymbol{\omega}_\beta \times b$.
2) $\aleph_\gamma^{\mathrm{Max}(\aleph_\alpha, \aleph_\beta)}$ is the number of all possible $\boldsymbol{b}$-relations $r \subseteq (\boldsymbol{\omega}_\alpha \times \boldsymbol{\omega}_\beta) \times b$.

CHAPTER VI

SECTION 1

Independence of the Axiom of Continuum

We shall now illustrate the general theory by means of some examples. We shall demonstrate among other things the independence of the Axiom of Continuum.

a) Generalized Cantor algebras.

6101. DEFINITION (**TSS**′). Let $\aleph_\alpha \leqq \aleph_\beta$ be cardinal numbers; for any $x \in \boldsymbol{\omega}_\beta$ let $\boldsymbol{b}_x = \mathbf{at}(2)$ (see 2520) and let l be the ideal of all subsets of $\boldsymbol{\omega}_\beta$ of power less than $\aleph_\alpha$. We denote by $\mathbf{Cant}(\aleph_\alpha, \aleph_\beta)$ (or more briefly $\mathbf{Cant}_\alpha^\beta$) the complete Boolean algebra with base $\boldsymbol{b}_\alpha^\beta = \prod_{x \in \boldsymbol{\omega}_\beta} {}^l\boldsymbol{b}_x$.

6102. In the sequel we assume that $\aleph_\alpha$ is regular and that the axiom (Cont) holds.

6103. LEMMA (**TSS**′, E1, Cont).

(a) $\mathbf{v}(\mathbf{Cant}_\sigma^\beta) \leqq \aleph_\alpha$,

(b) $\boldsymbol{\mu}(\mathbf{Cant}_\alpha^\beta) \leqq \aleph_{\alpha+1}$.

Proof. (a) by Theorem 2523.

(b) by Theorems 2524, 2525, 2526 and the axiom Cont.

6104. THEOREM (**TSS**′, E1, Cont). Let Z be a total Boolean support on $\mathbf{Cant}_\alpha^\beta$, where $\alpha \leqq \beta$ and $\aleph_\alpha$ is regular.

(a) If $\gamma < \alpha$ then Appr $(\aleph_\gamma, \aleph_\delta, 2)$ for all δ,

(b) If $\gamma > \alpha$ then Appr $(\aleph_\delta, \aleph_\gamma, \aleph_\gamma)$ for all δ.

(c) Every $\aleph_\gamma$ is an absolute cardinal.

(d) For every γ $\mathbf{cf}(\aleph_\gamma) = \widehat{\mathbf{cf}}(\aleph_\gamma)$.

Proof. (a) See 5331 (1).

(b) See 5329 (1).

(c) See 5329 (3), 5331 (3).

(d) See 5329 (2), 5331 (4).

First we shall examine the case $\aleph_\alpha = \aleph_\beta$ for regular $\aleph_\alpha$.

6105. LEMMA (**TSS′**, E1, Cont). (1) If $\aleph_\alpha$ is regular then $\mathbf{Cant}_\alpha^\alpha \mid u$ is not $\aleph_\alpha - \aleph_\alpha - \aleph_\alpha$-distributive for any $u \neq 0_{\mathbf{Cant}}$.

(2) $\mathbf{Cant}_\alpha^\alpha$ has power $\aleph_{\alpha+1}$.

Proof. (1) Let f be a 1-1-mapping of $\boldsymbol{\omega}_\alpha \times \boldsymbol{\omega}_\alpha$ onto $\boldsymbol{\omega}_\alpha$; for $\xi, \eta < \boldsymbol{\omega}_\alpha$ let $u_{\xi\eta}$ be the function which assumes the value one at $f`\langle\xi, \eta\rangle$ and the value zero at $f`\langle\delta, \eta\rangle$ $(\delta < \xi)$, and is undefined otherwise. Clearly $u_{\xi,\eta} \in b_\alpha^\alpha$. We let $r = \{\langle\langle\xi, \eta\rangle, u_{\xi\eta}\rangle;\ \xi, \eta < \boldsymbol{\omega}_\alpha\}$; r is a $\mathbf{Cant}_\alpha^\alpha$-relation and for every $v \in b_\alpha^\alpha$ there exists η such that $f`\langle\xi, \eta\rangle$ is not a fixing point of v for any ξ. For such an η we have $v \wedge u_{\xi\eta} \neq 0_{\mathbf{Cant}_\alpha^\alpha}$ for every ξ. Hence no basic element is $\aleph_\alpha$-layered by r, and consequently no nonzero element is $\aleph_\alpha$-layered.

(2) The power of b_α^α is $\aleph_\alpha$; each element of $\mathbf{Cant}_\alpha^\alpha$ is a join of not more than $\aleph_\alpha$ disjoint basic elements, so that $\mathbf{Cant}_\alpha^\alpha \preccurlyeq \aleph_{\alpha+1}$. We have found a disjointed system of power $\aleph_\alpha$ of b_α^α; distinct subsets of this system determine distinct elements of $\mathbf{Cant}_\alpha^\alpha$ and hence $\mathbf{Cant}_\alpha^\alpha \approx \aleph_{\alpha+1}$.

6106. THEOREM (**TSS′**, E1, Cont). If $\aleph_\alpha$ is regular and Z is a total Boolean support on $\mathbf{Cant}_\alpha^\alpha$ then

1) $\gamma = \alpha \equiv \neg\, \mathrm{Appr}(\aleph_\gamma, \aleph_\gamma, \aleph_\gamma)$ for any γ ;

2) $\hat{\beth}(2, \aleph_\gamma) = \aleph_{\gamma+1}$ for any γ.

Proof. 1) By the preceding lemma and 5327 we have $\neg\, \mathrm{Appr}(\aleph_\alpha, \aleph_\alpha, \aleph_\alpha)$; if $\gamma \neq \alpha$ then we have $\mathrm{Appr}(\aleph_\gamma, \aleph_\gamma, \aleph_\gamma)$ by Theorem 6104.

2) By Theorem 6104, all cardinals are absolute; since $\mathbf{v}(\mathbf{Cant}_\alpha^\alpha) \geq \aleph_\alpha$, every semiset of power $< \aleph_\alpha$ is a set, and so $\hat{\beth}(2, \aleph_\gamma) = \beth(2, \aleph_\gamma) = \aleph_{\gamma+1}$ for each $\gamma < \alpha$. If $\gamma \geq \alpha$ then $\hat{\beth}(2, \aleph_\gamma) \geq \aleph_{\gamma+1}$; on the other hand, every part of $\aleph_\gamma$ is determined by a $\mathbf{Cant}_\alpha^\alpha$-set whose range is included in $\boldsymbol{\omega}_\gamma$. Since $\mathbf{Cant}_\alpha^\alpha \approx \aleph_{\alpha+1}$ and $\aleph_{\alpha+1}^{\aleph_\gamma} = \aleph_{\gamma+1}$, there are only $\aleph_{\gamma+1}$ such $\mathbf{Cant}_\alpha^\alpha$-sets. Consequently $\hat{\beth}(2, \aleph_\gamma) \leqq \aleph_{\gamma+1}$.

Suppose now that $\aleph_\alpha < \mathbf{cf}(\aleph_\beta)$ and that $\aleph_\alpha$ is regular.

6107. LEMMA (**TSS**′, E1, Cont). If $\aleph_\alpha < \mathbf{cf}(\aleph_\beta)$ and if $\aleph_\alpha$ is regular then

(a) $\aleph_\alpha \leqq \aleph_\gamma < \mathbf{cf}(\aleph_\beta) \rightarrow \text{Beth}_{\mathbf{Cant}_\alpha{}^\beta}(2, \aleph_\gamma, \aleph_\beta)$;
(b) $\mathbf{cf}(\aleph_\beta) \leqq \aleph_\gamma \leqq \aleph_\beta \rightarrow \text{Beth}_{\mathbf{Cant}_\alpha{}^\beta}(2, \aleph_\gamma, \aleph_{\beta+1})$.

Proof. (a) Let f be a 1-1-mapping of $\boldsymbol{\omega}_\gamma \times \boldsymbol{\omega}_\beta$ onto $\boldsymbol{\omega}_\beta$ and for each $\xi < \boldsymbol{\omega}_\gamma$ and $\eta < \boldsymbol{\omega}_\beta$, let $u_{\xi\eta}$ be a function which assumes the value 1 at $f`\langle\xi, \eta\rangle$ and is undefined elsewhere. Let $r = \{\langle\langle\xi, \eta\rangle, u_{\xi\eta}\rangle; \xi < \boldsymbol{\omega}_\gamma, \eta < \boldsymbol{\omega}_\beta\}$. It suffices to prove that if $\eta_1 \neq \eta_2$ then $\text{sSim}_{\mathbf{Cant}_\alpha{}^\beta}(\mathbf{D}(r[\eta_1]), \mathbf{D}(r[\eta_2])) = 0_{\mathbf{Cant}_\alpha{}^\beta}$. If $v \in b_\alpha^\beta$ then for some ξ neither $f`\langle\xi\eta_1\rangle$ nor $f`\langle\xi_1\eta_2\rangle$ is a fixing point of v; it follows that $v \wedge u_{\xi\eta_1} \neq v \wedge u_{\xi\eta_2}$, since $f`\langle\xi, \eta_1\rangle$ is a fixing point of the left-hand element but not of the right-hand element. Hence v distinguishes $r[\eta_1]$ and $r[\eta_2]$, and so $\text{sSim}_{\mathbf{Cant}_\alpha{}^\beta}(\mathbf{D}(r[\eta_1]), \mathbf{D}(r[\eta_2])) = 0_{\mathbf{Cant}_\alpha{}^\beta}$.

(b) Let c be the set of all subsets of $\boldsymbol{\omega}_\beta$ of power $\aleph_\gamma$; clearly $c \sim \aleph_{\beta+1}$. For $u, v \in c$ let $\langle u, v\rangle \in t \equiv (u - v) \cup (v - u) \prec \aleph_\alpha$; t is an equivalence and for every $u \in c$ we have $\mathbf{Ext}_t(u) \approx \aleph_\beta$. Hence there exists a set d of pairwise nonequivalent elements of c such that $d \approx \aleph_{\beta+1}$. We choose a well-ordering of d of type $\boldsymbol{\omega}_{\beta+1}$ and for every $u \in d$ we chose a well-ordering of u of type $\boldsymbol{\omega}_\gamma$; let $f`\langle\xi, \eta\rangle$ be the $\xi$-th element of the $\eta$-th set in $d$. If we let $u_{\xi\eta} = \{\langle 1, f`\langle\xi, \eta\rangle\rangle\}$ and $r = \{\langle\langle\xi, \eta\rangle, u_{\xi\eta}\rangle; \xi < \boldsymbol{\omega}_\gamma, \eta < \boldsymbol{\omega}_{\beta+1}\}$ then for $\eta_1 \neq \eta_2$ and $v \in b_\alpha^\beta$ there exists ξ such that $f`\langle\xi, \eta_1\rangle \neq f`\langle\xi, \eta_2\rangle$ and such that none of these elements is a fixing point of v. The rest of argument is similar to that sub (a).

6108. THEOREM (**TSS**′, E1, Cont). Let $\aleph_\alpha$ be a regular cardinal such that $\aleph_\alpha < \mathbf{cf}(\aleph_\beta)$. If Z is a total Boolean support on $\mathbf{Cant}_\alpha^\beta$ then

(a) $\aleph_\gamma < \aleph_\alpha \rightarrow \hat{\beth}(2, \aleph_\gamma) = \aleph_{\gamma+1}$,
(b) $\aleph_\alpha \leqq \aleph_\gamma < \mathbf{cf}(\aleph_\beta) \rightarrow \hat{\beth}(2, \aleph_\gamma) = \aleph_\beta$,
(c) $\mathbf{cf}(\aleph_\beta) \leqq \aleph_\gamma \leqq \aleph_\beta \rightarrow \hat{\beth}(2, \aleph_\gamma) = \aleph_{\beta+1}$,
(d) $\aleph_\beta \leqq \aleph_\gamma \rightarrow \hat{\beth}(2, \aleph_\gamma) = \aleph_{\gamma+1}$.

Proof. (a) is proved in the same way as (2) in Theorem 6106.

(b) Since $b_\alpha^\beta \approx \aleph_\beta$ and since every element of $\mathbf{Cant}_\alpha^\beta$ is a join of at most $\aleph_\alpha$ disjoint basic elements we have $\mathbf{Cant}_\alpha^\beta \approx \aleph_\beta$. By the lemma 6107(a) $\hat{\beth}(2, \aleph_\gamma) \geq \aleph_\beta$. Since $\aleph_\beta^{\aleph_\gamma} = \aleph_\beta$ it follows from $\mathbf{Cant}_\alpha^\beta \approx \aleph_\beta$ that there are $\aleph_\beta$ $(\mathbf{Cant}_\alpha^\beta)$-sets with range $\subseteq \boldsymbol{\omega}_\gamma$; hence $\hat{\beth}(2, \aleph_\gamma) \leqq \aleph_\beta$.

(c) By Lemma 6107(b) we have $\hat{\beth}(2, \aleph_\gamma) \geq \aleph_{\beta+1}$. Since $\aleph_\beta^{\aleph_\gamma} = \aleph_{\beta+1}$ it follows from $\mathbf{Cant}_\alpha^\beta \approx \aleph_\beta$ that $\hat{\beth}(2, \aleph_\gamma) = \aleph_{\beta+1}$.

(d) If $\aleph_\beta \leqq \aleph_\gamma$ then $\hat{\beth}(2, \aleph_\gamma) \geq 2^{\aleph_\gamma} = \aleph_{\gamma+1}$; since $\aleph_\beta^{\aleph_\gamma} = \aleph_{\gamma+1}$ we have $\hat{\beth}(2, \aleph_\gamma) = \aleph_{\gamma+1}$.

6109. Theorem (**TSS′**, E1, Cont). If $\aleph_\alpha$ and $\aleph_\beta$ satisfy the hypothesis of the preceding theorem then

(a) $\mathbf{cf}(\aleph_\gamma) < \aleph_\alpha \rightarrow \hat{\gimel}(\aleph_\gamma) = \gimel(\aleph_\gamma) = \aleph_{\gamma+1}$,
(b) $\aleph_\alpha \leqq \mathbf{cf}(\aleph_\gamma) < \mathbf{cf}(\aleph_\beta) \rightarrow \hat{\gimel}(\aleph_\gamma) = \mathbf{Max}(\aleph_\beta, \aleph_{\gamma+1})$,
(c) $\mathbf{cf}(\aleph_\beta) \leqq \mathbf{cf}(\aleph_\gamma) \rightarrow \hat{\gimel}(\aleph_\gamma) = \mathbf{Max}(\aleph_{\beta+1}, \aleph_{\gamma+1})$.

Proof. (a) is obvious, since every semiset of power $<\aleph_\alpha$ is a set.

(b) $\hat{\gimel}(\aleph_\gamma) \geq \gimel(\aleph_\gamma) = \aleph_{\gamma+1}$; $\hat{\gimel}(\aleph_\gamma) \geq \hat{\beth}(2, \mathbf{cf}(\aleph_\gamma)) = \aleph_\beta$. Hence $\hat{\gimel}(\aleph_\gamma) \geq \mathbf{Max}(\aleph_\beta, \aleph_{\gamma+1})$. Conversely, for every semiset σ which maps $\mathbf{cf}(\aleph_\gamma)$ into $\aleph_\gamma$ there exists a $\mathbf{Cant}_\alpha^\beta$-function $r \subseteq (\boldsymbol{\omega}_\gamma \times \mathbf{cf}(\boldsymbol{\omega}_\gamma)) \times \mathbf{Cant}_\alpha^\beta$ such that $\sigma = r``Z$. Since $\boldsymbol{\mu}(\mathbf{Cant}_\alpha^\beta) \leqq \aleph_{\alpha+1}$ we have $\mathbf{D}(r) \preccurlyeq \mathbf{Max}(\aleph_\alpha, \mathbf{cf}(\aleph_\gamma)) = \mathbf{cf}(\aleph_\gamma)$. Each $u \in \mathbf{D}(r)$ is a join of at most $\aleph_\alpha$ disjoint basic elements, each having less than $\aleph_\alpha$ fixing points. If we choose a set q_u of at most $\aleph_\alpha$ basic elements for every $u \in \mathbf{D}(r)$ such that $u = \bigvee q_u$, and if we let t be the set of all fixing points in $\bigcup_{u \in \mathbf{D}(r)} q_u$, then $t \preccurlyeq \mathbf{cf}(\aleph_\gamma)$ and r is a $\boldsymbol{b}^{(t)}$-function, where $\boldsymbol{b}^{(t)}$ is the complete Boolean algebra generated by the base $\prod_{x \in \boldsymbol{\omega}_\beta}^{1/t} \mathbf{at}(2)$. The cardinality of $b^{(t)}$ is at most $\mathbf{cf}(\boldsymbol{\omega}_\gamma)$, so that the number of all $\boldsymbol{b}^{(t)}$-functions $r \subseteq (\boldsymbol{\omega}_\gamma \times \mathbf{cf}(\boldsymbol{\omega}_\gamma)) \times b^{(t)}$ is at most $\mathbf{cf}(\aleph_\gamma)^{\aleph_\gamma} = \aleph_{\gamma+1}$; the number of all subsets of $\boldsymbol{\omega}_\beta$ of power at most $\mathbf{cf}(\aleph_\gamma)$ is $\aleph_\beta^{\mathbf{cf}(\aleph_\gamma)} = \aleph_\beta$. Hence we have found a set of $\mathbf{cant}_\alpha^\beta$-functions of power at most $\mathbf{Max}(\aleph_\beta, \aleph_{\gamma+1})$ such that every semiset-mapping of $\mathbf{cf}(\boldsymbol{\omega}_\gamma)$ into $\boldsymbol{\omega}_\gamma$ is coded by at least one of these $\mathbf{Cant}_\alpha^\beta$-functions. Hence

$$\hat{\gimel}(\aleph_\gamma) \leqq \mathbf{Max}(\aleph_\beta, \aleph_{\gamma+1}).$$

(c) Supposing $\mathbf{cf}(\aleph_\beta) \leqq \mathbf{cf}(\aleph_\gamma)$ we have

$$\hat{\gimel}(\aleph_\gamma) \geq \gimel(\aleph_\gamma) = \aleph_{\gamma+1},$$
$$\hat{\gimel}(\aleph_\gamma) \geq \hat{\beth}(2, \mathbf{cf}(\aleph_\gamma)) \geq \aleph_{\beta+1},$$

and $\hat{\gimel}(\aleph_\gamma) \leqq \hat{\beth}(\aleph_\gamma, \aleph_\gamma) = \mathbf{Max}(\aleph_{\beta+1}, \aleph_{\gamma+1})$. Consequently $\hat{\gimel}(\aleph_\gamma) = \mathbf{Max}(\aleph_{\beta+1}, \aleph_{\gamma+1})$.

6110. Definition (**TSS′**).

$$\mathrm{Ap}(X, \aleph_\alpha) \equiv (\forall f)(\mathrm{Un}(f, \boldsymbol{\omega}_\alpha, \boldsymbol{\omega}_\alpha) \rightarrow$$
$$\rightarrow (\exists r \in X)(f \subseteq r \,\&\, (\forall \iota)(\mathbf{Ext}_r(\iota) \prec \aleph_\alpha)).$$

The reader may verify as an exercise that the following is provable in **TS**, D3, $\mathrm{Mcl}_{\mathrm{cn}}(\mathbf{M})$, $\partial\mathbf{TSS}/\partial\mathfrak{St}(\mathbf{M})$:

$$\mathrm{Ap}(\mathbf{M}, \aleph_\alpha) \equiv \mathrm{Appr}^{\mathfrak{St}(\mathbf{M})}(\aleph_\alpha, \aleph_\alpha, \aleph_\alpha)\,.$$

Metamathematical cvnsequences.

6111. METATHEOREM. Let λ, μ be good definitions of cardinal numbers in **TSS**, D3, E2; denote the cardinal numbers satisfying λ and μ by $\aleph_\alpha$ and $\aleph_\beta$ respectively.

(1) Suppose TSS, D3, E2 $\vdash$ ($\aleph_\alpha$ is regular). Then the axiom

(Appr_α) $\mathrm{Mcl}_{\mathrm{cn}}(\mathbf{M})$ & ($\aleph_\alpha$ is the only cardinal such that $\neg\,\mathrm{Ap}(\mathbf{M}, \aleph_\alpha)$)

is consistent with **TS**, D3, E2, Cont; in fact, $\mathfrak{Up}$ is a model in the following diagram:

TS, D3, E2, Cont, Mcl (M), (Appr_α)	⟶	TSS, D3, E2, Cont, $(S\,\mathrm{Cant}_\alpha^\alpha)$ + upw

In particular the following is consistent with **TS**, D3, E2, Cont:

$$\mathrm{Mcl}_{\mathrm{cn}}(\mathbf{L})\ \&\ (\aleph_\alpha \text{ is the only cardinal such that } \neg\,\mathrm{Ap}(\mathbf{L}, \aleph_\alpha))\,.$$

(2) Suppose **TSS**, D3, E2 $\vdash$ ($\aleph_\alpha$ is regular and $\mathbf{cf}(\aleph_\beta) > \aleph_\alpha$). Then the axiom

$$(\mathrm{Gim}_\alpha^\beta)\quad \left\{\begin{array}{l} \text{(a) } \mathbf{cf}(\aleph_\gamma) < \aleph_\alpha \rightarrow \gimel(\aleph_\gamma) = \aleph_{\gamma+1}\,, \\ \text{(b) } \aleph_\alpha \leqq \mathbf{cf}(\aleph_\gamma) < \mathbf{cf}(\aleph_\beta) \rightarrow \gimel(\aleph_\gamma) = \\ \qquad = \mathbf{Max}(\aleph_\beta, \aleph_{\gamma+1})\,, \\ \text{(c) } \mathbf{cf}(\aleph_\beta) \leqq \mathbf{cf}(\aleph_\gamma) \rightarrow \gimel(\aleph_\gamma) = \\ \qquad = \mathbf{Max}(\aleph_{\beta+1}, \aleph_{\gamma+1}) \end{array}\right\} \text{for each } \gamma$$

is consistent with **TS**, D3, E2; in fact, $\mathfrak{Up}$ is a model in the following diagram:

TS, D3, E2, $(\mathrm{Gim}_\alpha^\beta)$	⟶	TSS, D3, E2, Cont $(S\,\mathrm{Cant}_\alpha^\beta)$ + upw

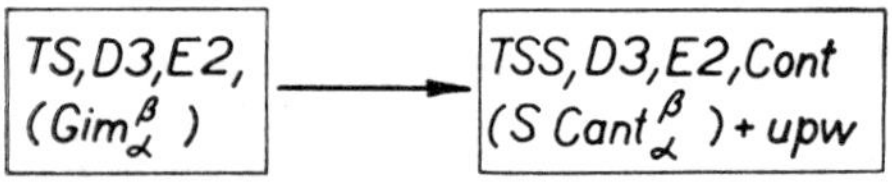

Demonstration. (1) We proceed in **TSS**, D3, E2, Cont, (S $\mathbf{Cant}_{\alpha}^{\alpha}$) + upw. By 6104 all cardinals are absolute and therefore $\mathrm{Mcl}_{\mathbf{cn}}^{*}$ (**Tor**). By 5142 and 5136 we have $\beth^*(2^*, \aleph_{\gamma*}^*) = \mathbf{Imb}''\,(\hat{\beth}(2, \aleph_\gamma)) = \aleph_{\gamma*+1*}^*$ for every γ and γ^* such that $\gamma^* = \mathbf{Imb}''\,\gamma$; i.e. (Cont)* holds. Set $\mathbf{Imb}''\,\aleph_\alpha = \aleph_{\alpha*}^*$. We have $\lambda(\aleph_\alpha)$, hence by 5126 we have $\lambda^{\square}(\aleph_{\alpha*}^*)$ (we use $\partial\mathbf{TSS}/\partial\mathfrak{St}^*$ (**Tor**) for the moment) but since λ is a good definition we have $(\mathrm{Mcl}_{\mathbf{cn}}^{*}(\mathbf{Tor})\ \&\ \lambda^{\square}(\aleph_{\alpha*}^*)) \to$ $\to \lambda^*(\aleph_{\alpha*}^*)$, i.e. $\lambda^*(\aleph_{\alpha*}^*)$ holds. Finally, for any γ and γ^* such that $\gamma^* = \mathbf{Imb}''\,\gamma$, $\mathrm{Ap}^*(\mathbf{Tor}^*, \aleph_{\gamma*}^*)$ is equivalent to

$$(1)\quad (\forall f^* \subseteq \mathbf{Tor})\,(\mathrm{Un}^*(f^*, \aleph_{\gamma*}^*, \aleph_{\gamma*}^*) \to (\exists r^* \in \mathbf{Tor})\,(f^* \subseteq r^*\ \&$$

$$\&\ (\forall \iota^*)\,(\mathbf{Ext}_{r*}^*(\iota^*) \prec^* \aleph_{\gamma*}^*)))$$

and (1) is obviously equivalent to

$$(2)\quad (\forall\sigma)\,(\mathrm{Un}(\sigma, \aleph_\gamma, \aleph_\gamma) \to (\exists r)\,(\sigma \subseteq r\ \&\ (\forall\iota)\,(\mathbf{Ext}_r(\iota)\ \hat{\precsim}\ \aleph_\gamma)))\,.$$

In (2) we may equivalently replace $\hat{\precsim}$ by $\prec$ because cardinals are absolute; hence (2) is equivalent to $\mathrm{Appr}(\aleph_\gamma, \aleph_\gamma, \aleph_\gamma)$ and hence by 6106 to $\aleph_\gamma \neq \aleph_\alpha$. But this is equivalent to $\aleph_{\gamma*}^* \neq \aleph_{\alpha*}^*$; this proves $(\mathrm{Appr}_{\alpha*}^*)$. We have thus demonstrated that $\mathfrak{Up}$ is a model in the diagram sub (1).

(2) We proceed in **TSS**, D3, E2, Cont, (S $\mathbf{Cant}_{\alpha}^{\beta}$) + upw. By 5138 and 6104 (D3)* and (E2)* hold and cardinals and cofinalities are absolute; hence if we set $\aleph_{\alpha*}^* = \mathbf{Imb}''\,\aleph_\alpha$ and $\aleph_{\beta*}^* = \mathbf{Imb}''\,\aleph_\beta$ we have $\lambda^*(\aleph_{\alpha*}^*)$ and $\mu^*(\aleph_{\beta*}^*)$ (cf. the proof of (1)). Let γ be an arbitrary ordinal number and set $\aleph_{\gamma*}^* =$ $= \mathbf{Imb}''\,\aleph_\gamma$. Then we have $\mathbf{cf}^*(\aleph_{\gamma*}^*) <^* \aleph_{\alpha*}^* \to \mathbf{cf}(\aleph_\gamma) < \aleph_\alpha \to \hat{\gimel}(\aleph_\gamma) =$ $= \aleph_{\gamma+1} \to \gimel^*(\aleph_{\gamma*}^*) = \aleph_{\gamma*+1*}^*$, and similarly for (b), (c) in $(\mathrm{Gim}_{\alpha*}^{*\beta*})$ (cf. 5144 and 6109). Thus we obtain $(\mathrm{Gim}_{\alpha*}^{*\beta*})$. This completes the demonstration. (For the corollary in (1) see 5138; **L** denotes the class **Cstr**.)

It is of some interest that using $\mathfrak{Up}$ as a model in **TSS**, D3, E2, (S $\mathbf{Cant}_1^1$) + + upw we can prove the consistency of $2^{\aleph_0} = \aleph_1$ without knowing anything about the model-class **Cstr** (0).

6112. METATHEOREM. $\mathfrak{Up}$ is a model in the following diagram:

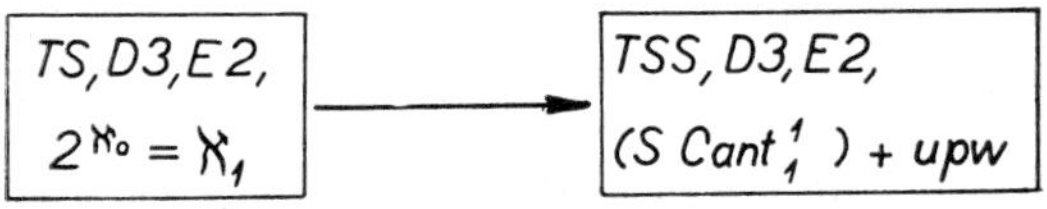

Demonstration. We proceed in the latter theory. Denote $\mathbf{Cant}_1^1$ by $\boldsymbol{b}$. Obviously $\mathbf{v}(\boldsymbol{b}) \leqq \aleph_1$; hence every subsemiset of $\boldsymbol{\omega}_0$ is a set and $\aleph_1$ is absolute. It suffices to prove the following: If $\varrho \subseteq \boldsymbol{\omega}_0 \times \boldsymbol{\omega}_\alpha$ is an exact functor then $\aleph_\alpha \hat{\lessapprox} \aleph_1$. First we prove $2^{\aleph_0} \hat{\approx} \aleph_1$. For this purpose let i be a one--one mapping of $\boldsymbol{\omega}_0 \times \boldsymbol{\omega}_1$ onto $\boldsymbol{\omega}_1$ and write $i_{n\xi}$ instead of $i'\langle n, \xi\rangle$. For $r \in \mathbf{exp}(2, \boldsymbol{\omega})$ we define $u_{\xi r}$ as the element of $\mathbf{Cant}_1^1$ such that $\mathbf{D}(u_{\xi r}) = i''(\boldsymbol{\omega} \times \{\xi\})$ and $u'_{\xi r} i_{n\xi} = r'n$.

Further define a $\boldsymbol{b}$-set f such that $\mathbf{D}(f) = \{u_{\xi r};\ \xi \in \boldsymbol{\omega}_1 \ \&\ r \in \mathbf{exp}(2, \boldsymbol{\omega})\}$ and $f'u_{\xi r} = \langle r, \xi\rangle$. If Z is a total Boolean support on $\boldsymbol{b}$ then f''Z is a semiset mapping of $\boldsymbol{\omega}_1$ onto $\mathbf{exp}(2, \boldsymbol{\omega})$ (exercise).

Now if σ is an exact functor which is a subsemiset of $\boldsymbol{\omega} \times \boldsymbol{\omega}_\alpha$ such that $\mathbf{D}(\sigma) = \boldsymbol{\omega}_\alpha$, then, for every ι, $\mathbf{Ext}_\sigma(\iota)$ is a subset of $\boldsymbol{\omega}$; thus the mapping ϱ which associates with every ι the set $\mathbf{Ext}_\sigma(\iota)$ is a one-to-one mapping of $\boldsymbol{\omega}_\alpha$ into $\mathbf{P}(\boldsymbol{\omega})$; hence $\aleph_\alpha \hat{\lessapprox} 2^{\aleph_0} \hat{\lessapprox} \aleph_1$, i.e. $\beth^*(2^*, \aleph_0^*) = \aleph_{1*}^*$, q.e.d.

6113. Remark. In the same way one can prove the following in **TSS**, D3, E2: if there is a total Boolean support on $\mathbf{Cant}_{\alpha+1}^{\alpha+1}$ then every subsemiset of $\boldsymbol{\omega}_\alpha$ is a set and $\hat{\beth}(2, \aleph_\alpha) = \aleph_{\alpha+1}$. To derive metamathematical consequences, one must be careful about the definition of $\aleph_\alpha$; e.g. it suffices that $\aleph_\alpha$ be defined by a definition λ such that **TS**, D3, E2, Mcl$(\mathbf{M})$, $\partial\mathbf{TSS}/\partial\mathfrak{St}(\mathbf{M}) \vdash (\forall\gamma)\,[(\forall\delta < \aleph_\gamma)(\mathrm{Card}(\delta) \equiv \mathrm{Card}^\square(\delta))] \rightarrow \lambda(\aleph_\gamma) \equiv \lambda^\square(\aleph_\gamma)]$.

6114. Remark. The reader may establish the following as an exercise: if $\aleph_\alpha$ is regular and $\aleph_\alpha = \mathbf{cf}(\aleph_\beta) < \aleph_\beta$ then $\hat{\beth}$ and $\hat{\gimel}$ satisfy the same equalities if we assume the existence of a total support on $\mathbf{Cant}_\alpha^{\beta+1}$ as if we assume the existence of a total support on $\mathbf{Cant}_\alpha^\beta$. However, if there is a total support on $\mathbf{Cant}_\alpha^{\beta+1}$ then there is no total support on any algebra with a base of power less than $\aleph_{\beta+1}$, whereas if there is a total support on $\mathbf{Cant}_\alpha^\beta$ then there is a Boolean support on an algebra with a base of power $\aleph_\beta$, since $\mathbf{Cant}_\alpha^\beta$ has a base of power $\aleph_\beta$.

b) *Simultaneous violation of the continuum hypothesis.*

6115. DEFINITION. Let f be a function whose domain consists of all ordinals α less than a fixed ordinal λ and such that $\aleph_\alpha$ is regular. If, for all $\alpha, \beta \in \mathbf{D}(f)$, we have

1) $f'\alpha \leqq f'\beta$ whenever $\alpha < \beta$,

2) $\aleph_\alpha < \mathbf{cf}(\aleph_{f'\alpha})$,

then we call f a *continuum function* on λ.

The algebra **PCant** (f) *corresponding* to the continuum function f is given by the base

$$\mathbf{b}(f) = \prod_{x \in \mathbf{D}(f)}^{l} b_{\alpha}^{f'\alpha},$$

where l is the ideal of all subsets of $\mathbf{D}(f)$ such that $x \cap \beta \prec \aleph_\beta$ for every regular cardinal $\aleph_\beta < \lambda$. Clearly l is an ideal; the condition $x \cap \beta \prec \aleph_\beta$ is nontrivial only if $\beta = \aleph_\beta$.*). For the definition of $b_\alpha^{f'\alpha}$ see 6101.

6116. We shall show that in **TSS'**, E1, Cont, the existence of a total support on **PCant** (f) implies $\hat{\beth}(2, \aleph_\alpha) = \aleph_{f'\alpha}$ for all $\alpha \in \mathbf{D}(f)$.

6117. LEMMA (**TSS'**, E1, Cont). Let f be a continuum-function on λ and let $\alpha \in \mathbf{D}(f)$. If

$$\mathbf{b}_1(f, \alpha) = \prod_{\substack{x \leqq \alpha \\ x \in \mathbf{D}(f)}}^{l/(\alpha+1)} b_x^{f'x},$$

and

$$\mathbf{b}_2(f, \alpha) = \prod_{\substack{x > \alpha \\ x \in \mathbf{D}(f)}}^{l/(\mathbf{D}(f)-(\alpha+1))} b_x^{f'x},$$

then

a) $\mathbf{b}(f)$ is isomorphic to $\mathbf{b}_1(f, \alpha) \odot \mathbf{b}_2(f, \alpha)$;
b) every exclusive system in $\mathbf{b}_1(f, \alpha)$ has power at most $\aleph_\alpha$;
c) $\mathbf{b}_2(f, \alpha)$ is $\aleph_\alpha$-multiplicative;
d) the algebra with base $\mathbf{b}_1(f, \alpha)$ has power at most $\aleph_{f'\alpha}$.

Proof. a) Obvious – see 2529.

b) Since $b_\alpha^{f'\alpha}$ is a product of $\aleph_{f'\alpha}$ copies of the algebra **at** (2), it follows by 2533 that $\mathbf{b}_1(f, \alpha)$ is a base of the same algebra as $\prod_{x \in s}^{j}$ **at** (2), where $s = \{\langle \gamma, \delta \rangle; \delta \leqq \alpha \,\&\, \delta \in \mathbf{D}(f) \,\&\, \gamma < \aleph_{f'\delta}\}$ and $x \in j \equiv .\, x \subseteq s \,\&\, \mathbf{D}(x) \in l/(\alpha + 1) \,\&\, (\forall \delta \in \mathbf{D}(x))\, \mathbf{Ext}_x(\delta) \prec \aleph_\delta)$. If $\alpha = \xi + 1$ then (b) follows from Theorem 2524, since $\mathbf{Norm}(j) = \aleph_{\xi+1}$ and since every exclusive system in **at** (2) has power at most 2; if α is a limit number then (b) follows from Theorem 2526, since $\mathbf{Norm}(j) = \aleph_\alpha$ and since every exclusive system in **at** (2) has power at most 2.

c) This follows from Theorem 2523, since the ideal $l/(\lambda - (\alpha + 1))$ is $\aleph_\alpha$-additive and since $b_\alpha^{f'\alpha}$ is $\aleph_\alpha$-multiplicative.

*) We mention that the existence of regular cardinals such that $\aleph_\beta = \beta$ is not provable in the theory of sets; such cardinals are called weakly inaccessible and we shall not treat them in this book.

d) By easy calculations, $\mathbf{b}_1(f, \alpha) \approx \aleph_{f'\alpha}$. Hence the assertion follows by b).

6118. THEOREM (**TSS′**, E1, Cont). Let f be a continuum function and let Z be a total Boolean support on $\mathbf{PCant}(f)$. Then

a) all cardinals are absolute,

b) $\hat{\beth}(2, \aleph_\alpha) = \aleph_{f'\alpha}$ for all $\alpha \in \mathbf{D}(f)$.

Proof. If $\aleph_\lambda$ is regular then, using the same argument as sub b) in the preceding lemma, we can prove that in $\mathbf{b}(f)$ – and hence in $\mathbf{PCant}(f)$ – the number of disjoint elements is at most $\aleph_\lambda$, so that the cardinals greater than $\aleph_\lambda$ are absolute. If $\aleph_\lambda$ is singular then, using the estimate 2524, we see that every exclusive system has power at most $\aleph_{\lambda+1}$, so that all cardinals greater than $\aleph_{\lambda+1}$ are absolute. Hence it is sufficient to prove that $\aleph_\alpha \mathrel{\hat{\approx}} \aleph_{\alpha+1}$ cannot hold for $\alpha \leqq \lambda$; the case $\alpha = \lambda$ is to be investigated only if $\aleph_\lambda$ is singular.

If $\alpha < \lambda$ and if $\aleph_\alpha$ is regular then $\boldsymbol{\mu}(\mathbf{b}_1(f, \alpha)) \leqq \aleph_{\alpha+1} \leqq \mathbf{v}(\mathbf{b}_2(f, \alpha))$, so that by 5333 $\mathrm{Un}_2(\sigma, \aleph_\alpha, \aleph_{\alpha+1})$ implies $\mathrm{Dep}(\sigma, Z_1)$ where $Z = Z_1 \odot Z_2$ and Z_1 is a support on $\mathbf{b}_1(f, \alpha)$. On the other hand, $\aleph_{\alpha+1}$ is absolute in the sense of $\mathfrak{Supp}(Z_1)$ (because $\boldsymbol{\mu}(\mathbf{b}_1(f, \alpha)) = \aleph_{\alpha+1}$), a contradiction.

If $\alpha \leqq \lambda$ and if $\aleph_\alpha$ is singular then let $\aleph_\xi = \hat{\mathbf{cf}}(\aleph_{\alpha+1})$. If $\aleph_{\alpha+1} \mathrel{\hat{\approx}} \aleph_\alpha$ then $\aleph_\xi < \aleph_\alpha$ by 5145 and 5134(a). Since $\boldsymbol{\mu}(\mathbf{b}_1(f, \xi)) \leqq \aleph_{\xi+1} \leqq \mathbf{v}(\mathbf{b}_2(f, \xi))$, we have $\hat{\mathrm{Conf}}(\aleph_\xi, \aleph_{\alpha+1})$ in the sense of $\mathfrak{Supp}(Z_1)$. Hence $\mathbf{b}_1(f, \xi)$ is not $\aleph_\xi - \aleph_{\alpha+1} - \aleph_{\alpha+1}$-distributive contradicting $\boldsymbol{\mu}(\mathbf{b}_1(f, \xi)) \leqq \aleph_{\xi+1}$ (see 5313 and 5328). This completes the proof of a).

b) If $\alpha \in \mathbf{D}(f)$ then every subsemiset of $\boldsymbol{\omega}_\alpha$ depends on Z_1 and we have $\hat{\beth}(2, \aleph_\alpha) \geq \aleph_{f'\alpha}$ in the sense of $\mathfrak{Supp}(Z_1)$; for, if ϱ is a support on $\boldsymbol{b}_\alpha^{f'\alpha}$ then $\hat{\beth}(2, \aleph_\alpha) \geq \aleph_{f'\alpha}$ holds in the sense of $\mathfrak{Supp}(\varrho)$. On the other hand, as in 6108 sub (b), Lemma 6117(d) yields $\hat{\beth}(2, \aleph_\alpha) \leqq \aleph_{f'\alpha}$ in the sense of $\mathfrak{Supp}(Z_1)$ and, by absoluteness of cardinals, we have $\hat{\beth}(2, \aleph_\alpha) \leqq \aleph_{f'\alpha}$.

6119. METATHEOREM. Let $\varphi(x)$ be a set formula such that

$$\mathbf{TSS}'' \vdash (\exists x)\, \varphi(x) \,\&\, (\forall x)\,(\varphi(x)) \to x \text{ is a continuum function})\,,$$

$$\mathbf{TSS}'', \mathrm{FMcl}(\mathbf{M}), \partial\mathbf{TSS}/\partial\mathfrak{St}(\mathbf{M}) \vdash (\forall x \in \mathbf{M})\,(\varphi(x) \equiv \varphi^{\square}(x))\,.$$

Then $\mathfrak{Up}$ is a model in the following diagram:

where (Powers$_{\mathbf{f}}$) is the axiom $(\forall\alpha \in \mathbf{D}(\mathbf{f}))\,(2^{\aleph_\alpha} = \aleph_{\mathbf{f}'\alpha})$. Hence (Powers$_{\mathbf{f}}$) is consistent with **TS**, D3, E2.

c) Collapsing algebras.

6120. DEFINITION (**TSS**$'$). Let $\aleph_\alpha \leqq \aleph_\beta$ be cardinal numbers, let $\boldsymbol{c}_x = \mathbf{at}(\aleph_\beta)$ for every $x \in \boldsymbol{\omega}_\alpha$ and let l be the ideal of all subsets of $\boldsymbol{\omega}_\alpha$ of power less than $\aleph_\alpha$. We denote by $\mathbf{Coll}(\aleph_\alpha, \aleph_\beta)$ or simply $\mathbf{Coll}_\alpha^\beta$, the complete Boolean algebra with base $\boldsymbol{c}_\alpha^\beta = \prod^l_{x \in \boldsymbol{\omega}_\alpha} \boldsymbol{c}_x$.

We shall investigate the case when $\aleph_\alpha$ is regular and less than or equal to $\mathbf{cf}(\aleph_\beta)$.

6121. LEMMA (**TSS**$'$, E1, Cont).

(a) $\mathbf{v}(\mathbf{Coll}_\alpha^\beta) \geq \aleph_\alpha$;

(b) $\boldsymbol{\mu}(\mathbf{Coll}_\alpha^\beta) \leq \aleph_{\beta+1}$.

Proof. (a) follows from Theorem 2523; (b) follows from the fact that $\boldsymbol{c}_\alpha^\beta$ has power $\aleph_\beta$, so that every disjointed system has power at most $\aleph_\beta$.

6122. LEMMA (**TSS**$'$, E1, Cont). $\mathbf{Coll}_\alpha^\beta$ is not $\aleph_\alpha - \aleph_\beta - \aleph_\beta$-distributive.

Proof. For $\xi < \boldsymbol{\omega}_\alpha$ and $\eta < \boldsymbol{\omega}_\beta$ let $u_{\eta\xi}{}'\xi = \eta$ and let $u_{\eta\xi}$ be undefined elsewhere; clearly, $u_{\eta\xi} \in \boldsymbol{c}_\alpha^\beta$. Furthermore, we let $r = \{\langle\langle\eta, \xi\rangle, u_{\eta\xi}\rangle; \xi \in \boldsymbol{\omega}_\alpha, \eta \in \boldsymbol{\omega}_\beta\}$. Since $u_{\eta_1\xi} \wedge u_{\eta_2\xi} = 0_{\mathrm{Coll}_\alpha^\beta}$ whenever $\eta_1 \neq \eta_2$, it follows that r is a $\mathbf{Coll}_\alpha^\beta$-mapping. Let $v \in \boldsymbol{c}_\alpha^\beta$ be a basic element and let ξ be such that $u_{\eta\xi}$ is not a fixing point of v for any η. Then $v \wedge u_{\eta\xi}$ $(= v \cup u_{\eta\xi})$ is nonzero for each $\eta \in \boldsymbol{\omega}_\beta$. Hence no element is $\aleph_\beta$-layered.

6123. THEOREM (**TSS**$'$, E1, Cont). Assuming the existence of a total Boolean support on $\mathbf{Coll}_\alpha^\beta$ we have $\aleph_\alpha \approx \aleph_\beta$; moreover, if $\gamma \leqq \alpha$ or $\gamma \geqq \beta + 1$ then $\aleph_\gamma$ is an absolute cardinal.

Proof. Let Z be a total Boolean support on $\mathbf{Coll}_\alpha^\beta$. Let r have the same meaning as in the previous proof. Put $\sigma = r''Z$. Then σ is a mapping of $\boldsymbol{\omega}_\alpha$ into $\boldsymbol{\omega}_\beta$; we prove that σ is a mapping onto $\boldsymbol{\omega}_\beta$. Let $\eta < \boldsymbol{\omega}_\beta$, $v \in \boldsymbol{c}_\alpha^\beta$. Then there exists $\xi < \boldsymbol{\omega}_\alpha$ such that $v \wedge u_{\eta\xi} \neq 0$ (see above). In other words, $\bigvee_{\xi \in \omega_\alpha} u_{\eta\xi} = 1_{\mathrm{Coll}_\alpha^\beta}$. Hence there is a ξ such that $u_{\eta\xi} \in Z$. Consequently $\langle\eta, \xi\rangle \in \sigma$.

The second part follows from Lemma 6121 and Corollaries 5329 and 5331.

6124. Metatheorem. Let λ, μ be definitions of infinite cardinal numbers in **TSS**, D3, E2 and suppose that **TSS**, D3, E2 $\vdash (\forall\alpha, \beta)\,(\lambda(\aleph_\alpha)\,\&\,\mu(\aleph_\beta)\,.\rightarrow .\,\aleph_\alpha \text{ regular } \&\, \aleph_\alpha \leqq \mathbf{cf}\,(\aleph_\beta))$. Let (Collaps$^\beta_\alpha$) be the following axiom (formulated in **TSS**, D3, E2, Mcl (**M**), ∂**TSS**/$\partial\mathfrak{S}$t (**M**)):

$$(\forall\alpha, \beta, \gamma)\,(\lambda^\square(\aleph^\square_\alpha)\,\&\,\mu^\square(\aleph^\square_\beta)\,.\rightarrow .\,\text{Card}\,(\aleph^\square_\gamma) \equiv (\gamma \leqq \alpha \vee \gamma > \beta))\,.$$

Then $\mathfrak{Up}$ is a model in the following diagram:

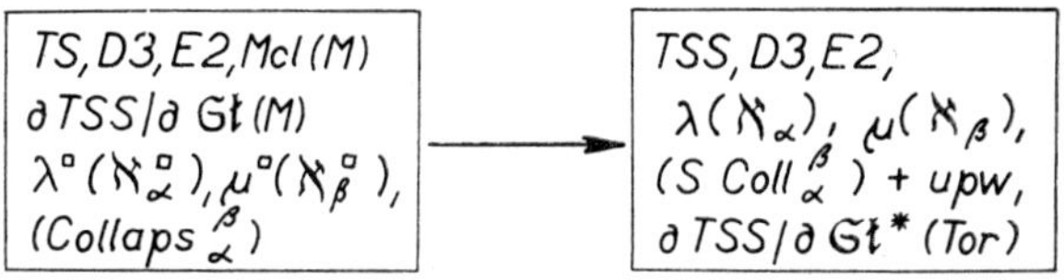

(In fact, the model in question is the extension of $\mathfrak{Up}$ which interpretes the symbols defined by ∂**TSS**/$\partial\mathfrak{S}$t (**M**) by the symbols defined by ∂**TSS**/$\partial\mathfrak{S}$t* (**Tor**) (cf. diagram 5112, arrow 8).

SECTION 2

Independence of the Axioms of Choice

The aim of the present section is to establish the consistency of the negation of the weakest axiom of choice E1 (and of certain stronger statements) w.r.t. the theory **TSS** + D3. Earlier we proved the consistency of ¬ E1 with **TSS** + U3 using the model $\mathfrak{St}(\mathbf{HS}(\mathbf{z}))$ with a suitable specification of the group-filter **z**; this model – and each specification of it – will be referred to as the *permutation model*. We also proved the consistency of the axiom of regularity D3 with **TSS** using the model $\mathfrak{St}(\mathbf{Ker})$, where **Ker** is the constant denoting the regular kernel; this model will be referred to as the *nuclear model*. Since the latter model is an essentially faithful model of **TSS** + D3 in **TSS**, $\partial\mathbf{TSS}/\partial\mathfrak{St}(\mathbf{Ker})$, the consistency of ¬E1 & D3 with **TSS** is equivalent to the consistency of $\neg(\mathrm{E1})^{\mathfrak{St}(\mathbf{Ker})}$ with **TSS**. To prove this consistency we cannot make use of permutation models, since **TSS**, U3, Sym (**z**), $\partial\mathbf{TSS}/\partial\mathfrak{St}(\mathbf{HS}(\mathbf{z})) \vdash \mathbf{Ker} = \mathbf{Ker}^{\mathfrak{St}(\mathbf{HS}(\mathbf{z}))}$; in order that the sentence $\neg(\mathrm{E1})^{\mathfrak{St}(\mathbf{Ker})*\mathfrak{St}(\mathbf{HS}(\mathbf{z}))}$ should be provable in some extension **T** of **TSS**, U3, Sym (**z**), $\partial\mathbf{TSS}/\partial\mathfrak{St}(\mathbf{HS}(\mathbf{z}))$ it is necessary that $\mathbf{T} \vdash \neg(\mathrm{E1})^{\mathfrak{St}(\mathbf{Ker})}$. Fortunately we have another method at hand. We shall start with some extension of **TSS** + ¬ E1 which can be shown to be consistent using the permutation model and consider the extending model in this theory. We shall first investigate how to extend $\mathbf{TSS}^{\mathrm{e}}$ so that $\neg(\mathrm{E1})^{\mathfrak{St}(\mathbf{Ker})}$ is provable in the sense of the model $\mathfrak{Up}$.

6201. Definition (**TSS**′). Let $\boldsymbol{b}$ be a complete Boolean algebra. A non-empty set j of subalgebras of $\boldsymbol{b}$ is called an *ideal* (*of subalgebras*) if the following hold:

1) if $\boldsymbol{c}_1 \in j$ and if $\boldsymbol{c}_2$ is a subalgebra of $\boldsymbol{c}_1$ then $\boldsymbol{c}_2 \in j$,

2) if $\boldsymbol{c}_1, \boldsymbol{c}_2 \in j$ and if $\boldsymbol{c}$ is the subalgebra of $\boldsymbol{b}$ generated by $c_1 \cup c_2$ then $\boldsymbol{c} \in j$,

3) if $\boldsymbol{c} \in j$ and f is an automorphism of $\boldsymbol{b}$ then $f''\boldsymbol{c} \in j$.

6202. Definition (**TSS'**). If $\boldsymbol{b}$ is a complete Boolean algebra then the set $\mathbf{Ki}(\boldsymbol{b})$ of all subalgebras $\boldsymbol{c}$ of $\boldsymbol{b}$ such that $(\exists d \in \mathbf{Ker})(c \approx d)$ is called the *ideal of nuclear subalgebras.*

6203. Lemma (**TSS'**). $\mathbf{Ki}(\boldsymbol{b})$ is an ideal of subalgebras (in the sense of Definition 6201).

Proof. $\mathbf{Ki}(\boldsymbol{b})$ is nonempty since it contains the algebra whose field is $\{0_{\boldsymbol{b}}, 1_{\boldsymbol{b}}\}$. Conditios 1) and 3) obviously hold; let us prove 2). It is sufficient to show that if $a \subseteq b$ can be mapped 1-1 into **Ker** then c can also be mapped 1-1 into **Ker**, where c is the field of the subalgebra of $\boldsymbol{b}$ generated by a. Let f_0 be a 1-1 mapping of some $d_0 \in \mathbf{Ker}$ onto a. Obviously, we can assume that all the elements of d_0 have the same rank. We define mappnigs f_α with domains d_α as follows:

$$f_{2\alpha+1} = \{\langle -f_{2\alpha}'x, \langle 1, x\rangle\rangle \; ; \; x \in d_{2\alpha}\} \cup f_{2\alpha} ,$$
$$f_{2\alpha+2} = \{\langle \bigwedge f_{2\alpha+1}'' z, \langle 0, z\rangle\rangle ; \; z \subseteq d_{2\alpha+1}\} \cup f_{2\alpha+1} ,$$
$$f_\lambda = \bigcup_{\alpha<\lambda} f_\alpha \quad (\lambda \text{ a limit}) .$$

There exists a least α such that $\mathbf{W}(f_\alpha) = c$; we define an equivalence e on $\mathbf{D}(f_\alpha)$ by setting $\langle u, v\rangle \in e \equiv f_\alpha'u = f_\alpha'v$. If d is the set of all maximal sets of equivalent elements and if we define $f'x = y \equiv (\exists u \in x)(f_\alpha'u = y)$ for each $x \in d$ then f is a 1-1 mapping of d onto c. In addition to the assertion of the lemma we have proved the following result:

6204. Lemma (**TSS'**). Let $\boldsymbol{b}$ be a complete Boolean algebra and let $\boldsymbol{c}$ be the subalgebra of $\boldsymbol{b}$ generated by a set $a \subseteq b$. If $a \approx q$ for some $q \in \mathbf{Ker}$ then $\boldsymbol{c} \in \mathbf{Ki}(\boldsymbol{b})$.

6205. Lemma (**TSS'**). If Z is a total Boolean support on a complete Boolean algebra $\boldsymbol{b}$ then for every $\sigma \subseteq \mathbf{Ker}$ there exists $\boldsymbol{c} \in \mathbf{Ki}(\boldsymbol{b})$ such that σ depends on $Z \cap c$.

Proof. We have $\sigma = r''Z$ for some $\boldsymbol{b}$-set r such that $\mathbf{W}(r) \in \mathbf{Ker}$. $\mathbf{D}(r)$ can be 1-1 mapped into **Ker** by assigning $r''\{x\}$ to each $x \in \mathbf{D}(r)$. Let $\boldsymbol{c}$ be the subalgebra of $\boldsymbol{b}$ generated by $\mathbf{D}(r)$; we have $\boldsymbol{c} \in \mathbf{Ki}(\boldsymbol{b})$ and since $\mathrm{Dep}(\sigma, \mathbf{D}(r) \cap Z)$ and $\mathrm{Dep}(\mathbf{D}(r) \cap Z, c \cap Z)$, we have $\mathbf{Dep}(\sigma, c \cap Z)$.

6206. Lemma (**TSS''**, (S6), ∂**TSS**/$\partial\mathfrak{S}\mathrm{t}$(**Ker**))

(a) $(\mathrm{St})^{\mathfrak{S}\mathrm{t}(\mathbf{Ker})}$,

(b) $(\mathrm{Bd})^{\mathfrak{S}\mathrm{t}(\mathbf{Ker})}$.

Proof. Each $\sigma \subseteq \mathbf{On}$ is a semiset in the sense of $\mathfrak{St}(\mathbf{Ker})$ and ordinal numbers are ordinal numbers in the sense of $\mathfrak{St}(\mathbf{Ker})$; hence (*a*) follows easily. We now prove (b). For each $\boldsymbol{c} \in \mathbf{Ki}(\boldsymbol{b})$ let $\boldsymbol{\alpha}(\boldsymbol{c})$ be the least ordinal number such that $c \preccurlyeq p_{\alpha(c)}$; let $\alpha = \mathbf{Sup}\{\boldsymbol{\alpha}(c); \boldsymbol{c} \in \mathbf{Ki}(\boldsymbol{b})\}$ and set $a = p_\alpha$. We prove that a is a bounding set. If $\sigma \subseteq \mathbf{Ker}$ then $\sigma = r``(Z \cap c)$ for some $r$ and some $\boldsymbol{c} \in \mathbf{Ki}(\boldsymbol{b})$. We may suppose that $\mathbf{W}(r) \in \mathbf{Ker}$ and $\mathbf{D}(r) \subseteq c$; if we let $q = \{r``\{x\}, x \in \mathbf{D}(r)\}$ then $q \approx \mathbf{D}(r) \preccurlyeq c \preccurlyeq a$; if we let $\varrho = \{r``\{x\}; x \in Z \cap c\}$ then $\varrho \subseteq q$ and $\bigcup \varrho = \sigma$.

6207. THEOREM ($\mathbf{TSS}^c$, S6, $\partial\mathbf{TSS}/\partial\mathfrak{St}(\mathbf{Ker})$, $\partial\mathbf{TSS}/\partial\mathfrak{St}^*(\mathbf{Ker}^*)$).

$$(\mathrm{E1})^{\mathfrak{St}^*(\mathbf{Ker}^*)} \rightarrow (\mathrm{S6})^{\mathfrak{St}(\mathbf{Ker})}.$$

Proof. By the preceding lemma it suffices to prove $(\mathrm{Pot})^{\mathfrak{St}(\mathbf{Ker})}$ (cf. 4134, 4241). We have $\mathbf{Imb}``\,\mathbf{Ker} \subseteq \mathbf{Ker}^*$ (where $*$ is used to represent the notions in the sense of $\mathfrak{Up}$); this follows from the fact that $\mathbf{Imb}``\{0\} \subseteq \mathbf{Ker}^*$ and that for every α, $\mathbf{Imb}``\,p_\alpha \subseteq \mathbf{Ker}^*$ implies $\mathbf{Imb}``\,p_{\alpha+1} \subseteq \mathbf{Ker}^*$. Suppose $a \in \mathbf{Ker}$ and $\langle b, a\rangle \in \mathbf{Imb}$. Clearly $y^* = \mathbf{Ext}_{\mathbf{FUpw}}(b) \in^* \mathbf{Ker}^*$ and there exists $x^* \in^* \mathbf{Ker}^*$ such that, in the sense of $\mathfrak{Up}$, x^* is an exact functor whose domain is an ordinal number and which codes all subclasses of y^*. We have $x^* \subseteq \mathbf{Tor}$; if we let $\varrho = [\mathbf{Cnv}(\mathbf{Imb})]``\,x^*$ then ϱ is an exact functor whose domain is an ordinal number and ϱ codes all semisets included in a. This proves $(\mathbf{Pot})^{\mathfrak{St}(\mathbf{Ker})}$ and hence also $(\mathrm{S6})^{\mathfrak{St}(\mathbf{Ker})}$.

6208. Remark. It follows that if we want (E1) to fail in the sense of the model $\mathfrak{St}(\mathbf{Ker})$ constructed within the model $\mathfrak{Up}$ it is sufficient to prove that there exists no total support in the sense of $\mathfrak{St}(\mathbf{Ker})$.

6209. DEFINITION ($\mathbf{TSS}'$). Let j be an ideal of subalgebras of $\boldsymbol{b}$. We call $\boldsymbol{c} \in j$ a *principal algebra* in j if for every $\boldsymbol{d} \in j$ and every $u \geqq 0_{\boldsymbol{b}}$ there exists $v \geqq 0_{\boldsymbol{b}}$ such that $v \leqq u$ and $d \mid_{\boldsymbol{b}} v \subseteq c \mid_{\boldsymbol{b}} v$; it follows that $\boldsymbol{d} \mid_{\boldsymbol{b}} v$ is a subalgebra of $\boldsymbol{c} \mid_{\boldsymbol{b}} v$. An ideal j is called *primary* if it has a principal algebra.

6210. THEOREM ($\mathbf{TSS}''$, $\partial\mathbf{TSS}/\partial\mathfrak{St}(\mathbf{Ker})$). If Z is a total Boolean support on $\boldsymbol{b}$ then the following conditions are equivalent:

(a) there exists $u \in Z$ such that $\mathbf{Ki}(\boldsymbol{b} \mid u)$ is primary;

(b) in the sense of $\mathfrak{St}(\mathbf{Ker})$ there exists a total (Boolean semiset) support.

Proof. Assume that (a) holds. We may suppose that $\mathbf{Ki}(\boldsymbol{b})$ is primary; for otherwise we may consider $\boldsymbol{b} \mid u$ and $Z \mid_{\boldsymbol{b}} u$. Let $\boldsymbol{c}$ be a principal algebra in $\mathbf{Ki}(\boldsymbol{b})$ and let $\boldsymbol{c}' \in \mathbf{Ker}$ be a complete Boolean algebra isomorphic to $\boldsymbol{c}$.

Let f be an isomorphism of $\boldsymbol{c}'$ and $\boldsymbol{c}$ and let $Z' = f''(Z \cap c)$. Since Z' is a complete ultrafilter on $\boldsymbol{c}'$, Z' is a support in the sense of $\mathfrak{St}$ (**Ker**). We prove that Z' is a total support in the sense of $\mathfrak{St}$ (**Ker**); thus we prove that Dep (σ, Z') for each $\sigma \subseteq$ **Ker**. We have Dep $(\sigma, Z \cap d)$ for some $\boldsymbol{d} \in$ **Ki** $(\boldsymbol{b})$; for every $\boldsymbol{u} \in \mathring{b}$ there exists $v \in \mathring{b}$ such that $v \leqq \boldsymbol{u}$ and $d \mid_{\boldsymbol{b}} v \subseteq c \mid_{\boldsymbol{b}} v$; hence there exists $v \in Z$ such that $d \mid_{\boldsymbol{b}} v \subseteq c \mid_{\boldsymbol{b}} v$. It follows that

$$\text{Dep}\,(Z \cap (d \mid_{\boldsymbol{b}} v), Z \cap (c \mid_{\boldsymbol{b}} v))$$

(here we use the relation $r = \mathbf{I} \upharpoonright (d \mid_{\boldsymbol{b}} v)$). We also have Sim $(Z \cap c, Z \cap \cap (c \mid_{\boldsymbol{b}} v))$, Sim $(Z \cap d, Z \cap (d \mid_{\boldsymbol{b}} v))$ and Sim $(Z \cap c, Z')$, so that Dep (σ, Z') which proves (b).

Conversely, let Z_1 be a total Boolean support on $\boldsymbol{b}_1$ in the sense of $\mathfrak{St}$ (**Ker**). By 4257, there exists a subalgebra $\boldsymbol{b}'$ of $\boldsymbol{b}$ such that Z_1 is similar to $b' \cap Z$; hence, by 4323, there exist $u_0 \in Z_1$ and $v_0 \in b' \cap Z$ such that $\boldsymbol{b}_1 \mid u_0$ is isomorphic to $\boldsymbol{b}' \mid v_0$. It follows that $\boldsymbol{b}' \mid v_0 \in$ **Ki** $(\boldsymbol{b} \mid v_0)$ and that $\boldsymbol{b}' \mid w_0 \in \in$ **Ki** $(\boldsymbol{b} \mid w_0)$ for each $w_0 \leqq v_0$. Let $\boldsymbol{c} \in$ **Ki** $(\boldsymbol{b})$; the subalgebra $\boldsymbol{b}''$ of $\boldsymbol{b}$ generated by $(b' \cup c)$ also belongs to **Ki** $(\boldsymbol{b})$. We shall prove that $Z \cap b'$ is similar to $Z \cap b''$. Since $b' \subseteq b''$ we have Dep $(Z \cap b', Z \cap b'')$; since $\boldsymbol{b}'' \in$ **Ki** $(\boldsymbol{b})$ there exists an algebra in **Ker** isomorphic to $\boldsymbol{b}''$. Let f be the isomorphism in question and let $\sigma = f''(Z \cap b'')$. We have $\sigma \subseteq$ **Ker**, Dep (σ, Z_1), Sim $(\sigma, Z \cap b'')$ and Sim $(Z_1, Z \cap b')$, so that Dep $(Z \cap b'', Z \cap b')$.

By Theorem 4328, there exists $v \in Z$ such that $b' \mid_{\boldsymbol{b}} v = b'' \mid_{\boldsymbol{b}} v$, and since $c \subseteq b''$ we have $c \mid_{\boldsymbol{b}} v \subseteq b' \mid_{\boldsymbol{b}} v$, and $c \mid_{\boldsymbol{b}} w \subseteq b' \mid_{\boldsymbol{b}} w$ for each $w \leqq v$. Hence we have

$$(\forall u \in Z)\,(\forall \boldsymbol{c} \in \mathbf{Ki}\,(\boldsymbol{b}))\,(\exists v \leqq u)\,(v \geqq 0_{\boldsymbol{b}}\ \&\ c \mid_{\boldsymbol{b}} v \subseteq b' \mid_{\boldsymbol{b}} v)\,.$$

By 4302, there exists $u \in Z$ such that

$$(\forall v \geqq 0_{\boldsymbol{b}})\,(v \leqq u \rightarrow (\forall \boldsymbol{c} \in \mathbf{Ki}\,(\boldsymbol{b}))\,(\exists w \leqq v)\,(w \geqq 0_{\boldsymbol{b}}\ \&\ c \mid_{\boldsymbol{b}} w \subseteq b' \mid_{\boldsymbol{b}} w)\,.$$

We may suppose $u \leqq v_0$, so that $\boldsymbol{b}' \mid_{\boldsymbol{b}} u \in$ **Ki** $(\boldsymbol{b} \mid u)$ and

$$(\forall \boldsymbol{c} \in \mathbf{Ki}\,(\boldsymbol{b} \mid u)\,(\forall v \geqq 0_{\boldsymbol{b}|u})\,(\exists w \leqq v)\,(w \geqq {}_{\boldsymbol{b}|u}\ \&\ c \mid_{\boldsymbol{b}} w \subseteq (b' \mid_{\boldsymbol{b}} u) \mid_{\boldsymbol{b}} w)\,;$$

for, if $\boldsymbol{c} \in$ **Ki** $(\boldsymbol{b} \mid u)$ then the subalgebra $\boldsymbol{d}$ of $\boldsymbol{b}$ generated by c belongs to **Ki** $(\boldsymbol{b})$ and $c \subseteq d \mid_{\boldsymbol{b}} u$. Hence **Ki** $(\boldsymbol{b} \mid u)$ is primary and $\boldsymbol{b}' \mid_{\boldsymbol{b}} u$ is the principal algebra.

6211. Remark. ($\mathbf{TSS}'$, E1, WSym (z), $\partial\mathbf{TSS}/\partial\mathfrak{St}(\mathbf{HS}(\mathbf{z}))$). Let $\boldsymbol{b}$, $\boldsymbol{c} \in \mathbf{HS}(\mathbf{z})$. Suppose $\boldsymbol{b}$ is a complete Boolean algebra in the sense of $\mathfrak{St}(\mathbf{HS}(\mathbf{z}))$ and $\boldsymbol{c}$ is a subalgebra of $\boldsymbol{b}$ in the sense of $\mathfrak{St}(\mathbf{HS}(\mathbf{z}))$. Then $\boldsymbol{c} \in \mathbf{Ki}^*(\boldsymbol{b})$ iff $\mathbf{PInv}(c) \in \mathbf{z}$. For, if $x \in \mathbf{HS}(\mathbf{z})$ then x can be 1-1 mapped in the sense of $\mathfrak{St}(\mathbf{HS}(\mathbf{z}))$ into **Ker** iff $\mathbf{PInv}(x) \in \mathbf{z}$.

6212. We shall now introduce some notations in $\mathbf{TSS}'$. Let $\mathbf{a} = \langle \mathrm{a}, \leqq \rangle$ be a separatively ordered set fruitful w.r.t. a symmetric **g**-filter **z** and let $\mathscr{P} = \mathbf{HS}(\mathbf{z})$. Let **b** be the complete Boolean algebra with base **a** and let $\mathbf{b}_1$ be the complete Boolean algebra in the sense of $\mathfrak{St}(\mathbf{HS}(\mathbf{z}))$ with base $\mathbf{a}_1 = \langle \mathrm{a} \cap \mathscr{P}, \leqq \cap \mathscr{P} \rangle$. Let **f** be the embedding of $\mathbf{b}_1$ into **b** described in 3352 and let **Z** be a complete ultrafilter on **b**. It follows that $(\mathbf{Cnv}(\mathbf{f}))$" **Z** is a complete ultrafilter on $\mathbf{b}_1$ in the sense of $\mathfrak{St}(\mathscr{P})$. This ultrafilter will be denoted by $\mathbf{Z}_1$. In the sequel we shall suppose **a**, **b**, **z** etc. to have the meanings given to them above. We shall also assume that the algebras **b** and $\mathbf{b}_1$ are constructed in the canonical manner from the saturated subsets of the corresponding bases (cf. 2442). Under these assumptions we shall investigate the model $\mathfrak{Supp}(\mathbf{Z}_1) * \mathfrak{St}(\mathbf{HS}(\mathbf{z}))$, whose sets are the elements of $\mathbf{HS}(\mathbf{z})$ and whose semisets are those subclasses σ of $\mathbf{HS}(\mathbf{z})$ which are, in the sense of $\mathfrak{St}(\mathbf{HS}(\mathbf{z}))$, dependent on $\mathbf{Z}_1$ (i.e. such that $\sigma = r$"$\mathbf{Z}_1$ for some $r \in \mathbf{HS}(\mathbf{z})$). We shall sometimes be interested in whether elements of $\mathbf{V} - \mathbf{HS}(\mathbf{z})$, i.e. nonsymmetric subsets of $\mathbf{HS}(\mathbf{z})$, can also be semisets in the sense of the above-mentioned model. The following theorem provides the answer in case **a** is hereditarily symmetric (and hence $\mathbf{a}_1 = \mathbf{a}$).

6213. THEOREM ($\mathbf{TSS}'$, **a**, **b**, **Z**, **z** etc. as above). Suppose $\mathbf{a} \in \mathscr{P}$, so that $\mathbf{a}_1 = \mathbf{a}$. If $r \in \mathscr{P}$ and $x = r$"$\mathbf{Z}_1$ then $x \in \mathscr{P}$.

Proof. We may suppose that r is a disjointed relation, since $\mathbf{Z}_1$ is, in the sense of $\mathfrak{St}(\mathscr{P})$, a disjointed support. If we let $y = (\mathbf{Cnv}(r))$" x then $y \subseteq \mathbf{Z}_1$, i.e. **f**"$y \subseteq \mathbf{Z}$. Let $u \in \mathrm{a} \cap \mathbf{Z}$, $u \leqq \bigwedge \mathbf{f}$"$y$. If we let $\mathrm{b}_1^u = \{v \in \mathrm{b}_1, v \geq u\}$ then we have $\mathrm{b}_1^u \in \mathscr{P}$, since $u \in \mathrm{a} = \mathrm{a}_1$ and $\mathbf{Inv}(\mathrm{b}_1^u) \supseteq \mathbf{Inv}(u) \cap \mathbf{Inv}(\mathbf{a})$. We shall prove that $x = r$"(b_1^u); this implies $x \in \mathscr{P}$ which completes the proof of the theorem. The inclusion $x \subseteq r$"b_1^u follows from $y \subseteq \mathrm{b}_1^u$; since $\mathrm{b}_1^u \subseteq \mathbf{Z}_1$ we have r"$\mathrm{b}_1^u \subseteq r$"$\mathbf{Z}_1 = x$ and we are finished.

We now consider products of complete Boolean algebras.

6214. THEOREM ($\mathbf{TSS}'$). For each $x \in \mathbf{Ur}$ let $\boldsymbol{b}_x \in \mathbf{Ker}$ be a complete atomless Boolean algebra. Let l be a cut on **Ur** having the singleton property. If Z is a total Boolean support on $\boldsymbol{a} = \prod^l_{x \in \mathbf{Ur}} \boldsymbol{b}_x$ then there exists an exact functor ϱ such that $\mathbf{D}(\varrho) = \mathbf{Ur}$ and $\mathbf{W}(\varrho) \subseteq \mathbf{Ker}$.

Proof. We define $\varrho = \{\langle u, x\rangle; x \in \mathbf{Ur} \;\&\; \{\langle u, x\rangle\} \in Z\}$. For each $x \in \mathbf{Ur}$ we have $0 \neq \varrho''\{x\} \subseteq \boldsymbol{b}_x$; it suffices to prove that ϱ is nowhere constant. If $x_i \in \mathbf{Ur}$ then $\varrho''\{x_i\} = r_i''Z$, where $r_i = \{\langle u, \{\langle u, x_i\rangle\}\rangle, \{\langle u, x_i\rangle\} \in a\}$. By 4320, it suffices to prove that $x_1 \neq x_2$ implies $\mathrm{sSim}_{\boldsymbol{b}}((\mathbf{D}(r_1), \mathbf{D}(r_2)) = 0_{\boldsymbol{b}}$ where $\boldsymbol{b}$ is the complete Boolean algebra with base $\boldsymbol{a}$. Let f be an element of a; we wish to find $g \leqq f$ such that $\mathbf{D}(r_1) |_{\boldsymbol{b}}\, g \neq \mathbf{D}(r_2) |_{\boldsymbol{b}}\, g$; we shall show that $g = f$ has the required property.

(a) If $x_1 \notin \mathbf{D}(f)$ then for every $u \in \dot{\boldsymbol{b}}_{x_2}$, $f \wedge \{\langle u, x_2\rangle\}$ is a basic element not defined at x_1 or 0_b whereas for every $v \in \dot{\boldsymbol{b}}_{x_1}$, $f \wedge \{\langle v, x_1\rangle\}$ is a basic element defined at x_1.

(b) If $x_1 \in \mathbf{D}(f)$ then, for every $u \in \dot{\boldsymbol{b}}_{x_2}$, $f \wedge \{\langle u, x_2\rangle\}$ is a basic element which assumes the value $f'x_1$ at x_1, or 0_b whereas for every $v \in \dot{\boldsymbol{b}}_{x_1}$ such that $v < f'x_1$, $f \wedge \{\langle v, x_1\rangle\}$ is a basic element which has the value v at x_1 (here we use the fact that $\boldsymbol{b}_{x_1}$ is atomless).

6215. EXAMPLE. Suppose $\mathbf{Ur} \approx \aleph_\beta$ where $\aleph_\beta$ is a regular cardinal number, and let $\mathbf{I}$ be the ideal of all subsets of $\mathbf{Ur}$ which have power less than $\aleph_\beta$. We shall use the ideal $\mathbf{I}$ in two different ways. First we define a $\mathbf{g}$-filter $\mathbf{z}$ on $\mathbf{g}(\mathbf{Ur})$ by letting $\mathbf{z} = \mathbf{z}_\mathbf{I}$; $\mathbf{z}$ determines the class $\mathscr{P} = \mathbf{HS}(\mathbf{z})$. Secondly, we define a separatively ordered set $\mathbf{a}$ by letting $\mathbf{a} = \prod^{\mathbf{I}}_{x \in \mathbf{Ur}} \dot{\mathbf{b}}_x$, where $\mathbf{b}_x = \mathbf{Cant}(\aleph_\beta, \aleph_\beta)$ for every x.

Clearly $\mathbf{a} \in \mathscr{P}$; let $\mathbf{b}$ be the complete Boolean algebra with base $\mathbf{a}$ and let $\mathbf{b}_1$ be the complete Boolean algebra with base $\mathbf{a}$ in the sense of $\mathscr{P}$; we suppose both algebras to be constructed in the canonical manner. Suppose there is a Boolean total support on $\mathbf{b}$ and let $\mathbf{Z}_1$ be its restriction to $\mathbf{b}_1$. We know that $\mathbf{Z}_1$ is a Boolean support on $\mathbf{b}_1$ in the sense of $\mathfrak{St}(\mathscr{P})$ (see 6212). Also suppose (E2). For the time being, $\mathbf{TSS}^{\mathscr{P}}$ will denote $\mathbf{TSS}'$ extended by the addition of the above-mentioned assumptions; the relative consistency of $\mathbf{TSS}^{\mathscr{P}}$ is given by 5215. It follows from the results of Chapter III, Sect. 3 that, in the sense of $\mathfrak{St}(\mathscr{P})$, $\aleph_\alpha \prec \mathbf{Ur}$ holds for every $\alpha < \beta$ but neither $\aleph_\beta \preccurlyeq \mathbf{Ur}$ nor $\mathbf{Ur} \preccurlyeq \aleph_\beta$; moreover, a subset of $\mathbf{Ur}$ either has power less than $\mathbf{Ur}$ or its complement has power less than $\mathbf{Ur}$. Every exclusive system q of elements of $\mathbf{b}_1$ which is in $\mathscr{P}$ is in the algebra with base $\prod^{\mathbf{I}/e}_{\mathbf{Ur}} \mathbf{b}_x$, where $e \in \mathbf{I}$ and $\mathbf{PInv}(e) \subseteq \mathbf{Inv}(q)$ (see 3357). This algebra is in $\mathscr{P}$, since $\mathbf{Ker}(e) \subseteq \mathscr{P}$. We now prove

6216. THEOREM ($\mathbf{TSS}^{\mathscr{P}}$, $\partial\mathbf{TSS}/\partial\mathfrak{St}(\mathscr{P})$, $\partial\mathbf{TSS}/\partial\mathfrak{Supp}^{\square}(\mathbf{Z}_1)$). The following holds in the sense of $\mathfrak{Supp}(\mathbf{Z}_1) * \mathfrak{St}(\mathscr{P})$:

(a) All cardinals are absolute.

(b) If $\alpha < \beta$ then $\aleph_\alpha \mathrel{\hat{\prec}} \mathbf{Ur}$.

(c) Neither $\aleph_\beta \mathrel{\hat{\preccurlyeq}} \mathbf{Ur}$ nor $\mathbf{Ur} \mathrel{\hat{\preccurlyeq}} \aleph_\beta$.

(d) There exists an exact functor ϱ such that $\mathbf{D}(\varrho) = \mathbf{Ur}$ and $\mathbf{W}(\varrho) \subseteq \boldsymbol{\omega}_\beta$.

Proof. As in the proof of 6117 (b) we may prove that $\mathbf{b}$ is isomorphic to $\mathbf{Cant}(\aleph_\beta, \aleph_\beta)$. It follows by 6104 that if there is a Boolean support on $\mathbf{b}$ then all cardinals are absolute; hence all cardinals are absolute in the sense of $\mathfrak{Supp}(\mathbf{Z}_1) * \mathfrak{St}(\mathscr{P})$.

(b) If $\alpha < \boldsymbol{\beta}$ then there exists a 1-1 mapping $f \in \mathscr{P}$ of $\aleph_\alpha$ into $\mathbf{Ur}$; hence $\aleph_\alpha \preccurlyeq \mathbf{Ur}$ holds in the sense of $\mathfrak{Supp}(\mathbf{Z}_1) * \mathfrak{St}(\mathscr{P})$, i.e. $\aleph_\alpha \mathrel{\hat{\preccurlyeq}} \mathbf{Ur}$. Since $\aleph_\alpha \mathrel{\hat{\approx}} \aleph_\beta$ is impossible by absoluteness of cardinals, it follows that $\aleph_\alpha \mathrel{\hat{\approx}} \mathbf{Ur}$ cannot hold in the sense of the model.

(c) Let σ be a semiset of the model which is a 1-1 mapping of $\aleph_\beta$ into $\mathbf{Ur}$. There exists a relation $r \in \mathscr{P}$ which is a $\boldsymbol{b}_1$-mapping of $\aleph_\beta$ into $\mathbf{Ur}$ and such that $\sigma = r\text{''}\mathbf{Z}_1$. Suppose that $\mathbf{Inv}(r) \supseteq \mathbf{PInv}(e)$ for some $e \in \mathbf{l}$, $e \approx \aleph_\alpha$. For $\iota \in \mathbf{Ur}$ let $h'\iota$ be the least $\gamma < \boldsymbol{\omega}_\beta$ such that $r_{\iota\gamma} \neq 0_{\boldsymbol{b}}$ (if such a γ exists). There exists $\iota \in \mathbf{Ur} - e$ such that $h'\iota$ is defined; for otherwise $\mathbf{W}(\sigma) \subseteq e$ and hence $\aleph_\alpha \mathrel{\hat{\approx}} \aleph_\beta$, a contradiction. Since $\iota \notin e$ we have $\kappa \in \mathbf{D}(h)$ and $h'\kappa = h'\iota$ whenever $\kappa \in \mathbf{Ur} - e$. If we set $h'\iota = \gamma$ then $\{r_{\kappa\gamma};\ \kappa \in \mathbf{D}(h) - e\}$ is a disjointed family of $\boldsymbol{b}_1$ which belongs to $\mathscr{P}$ and which cannot be well-ordered in the sense of $\mathscr{P}$. This is a contradiction, since by 3357 the family $\{r_{\kappa\gamma};\ \kappa \in \mathbf{D}(h) - e\}$ belongs to the subalgebra with base $\prod_{\mathbf{Ur}}^{1/e} \mathbf{b}_\kappa$ and consequently to the class $\mathbf{Ker}^{\mathfrak{St}(\mathscr{P})}(e)$, which can be well-ordered in the sense of $\mathfrak{St}(\mathscr{P})$.

Similarly, let σ be a semiset of the model and suppose that σ is a 1-1 mapping of $\mathbf{Ur}$ into $\aleph_\beta$; as before, $h'\iota$ is defined for every $\iota \in \mathbf{Ur}$ so that there exists $\iota \in \mathbf{D}(h) - e$ (where r and e are defined as above). The rest is obvious.

(d) For each $\gamma < \boldsymbol{\omega}_\beta$ let u_γ be the basic element of $\mathbf{Cant}(\aleph_\beta, \aleph_\beta)$ which assumes the value 1 at γ and is undefined elsewhere, and define r by $r = \{\langle\gamma, \iota\rangle, \langle u_\gamma, \iota\rangle\}$. We let $\varrho = r\text{''}\mathbf{Z}$; as in the proof of Theorem 6107 (a) the reader may verify that ϱ has the required property.

6217. COROLLARY. Let $\aleph_\beta$ be a constant for a regular cardinal introduced by a good definition. The theory $\mathbf{TSS}^e$ extended by adding (a)–(d) of the preceding theorem is consistent relative to $\mathbf{TSS}'$; moreover, the following holds in the sense of the model $\mathfrak{Up}$ as a model in this theory;

There exists a family c of subsets of $\boldsymbol{\omega}_\beta$ such that $\aleph_\alpha \prec c$ for every $\aleph_\alpha \prec \aleph_\beta$ and neither $\aleph_\beta \preccurlyeq c$ nor $c \preccurlyeq \aleph_\beta$.

In particular, for $\beta = 0$, we obtain the consistency of the following statement:

There exists a family of subsets of ω_0 which is infinite (in the sense of 2216) but does not contain any countable subset.

6218. We shall now continue from Chapter III, Section 3 our investigations concerning α-boundable formulas and their permutation models. Suppose that $\varphi(x)$ is α-boundable where α is a constant introduced by a good definition and that $\varphi(x)$ has a permutation model given by a formula $\omega(x, z)$ where $\omega(x, z)$ well describes a symmetric **g**-filter on x (everything w.r.t. **TSS**, U3). We proved in 3348 that the theory

$$\mathbf{TSS}, \text{U3, E2}, (\text{col}(\mathbf{q}, \alpha) \,\&\, \omega(\mathbf{q}, \mathbf{z}))$$

is consistent.

Let $\aleph_\beta$ be a constant for the least regular cardinal number $\succ \bar{p}^{\mathbf{q}}_{\alpha}$; recall that col $(\mathbf{q}, \alpha)$ implies that $(\forall u \in \mathbf{q})\,(u \approx \aleph_\beta)$. Let **b** be the algebra with the base $\mathbf{a} = \prod^{\mathbf{I}}_{\mathbf{Ur}} b^{\beta}_{\beta}$ (cf. 6101), where **I** is the ideal of all subsets of **Ur** of power $\prec \aleph_\beta$. It follows from 5215 that the theory

$$\mathbf{TSS}, \text{U3, E2, col}\,(\mathbf{q}, \alpha) \,\&\, \omega(\mathbf{q}, \mathbf{z})\; \vartheta(\mathbf{b}, \text{Z}) \,\&\, \text{TSupp}\,(\mathbf{Z})$$

is consistent (relative to **TSS**). In this theory we can define the filter $\mathbf{z}\downarrow$ (see 3349) and the model class $\mathscr{P} = \mathbf{HS}\,(\mathbf{z}\downarrow)$; as usual, $\mathbf{a}_1$ denotes the ordered set $\langle \mathrm{a} \cap \mathscr{P}, \leqq \cap \mathscr{P} \rangle$ and $\mathbf{b}_1$ is the complete Boolean algebra in the sense of $\mathfrak{St}\,(\mathscr{P})$ with base $\mathbf{a}_1$. $\mathbf{Z}_1$ denotes a Boolean support on $\mathbf{b}_1$ in the sense of $\mathfrak{St}\,(\mathscr{P})$ which is a restriction of **Z**. The latter theory together with all defined notations will be denoted by $\mathbf{TSS}^{\mathscr{P}}$ as before.

The following theorem is crucial for the demonstration of the consistency of **TS** + D3 + $(\exists x)\, \varphi(x)$:

6219. THEOREM ($\mathbf{TSS}^{\mathscr{P}}$, $\partial\mathbf{TSS}/\partial\mathfrak{St}\,(\mathscr{P})$, $\partial\mathbf{TSS}/\partial\mathfrak{Supp}^{\square}\,(\mathbf{Z}_1)$). In the sense of $\mathfrak{Supp}\,(\mathbf{Z}_1) * \mathfrak{St}\,(\mathscr{P})$, every subclass of $\bar{p}^{\mathbf{q}}_{\alpha}$ is a set.

Proof. We shall say "in the sense of the model" instead of "in the sense of $\mathfrak{Supp}\,(\mathbf{Z}_1) * \mathfrak{St}\,(\mathscr{P})$". Let us recall that the sets in the sense of the model are the elements of $\mathscr{P}$ and the semisets in the sense of the model are the semisets σ such that $\sigma = r''\mathbf{Z}_1$ for some $r \in \mathscr{P}$. The power of $\bar{p}^{\mathbf{q}}_{\alpha}$ is less than $\aleph_\beta$, so by 5331 (2) every part of $\bar{p}^{\mathbf{q}}_{\alpha}$ is a set (and, consequently, every part of $\bar{p}^{\mathbf{q}}_{\alpha} \cap \mathbf{HS}\,(\mathbf{z}\downarrow)$ is a set). It remains to prove that every semiset in the model which is included in $\bar{p}^{\mathbf{q}}_{\alpha} \cap \mathbf{HS}\,(\mathbf{z}\downarrow)$ is a symmetric set. We shall prove the following stronger statement: if $x \subseteq \mathbf{HS}'(\mathbf{z}, \mathbf{q}) = \mathbf{HS}(\mathbf{z}\downarrow) \cap \mathbf{Ker}'(\mathbf{q})$ and if $x = r''\mathbf{Z}_1$ for some $r \in \mathscr{P}$ then $x \in \mathscr{P}$. Suppose $k \subseteq \mathbf{Inv}_{\mathbf{Ur}}(r) \cap \mathbf{Inv}_{\mathbf{Ur}}(\mathbf{z}\downarrow)$, $k = h\downarrow \cap \mathbf{PInv}_{\mathbf{Ur}}(e)$, $h \in \mathbf{z}$, $e \in \mathbf{P_{fin}}(\mathbf{Ur})$; we prove that $k \subseteq \mathbf{Inv}_{\mathbf{Ur}}(x)$. We may suppose that r is disjointed; if f is the canonical embedding of $\mathbf{b}_1$ into $\mathbf{b}$ and if we let $\langle t, f'u\rangle \in \bar{r} \equiv \langle t, u\rangle \in r$ then $\mathbf{Inv}_{\mathbf{Ur}}(f) \supseteq \mathbf{Inv}_{\mathbf{Ur}}(\mathbf{z}\downarrow)$ and $k \subseteq \subseteq \mathbf{Inv}_{\mathbf{Ur}}(\bar{r})$. Since $\bar{r}''\mathbf{Z}$ is a set there exists an exclusive system m of elements of $\mathbf{b}$ with the following properties:

1) $\bigvee m = \mathrm{Thru}(\mathbf{D}(\bar{r})) \in \mathbf{Z}$.

2) If $u \in m$ then u is a maximal element such that

$$(\forall y \in \mathbf{D}(\bar{r}))(y \wedge u = 0_b \vee u \leqq y).$$

Clearly $\mathbf{Inv}_{\mathbf{Ur}}(\bar{r}) \subseteq \mathbf{Inv}_{\mathbf{Ur}}(\mathbf{D}(\bar{r})) \subseteq \mathbf{Inv}_{\mathbf{Ur}}(m)$, so that $k \subseteq \mathbf{Inv}_{\mathbf{Ur}}(m)$. Let $u \in m \cap \mathbf{Z}$ and for each $p \in k$ let p_1 be an element of k such that p_1 and p have the same extension onto q (see 3350) such that $\hat{p}_1'u \wedge u \neq 0_b$. Since m is an exclusive system it follows that $\hat{p}_1'u = u$ and hence $\hat{p}_1'(\mathrm{b}^u) = \mathrm{b}^u$, where $\mathrm{b}^u = \{v \in \mathrm{b}, v \geq u\}$. On the other hand we have $\bar{r}''\mathbf{Z} = \bar{r}''(\mathrm{b}^u)$, since $(\mathbf{Cnv}(\bar{r}))''\, x \subseteq \mathrm{b}^u \subseteq \mathbf{Z}$, i.e. $x \subseteq \bar{r}''\mathrm{b}^u \subseteq r''Z_1 = x$. Hence

$$\hat{p}_1'x = \hat{p}_1(\bar{r}''\mathrm{b}^u) = (\hat{p}_1'\bar{r})''\,(\hat{p}_1'b^u) = \bar{r}''b^u = x\,.$$

Since $\hat{p} \upharpoonright \mathbf{Ker}'(\mathbf{q}) = \hat{p}_1 \upharpoonright \mathbf{Ker}'(\mathbf{q})$, we have $\hat{p}'x = x$ and hence $p \in \mathbf{Inv}_{\mathbf{Ur}}(x)$.

The following theorem can be proved as 6216 (d):

6220. THEOREM ($\mathsf{TSS}^{\mathscr{P}}$, $\partial\mathsf{TSS}/\partial\mathfrak{St}(\mathscr{P})$, $\partial\mathsf{TSS}/\partial\mathfrak{Supp}^{\square}(\mathbf{Z}_1)$). The following statement holds in the sense of $\mathfrak{Supp}(\mathbf{Z}_1) * \mathfrak{St}(\mathscr{P})$:

There exists an exact functor ϱ such that $\mathbf{D}(\varrho) = \mathbf{Ur}$ and $\mathbf{W}(\varrho) \subseteq \boldsymbol{\omega}_{\beta}$; moreover, if $\iota \in \mathbf{Ur}$ then $\bigcup(\varrho''\{\iota\}) = \boldsymbol{\omega}_{\beta}$.

6221. METACOROLLARY. Let $\boldsymbol{\alpha}$ be well-defined, let $\varphi(x)$ be an $\boldsymbol{\alpha}$-boundable formula and suppose that $\omega(x, z)$ well describes a symmetric $\mathbf{g}$-filter x and determines a permutation model for $\varphi(x)$; (everything w. r. t. **TSS**, U3); then the following theory is consistent (relative to **TSS**):

$\mathsf{TSS}^{\mathrm{e}}$, $(\exists q)(\bigcup(q) = \mathbf{Ur}\ \&\ q$ is disjointed and all elements of q have the same power as $\mathbf{Ur}\ \&\ \varphi(q)\ \&\ (\forall\sigma \subseteq \bar{p}^q_{\alpha})(\mathrm{M}(\sigma))\ \&\ (\exists\varrho)(\exists\beta)(\varrho$ is an exact functor $\&\ \mathbf{D}(\varrho) = \mathbf{Ur}\ \&\ (\forall\iota \in \mathbf{Ur})(\varrho''\{\iota\}$ is a cofinal part of $\boldsymbol{\omega}_{\beta})).$

The following statement holds in the model $\mathfrak{Up}$ as a model in the above theory:

($*$) The model class **M** contains all urelements; there is a partition q of **Ur** whose elements have the same power as **Ur**, such that

1) $\varphi^{\square}(q)$,

2) $p_{\alpha}^{q} \subseteq \mathbf{M}$,

3) there exists a 1-1 mapping of **Ur** onto a certain set of cofinal subsets of some cardinal $\boldsymbol{\omega}_{\beta}$.

6222. Proceeding in the theory $\mathbf{TS}^{c}$ extended by ($*$) we prove that $(\exists\beta)\,(\exists x \in \mathbf{PPP}\,(\boldsymbol{\omega}_{\beta}))\;\varphi(x)$.

Let f be the mapping of **Ur** onto cofinal subsets of $\boldsymbol{\omega}_{\beta}$ given by 3); clearly f'' **Ur** is a set of individuals (in the sense of 3339), so that $x = \{f''u,\, u \in q\}$ is also a set of individuals and $x \in \mathbf{Ker}$. The mapping f can be extended to an isomorphism between $\bar{p}_{\alpha}^{q}$ and $\bar{p}_{\alpha}^{x}$; since $\bar{p}_{\alpha}^{x}$ is the same as $\bar{p}_{\alpha}^{x}$ in the sense of **Ker** and since φ is $\boldsymbol{\alpha}$-boundable, we have $\varphi(x)$ and $\varphi^{\mathfrak{St}(\mathbf{Ker})}(x)$. It follows that $(\exists\beta)\,(\exists x \subseteq \mathbf{PP}\,(\boldsymbol{\omega}_{\beta}))\;\varphi(x)$ holds in the sense of $\mathfrak{St}\,(\mathbf{Ker})$; thus we have demonstrated the following result:

6223. METATHEOREM. Let $\boldsymbol{\alpha}$ be well-defined, let $\varphi(x)$ be $\boldsymbol{\alpha}$-boundable and suppose that $\omega(x, z)$ well describes a symmetric **g**-filter on x and determines a permutation model for $\varphi(x)$. Then the formula

$$(\exists x \in \mathbf{PPP}\,(\mathbf{On}))\;\varphi(x)$$

is consistent with $\mathbf{TS}$ + D3.

SECTION 3

Support of one model-class over another

The notion of a support has interesting applications in the study of model-classes, especially in the study of a pair of model-classes X, Y such that $X \subseteq Y$. In this section we discuss these applications. For the sake of brevity we consider the theory **TSS**, D3 (denoted here by **TSS‴**); however it would be easy to reformulate the main results of this section for full model-classes in **TSS″**.

First consider two possibilities of composing direct models determined by the F-definitions $\mathfrak{St}$ (**M**) and $\mathfrak{Real}$. We have the two following diagrams: (**TS‴** denotes **TS**, D3)

6301.

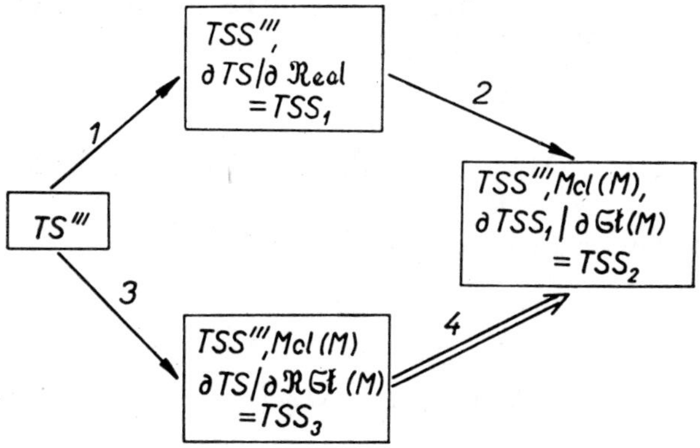

6302.

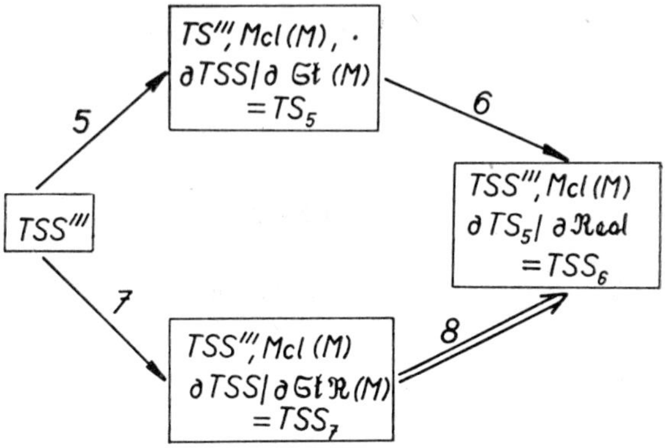

Consider the first diagram. Arrow 1 is $\mathfrak{Dir}\,(\partial\mathbf{TS}/\partial\mathfrak{Real})$ which is a model of $\mathbf{TS}'''$ in $\mathbf{TSS}_1$. Arrow 2 is $\mathfrak{Dir}\,(\partial\mathbf{TSS}_1/\partial\mathfrak{St}(\mathbf{M}))$ which is a model of $\mathbf{TSS}_1$ in $\mathbf{TSS}_2$. The axioms of $\mathbf{TSS}_2$ are as follows:

(a) the axioms of $\mathbf{TSS}'''$ and the fixing Mcl $(\mathbf{M})$;

(b) $*$-class-variables for subclasses of $\mathbf{M}$ and other $*$-notions corresponding to **TSS**-notions are defined by $\partial\mathbf{TSS}/\partial\mathfrak{St}\,(\mathbf{M})$;

c) $\bigcirc$-class variables are defined for those $*$-classes which are real in the $*$-sense;

(d) $X^{\bigcirc} \in^{\bigcirc} Y^{\bigcirc}$ is defined as $X^{\bigcirc} \in^{*} Y^{\bigcirc}$;

(e) all other $\bigcirc$-notions corresponding to **TS**-notions are defined by their definitions from $\partial\mathbf{TSS}/\partial\mathfrak{Real}^*$. Obviously, the following are provable in $\mathbf{TSS}'''$, Mcl $(\mathbf{M})$, $\partial\mathbf{TSS}/\partial\mathfrak{St}\,(\mathbf{M})$:

$$\text{Real}^*\,(X^*) \equiv_{\bullet} X^* \subseteq \mathbf{M}\,\&\,(\forall y \in \mathbf{M})\,(X^* \cap y \in \mathbf{M}) \text{ and } X^* \in^* Y^* \equiv X^* \in Y^*\,.$$

Hence the following are provable in $\mathbf{TSS}_2$:

6303.

(1) $(\forall X)\,[(\exists X^{\bigcirc})\,(X = X^{\bigcirc}) \equiv_{\bullet} X \subseteq \mathbf{M}\,\&\,(\forall y \in \mathbf{M})\,(X \cap y \in \mathbf{M})]\,\&\,sec,$

(2) $(\forall X^{\bigcirc}, Y^{\bigcirc})\,(X^{\bigcirc} \in^{\bigcirc} Y^{\bigcirc} = X^{\bigcirc} \in Y^{\bigcirc})\,\&\,sec.$

The pair of formulas (1), (2) is an F-definition which we denote by $\mathfrak{R}\mathfrak{St}\,(\mathbf{M})$.

Now consider arrow 3 and its terminal theory $\mathbf{TSS}_3$ and let the notions defined by $\partial\mathbf{TS}/\partial\mathfrak{R}\mathfrak{S}\mathrm{t}(\mathbf{M})$ be denoted by the superscript ○. If we add to $\mathbf{TSS}_3$ the derivation $\partial\mathbf{TSS}/\partial\mathfrak{S}\mathrm{t}(\mathbf{M})$ (where new notions are denoted by superscript $*$) we obtain a conservative extension $\mathbf{TSS}_4$ of $\mathbf{TSS}_3$. The axioms of $\mathbf{TSS}_2$ which occur sub (a) and (b) above are axioms of $\mathbf{TSS}_4$, those sub (c) and (d) are provable from (1) and (2) and those sub (e) coincide with the corresponding axioms from $\partial\mathbf{TSS}/\partial\mathfrak{R}\mathfrak{S}\mathrm{t}(\mathbf{M})$. Hence $\mathbf{TSS}_2$ is equivalent to $\mathbf{TSS}_4$ and is therefore a conservative extension of $\mathbf{TSS}_3$. Thus the diagram commutes and 4 is a faithful model; hence 3 is a model and a **TSS**-formula φ holds in $1 * 2$ iff it holds in 3.

Now consider the second diagram. As in the preceding discussion we verify that the diagram commutes and that 8 is a faithful identity where the F-definition $\mathfrak{S}\mathrm{t}\mathfrak{R}(\mathbf{M})$ is given by:

6304.

(3) $(\forall X)\,[(\exists X^*)(X^* = X) \equiv .\ \mathrm{Real}(X)\,\&\, X \subseteq \mathbf{M}]\,\&\, sec,$

(4) $(\forall X^*, Y^*)(X^* \in^* Y^* \equiv X^* \in Y^*)\,\&\, sec.$

We shall often have occasion to consider two model-classes X, Y such that $X \subseteq Y$.

6305. DEFINITION ($\mathbf{TSS}'''$).

$$2\mathrm{Mcl}(X, Y) \equiv .\ \mathrm{Mcl}(X)\,\&\,\mathrm{Mcl}(Y)\,\&\, X \subseteq Y.$$

Consider the theory $\mathbf{TSS}'''$, $2\mathrm{Mcl}(\mathbf{M}, \mathbf{N})$ and ask if $2\mathrm{Mcl}(\mathbf{M}, \mathbf{N})$ is a specification for a reasonable F-definition. $2\mathrm{Mcl}(\mathbf{M}, \mathbf{N})$ is certainly a specification for $\mathfrak{S}\mathrm{t}(\mathbf{N})$ stronger than $\mathrm{Mcl}(\mathbf{N})$ and if we denote the notions defined by $\partial\mathbf{TSS}/\partial\mathfrak{S}\mathrm{t}(\mathbf{N})$ by the superscript □ then by 3228 we have

$$\mathbf{TSS}''',\ 2\mathrm{Mcl}(\mathbf{M}, \mathbf{N}),\ \partial\mathbf{TSS}/\partial\mathfrak{S}\mathrm{t}(\mathbf{N}) \vdash \mathrm{Mcl}^{\square}(\mathbf{M})\,;$$

i.e. **M** is a class of the model and a model-class in the sense of the model. The easiest thing we could do is to construct $\mathfrak{S}\mathrm{t}(\mathbf{M})$ in $\mathfrak{S}\mathrm{t}(\mathbf{N})$, i.e. to consider the composition $9 * 10$ in the following diagram:

6306.

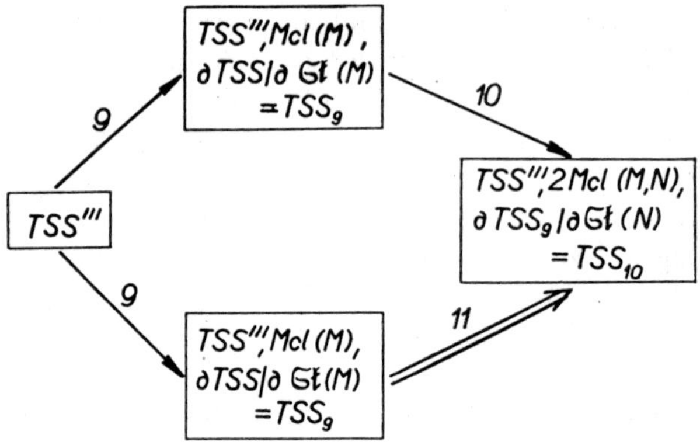

But this is of no use because $9 * 10$ is equivalent to $9 * 11$ and 11 is a faithful identity. Indeed, consider the theory $\mathbf{TSS}_{10}$ in the diagram. We have constants **M**, **N**. First, □-class variables are defined for subclasses of **N** and all other notions of $\partial\mathbf{TSS}/\partial\mathfrak{St}(\mathbf{N})$ are defined. We have $\mathrm{Mcl}^{\square}(\mathbf{M})$; we define $*$-class variables for those □-classes which are □-subclasses of **M** and define $X^* \in^* Y^*$ as $X^* \in^{\square} Y^*$, etc. But one shows as in the preceding diagrams that 11 is a model and that the diagram commutes. Moreover, $\mathbf{TSS}_{10}$ is a conservative extension of $\mathbf{TSS}_9$ so that 11 is a faithful identity. Hence a **TSS**-formula φ holds in $9 * 10$ iff it holds in 9 and therefore the composition $9 * 10$ is of no interest.

A slightly more complicated procedure is required. We recall 1444 telling us that

$$\mathbf{TSS}''', 2\mathrm{Mcl}(\mathbf{M}, \mathbf{N}), \partial\mathbf{TSS}/\partial\mathfrak{St}(\mathbf{N}) \vdash \mathrm{Real}^{\square}(\mathbf{M})$$

(as a consequence of $\mathrm{Mcl}^{\square}(\mathbf{M})$), i.e. if we construct the real model in $\mathfrak{Dir}(\partial\mathbf{TSS}/\partial\mathfrak{St}(\mathbf{N}))$, **M** will still be a class of this model. Hence it seems to be reasonable to consider the composition $12 * 13 * 14$ in the following diagram:

6307.

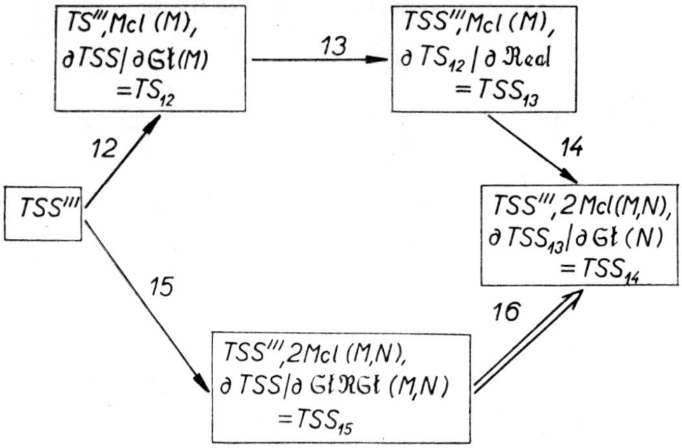

Consider the theory denoted by $\mathbf{TSS}_{14}$. In this theory, □-class-variables are variables for subclasses of **N** and □-notions are defined by $\partial\mathbf{TSS}/\partial\mathfrak{St}(\mathbf{N})$. $\mathfrak{Real}$ is constructed in the □-sense; in particular, ○-class variables are defined for □-classes which are □-real. We have $\mathbf{TSS}_{14} \vdash \mathrm{Mcl}^{\circ}(\mathbf{M})$ and we continue to define ∗-notions constructing $\mathfrak{St}(\mathbf{M})$ in the ○-sense. Thus ∗-classes are defined as ○-subclasses of **M**, or, equivalently, as those subclasses $X \subseteq \mathbf{M}$ for which $(\forall y \in \mathbf{N})(X \cap y \in \mathbf{N})$.

This leads to the following F-definition (called $\mathfrak{St}\mathfrak{R}\mathfrak{St}(\mathbf{M}, \mathbf{N})$):

6308.

(5) $(\forall X)\,[(\exists X^*)\,(X = X^*) \equiv .\, X \subseteq \mathbf{M} \,\&\, (\forall y \in \mathbf{N})\,(X \cap y \in \mathbf{N})] \,\&\, sec,$

(6) $(\forall X^*, Y^*)\,(X^* \in^* Y^* \equiv X^* \in Y^*) \,\&\, sec.$

We see that the diagram commutes (since 16 is an identity). It is easy to show that 16 is a faithful identity; i.e. a formula φ holds in $12 * 13 * 14$ iff it holds in 15. Moreover, if a formula holds in 12 then it obviously holds in $12 * 13 * 14$ and hence in 15. In particular, by 5123, the following holds in 15:

$$\text{(i) (St)}; \quad \text{(ii) (S1)} \to \text{(S5)}; \quad \text{(iii) (S1 \& Bd)} \to \text{(S6)}.$$

We consider one more diagram which will be useful later:

6309.

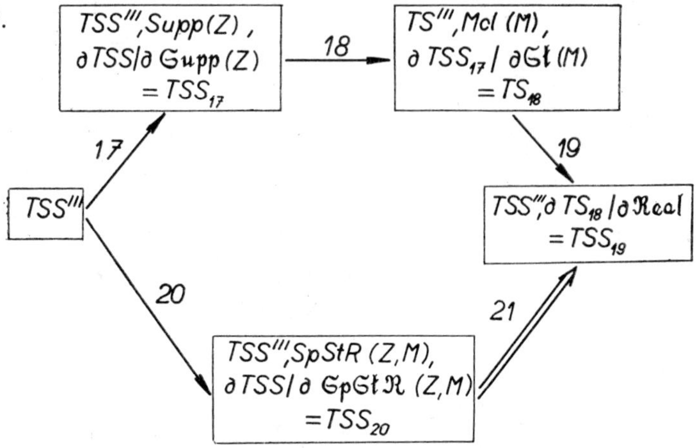

Here SpStR $(\mathbf{Z}, \mathbf{M})$ is defined as

$$\mathrm{Mcl}(\mathbf{M}) \,\&\, \mathrm{Real}(\mathbf{Z}) \,\&\, \mathbf{Z} \subseteq \mathbf{M} \,\&$$

$$\&\, (\forall u, v \subseteq \mathbf{M})\,(\mathrm{Dep}(\mathbf{M}, u, \mathbf{Z}) \,\&\, \mathrm{Dep}(\mathbf{M}, v, \mathbf{Z}) \rightarrow$$

$$\rightarrow \mathrm{Dep}(\mathbf{M}, u - v, \mathbf{Z}))\,,$$

where $\mathrm{Dep}(\mathbf{M}, u, \mathbf{Z}) \equiv (\exists r \in \mathbf{M})\,(u = r\text{''}\mathbf{Z})$.

$\mathfrak{SpStR}(\mathbf{Z}, \mathbf{M})$ is the following F-definition:

6310.

(7) $(\forall X)\,[(\exists X^{\triangle})\,(X = X^{\triangle}) \equiv .\, X \subseteq \mathbf{M} \,\&\, (\forall y \in \mathbf{M})\,(\mathrm{Dep}(\mathbf{M}, y \cap X, \mathbf{Z})] \,\&$
$\&\ sec$;

(8) $(\forall X^{\triangle}, Y^{\triangle})\,(X^{\triangle} \in^{\triangle} Y^{\triangle} \equiv X^{\triangle} \in Y^{\triangle}) \,\&\ sec$.

The reader may easily show that in the last diagram all arrows are models, that the diagram commutes and that 21 is a faithful identity.

We now discuss the notion of a support in the sense of some of the models just discussed.

6311. THEOREM ($\mathbf{TSS}'''$, Mcl ($\mathbf{A}$), $\partial\mathbf{TSS}/\partial\mathfrak{StR}(\mathbf{A})$). If Z is a total support in the sense of $\mathfrak{StR}(\mathbf{A})$ then $\mathbf{V} = \mathbf{Cstr}(Z \times \mathbf{A})$.

Proof. Let $\mathbf{Z}$ be a fixed total support in the sense of $\mathfrak{StRSt}(\mathbf{A})$ and let R be the full upward extension $\mathbf{Fup}$ in the sense of $\mathfrak{StR}(\mathbf{A})$. (The existence of such an R follows from the fact that $\mathbf{Z}$ is a total support in the sense of $\mathfrak{StR}(\mathbf{A})$.) We set $\mathbf{B} = \mathbf{Cstr}(\mathbf{Z} \times \mathbf{A})$ and observe that $R \subseteq \subseteq \mathbf{A} \subseteq \mathbf{B}$ and R is a real class in the sense of $\mathfrak{StR}(\mathbf{B})$. We define, in the sense of $\mathfrak{StR}(\mathbf{B})$, a real morphism of $\mathbf{C}(R)$ onto a complete class C as follows:

$$H``\{0\} = \mathbf{Cnv}(\mathbf{Imb}^{\mathfrak{StR}(\mathbf{A})}),$$

$$\alpha > 0 \rightarrow H``\{\alpha\} = \{\langle x, y\rangle; y \in \mathbf{C}(R) \;\&\; \mathbf{Ext}_R(y) \subseteq \mathbf{D}(H``\alpha) \;\&\; x = = (H``\alpha)``\,\mathbf{Ext}_R(y)\}.$$

Further we let $H``\,\mathbf{On} = F$ and $\mathbf{W}(F) = C$. The class C is complete (it is even a model-class) and we have $C \subseteq \mathbf{Cstr}(\mathbf{Z} \times \mathbf{A})$ and $\mathbf{D}(F) = \mathbf{C}(R)$. We prove $C = \mathbf{V}$. If, on the contrary, $\mathbf{V} \neq C$, then we may choose an element x of least rank in $\mathbf{V} - C$; clearly $x \subseteq C$. We let $(\mathbf{Cnv}(F))``\,x = y$. Since y is a semiset in the sense of $\mathfrak{StR}(\mathbf{A})$, and since R is saturated-universal in the sense of $\mathfrak{StR}(\mathbf{A})$, there exists some x_0 such that $\mathbf{Ext}_R(x_0) = y$. Hence we have $\langle x, x_0\rangle \in F$ and $x \in C$, a contradiction. We have proved $\mathbf{V} \subseteq \mathbf{Cstr}(\mathbf{Z} \times \mathbf{A})$, so that $\mathbf{V} = \mathbf{Cstr}(\mathbf{Z} \times \mathbf{A})$.

6312. COROLLARY ($\mathbf{TSS}'''$, 2Mcl ($\mathbf{A}, \mathbf{B}$), $\partial\mathbf{TSS}/\partial\mathfrak{StRSt}(\mathbf{A}, \mathbf{B})$). If Z is a total support in the sense of $\mathfrak{StRSt}(\mathbf{A}, \mathbf{B})$ then $\mathbf{B} = \mathbf{Cstr}(Z \times \mathbf{A})$.

Proof. We recall that the classes of the model $\mathfrak{Dir}(\partial\mathbf{TSS}/\partial\mathfrak{StRSt}(\mathbf{A}, \mathbf{B}))$ are the subclasses of $\mathbf{A}$ which are real in the sense of $\mathfrak{St}(\mathbf{B})$; in particular, the semisets of the model are the subclasses of $\mathbf{A}$ which are elements of $\mathbf{B}$ and the sets of the model are elements of $\mathbf{A}$. By the preceding theorem we have $[\mathbf{V} = \mathbf{Cstr}(Z \times \mathbf{A})]^{\mathfrak{StR}(\mathbf{B})}$ and hence by 3525 we have $\mathbf{B} = = \mathbf{Cstr}(Z \times \mathbf{A})$ (see Diagram 6307).

6313. THEOREM ($\mathbf{TSS}'''$, SpStR ($\mathbf{Z}, \mathbf{A}$), $\partial\mathbf{TSS}/\partial\mathfrak{SpStR}(\mathbf{Z}, \mathbf{A})$, $\partial\mathbf{TSS}/\partial\mathfrak{StRSt}(\mathbf{A}, \mathbf{Cstr}(\mathbf{Z} \times \mathbf{A}))$).

(1) $\mathbf{Z}$ is a total support in the sense of $\mathfrak{StRSt}(\mathbf{A}, \mathbf{Cstr}(\mathbf{Z} \times \mathbf{A}))$;

(2) For any X, X is a class in the sense of $\mathfrak{SpStR}(\mathbf{Z}, \mathbf{A})$ iff it is a class in the sense of $\mathfrak{StRSt}(\mathbf{A}, \mathbf{Cstr}(\mathbf{Z} \times \mathbf{A}))$.

Proof. (1) Let $\mathbf{R}^{\triangle}$ be the full upward extension of $\mathbf{E}$ in the sense of $\mathfrak{SpStR}(\mathbf{Z}, \mathbf{A})$ and let $\mathbf{Imb}^{\triangle}$ be the embedding of the universal class into $\mathbf{R}^{\triangle}$ in the sense of this model. Clearly $\mathbf{R}^{\triangle}$ is an almost universal, extensional and regular relation. We define a relation H by the same definition as in the preceding proof and let $F = H\text{''}\,\mathbf{On}$, $\mathbf{W}(F) = \mathbf{B}$. $\mathbf{B}$ is a model-class: it is closed since it is morphic to $\mathbf{C}(\mathbf{R}^{\triangle})$ w.r.t. $\mathbf{E}$, $\mathbf{R}^{\triangle}$, it is obviously complete and it is an almost universal class because $\mathbf{R}^{\triangle}$ is an almost universal relation. Suppose that x is a semiset in the sense of $\mathfrak{S}\mathfrak{t}\mathfrak{R}\mathfrak{S}\mathfrak{t}(\mathbf{A}, \mathbf{B})$, i.e. $x \in \mathbf{P}(\mathbf{A}) \cap \mathbf{B}$; we prove $\mathrm{Dep}^{\mathfrak{S}\mathfrak{t}\mathfrak{R}(\mathbf{A})}(x, \mathbf{Z})$. For some $y \in \mathbf{C}(\mathbf{R}^{\triangle})$ we have $\langle x, y\rangle \in F$ and since the composition of $\mathbf{Imb}^{\triangle}$ and F is an identity on $\mathbf{A}$ we have $\mathbf{Ext}_{\mathbf{R}^{\triangle}}(y) \subseteq \mathbf{W}(\mathbf{Imb}^{\triangle})$, $\mathbf{Ext}_{\mathbf{R}^{\triangle}}(y) = \mathbf{Imb}^{\triangle}\text{''}\,x$ and $x = (\mathbf{Cnv}(\mathbf{Imb}^{\triangle}))\text{''}\,\mathbf{Ext}_{\mathbf{R}^{\triangle}}(y)$. Since $y \in \mathbf{C}(\mathbf{R}^{\triangle})$ and both $\mathbf{R}^{\triangle}$ and $\mathbf{Imb}^{\triangle}$ are classes in the sense of $\mathfrak{S}\mathfrak{p}\mathfrak{S}\mathfrak{t}\mathfrak{R}(\mathbf{Z}, \mathbf{A})$, both $\mathbf{Ext}_{\mathbf{R}^{\triangle}}(y)$ and x are semisets of $\mathfrak{S}\mathfrak{p}\mathfrak{S}\mathfrak{t}\mathfrak{R}(\mathbf{Z}, \mathbf{A})$, i.e. x is dependent on $\mathbf{Z}$ in the sense of $\mathfrak{S}\mathfrak{t}\mathfrak{R}(\mathbf{A})$. Consequently, $\mathbf{Z}$ is a total support in the sense of $\mathfrak{S}\mathfrak{t}\mathfrak{R}\mathfrak{S}\mathfrak{t}(\mathbf{A}, \mathbf{B})$. Using the above corollary we obtain $\mathbf{B} = \mathbf{Cstr}(\mathbf{Z} \times \mathbf{A})$, which completes the proof of (1).

(2) In both models the sets are the elements of $\mathbf{A}$; the semisets of the first are the subsets x of $\mathbf{A}$ such that $x = r\text{''}\mathbf{Z}$ for some $r \in \mathbf{A}$ and the semisets of the second are the subsets of $\mathbf{A}$ which are elements of $\mathbf{Cstr}(\mathbf{Z} \times \mathbf{A})$. Clearly every semiset of the first model is a semiset of the second, since $\mathbf{Z}$ is a class of $\mathfrak{S}\mathfrak{t}\mathfrak{R}(\mathbf{Cstr}(\mathbf{Z} \times \mathbf{A}))$. As for the converse, we have proved in (1) that every semiset of the second model is a semiset of the first. The classes of the first model are the subclasses of $\mathbf{A}$ which intersect every semiset of the model in a semiset of the model; the classes of the second model are the subclasses of $\mathbf{A}$ which are real in $\mathfrak{S}\mathfrak{t}(\mathbf{Cstr}(\mathbf{Z} \times \mathbf{A}))$, i.e. which intersect every semiset of the second model in a semiset of that model. It follows that both models have the same classes. Since the membership predicate in both models coincides with the actual membership relation we obtain the following

6314. METALEMMA. The following diagram commutes and both 24 and 25 are faithful identities.

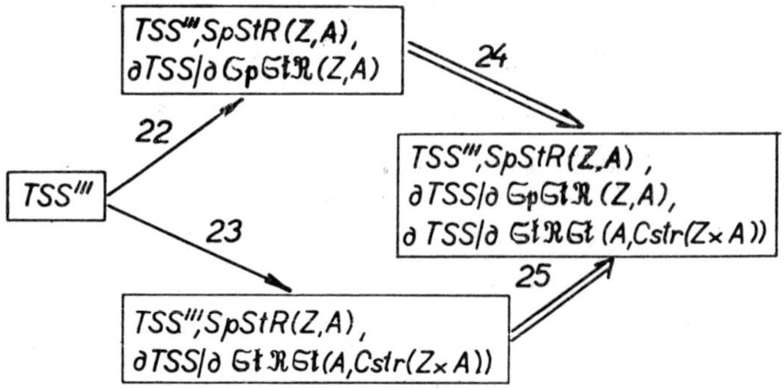

6315. COROLLARY. (1) (**TSS**''', Mcl (**A**), ∂**TSS**/$\partial$$\mathfrak{S}t\mathfrak{R}$ (**A**)). If Z is a support in the sense of $\mathfrak{S}$t$\mathfrak{R}$ (**A**) and $x \subseteq$ **A** then $\mathrm{Dep}^{\mathfrak{S}t\mathfrak{R}(\mathbf{A})}(x, Z)$ iff $x \in$ $\in$ **Cstr** (Z × **A**).

(2) (**TS**''', ∂**TSS**/$\partial$$\mathfrak{S}$t (**L**)). If Z is a support in the sense of $\mathfrak{S}$t (**L**) and $x \subseteq$ **L** then $\mathrm{Dep}^{\mathfrak{S}t(\mathbf{L})}(x, Z)$ iff $x \in$ **Cstr** (Z).

Indeed, we proved (1) for fixed **A** and **Z** in the theory indicated in the preceding theorem (cf. the proof of that theorem); it follows that our statement (1) is provable in the theory indicated sub (1). (2) follows from (1) because **TSS**''' $\vdash$ **Cstr** (Z) = **Cstr** (Z × **L**).

6316. DEFINITION (**TSS**'''). Let X, Y be model-classes and let $X \subseteq Y$. We say that $Z \subseteq X$ is a *support for Y over X* (Support (Z, X, Y)) iff

(1) $(\forall y \in Y)(Z \cap y \in Y)$ (consequently, Z is a real class),

(2) $(\forall x \in \mathbf{P}(X) \cap Y)(\exists r \in X)(x = r\text{''}Z)$.

If Z is a set then it is called a *set support for Y over X* ((1) can then be replaced by $Z \in Y$).

6317. Remark (**TSS**''', 2Mcl (**A**, **B**), ∂**TSS**/$\partial$$\mathfrak{S}t\mathfrak{R}$$\mathfrak{S}$t (**A**, **B**)). For any Z, (1) Z is a support for **B** over **A** iff Z is a total support in the sense of $\mathfrak{S}$t$\mathfrak{R}$$\mathfrak{S}$t (**A**, **B**); (2) Z is a set support for **B** over **A** iff Z is a semiset total support in the sense of $\mathfrak{S}$t$\mathfrak{R}$$\mathfrak{S}$t (**A**, **B**).

6318. THEOREM (**TSS**'''). Let $A \subseteq B \subseteq C$ be model-classes and let B be a model-class with E1. If Z is a support for C over A then Z is a support for C over B.

Proof. Clearly $Z \subseteq B$ and $(\forall y \in C)(Z \cap y \in C)$.

Suppose that $x \subseteq B$ and $x \in C$. We must find $r \in B$ such that $r\text{''}Z = x$. There exists some $a \in B$ such that $x \subseteq a$; we let $f \in B$ be some one-one mapping of a onto an ordinal α and $x_0 = f\text{''}x$. Clearly $x_0 \subseteq A$ and $x_0 \in C$, so that $x_0 = s\text{''}Z$ for some $s \in A$. If we denote by r the composition of **Cnv** (f) and s we obtain $r\text{''}Z = x$; hence Z is a support for C over B.

We now derive some theorems on set-supports.

6319. THEOREM (**TSS**'''). Let $A \subseteq B \subseteq C$ be model-classes and let B be a model-class with E1. If z_1 is a support for B over A and z_2 is a support for C over B then there exists a set-support for C over A.

Proof. Fix **A**, **B**, **C**. Clearly $z_1 \subseteq \mathbf{A}$ and $z_1 \in \mathbf{B}$; similarly $z_2 \subseteq \mathbf{B}$ and $z_2 \in \mathbf{C}$. We may suppose that $z_2 \subseteq \mathbf{A}$; for if $z_2 \subseteq a \in \mathbf{B}$ and if $f \in \mathbf{B}$ is a one-one mapping of a onto an ordinal then $\bar{z}_2 = f\text{''}z_2$ is a subset of **A** and, since

$\mathrm{Sim}^{\mathfrak{StRSt}(\mathbf{B},\mathbf{C})}(z_2, \bar{z}_2)$, $\bar{z}_2$ is a support for $\mathbf{C}$ over $\mathbf{B}$. If we let $z = z_2 \times z_1$ then $z \subseteq \mathbf{A}$ and $z \in \mathbf{C}$. If $x \subseteq \mathbf{A}$ and $x \in \mathbf{C}$ then $x = r_2'' z_2$ for some r_2 such that $r_2 \in \mathbf{B}$, $r_2 \subseteq \mathbf{A}$. Since $r_2 \in \mathbf{B} \cap \mathbf{P}(\mathbf{A})$ there exists $r_1 \in \mathbf{A}$ such that $r_2 = r_1'' z_1$. If we let $\langle u, v, w\rangle \in r \equiv \langle\langle u, v\rangle, w\rangle \in r_1$ then $r \in \mathbf{A}$. If $u \in x$ then $\langle u, v\rangle \in r_2$ for some $v \in z_2$ and hence $\langle\langle u, v\rangle, w\rangle \in r_1$ for some $w \in z_1$. Hence we have $\langle u, v, w\rangle \in r$, $\langle v, w\rangle \in z$ and $u \in r''z$, so that $x \subseteq r''z$. Conversely, if $u \in r''z$ then $\langle u, v, w\rangle \in r$ for some $\langle v, w\rangle \in z$; hence $\langle u, v\rangle \in$ $\in r_2$, $v \in z_2$ and we have $u \in x$. We have proved $x = r''z$, i.e. z is a support for $\mathbf{C}$ over $\mathbf{A}$.

6320. Theorem (**TSS‴**). Let $A \subseteq B \subseteq C$ be model-classes and suppose that B is a model-class with E1. If there exists a set-support for C over A then there exists a set-support for B over A.

Proof. Again fix $\mathbf{A}$, $\mathbf{B}$, $\mathbf{C}$. By 4134 it is sufficient to show that (Pot) and (Bd) hold in the sense of $\mathfrak{StRSt}(\mathbf{A}, \mathbf{B})$. If $a \in \mathbf{A}$ then by $(\mathrm{E1})^{\mathfrak{St}(\mathbf{B})}$ there exists $f \in \mathbf{B}$ which is a one-one mapping f of some ordinal α onto $\mathbf{P}(a) \cap \mathbf{B}$. For $\gamma < \alpha$ we let $\langle y, \gamma\rangle \in r \equiv y \in f'\gamma$; in the sense of $\mathfrak{StRSt}(\mathbf{A}, \mathbf{B})$, r is an exact functor which codes all subsemisets of a. Hence (Pot) holds in the sense of $\mathfrak{StRSt}(\mathbf{A}, \mathbf{B})$. Suppose that z is a disjointed support for $\mathbf{C}$ over $\mathbf{A}$ and that $z \subseteq a \in \mathbf{A}$. If $x \subseteq \mathbf{A}$ and $x \in \mathbf{B}$ then $x = r''z$ for some disjointed relation $r \in \mathbf{A}$ with $\mathbf{D}(r) \subseteq a$. For $u \in a$ we let $f'u = r''\{u\}$, $b =$ $= \mathbf{W}(f)$ and $y = f''z$. We have $y \subseteq b$, $\bigcup(y) = x$ and $b \preccurlyeq^{\mathfrak{StRSt}(\mathbf{A},\mathbf{B})} a$; hence (Bd) holds in the sense of $\mathfrak{StRSt}(\mathbf{A}, \mathbf{B})$.

6321. Corollary. (**TSS‴**). Let $A \subseteq B \subseteq C$ be model-classes and suppose that B is a model-class with E1. A set-support for C over A exists iff there exist set-supports for C over B and for B over A.

We now state a theorem showing the significance of the class **HDf** of all hereditarily definable sets in the theory of semisets with the axiom (SConstr).

6322. Theorem (**TSS‴**). If $\mathbf{V} = \mathbf{Cstr}(x)$ for some $x \subseteq \mathbf{On}$ then $\mathbf{V}$ has a set-support over **HDf**.

Proof. Let $x \subseteq \xi \in \mathbf{On}$. To begin with we construct a certain definable (not necessarily hereditarily definable) "definably complete" Boolean algebra and an ultrafilter on it which is closed under definable intersections. If we now let $b_0 = \mathbf{P}(\mathbf{P}(\xi)) \cap \mathbf{Df}$ (the set of all definable sets of subsets of ξ) then $b_0 \in \mathbf{Df}$. For $u, v \in b_0$ we let $\langle u, v\rangle \in c_0 \equiv u = \mathbf{P}(\xi) - v$; clearly $c_0 \in$ $\in \mathbf{Df}$. For $q \subseteq b_0$, $q \in \mathbf{Df}$, $u \in b_0$ we let $\langle u, q\rangle \in f_0 \equiv u = \bigcap q$; clearly $f_0 \in \mathbf{Df}$. If we define $\boldsymbol{b}_0 = \langle b_0, c_0, f_0\rangle$ and $u \in z_0 \equiv x \in u$ for $u \in b_0$ then we have $u \in z_0 \equiv \mathbf{P}(\xi) - u \notin z_0$ and $q \subseteq z_0 \equiv \bigcap q \in z_0$ for $q \neq 0$.

Let F be the one-one mapping of **Df** onto **On** which was defined in 3406. F has the following property: for any $x \subseteq \mathbf{Df}$, $x \in \mathbf{Df} \equiv F\text{''}x \in \mathbf{Df}$. We let $b = F\text{''}b_0$ and define the operations on b in such a way that F becomes an isomorphism of $\boldsymbol{b}_0$ and $\boldsymbol{b}$. We have $\boldsymbol{b} \in \mathbf{Df}$ and since the elements of $\boldsymbol{b}$ are ordinal numbers, it follows that $\boldsymbol{b} \in \mathbf{HDf}$. If we let $z = F\text{''}z_0$ then, in the sense of $\mathfrak{StR}(\mathbf{HDf})$, $\boldsymbol{b}$ is a complete Boolean algebra and z is a complete ultrafilter on $\boldsymbol{b}$. Let $\mathbf{A} = \mathbf{Cstr}(\mathbf{HDf} \times z)$.

By Theorem 6313 z is a total support in the sense of $\mathfrak{StRSt}(\mathbf{HDf}, \mathbf{A})$ hence z is a set-support for $\mathbf{A}$ over $\mathbf{HDf}$. To prove that $\mathbf{A} = \mathbf{V}$ it suffices to show that $x \in \mathbf{A}$. We define a function g by setting $g\text{'}u = \bigcup((\mathbf{Cnv}(F)\text{''}\, u)$ for all $u \in b$. Since the sum of a definable set is definable, g is definable and, since g assigns to each ordinal number in its domain a definable set of ordinal numbers, g is hereditarily definable. Hence $g\text{''}z \in \mathbf{A}$ and $\bigcap(g\text{''}z) \in \mathbf{A}$. We also have $\bigcap(g\text{''}z) = x$, since $\alpha \in x$ iff for all $u \in z_0$ there exists $a \in u$ such that $x \in a$; one implication is obvious while if $\alpha \notin x$ then $\mathbf{P}(\xi - \{\alpha\}) \in$ $\in z_0$ and x is in no $a \in \mathbf{P}(\xi - \{\alpha\})$. Hence we have $\alpha \in x \equiv \alpha \in \bigcap_{u \in z_0} \bigcup(u) \equiv$ $\equiv \alpha \in \bigcap(g\text{''}z)$.

6323. COROLLARY ($\mathbf{TSS}'''$, $\mathbf{x} \subseteq \mathbf{On}$, $\partial\mathbf{TSS}/\partial\mathfrak{St}(\mathbf{Cstr}(\mathbf{x}))$). $\mathbf{Cstr}(\mathbf{x})$ has a set-support over $\mathbf{HDf}^{\mathfrak{St}(\mathbf{Cstr}(\mathbf{x}))}$.

6324. THEOREM ($\mathbf{TSS}'''$). (a) If $M = \mathbf{Cstr}(x)$ and if a model-class N has a set-support over M then $N = \mathbf{Cstr}(y)$ for some y.

(b) If $x \subseteq \mathbf{On}$ and if $M = \mathbf{Cstr}(x)$ has a set-support over a model-class N with E1 then $N = \mathbf{Cstr}(y)$ for some y.

Proof. (a) follows from Theorem 6312; if y is a set-support for N over M then $N = \mathbf{Cstr}(\mathbf{Cstr}(x) \times y) = \mathbf{Cstr}(x \times y)$.

(b) We may suppose $M = \mathbf{V}$ and fix $\mathbf{N}$. By Theorems 6318, 6312 it suffices to find a model-class $\mathbf{N}_0 \subseteq \mathbf{N}$ such that $\mathbf{N}_0 = \mathbf{Cstr}(q)$ for some q and such that $\mathbf{V}$ has a support over $\mathbf{N}_0$. By assumption there exists, in the sense of $\mathfrak{StR}(\mathbf{N})$, a complete Boolean algebra $\boldsymbol{b}$ and a complete ultrafilter z such that z is a total support in the sense of $\mathfrak{StR}(\mathbf{N})$. We have $x = r\text{''}z$ for some $r \in \mathbf{N}$. If we let $\mathbf{N}_0 = \mathbf{Cstr}(\boldsymbol{b} \times r)$ then $\mathbf{N}_0 \subseteq \mathbf{N}$ and z is a support for $\mathbf{V}$ over $\mathbf{N}_0$. For, z is a complete ultrafilter on $\boldsymbol{b}$ in the sense of $\mathfrak{StR}(\mathbf{N}_0)$; hence z is a support and, since $x = r\text{''}z$, $\mathbf{Cstr}(x)$ is included in a unique model-class for which z is a support over $\mathbf{N}_0$; consequently this model-class is $\mathbf{V}$.

6325. COROLLARY ($\mathbf{TSS}'''$). If $\mathbf{V} = \mathbf{Cstr}(x)$ then $\mathbf{HDf} = \mathbf{Cstr}(y)$ for some y.

We shall now continue the discussion begun at the end of Chapter III, Sect. 5; we shall demonstrate the independence of the axiom of constructibility from the axiom of definability. (See also Section 4 of the present Chapter.)

6326. We define an algebra **Comb** as follows. Let ϑ be the smallest cardinal $\xi > 0$ such that $\alpha < \xi$ implies $2^{\aleph_\alpha} < \xi$, i.e. $\vartheta = \bigcup_{\omega} h'n$ where $h'0 = 0$, $h'(n+1) = 2^{\aleph_{h'n}}$, **Comb** is the algebra with the base $\boldsymbol{a} = \prod^{l}_{\iota \in \vartheta} \boldsymbol{b}^{\iota+1}_{\iota+1}$ where $l = \mathbf{P}(\vartheta)$. (See 6101 for the definition of $\boldsymbol{b}^{\iota+1}_{\iota+1}$.)

Consider **TSS**, D3, (Constr), (S **Comb**). We put $\boldsymbol{a}_{<\alpha} = \prod^{l/\alpha}_{\iota<\alpha} \boldsymbol{b}^{\iota+1}_{\iota+1}$, similarly for $>\alpha$, $\geq\alpha$ and $\neq\alpha$. Clearly $\boldsymbol{a}$ is isomorphic to $\boldsymbol{a}_{<\alpha} \odot \boldsymbol{a}_{\geq\alpha}$ and also to $\boldsymbol{a}_{<\alpha} \odot \boldsymbol{b}^{\alpha+1}_{\alpha+1} \odot \boldsymbol{a}_{>\alpha}$.

6327. Lemma. (**TSS′**, E1, Cont).

(a) Every exclusive system in $\boldsymbol{a}_{<\alpha}$ has power at most $\aleph_\alpha$;

(b) $\boldsymbol{a}_{\geqq\alpha}$ is $\aleph_\alpha$-multiplicative.

Proof. (a) As in the proof of Lemma 6117, $\boldsymbol{a}_{<\alpha}$ may be represented by $\prod^{j}_{x \in s} \mathbf{at}(2)$ where $s = \{\langle\gamma, \delta\rangle; \delta < \alpha \,\&\, \gamma < \omega_{\delta+1}\}$ and

$$x \in j \equiv .\, x \subseteq s \,\&\, \mathbf{D}(x) \in l \,\&\, (\forall\delta)\,(\mathbf{Ext}_x(\delta) \prec \omega_{\delta+1})\,.$$

We have $\mathbf{Norm}(j) \leqq \aleph_\alpha$, so that every exclusive system in $\boldsymbol{a}_{<\alpha}$ has power at most $\aleph_\alpha$, by 2524 or by 2526.

(b) If $\iota \geq \alpha$ then $\mathbf{v}(\boldsymbol{b}^{\iota+1}_{\iota+1}) \geq \aleph_{\iota+1} \geq \aleph_{\alpha+1}$ and, since the ideal $l \,|\, (\lambda - \alpha)$ is $\aleph_\alpha$-additive, $\boldsymbol{a}_{>\alpha}$ is $\aleph_\alpha$-multiplicative.

6328. Lemma (**TSS′**, E1). If Z is a Boolean support on $\boldsymbol{a}$ and if $\varrho = \bigcup(Z)$ then $Z = \mathbf{P}(\varrho) \cap a$ and for $\alpha < \vartheta$ we have

$$Z \cap a_\alpha = \mathbf{P}(\varrho \upharpoonright \{\alpha\}) \cap a_\alpha\,,$$
$$Z \cap a_{\neq\alpha} = \mathbf{P}(\varrho \upharpoonright (\mathbf{D}(\varrho) - \{\alpha\}) \cap a_{\neq\alpha}\,.$$

Proof. See 4329. (a_α is $\{\{\langle u, \alpha\rangle\}; u \in b^{\alpha+1}_{\alpha+1}\}$.)

6329. Lemma (**TSS**, D3, Constr, (S **Comb**)).

(a) All cardinals are absolute.

(b) For every α, $\hat{\beth}(2, \aleph_\alpha) = \aleph_{\alpha+1}$.

This can be proved in analogy with the proof of 6118.

6330. METALEMMA. Let $\mathbf{T}(X)$ be a gödelian term such that **TSS**, D3 $\vdash$ $\vdash \mathbf{T}(\mathbf{V}) = \{\alpha; \neg \mathrm{Ap}(\mathbf{L}, \aleph_{\alpha+1})\}$. Then the following is provable in **TS**, D3, $\mathrm{Mcl}_{\mathbf{cn}}(\mathbf{L})$: for every model-class M,

$$\mathbf{T}(M) = \{\alpha; \neg(\forall f \in M)(\mathrm{Un}(f, \boldsymbol{\omega}_{\alpha+1}, \boldsymbol{\omega}_{\alpha+1}) \rightarrow \rightarrow (\exists s \in \mathbf{L})(f \subseteq s \,\&\, (\forall \iota)(\mathbf{Ext}_s(\iota) \prec \aleph_{\alpha+1})))\} .$$

Demonstration. It suffices to show the following in **TS**, D3, $\mathrm{Mcl}_{\mathbf{cn}}(\mathbf{M})$, $\partial\mathbf{TSS}/\partial\mathfrak{St}(\mathbf{M})$:

$$(*)\quad \mathrm{Ap}^{\mathfrak{St}(\mathbf{M})}(\mathbf{L}, \boldsymbol{\omega}_{\alpha+1}) \equiv (\forall f \in \mathbf{M})(\mathrm{Un}(f, \boldsymbol{\omega}_{\alpha+1}, \boldsymbol{\omega}_{\alpha+1}) \rightarrow \rightarrow (\exists s \in \mathbf{L})(f \subseteq s \,\&\, (\forall \iota)(\mathbf{Ext}_s(\iota) \prec \aleph_{\alpha+1}))) .$$

Since semisets in the sense of $\mathfrak{St}(\mathbf{M})$ are subsets of $\mathbf{M}$ (we are working in a theory stronger than **TS**), $\mathrm{Ap}^{\mathfrak{St}(\mathbf{M})}(\mathbf{L}, \boldsymbol{\omega}_{\alpha+1})$ is equivalent to

$$(\forall f \in \mathbf{M})(\mathrm{Un}^{\square}(f, \boldsymbol{\omega}^{\square}_{\alpha+1}, \boldsymbol{\omega}^{\square}_{\alpha+1}) \rightarrow (\exists s \in \mathbf{L})(f \subseteq s \,\&\, (\forall \iota)(\mathbf{Ext}_s(\iota) \prec^{\square} \boldsymbol{\omega}^{\square}_{\alpha+1}))) .$$

But $\mathbf{L}$ is absolute, Un is absolute and, since $\mathrm{Mcl}_{\mathbf{cn}}(\mathbf{M})$ holds, $\prec$ is also absolute and $(*)$ follows.

6331. METATHEOREM. Let $\mathbf{T}(X)$ be a gödelian term such that **TSS**, D3 $\vdash \mathbf{T}(\mathbf{V}) = \{\alpha; \neg \mathrm{Ap}(\mathbf{L}, \aleph_{\alpha+1})\}$ and denote by $(\mathrm{Det}_{\mathbf{T}})$ the following axiom:

$$(\exists r, \bar{r})(\mathrm{Det}_{\mathbf{T}}(r) \,\&\, \mathrm{Est}(r, \bar{r}) \,\&\, r \upharpoonright \{0\} \notin \mathbf{Cstr}(0)) .$$

(Cf. 3532 and 3533.)

$\mathfrak{Up}$ is a model in the following diagram:

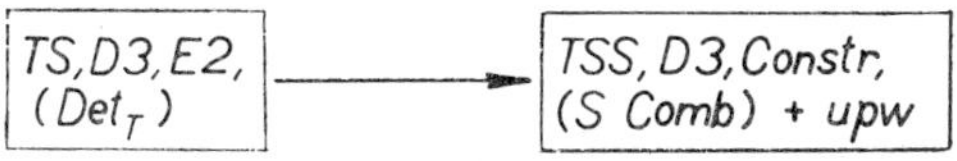

Demonstration. We proceed in the latter theory. By 6329 all cardinals are absolute so that we have $\mathrm{Mcl}^*_{\mathbf{cn}}(\mathbf{Tor})$; $\mathbf{V} = \mathbf{L}$ implies $\mathbf{Tor} = \mathbf{L}^*$ (see 5138 (e)). Therefore $\mathrm{Mcl}^*_{\mathbf{cn}}(\mathbf{L}^*)$ holds. Secondly, the axiom of continuum holds in the $*$-sense (cf. 6118). Consequently if ϑ^* denotes the cardinal of the model defined in the same way as ϑ but in the sense of the model then we have $\vartheta^* = \mathbf{Imb}''\, \vartheta$.

Arrow 1 in the following diagram is the model $\mathfrak{Up}$:

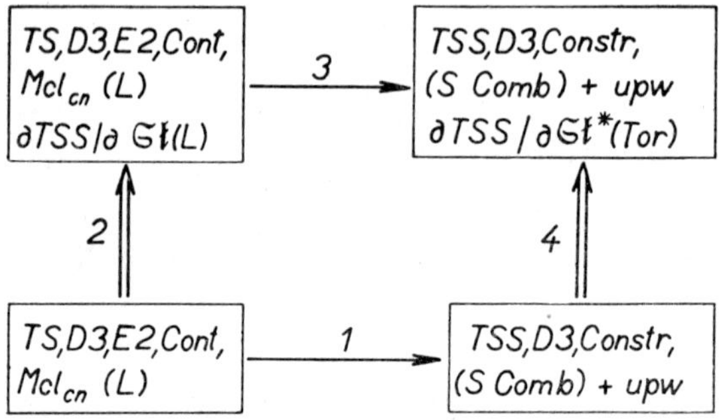

We extend $\mathfrak{Up}$ by defining □-notions in the terminal theory of 1 by ∂**TSS**/$\partial\mathfrak{S}\mathfrak{t}^*$ (**Tor**) (arrow 4) and we know that in this way the whole diagram becomes commutative. (Cf. arrows 5, 6, 7, 8 in Diagram 5112). By 5126 (S **Comb**)$^\square$ holds in 3. Hence let us proceed in **TS**, D3, E2, Cont, $\mathrm{Mcl}_{\mathbf{cn}}$ (**L**), ∂**TSS**/$\partial\mathfrak{S}\mathfrak{t}$ (**L**), (S **Comb**)$^\square$. Denote by z a complete ultrafilter on **Comb**$^\square$ in the sense of $\mathfrak{S}\mathfrak{t}$ (**L**).

Defining $r = \bigcup z$ we have by 4329

$$(1) \begin{cases} r \upharpoonright \{\alpha\} = \bigcup(z \cap a_\alpha^\square),\ r \upharpoonright (\mathbf{D}(r) - \{\alpha\}) = \bigcup(z \cap a_{\neq\alpha}^\square), \\ z \cap a_\alpha^\square = \mathbf{P}^\square(r \upharpoonright \{\alpha\}) \cap a_\alpha^\square,\quad z \cap a_{\neq\alpha}^\square = \mathbf{P}(r \upharpoonright \mathbf{D}(r) - \{\alpha\}) \cap \alpha_{\neq\alpha}^\square . \end{cases}$$

We wish to prove that r is a **T**-determining well estimated relation. In particular, we want to prove

$$(2)\quad (\forall\alpha < \vartheta)(\alpha \in \mathbf{T}(\mathbf{Cstr}(r \upharpoonright \{\alpha\})) \,\&\, \alpha \notin \mathbf{T}(\mathbf{Cstr}(r \upharpoonright (\mathbf{D}(r) - \{\alpha\})))) .$$

L is a model-class with absolute cardinals and therefore, by 6330, (2) is equivalent to

$$(3)\quad (\forall\alpha < \vartheta)\,[\neg(\forall f \in \mathbf{Cstr}(r \upharpoonright \{\alpha\}))\, \chi(f, \alpha)) \,\&$$
$$\& \,(\forall f \in \mathbf{Cstr}(r \upharpoonright (\mathbf{D}(r) - \{\alpha\})))\, \chi(f, \alpha)]$$

where $\chi(f, \alpha)$ is the formula

$$\mathrm{Un}(f, \omega_{\alpha+1}, \omega_{\alpha+1}) \to (\exists s \in \mathbf{L})\,(f \subseteq s \,\&\, (\forall\iota)\,(\mathbf{Ext}_s(\iota) \prec \omega_{\alpha+1})) .$$

By (1), $\mathbf{Cstr}(r \upharpoonright \{\alpha\}) = \mathbf{Cstr}(z \cap a_\alpha^\square)$ and $\mathbf{Cstr}(r \upharpoonright (\mathbf{D}(r) - \{\alpha\})) =$

$= \mathbf{Cstr}\,(z \cap a^{\square}_{\neq\alpha})$; hence the quantifiers in (3) may be replaced by

$$(\forall f \in \mathbf{Cstr}\,(z \cap a^{\square}_{\alpha})) \quad \text{and} \quad (\forall f \in \mathbf{Cstr}\,(z \cap a^{\square}_{\neq\alpha}))$$

respectively. Since $\mathrm{Un}\,(f, \boldsymbol{\omega}_{\alpha+1}, \boldsymbol{\omega}_{\alpha+1})$ implies $f \subseteq \mathbf{L}$ we may use 6315 and so these quantifiers may be replaced by $(\forall f)\,(\mathrm{Dep}^{\square}\,(f, z \cap a^{\square}_{\alpha}) \rightarrow$ and $(\forall f)\,(\mathrm{Dep}^{\square}\,(f, z \cap a^{\square}_{\neq\alpha}) \rightarrow$ respectively.

Thus we must prove the following:

$$\begin{aligned}(4)\quad & (\forall \alpha < \vartheta)\,(\neg[(\forall \sigma^{\square})\,(\mathrm{Dep}^{\square}\,(\sigma^{\square}, z \cap a^{\square}_{\alpha}) \rightarrow \\ & \rightarrow (\exists s^{\square})\,(\sigma^{\square} \subseteq s^{\square} \,\&\, (\forall \iota)\,(\mathbf{Ext}_{s^{\square}}\,(\iota) \prec \boldsymbol{\omega}_{\alpha+1})))] \,\& \\ & \& \;[(\forall \sigma^{\square})\,(\mathrm{Dep}^{\square}\,(\sigma^{\square}, z \cap a^{\square}_{\neq\alpha}) \rightarrow \\ & \rightarrow (\exists s^{\square})\,(\sigma^{\square} \subseteq s^{\square} \,\&\, (\forall \iota)\,(\mathbf{Ext}_{s^{\square}}\,(\iota) \prec \boldsymbol{\omega}_{\alpha+1})))]) \,.\end{aligned}$$

Instead of proving this it suffices to prove the following in **TSS**, D3, Constr, (S **Comb**) + upw: if Z is a total Boolean support on **Comb** then

$$\begin{aligned}(5)\quad & (\forall \alpha < \vartheta)\,(\neg[(\forall \sigma)\,(\mathrm{Dep}\,(\sigma, \mathrm{Z} \cap a_{\alpha}) \rightarrow (\exists s)\,(\sigma \subseteq s \,\& \\ & \&\; (\forall \iota)\,(\mathbf{Ext}_{\sigma}\,(\iota) \prec \boldsymbol{\omega}_{\alpha+1})))] \,\&\, [(\forall \sigma)\,(\mathrm{Dep}\,(\sigma, \mathrm{Z} \cap a_{\neq\alpha}) \rightarrow \\ & \rightarrow (\exists s)\,(\sigma \subseteq s \,\&\, (\forall \iota)\,(\mathbf{Ext}_{\sigma}\,(\iota) \prec \boldsymbol{\omega}_{\alpha+1})))]) \,.\end{aligned}$$

In **TSS**, D3, Constr, (S **Comb**) + upw, fix **Z** and $\boldsymbol{\alpha} < \vartheta$, and add the definitions $\partial\mathbf{TSS}/\partial\mathfrak{Supp}\,(\mathbf{Z} \cap a_{\alpha})$ and $\partial\mathbf{TSS}/\partial\mathfrak{Supp}\,(\mathbf{Z} \cap a_{\neq\alpha})$. Our problem reduces to proving $\neg\,\mathrm{Appr}\,(\boldsymbol{\omega}_{\alpha+1}, \boldsymbol{\omega}_{\alpha+1}, \boldsymbol{\omega}_{\alpha+1})$ in the sense of $\mathfrak{Supp}\,(\mathbf{Z} \cap a_{\alpha})$ and $\mathrm{Appr}\,(\boldsymbol{\omega}_{\alpha+1}, \boldsymbol{\omega}_{\alpha+1}, \boldsymbol{\omega}_{\alpha+1})$ in the sense of $\mathfrak{Supp}\,(\mathbf{Z} \cap a_{\neq\alpha})$. In the sense of $\mathfrak{Supp}\,(\mathbf{Z} \cap a_{\alpha})$ we have a total Boolean support on $\mathbf{Cant}^{\alpha+1}_{\alpha+1}$ and therefore by 6106 $\neg\,\mathrm{Appr}\,(\boldsymbol{\omega}_{\alpha+1}, \boldsymbol{\omega}_{\alpha+1}, \boldsymbol{\omega}_{\alpha+1})$ holds in the sense of $\mathfrak{Supp}\,(\mathbf{Z} \cap a_{\alpha})$. To prove $\mathrm{Appr}\,(\boldsymbol{\omega}_{\alpha+1}, \boldsymbol{\omega}_{\alpha+1}, \boldsymbol{\omega}_{\alpha+1})$ in the sense of $\mathfrak{Supp}\,(\mathbf{Z} \cap a_{\neq\alpha})$, suppose that $\mathrm{Appr}\,(\aleph_{\alpha+1}, \aleph_{\alpha+1}, \aleph_{\alpha+1})$ fails in the sense of $\mathfrak{Supp}\,(\mathbf{Z} \cap \cap\, (a_{<\alpha} \odot a_{\alpha<}))$; then there exists some $\sigma \subseteq \boldsymbol{\omega}_{\alpha+1}$ which is not approximable. Observing that $a_{>\alpha}$ is $\aleph_{\alpha+1}$-multiplicative, we see that σ is dependent on $\mathrm{Z} \cap a_{<\alpha}$. But since $\boldsymbol{\mu}(a_{<\alpha}) \leqq \aleph_{\alpha+1}$, it follows that $\mathrm{Appr}\,(\aleph_{\alpha+1}, \aleph_{\alpha+1}, \aleph_{\alpha+1})$ holds in the sense of $\mathfrak{Supp}\,(\mathbf{Z} \cap a_{<\alpha})$, a contradiction.

Hence in **TS**, D3, E2, Cont, $\partial\mathbf{TSS}/\partial\mathfrak{St}\,(\mathbf{L})$, $(\mathrm{S}\;\mathbf{Comb})^{\square}$ we have shown that r is a T-determining relation. It remains to show that r is estimated. Define $\bar{r} = \bigcup_{\alpha<\delta} \left(b^{\alpha+1}_{\alpha+1}\right)^{\square} \times \{\alpha\}$; evidently $r \subseteq \bar{r}$; since (Cont) holds $\bar{r}$ is a good estimation for r. Evidently, $r \restriction \{0\} \notin \mathbf{L}$. This completes the demonstration.

By 6331 and 3534, we have the following

6332. METACOROLLARY. The axiom (E3) & $\neg$ (Constr) is consistent relative to **TSS**, D3.

SECTION 4

Boolean values and forcing

This final section is concerned with a metamathematical notion which corresponds to the intuitive notion "generalized truth value of a formula". We give an algorithm which enables us to define, for each closed **TSS**-formula φ of a certain type, a gödelian term $\mathbf{F}_\varphi(\boldsymbol{b})$ in **TSS**′ such that the following is provable in **TSS**′:

1) If $\boldsymbol{b}$ is a complete Boolean algebra then $\mathbf{F}_\varphi(\boldsymbol{b}) \in b$.

2) If $\boldsymbol{b}$ is a complete Boolean algebra and Z is a total Boolean support on $\boldsymbol{b}$ then $\varphi \equiv \mathbf{F}_\varphi(\boldsymbol{b}) \in Z$. The element $\mathbf{F}_\varphi(\boldsymbol{b})$ will be called the φ-value in $\boldsymbol{b}$.

With the aid of this notion we can, for example, give a uniform reduction of consistency problems w.r.t. the theory **TS**, D3, E2, ∂**TSS**$/\partial\mathfrak{St}(\mathbf{L})$, $(\mathrm{Bd})^{\mathfrak{St}(\mathbf{L})}$ to consistency problems w.r.t. **TSS**‴, (**Constr**). We shall show how to find for any SF φ another SF $\hat{\varphi}$ such that φ is consistent with **TS**, E2, ∂**TSS**$/\partial\mathfrak{St}(\mathbf{L})$, $(\mathrm{Bd})^{\mathfrak{St}(\mathbf{L})}$ if and only if $\hat{\varphi}$ is consistent with **TS**, (Constr). In case $\hat{\varphi}$ is provable in **TSS**, (Constr) it follows that φ is consistent with **TS**, E2, ∂**TSS**$/\partial\mathfrak{St}(\mathbf{L})$, $(\mathrm{Bd})^{\mathfrak{St}(\mathbf{L})}$; the results of Section 1 and other similar results may be obtained in this way. The formula $\hat{\varphi}$ always has the form "there exists a complete Boolean algebra with such and such property"; in other words, we give here a "canonical form" for conditions to be imposed on a complete Boolean algebra in order that the existence of a total Boolean support should imply that a particular formula φ is valid in the sense of the model $\mathfrak{Up}$. The "canonical form" in question is the form $(\exists\boldsymbol{b})(\mathbf{F}_\varphi(\boldsymbol{b}) = 1_{\boldsymbol{b}})$. We shall also investigate properties of operations on φ-values and we shall then consider the matter more generally in the case when the arguments of these operations are not complete Boolean algebras but separatively ordered sets. Thus we finally arrive back, in a modified form, at the original notion of forcing due to Cohen.

In 4311 a disjointed relation which is a set and whose domain is included in a complete Boolean algebra $\boldsymbol{b}$ was called a $\boldsymbol{b}$-set. If Z is a total Boolean support on $\boldsymbol{b}$ then for every semiset σ there exists a $\boldsymbol{b}$-set r such that $\sigma = r``Z$. In particular, if $x$ is a set and if we let $x^{(\boldsymbol{b})} = \{\langle y, 1_{\boldsymbol{b}}\rangle, y \in x\}$ then $x^{(\boldsymbol{b})}$ is a $\boldsymbol{b}$-set and we have $x = x^{(\boldsymbol{b})}``Z$. Hence we make the following

6401. DEFINITION (**TSS**′). Let $\boldsymbol{b}$ be a complete Boolean algebra. A $\boldsymbol{b}$-set r is called a *$\boldsymbol{b}$-constant* if $r = x \times \{1_{\boldsymbol{b}}\}$ for some x.

6402. METADEFINITION. Let x and y be set-variables and let σ be a semiset-variable.

(a) $x \in y$ and $x \in \sigma$ are *pure* SNF ;

(b) if φ and ψ are pure SNF then all formulas of the forms $\neg\, \varphi$, $(\varphi \,\&\, \psi)$, $(\varphi \vee \psi)$ are *pure* SNF ;

(c) if φ is a pure SNF then all formulas of the forms $(\forall x)\, \varphi$, $(\forall \sigma)\, \varphi$, $(\exists x)\, \varphi$ and $(\exists \sigma)\, \varphi$ are *pure* SNF ;

(d) every pure SNF can be obtained from pure SNF sub (a) by a finite number of operations sub (b) and (c).

We now give an algorithm which associates, with each pure SNF φ with n free set-variables and m free semiset variables, a normal constant $[\![\varphi]\!]$ in **TSS**′ such that it is provable in **TSS**′ that $[\![\varphi]\!]$ is a mapping which assigns to every complete Boolean algebra $\boldsymbol{b}$, to every n-tuple of $\boldsymbol{b}$-constants and to every m-tuple of $\boldsymbol{b}$-sets an element of $\boldsymbol{b}$.

6403. METADEFINITION. We define in **TSS**′:

$$[\![x \in y]\!]` \langle z_1^{(\boldsymbol{b})}, z_2^{(\boldsymbol{b})}, \boldsymbol{b}\rangle = \begin{cases} 1_{\boldsymbol{b}} \text{ if } z_1 \in z_2 \,, \\ 0_{\boldsymbol{b}} \text{ if } z_1 \notin z_2 \,, \end{cases}$$

$$[\![x \in \sigma]\!]` \langle z_1^{(\boldsymbol{b})}, r, \boldsymbol{b}\rangle = r_{z_1} \text{ if } r \text{ is a } \boldsymbol{b}\text{-set} .$$

(Hence $[\![x \in y]\!]$ is a restriction of $[\![x \in \sigma]\!]$.)

Suppose we have already defined $[\![\varphi]\!]$ and $[\![\psi]\!]$ and that $\varphi \,\&\, \psi$ is a formula; we define

$$[\![\neg \varphi]\!]` \langle r_1, \bullet, \boldsymbol{b}\rangle = -\left([\![\varphi]\!]` \langle r_1, \bullet, \boldsymbol{b}\rangle\right),$$

$$[\![\varphi \,\&\, \psi]\!]` \langle r_1, \bullet, \boldsymbol{b}\rangle = [\![\varphi]\!]` \langle r_1, \bullet, \boldsymbol{b}\rangle \wedge [\![\psi]\!]` \langle r_1, \bullet, \boldsymbol{b}\rangle ,$$

$$[\![\varphi \vee \psi]\!]` \langle r_1, \bullet, \boldsymbol{b}\rangle = [\![\varphi]\!]` \langle r_1, \bullet, \boldsymbol{b}\rangle \vee [\![\psi]\!]` \langle r_1, \bullet, \boldsymbol{b}\rangle .$$

(More exactly, if φ has free variables $u_1, \bullet, v_1, \bullet$ for sets or semisets and ψ has free variables $u_1, \bullet, w_1, \bullet$, then

$$[\![\varphi \,\&\, \psi]\!]\text{'}\, \langle r_1, \bullet, s_1, \bullet, t_1, \bullet, \boldsymbol{b}\rangle = [\![\varphi]\!]\text{'}\, \langle r_1, \bullet, s_1, \bullet, \boldsymbol{b}\rangle \wedge$$
$$\wedge\; [\![\psi]\!]\text{'}\, \langle r_1, \bullet, t_1, \bullet, \boldsymbol{b}\rangle\,;$$

similarly for disjunction.)

Suppose we have already defined $[\![\varphi]\!]$ and that $(\forall x)\,\varphi$, resp. $(\forall \sigma)\,\varphi$ are formulas; we define

$$[\![(\forall x)\,\varphi]\!]\text{'}\, \langle r_1, \bullet, \boldsymbol{b}\rangle = \bigwedge\{[\![\varphi]\!]\text{'}\, \langle z^{(\boldsymbol{b})}, r_1, \bullet, \boldsymbol{b}\rangle;\ z \text{ is a set}\}\,;$$

$$[\![(\forall \sigma)\,\varphi]\!]\text{'}\, \langle r_1, \bullet, \boldsymbol{b}\rangle = \bigwedge\{[\![\varphi]\!]\text{'}\, \langle s, r_1, \bullet, \boldsymbol{b}\rangle;\ s \text{ is a } \boldsymbol{b}\text{-set}\}\,;$$

$$[\![(\exists x)\,\varphi]\!]\text{'}\, \langle r_1, \bullet, \boldsymbol{b}\rangle = \bigvee\{[\![\varphi]\!]\text{'}\, \langle z^{(\boldsymbol{b})}, r_1, \bullet, \boldsymbol{b}\rangle;\ z \text{ is a set}\}\,;$$

$$[\![(\exists \sigma)\,\varphi]\!]\text{'}\, \langle r_1, \bullet, \boldsymbol{b}\rangle = \bigvee\{[\![\varphi]\!]\text{'}\, \langle s, r_1, \bullet, \boldsymbol{b}\rangle;\ s \text{ is a } \boldsymbol{b}\text{-set}\}\,.$$

6404. Remark. If $k, \bullet$ are $\boldsymbol{b}$-constants and $r, \bullet$ are $\boldsymbol{b}$-sets we shall often write $[\![\varphi(k, \bullet, r, \bullet)]\!]_{\boldsymbol{b}}$ instead of $[\![\varphi(x, \bullet, \sigma, \bullet)]\!]\text{'}\, \langle k, \bullet, r, \bullet, \boldsymbol{b}\rangle$, i.e. we "substitute k_1 for x_1, r_1 for σ_1" etc. Notice however that $[\![\varphi(x, \bullet, \sigma, \bullet)]\!]$ is the name of a mapping which has $\langle k, \bullet, r, \bullet, \boldsymbol{b}\rangle$ in its domain. All constants $[\![\varphi]\!]$ are defined in **TSS'** by normal formulas; hence it is provable in **TSS'** that $[\![\varphi]\!]$ is a real class. If φ is closed then we may define an operation $\mathbf{F}_\varphi(\boldsymbol{b})$ by setting $\mathbf{F}_\varphi(\boldsymbol{b}) = [\![\varphi]\!]\text{'}\ \boldsymbol{b}$. In **TSS'** we call $[\![\varphi]\!]\text{'}\, \langle k, \bullet, r, \bullet, \boldsymbol{b}\rangle$ "the φ-value of $k, \bullet, r, \bullet$ in $\boldsymbol{b}$".

6405. Example (**TSS'**). If $\boldsymbol{b}$ is a complete Boolean algebra and r is a $\boldsymbol{b}$-set then

$$\mathbf{Thru}_{\boldsymbol{b}}\,(\mathbf{D}\,(r)) = [\![(\exists x)\,(x = \sigma)]\!]\text{'}\, \langle r, \boldsymbol{b}\rangle\,.$$

Proof. To be precise, the formula $(\exists x)\,(x = \sigma)$ is not a pure SNF since it contains $=$. By $[\![(\exists x)\,(x = \sigma)]\!]$ we mean $[\![(\exists x)\,(\forall y)\,((y \in x \,\&\, y \in \sigma) \vee \vee\, (y \notin x \,\&\, y \notin \sigma))]\!]$. We have

$$[\![(\exists x)\,(\forall y)\,(y \in x \,\&\, y \in \sigma\,.\, \vee\,.\, y \notin x \,\&\, y \notin \sigma)]\!]\text{'}\, \langle r, \boldsymbol{b}\rangle =$$
$$= \bigvee_{x} \bigwedge_{y} ([\![y \in x \,\&\, y \in \sigma]\!]\text{'}\, \langle x^{(\boldsymbol{b})}, y^{(\boldsymbol{b})}, r, \boldsymbol{b}\rangle \vee$$
$$\vee\; [\![y \notin x \,\&\, y \notin \sigma]\!]\text{'}\, \langle x^{(\boldsymbol{b})}, y^{(\boldsymbol{b})}, r, \boldsymbol{b}\rangle) = \bigvee_{x}(\bigwedge_{y \in x} r_y \wedge \bigwedge_{y \notin x} - r_y)\,.$$

Let $u_x = \bigwedge_{y \in x} r_y \wedge \bigwedge_{y \notin x} - r_y$. We prove the following:

(1) for every x, u_x is layered by $\mathbf{D}(r)$,

(2) for every v layered by $\mathbf{D}(r)$ there is an x such that $v \leqq u_x$.

(By (1) and (2), $[\![(\exists x)(x = \sigma)]\!]$ is the sum of all elements layered by $\mathbf{D}(r)$.)

(1) is evident; for, if $y \in x$ then $r_y \geqq u_x$ otherwise $r_y \leqq -u_x$.

(2) Put $x = \{y; v \leqq r_y\}$, then $v \leqq u_x$.

6406. METATHEOREM. If $\varphi(x, \bullet, \sigma, \bullet)$ is a pure SNF then the following statement is provable in **TSS'**:

Let $\boldsymbol{b}$ be a complete Boolean algebra and let Z be a total Boolean support on $\boldsymbol{b}$; if $k = x^{(\boldsymbol{b})}, \bullet$ and if $\sigma = r\text{''}Z, \bullet$ then

$$([\![\varphi(x, \bullet, \sigma, \bullet)]\!]' \langle k, \bullet, r, \bullet, \boldsymbol{b}\rangle \in Z) \equiv \varphi(x, \bullet, \sigma, \bullet)\,.$$

Demonstration. By induction on the length of φ. We give here the proof for $(\forall \varrho)\, \varphi(\varrho, x)$ under the assumption that the statement holds for $\varphi(\varrho, x)$. We proceed in **TSS'**. If $(\forall \varrho)\, \varphi(\varrho, x)$ for some x then $\varphi(r\text{''}Z, x)$ for any $\boldsymbol{b}$-set r, so that $(\forall r)(r$ is a $\boldsymbol{b}$-set $\rightarrow [\![\varphi(r, x^{(\boldsymbol{b})})]\!] \in Z)$ and hence $\bigwedge_r [\![\varphi(r, x^{(\boldsymbol{b})})]\!] \in Z$. The converse is proved in the same way.

6407. METATHEOREM. The following statement is provable in **TSS'** for any pure SNF φ:

If $\boldsymbol{b}$ is a complete Boolean algebra and if $u \in \boldsymbol{b}$ then

$$u \wedge [\![\varphi]\!]' \langle k, \bullet, r, \bullet, \boldsymbol{b}\rangle = [\![\varphi]\!]' \langle k \restriction_{\boldsymbol{b}} u, \bullet, r \restriction_{\boldsymbol{b}} u, \bullet, \boldsymbol{b} \mid u\rangle\,.$$

Demonstration. By induction on the length of φ. We give the induction step for $(\forall \sigma)\, \varphi$. We have

$$\begin{aligned} u \wedge [\![(\forall \sigma)\, \varphi]\!]' \langle k, r, \boldsymbol{b}\rangle &= u \wedge \bigwedge\{[\![\varphi]\!]' \langle s, k, r, \boldsymbol{b}\rangle; s \text{ is a } \boldsymbol{b}\text{-set}\} = \\ &= \bigwedge\{u \wedge [\![\varphi]\!]' \langle s, k, r, \boldsymbol{b}\rangle; s \text{ is a } \boldsymbol{b}\text{-set}\} = \\ &= \bigwedge\{[\![\varphi]\!]' \langle s \mid u, k \mid u, r \mid u, \boldsymbol{b} \mid u; s \text{ is a } \boldsymbol{b}\text{-set}\} = \\ &= \bigwedge\{[\![\varphi]\!]' \langle t, k \mid u, r \mid u, \boldsymbol{b} \mid \boldsymbol{u}\rangle, t \text{ is a } (\boldsymbol{b} \mid u)\text{-set}\} = \\ &= [\![(\forall \sigma)\, \varphi]\!]' \langle k \mid u, r \mid u, \boldsymbol{b} \mid u\rangle\,. \end{aligned}$$

6408. METATHEOREM. Let φ be a closed pure SNF. The formula $\varphi^{\mathfrak{St}(\mathbf{L})}$ is consistent with **TS**, D3, E2, ∂**TSS**$/\partial\mathfrak{St}(\mathbf{L})$, $(\mathrm{Bd})^{\mathfrak{St}(\mathbf{L})}$ iff the formula $(\exists \boldsymbol{b})(\llbracket\varphi\rrbracket' \boldsymbol{b} \neq 0_{\boldsymbol{b}})$ is consistent with **TSS**, D3, Constr.

Demonstration. (1) Suppose that **TSS**, D3, Constr, $(\exists \boldsymbol{b})(\llbracket\varphi\rrbracket' \boldsymbol{b} \neq 0_{\boldsymbol{b}})$ is consistent. In this theory $(\exists \boldsymbol{b})(\llbracket\varphi\rrbracket' \boldsymbol{b} = 1_{\boldsymbol{b}})$ is provable. (Take $\bar{\boldsymbol{b}} = \boldsymbol{b} \mid \mid (\llbracket\varphi\rrbracket' \boldsymbol{b})$ where $\llbracket\varphi\rrbracket' \boldsymbol{b} \neq 0_{\boldsymbol{b}}$; then $\llbracket\varphi\rrbracket' \bar{\boldsymbol{b}} = 1_{\bar{\boldsymbol{b}}}$.) Denote by $\mathrm{Val}_{\varphi}(\boldsymbol{b})$ the formula $\llbracket\varphi\rrbracket' \boldsymbol{b} = 1_{\boldsymbol{b}}$. By the Support Principle 5215, the theory **TSS**, D3, Constr, $(\mathrm{S}\ \mathrm{Val}_{\varphi}(\boldsymbol{b})) + \mathrm{upw}$ is consistent and by 6406 φ is provable in this theory. Using the model $\mathfrak{Up}$ as a model in the last theory we see that **TS**, D3, E2, ∂**TSS**$/\partial\mathfrak{St}(\mathbf{L})$, $(\mathrm{Bd})^{\mathfrak{St}(\mathbf{L})}$, $\varphi^{\mathfrak{St}(\mathbf{L})}$ is consistent.

(2) Suppose conversely that the theory **TS**, D3, E2, ∂**TSS**$/\partial\mathfrak{St}(\mathbf{L})$, $(\mathrm{Bd})^{\mathfrak{St}(\mathbf{L})}$, $\varphi^{\mathfrak{St}(\mathbf{L})}$ is consistent; we investigate the model $\mathfrak{St}(\mathbf{L})$ in this theory. In the sense of this model there exists a total support (by E2); hence, by $(\mathrm{Bd})^{\mathfrak{St}(\mathbf{L})}$, there exists a total Boolean support Z on an algebra $\boldsymbol{b}$ (in the sense of $\mathfrak{St}(\mathbf{L})$). Obviously, D3 and (Constr) hold in the sense of the model. By 6406 we have $\llbracket\varphi\rrbracket' \boldsymbol{b} \in Z$ in the sense of the model and hence $\llbracket\varphi\rrbracket' \boldsymbol{b} \neq 0_{\boldsymbol{b}}$ in the sense of the model. Consequently the theory **TSS**, D3, Constr, $(\exists \boldsymbol{b})(\llbracket\varphi\rrbracket' \boldsymbol{b} \neq 0_{\boldsymbol{b}})$ is consistent.

The demonstration of the following slightly more general metatheorem is left to the reader:

6409. METATHEOREM. Let φ be a closed pure SNF. The formula $\varphi^{\mathfrak{St}(\mathbf{M})}$ is consistent with **TS**″, FMcl $(\mathbf{M})$, ∂**TSS**$/\partial\mathfrak{St}(\mathbf{M})$, $(\mathrm{E2})^{\mathfrak{St}(\mathbf{M})}$, $(\mathrm{S6})^{\mathfrak{St}(\mathbf{M})}$ iff the formula $(\exists \boldsymbol{b})(\llbracket\varphi\rrbracket' \boldsymbol{b} \neq 0_{\boldsymbol{b}})$ is consistent with **TSS**, E2.

Thus we obtain the desired reduction of consistency problems for formulas of the form $\varphi^{\mathfrak{St}(\mathbf{M})}$. We shall show that in **TS**″, FMcl $(\mathbf{M})$, ∂**TSS**$/\partial\mathfrak{St}(\mathbf{M})$, $(\mathrm{S6})^{\mathfrak{St}(\mathbf{M})}$ every closed set formula is equivalent to some formula of this form.

6410. METATHEOREM. If φ is a closed set formula then there is a pure SNF ψ such that $\varphi \equiv \psi^{\mathfrak{St}(\mathbf{M})}$ is provable in **TS**″, FMcl $(\mathbf{M})$, ∂**TSS**$/\partial\mathfrak{St}(\mathbf{M})$, $(\mathrm{S6})^{\mathfrak{St}(\mathbf{M})}$.

Demonstration. We investigate the theory **TS**″, FMcl $(\mathbf{M})$, ∂**TSS**$/\partial\mathfrak{St}(\mathbf{M})$ $(\mathrm{S6})^{\mathfrak{St}(\mathbf{M})}$. This theory will be denoted by **TS**! in the course of the demonstration. Recall the gödelian term **Fupw** (R, S) defined in 5109. By the proof of 4119, there is a gödelian term **TU** (σ) such that **TSS**′ $\vdash$ TSupp $(\sigma) \rightarrow$ $\rightarrow$ Totunvr $(\mathbf{TU}(\sigma))$. Hence fix a constant $\boldsymbol{\sigma}^{\square}$ in **TS**! by $\mathrm{TSupp}^{\square}(\boldsymbol{\sigma}^{\square})$, define a constant $\mathbf{R}^{\square}$ by $\mathbf{Fupw}^{\square}(\mathbf{E}^{\square}, \mathbf{TU}^{\square}(\boldsymbol{\sigma}^{\square}))$ and add the definitions ∂**TS**$/\partial\mathfrak{N}^{\square}(\mathbf{R}^{\square})$ (cf. arrows 4, 10 in 5112). We denote the notions defined by these definitions by the superscript $*$. Let φ be an arbitrary SF (not necessarily closed). We show by induction that there is a pure SNF ψ which is normal and such that

$$(x^* = \mathbf{Ext}_{\mathbf{R}^\square}(x^\square) \,\&\, \bullet) \to [\varphi^*(x^*, \bullet) \equiv \psi^\square(x^\square, \bullet, \boldsymbol{\sigma}^\square)]$$

is provable in **TS**!, $\mathrm{TSupp}^\square(\boldsymbol{\sigma}^\square)$, $\partial\mathbf{TS}/\partial\mathfrak{N}^\square(\mathbf{Fupw}^\square(\mathbf{E}^\square, \mathbf{TU}^\square(\boldsymbol{\sigma}^\square))$. For φ atomic we have $x^* \in^* y^* \equiv \langle x^\square, y^\square\rangle \in \mathbf{Fupw}^\square(\mathbf{E}^\square, \mathbf{TU}^\square(\boldsymbol{\sigma}^\square))$. Since both **Fupw** and **TU** are gödelian terms, the last formula is equivalent to a formula $\psi^\square(x^\square, y^\square, \boldsymbol{\sigma}^\square)$ of the desired form. For the induction step cf. 1146.

In particular, if φ is closed, then $\varphi \equiv \varphi^*$ is provable in our theory (by 5128) and therefore $\varphi \equiv \psi^\square(\boldsymbol{\sigma}^\square)$ is provable in **TS**!, $\mathrm{T\,Supp}^\square(\boldsymbol{\sigma}^\square)$. By 1245 we obtain

$$\mathbf{TS}! \vdash (\forall\sigma^\square)(\mathrm{TSupp}^\square(\sigma^\square) \to [\varphi \equiv \psi^\square(\sigma^\square)]) .$$

Consequently, φ is equivalent in **TS**! to either of the following formulas:

$$(\forall\sigma^\square)(\mathrm{TSupp}^\square(\sigma^\square) \to \psi^\square(\sigma^\square)) , \quad (\exists\sigma^\square)(\mathrm{TSupp}^\square(\sigma) \,\&\, \psi^\square(\sigma^\square)) .$$

Since $\mathrm{TSupp}^\square(\sigma^\square)$ is equivalent to $(\forall\varrho^\square)\,\mathrm{Dep}^\square(\varrho^\square, \sigma^\square)$ which is obviously equivalent to a pure SNF, the demonstration is complete. Moreover, note that in fact we have found two formulas of a relatively simple form equivalent to φ: the former has the form $[(\forall\sigma)(\exists\varrho)\,\overline{\psi}(\sigma, \varrho)]^\square$ and the latter the form $[(\exists\sigma)(\forall\varrho)\,\overline{\overline{\psi}}(\sigma, \varrho)]^\square$ where both $\overline{\psi}$ and $\overline{\overline{\psi}}$ are normal pure SNF's.

The following metatheorem is analogous to Metatheorem 6409 but refers to provability instead of consistency.

6411. METATHEOREM. Let φ be a closed pure SNF and suppose that $\pi(\boldsymbol{b})$ is an SF such that $\mathbf{TSS}' \vdash (\pi(\boldsymbol{b}) \to \boldsymbol{b}$ is a complete Boolean algebra). Then $\mathbf{TSS}'$, (E2) $\vdash (\pi(\boldsymbol{b}) \,\&\, Z$ is a total Boolean support on $\boldsymbol{b}) \to \varphi$ iff $\mathbf{TSS}'$, (E2) $\vdash \pi(\boldsymbol{b}) \to [\![\varphi]\!]'\,\boldsymbol{b} = 1_{\boldsymbol{b}}$.

Demonstration. The first expression follows from the second by Metatheorem 6406. Suppose the first expression holds; we shall prove $\pi(\boldsymbol{b}) \to$ $\to [\![\varphi]\!]'\,\boldsymbol{b} = 1_{\boldsymbol{b}}$ in $\mathbf{TSS}'$, (E2) (or in **TS**, E2, since the formula in question is an SF). Suppose that $\boldsymbol{b}$ is a complete Boolean algebra and that $\pi(\boldsymbol{b})\,\&$ $\&\, [\![\varphi]\!]'\,\boldsymbol{b} \neq 1_{\boldsymbol{b}}$. Let z be an ultrafilter on $\boldsymbol{b}$ which contains the element $-([\![\varphi]\!]'\,\boldsymbol{b})$ and let **R** be the ultraproduct relation given by $\boldsymbol{b}$ and z. In the sense of $\mathfrak{N}(\mathbf{R})$, $\bar{\boldsymbol{b}}$ is a complete Boolean algebra and we have $\pi^{\mathfrak{N}(\mathbf{R})}(\bar{\boldsymbol{b}})$ (notation as in 5203); there exists a complete ultrafilter Z on $\bar{b}$ such that $[\![\varphi]\!]$' $\bar{b} \notin Z$. Let $\mathbf{Z}$ be a fixed ultrafilter with these properties. By (i) the formula φ holds in the sense of $\mathfrak{Supp}(\mathbf{Z}) * \mathfrak{R}(\mathbf{R})$; the formula $\neg\varphi$ also holds, since in the sense of this model, the φ-value in the algebra $\bar{b}$ does not belong to the total Boolean support. This contradiction completes the proof.

6412. COROLLARY. If a formula φ is provable in **TSS′**, (E2) then

$$\mathbf{TSS}', (\mathrm{E2}) \vdash (\forall \boldsymbol{b}) ([\![\varphi]\!]\text{'}\, \boldsymbol{b} = 1_{\boldsymbol{b}}) .$$

Recall that if p is an isomorphism between two complete Boolean algebras $\boldsymbol{b}_1$ and $\boldsymbol{b}_2$ then we may define a 1-1 mapping $\hat{p}$ between $\boldsymbol{b}_1$-sets and $\boldsymbol{b}_2$-sets as follows (see 4324):

$$r_2 = \hat{p}\text{'} r_1 \equiv (\forall x)(\forall u \in b_1)(\langle x, u\rangle \in r_1 \equiv \langle x, p\text{'}u\rangle \in r_2) .$$

6413. METATHEOREM. If φ is a pure SNF then the following is provable in **TSS′**:

If p is an isomorphism between complete Boolean algebras $\boldsymbol{b}_1$ and $\boldsymbol{b}_2$ and if r_1, • are $\boldsymbol{b}_1$-sets then

$$p\text{'}([\![\varphi]\!]\text{'}\langle r_1, \bullet, \boldsymbol{b}_1\rangle) = [\![\varphi]\!]\text{'}\langle \hat{p}\text{'} r_1, \bullet, \boldsymbol{b}_2\rangle .$$

Demonstration. By induction (exercise).

6414. METADEFINITION. A constant $\mathbf{F}$ defined in **TSS′** by a gödelian term is called a Boolean-valued function if the following is provable in **TSS′** for suitable natural numbers m and n:

1) (a) $\mathbf{D}(\mathbf{F})$ is the class of all $(m + n + 1)$-tuples whose last element is some complete Boolean algebra $\boldsymbol{b}$, whose first m elements are $\boldsymbol{b}$-constants and whose remaining n elements are $\boldsymbol{b}$-sets;

(b) $\mathbf{F}\text{'}\langle k, \bullet, r, \bullet, \boldsymbol{b}\rangle$ is always in $\boldsymbol{b}$.

2) If $\boldsymbol{b}$ is a complete Boolean algebra, and if k, • are $\boldsymbol{b}$-constants, r, • s, • are $\boldsymbol{b}$-sets and $u \in \boldsymbol{b}$, then

(a) $$u \wedge \mathbf{F}\text{'}\langle k, \bullet, r, \bullet, \boldsymbol{b}\rangle = \mathbf{F}\text{'}\langle k \restriction_{\boldsymbol{b}} u, \bullet, r \restriction_{\boldsymbol{b}} u, \boldsymbol{b} \mid u\rangle ,$$

(b) $$r \doteq s \,\&\, \bullet \rightarrow \mathbf{F}\text{'}\langle k, \bullet, r, \bullet, \boldsymbol{b}\rangle = \mathbf{F}\text{'}\langle k, \bullet, s, \bullet, \boldsymbol{b}\rangle .$$

3) If p is an isomorphism between complete Boolean algebras $\boldsymbol{b}_1$ and $\boldsymbol{b}_2$ and if k, • are $\boldsymbol{b}$-constants and r, • are $\boldsymbol{b}$-sets then

$$p\text{'}(\mathbf{F}\text{'}\langle k, \bullet, r, \bullet, \boldsymbol{b}_1\rangle) = \mathbf{F}\text{'}\langle \hat{p}\text{'}k, \bullet, \hat{p}\text{'}r, \bullet, \boldsymbol{b}_2\rangle .$$

We have shown that for every pure SNF φ the function $[\![\varphi]\!]$ is a Boolean-valued function. We now prove the converse.

6415. METATHEOREM. If $\mathbf{G}$ is a Boolean-valued function then there exists a pure SNF φ such that

$$\mathbf{TSS}' \vdash \mathbf{G} = [\![\varphi]\!] .$$

Demonstration. We restrict ourselves to the case $m = n = 1$, since the other cases are analogous. The domain of $\mathbf{G}$ consists of triples $\langle k, \boldsymbol{r}, \boldsymbol{b}\rangle$ where k is a $\boldsymbol{b}$-constant and $\boldsymbol{r}$ is a $\boldsymbol{b}$-set. We define a formula $\varphi(x, \sigma)$ as follows:

$(*)$ $(\forall \boldsymbol{b}, \varrho, r)\,(\vartheta(\boldsymbol{b}, \varrho)\,\&\, r \text{ is a } \boldsymbol{b}\text{-set} \,\&\, r``\varrho = \sigma \rightarrow \mathbf{G}`\langle x^{(\boldsymbol{b})}, r, \boldsymbol{b}\rangle \in \varrho)$

where $\vartheta(\boldsymbol{b}, Z)$ is the formula "$\boldsymbol{b}$ is a complete Boolean algebra and Z is a total Boolean support on $\boldsymbol{b}$".

First we prove in $\mathbf{TSS}'$ that $(*)$ is implied by

$(**)$ $(\exists \boldsymbol{b}, \varrho, r)\,(\vartheta(\boldsymbol{b}, \varrho)\,\&\, r \text{ is a } \boldsymbol{b}\text{-set} \,\&\, r``\varrho = \sigma \,\&\, \mathbf{G}`\langle x^{(\boldsymbol{b})}, r, \boldsymbol{b}\rangle \in \varrho)$.

Suppose $(**)$ holds and let $\bar{\boldsymbol{b}}$, $\bar{\varrho}$ and $\bar{r}$ be such that $\vartheta(\bar{\boldsymbol{b}}, \bar{\varrho})$ and $\bar{r}``\bar{\varrho} = \sigma$. We prove that $\mathbf{G}`\langle x^{(\bar{\boldsymbol{b}})}, \bar{r}, \bar{\boldsymbol{b}}\rangle \in \bar{\varrho}$. Since both ϱ and $\bar{\varrho}$ are total supports, there exist $u \in \varrho$, $\bar{u} \in \bar{\varrho}$ and an isomorphism p between $\boldsymbol{b} \mid u$ and $\bar{\boldsymbol{b}} \mid \bar{u}$ such that $p``(\varrho \mid_{\boldsymbol{b}} u) = \bar{\varrho} \mid_{\bar{\boldsymbol{b}}} \bar{u}$ and $\hat{p}`(r \mid_{\boldsymbol{b}} u) \doteq \bar{r} \mid_{\bar{\boldsymbol{b}}} \bar{u}$ (cf. 4326). We have

$$\mathbf{G}`\langle x^{(\boldsymbol{b}\mid u)}, r \mid_{\boldsymbol{b}} u, \boldsymbol{b} \mid u\rangle = \mathbf{G}`\langle x^{(\boldsymbol{b})}, r, \boldsymbol{b}\rangle \wedge u \in \varrho ,$$
$$\mathbf{G}`\langle x^{(\bar{\boldsymbol{b}}\mid \bar{u})}, \bar{r} \mid_{\bar{\boldsymbol{b}}} \bar{u}, \bar{\boldsymbol{b}} \mid \bar{u}\rangle = p`(\mathbf{G}`\langle x^{(\boldsymbol{b}\mid u)}, r \mid_{\boldsymbol{b}} u, \boldsymbol{b} \mid u\rangle) \in \bar{\varrho} ,$$
$$\mathbf{G}`\langle x^{(\bar{\boldsymbol{b}})}, \bar{r}, \bar{\boldsymbol{b}}\rangle \geq \mathbf{G}`\langle x^{(\bar{\boldsymbol{b}}\mid \bar{u})}, \bar{r}, \mid_{\bar{\boldsymbol{b}}} \bar{u}, \bar{\boldsymbol{b}} \mid u\rangle \in \bar{\varrho} .$$

Suppose now that $[\![\varphi]\!] \neq \mathbf{G}$, i.e. that there exist $\boldsymbol{b}$, x and r such that

$$[\![\varphi]\!]`\,\langle x^{(\boldsymbol{b})}, r, \boldsymbol{b}\rangle \neq \mathbf{G}`\langle x^{(\boldsymbol{b})}, r, \boldsymbol{b}\rangle .$$

1) First suppose that there exist $\boldsymbol{b}$, x and r such that

$$[\![\varphi]\!]`\,\langle x^{(\boldsymbol{b})}, r, \boldsymbol{b}\rangle - \mathbf{G}`\langle x^{(\boldsymbol{b})}, r, \boldsymbol{b}\rangle \neq 0_{\boldsymbol{b}} .$$

Then there exist $\boldsymbol{b}$, x and r such that

$(***)$
$$[\![\varphi]\!]`\,\langle x^{(\boldsymbol{b})}, r, \boldsymbol{b}\rangle = 1_{\boldsymbol{b}} ,$$
$$\mathbf{G}`\langle x^{(\boldsymbol{b})}, r, \boldsymbol{b}\rangle = 0_{\boldsymbol{b}} .$$

We shall show that (∗∗∗) leads to a contradiction. By Metatheorem 5218 there is no loss of generality in supposing that $\boldsymbol{b}$ has a total Boolean support ϱ. Under this assumption we have $[\![\varphi]\!]'\langle x^{(\boldsymbol{b})}, r, \boldsymbol{b}\rangle \in \varrho$ and hence $\varphi(x, r''\varrho)$; if we set $r''\varrho = \sigma$ then $\varphi(x, \sigma)$ and consequently we have $\vartheta(\boldsymbol{b}, \varrho)$, $r''\varrho = \sigma$ and $\mathbf{G}'\langle x^{(\boldsymbol{b})}, r, \boldsymbol{b}\rangle = 0_b$, a contradiction.

2) Suppose conversely that there exist $\boldsymbol{b}$, x and r such that

$$\mathbf{G}'\langle x^{(\boldsymbol{b})}, r, \boldsymbol{b}\rangle - [\![\varphi]\!]'\langle x^{(\boldsymbol{b})}, r, \boldsymbol{b}\rangle \neq 0_{\boldsymbol{b}}\,.$$

Then there exist $\boldsymbol{b}$, x and r such that

$$\mathbf{G}'\langle x^{(\boldsymbol{b})}, r, \boldsymbol{b}\rangle = 1_{\boldsymbol{b}}\,,$$

$$[\![\varphi]\!]'\langle x^{(\boldsymbol{b})}, r, \boldsymbol{b}\rangle = 0_{\boldsymbol{b}}\,.$$

We shall show that this leads to a contradiction; as before there is no loss of generality in supposing that $\boldsymbol{b}$ has a total Boolean support ϱ. Under this assumption we have $\neg\varphi(x, r''\varrho)$ and setting $r''\varrho = \sigma$ we see that (∗) does not hold. Hence (∗∗) does not hold and so there exist no $\bar{\boldsymbol{b}}$, $\bar{r}$ and $\bar{\varrho}$ such that $\vartheta(\bar{\boldsymbol{b}}, \bar{\varrho})$, $\bar{r}''\bar{\varrho} = \sigma$ and $\mathbf{G}'\langle x^{(\bar{\boldsymbol{b}})}, \bar{r}, \bar{\boldsymbol{b}}\rangle \in \bar{\varrho}$. On the other hand, we have $\vartheta(\boldsymbol{b}, \varrho)$, $r''\varrho = \sigma$ and $\mathbf{G}'\langle x^{(\boldsymbol{b})}, r, \boldsymbol{b}\rangle \in \varrho$, a contradiction. This completes the demonstration.

By Metatheorem 6413 we have

6416. METATHEOREM. If φ is a closed pure SNF then the following is provable in **TSS′**:

If p is an automorphism of a complete Boolean algebra then $p'([\![\varphi]\!]'\,\boldsymbol{b}) = [\![\varphi]\!]'\,\boldsymbol{b}$.

6417. DEFINITION (**TSS′**). An element u of a complete Boolean algebra $\boldsymbol{b}$ is called *rigid* in $\boldsymbol{b}$ if $p'u = u$ for every automorphism p of $\boldsymbol{b}$. Clearly the rigid elements of $\boldsymbol{b}$ form a subalgebra of $\boldsymbol{b}$ which we denote by $\boldsymbol{b}_{\mathrm{rig}}$. We have shown that $[\![\varphi]\!]_{\boldsymbol{b}} \in \boldsymbol{b}_{\mathrm{rig}}$ is provable for every closed pure SNF φ.

We shall now investigate the algebra $\boldsymbol{b}_{\mathrm{rig}}$ and the $\boldsymbol{b}_{\mathrm{rig}}$-sets. First we prove two useful lemmas.

6418. LEMMA (**TSS′**). Let r and s be $\boldsymbol{b}$-sets and let $u \in b$; $u \leqq [\![r = s]\!]_b$ holds iff $r_x \wedge u = s_x \wedge u$ for every x.

Proof. We have

$$[\![r = s]\!] = \bigwedge_x((r_x \wedge s_x) \vee (-r_x \wedge -s_x)) = \bigwedge_x((r_x \vee -s_x) \wedge (-r_x \vee s_x)).$$

The rest of the proof is left as an exercise.

6419. LEMMA (**TSS′**). Let $\boldsymbol{b}$ be a complete Boolean algebra and let q be a partition of $\boldsymbol{b}$. If $h'u$ is a $\boldsymbol{b}$-set for each $u \in q$ then there exists a $\boldsymbol{b}$-set r such that $u \leqq [\![r = h'u]\!]_{\boldsymbol{b}}$ for every $u \in q$. ($[\![r = h'u]\!]_{\boldsymbol{b}}$ is an abbreviation for $[\![(\forall x)((x \in \sigma \,\&\, x \in \varrho) \vee (x \notin \sigma \,\&\, x \notin \varrho)]\!]' \langle r, h'u, \boldsymbol{b}\rangle$.)

Proof. Set $r_x = \bigvee_{u \in q} (u \wedge (h'u)_x)$ for each $x \in \bigcup_{u \in q} \mathbf{W}(h'u)$ and use the preceding lemma.

6420. METATHEOREM. If φ is a pure SNF with one free semiset-variable then the following is provable in **TSS**, (E2):

For every $\boldsymbol{b}$ and every $\boldsymbol{b}$-set $\boldsymbol{r}$ there exists a $\boldsymbol{b}$-set $\bar{\boldsymbol{r}}$ such that

$$[\![\varphi(r)]\!]_{\boldsymbol{b}} \leqq [\![r = \bar{r}]\!]_{\boldsymbol{b}},\ [\![\varphi(r)]\!]_{\boldsymbol{b}} \leqq [\![\varphi(\bar{r})]\!]_{\boldsymbol{b}},\ [\![\varphi(\bar{r})]\!]_{\boldsymbol{b}} \in b_{\mathbf{rig}}.$$

Demonstration. We prove the above statements in the theory **TSS′**, (E2). Let r be a $\boldsymbol{b}$-set. If $u = [\![\varphi(r)]\!] = 0_{\boldsymbol{b}}$ then we are finished, since $0_{\boldsymbol{b}} \in b_{\mathbf{rig}}$. Otherwise let $\bar{u} = \bigvee\{p'u;\, p$ is an automorphism of $\boldsymbol{b}\}$ so that $\bar{u} \in b_{\mathbf{rig}}$. If $\{p_\alpha;\, \alpha < \lambda\}$ is an enumeration of all automorphisms of $\boldsymbol{b}$ we define $v_0 = u$ and $v_\alpha = p_\alpha'u - \bigvee_{\beta<\alpha} v_\beta$. By Lemma 6419 there exists a $\boldsymbol{b}$-set $\bar{r}$ such that $v_\alpha \leqq [\![\bar{r} = \hat{p}_\alpha' r]\!]$ for every α. Since $v_\alpha \leqq [\![\varphi(\hat{p}_\alpha' r)]\!]$ we have $\bar{u} \leqq [\![\varphi(\bar{r})]\!]$. In addition, we may suppose $-\bar{u} \leqq [\![\bar{r} = 0^{(\boldsymbol{b})}]\!]$. Since $\varphi(0)$ is a closed pure SNF (or is equivalent to such a formula), we have $[\![\varphi(0^{(\boldsymbol{b})})]\!] \in b_{\mathbf{rig}}$ by Metatheorem 6416. Further we have $[\![\bar{r} = 0^{(\boldsymbol{b})}]\!] \wedge [\![\varphi(\bar{r})]\!] = [\![\bar{r} = 0^{(\boldsymbol{b})}]\!] \wedge \wedge [\![\varphi(0^{(\boldsymbol{b})})]\!]$; hence

$$[\![\varphi(\bar{r})]\!] = ([\![\varphi(\bar{r})]\!] \wedge \bar{u}) \vee ([\![\varphi(\bar{r})]\!] \wedge -\bar{u}) =$$

$$= \bar{u} \vee ([\![\varphi(\bar{r})]\!] \wedge -\bar{u} \wedge [\![\bar{r} = 0^{(\boldsymbol{b})}]\!]) = \bar{u} \vee ([\![\varphi(0^{(\boldsymbol{b})})]\!] \wedge -\bar{u} \wedge$$

$$\wedge [\![\bar{r} = 0^{(\boldsymbol{b})}]\!]) = \bar{u} \vee (-\bar{u} \wedge [\![\varphi(0)^{(\boldsymbol{b})}]\!]) = \bar{u} \vee [\![\varphi(0^{(\boldsymbol{b})})]\!] \in b_{\mathbf{rig}},$$

since both $\bar{u}$ and $[\![\varphi(0^{(\boldsymbol{b})})]\!]$ are in $b_{\mathbf{rig}}$. (Cf. 6406 and 5218.)

6421. METATHEOREM. If $\varphi(\sigma)$ is a pure SNF then the following is provable in **TSS′**, (E2):

Let $\boldsymbol{b}$ be a complete Boolean algebra and let Z be a total Boolean support on $\boldsymbol{b}$. If $(\exists!\sigma)\ \varphi(\sigma)$ then there exists a $\boldsymbol{b}$-set r such that $\mathbf{D}(r) \subseteq b_{\mathbf{rig}}$ and $\sigma = r\text{''}Z$.

Demonstration. We prove the above statement in the theory **TSS′**, (E2). Let r be a $\boldsymbol{b}$-set such that $\varphi(r\text{''}Z)$ and $[\![\varphi(r)]\!]_{\boldsymbol{b}} \in b_{\mathbf{rig}}$. If we set $[\![\varphi(r)]\!]_{\boldsymbol{b}} = u$ then we can assume that $\mathbf{D}(r) \subseteq b \mid u$. For each automorphism p of $\boldsymbol{b}$ we have $[\![\varphi(\hat{p}\text{'}r)]\!]_b = u \in Z$ so that $\varphi((\hat{p}\text{'}r)\text{''}\ Z)$, i.e. $(\hat{p}\text{'}r)\text{''}\ Z = r\text{''}Z$. If we let $r_1 = \bigcup\{\hat{p}\text{'}r;\ p$ is an automorphism of $\boldsymbol{b})\}$ we have $r_1\text{''}Z = r\text{''}Z$ if in addition we let $\bar{r}_x = \bigvee\{v;\ \langle x, v\rangle \in r_1\}$ then $\bar{r}$ is a $\boldsymbol{b}$-set, $\bar{r}\text{''}Z = r\text{''}Z = \sigma$.

As a corollary we have the following

6422. Theorem (**TS**, ∂**TSS**$/\partial\mathfrak{St}(\mathbf{L})$). If z is a set support for **V** over **L** and if z is a complete ultrafilter on an algebra $\boldsymbol{b}$ in the sense of $\mathfrak{St}(\mathbf{L})$, then $z \cap (b_{\mathbf{rig}})^{\mathfrak{St}(\mathbf{L})}$ is a set support for **HDf** over **L**.

Proof. We denote $(\boldsymbol{b}_{\mathbf{rig}})^{\mathfrak{St}(\mathbf{L})}$ by $\boldsymbol{b}_0$.

1) We have $(z \cap b_0) \in \mathbf{HDf}$; for, by 4327 we have $z \cap b_0 = b_0 \cap \bigcap\{q;\ q$ is a Boolean support for **V** over **L** on $\boldsymbol{b}\} = \boldsymbol{b}_0 \cap \mathbf{T}(\mathbf{V}, \boldsymbol{b}_0) = \boldsymbol{b}_0 \cap \mathbf{T}(p_\alpha, \boldsymbol{b}_0)$ for some α, where $\mathbf{T}(\mathbf{V}, \boldsymbol{b}_0)$ is the gödelian term corresponding to $\bigcap\{q;\ q$ is a Boolean support for **V** over **L** on $\boldsymbol{b}\}$. Hence $z \cap b_0$ is a Boolean support in the sense of $\mathfrak{St}\mathfrak{R}\mathfrak{St}(\mathbf{L}, \mathbf{HDf})$.

2) It remains to prove that $z \cap b_0$ is a total support in the sense of $\mathfrak{St}\mathfrak{R}\mathfrak{St}$ $(\mathbf{L}, \mathbf{HDf})$. Let $\varphi(q)$ be the formula which says that "q is the first total support for **HDf** over **L** in the definable well-ordering of **HDf**" (cf. 3406 and 6320). There is a unique q such that $\varphi(q)$ and so by Metatheorem 6421 there exists $\bar{r} \in \mathbf{L}$ such that $\mathbf{D}(\bar{r}) \subseteq b_0$ and $\bar{r}\text{''}z = \bar{r}\text{''}(z \cap b_0) = q$. Hence $\mathrm{Dep}^{\mathfrak{St}(\mathbf{L})}(q, z \cap b_0)$ and $z \cap b_0$ is a total support in the sense of $\mathfrak{St}\mathfrak{R}\mathfrak{St}(\mathbf{L}, \mathbf{HDf})$.

Remark. Using this theorem we can easily establish the consistency of the equality or inequality of the classes **L**, **HDf** and **V**. We have **TS**, D3 $\vdash$ $\vdash \mathbf{L} \subseteq \mathbf{HDf} \subseteq \mathbf{V}$. In Chapter III we showed that $\mathbf{L} = \mathbf{HDf} = \mathbf{V}$ is consistent and in 6332 we showed that $\mathbf{L} \neq \mathbf{HDf} = \mathbf{V}$ is consistent. In **TSS‴**, E1 we can prove that the only rigid elements of $\mathbf{Cant}_0^0$ are the zero and the unit. Hence in the model $\mathfrak{Up}$ as a model of **TS**, D3, E2 in **TSS**, D3, (Constr), (S $\mathbf{Cant}_0^0$) we have $\mathbf{L} = \mathbf{HDf}$ (since **HDf** has a support over **L** which is a singleton). Thus $\mathbf{L} = \mathbf{HDf} \neq \mathbf{V}$ is consistent. Finally, starting with the theory

$$\mathbf{TSS}''', \mathbf{L} \neq \mathbf{HDf} = \mathbf{V}, (\mathrm{S}\ \mathbf{Cant}_0^0)$$

which is relatively consistent w.r.t. **TSS**, and using the model $\mathfrak{Up}$, we obtain the consistency of the theory **TS**, D3, E2, Mcl (**M**), (**HDf** has a singleton support over **M**), ∂**TSS**/$\partial\mathfrak{St}$ (**M**), $\mathbf{L} \mp \mathbf{HDf}^{\mathfrak{St}(\mathbf{M})} = \mathbf{M} \mp \mathbf{V}$. In this theory, $\mathbf{HDf}^{\mathfrak{St}(\mathbf{M})} = \mathbf{HDf}$ is provable, so that we have $\mathbf{L} \mp \mathbf{HDf} \mp \mathbf{V}$.

In the following we shall consider separatively ordered sets instead of complete Boolean algebras. Disjoint relations are replaced by antimonotone relations (cf. 4253).

6423. DEFINITION (**TSS'**). Let $\boldsymbol{a}$ be a separatively ordered set. We say that r is an $\boldsymbol{a}$-set if it is an antimonotone relation and if $\mathbf{D}(r) \subseteq a$; we say that an $\boldsymbol{a}$-set is an *$\boldsymbol{a}$-constant* if $x, y \in a$ implies $\mathbf{Ext}_r(x) = \mathbf{Ext}_r(y)$. If z is a set then $z^{(\boldsymbol{a})}$ is the unique $\boldsymbol{a}$-constant r such that $\mathbf{W}(r) = z$, i.e. $z^{(\boldsymbol{a})} = z \times a$.

6424. METADEFINITION.

We define in **TSS'**: Let $\boldsymbol{a}$ be a separatively ordered set and let $u \in a$. If z_1 and z_2 are sets such that $z_1 \in z_2$ then we write

$$\langle u, a, z_1, z_2 \rangle \in \mathbf{Forc}_{x \in y} .$$

If z is a set and r is an $\boldsymbol{a}$-set such that $\langle z, u \rangle \in r$ then we write

$$\langle u, a, z, r \rangle \in \mathbf{Forc}_{x \in \sigma} .$$

If φ and ψ are pure SNF's for which $\mathbf{Forc}_\varphi$ and $\mathbf{Forc}_\psi$ are already defined then we let

$$\langle u, a, r, \bullet \rangle \in \mathbf{Forc}_{\neg\varphi} \equiv \neg (\exists v \leq u)(\langle v, a, r, \bullet \rangle \in \mathbf{Forc}_\varphi) ,$$

$$\langle u, a, r, \bullet \rangle \in \mathbf{Forc}_{\varphi \& \psi} \equiv . \langle u, a, r, \bullet \rangle \in \mathbf{Forc}_\varphi \& \langle u, a, r, \bullet \rangle \in \mathbf{Forc}_\psi ,$$

$$\langle u, a, r, \bullet \rangle \in \mathbf{Forc}_{\varphi \vee \psi} \equiv . \langle u, a, r, \bullet \rangle \in \mathbf{Forc}_\varphi \vee \langle u, a, r, \bullet \rangle \in \mathbf{Forc}_\psi .$$

(Here we should make a remark similar to that made in connection with the definition of φ-values of conjunctions and disjunctions in the Boolean case)

$$\langle u, a, r, \bullet \rangle \in \mathbf{Forc}_{(\forall x)\varphi} \equiv (\forall z)(\langle u, a, z^{(\boldsymbol{a})}, r, \bullet \rangle \in \mathbf{Forc}_\varphi) ,$$

$$\langle u, a, r, \bullet \rangle \in \mathbf{Forc}_{(\forall \sigma)\varphi} \equiv (\forall s)(s \text{ is an } \boldsymbol{a}\text{-set} \rightarrow \langle u, a, s, r, \bullet \rangle \in \mathbf{Forc}_\varphi) ,$$

$$\langle u, a, r, \bullet \rangle \in \mathbf{Forc}_{(\exists x)\varphi} \equiv (\exists z)(\langle u, a, z^{(\boldsymbol{a})}, r, \bullet \rangle \in \mathbf{Forc}_\varphi) ,$$

$$\langle u, a, r, \bullet \rangle \in \mathbf{Forc}_{(\exists \sigma)\varphi} \equiv (\exists s)(s \text{ is an } \boldsymbol{a}\text{-set} \,\&\, \langle u, a, s, r, \bullet \rangle \in \mathbf{Forc}_\varphi) .$$

Remark. The formula $\langle u, a, r, \bullet\rangle \in \mathbf{Forc}_\varphi$ is read "*u forces* φ *at* r, $\bullet$ *in a*".

6425. METALEMMA. For each pure SNF φ the following is provable in **TSS**′:

If u forces φ and $v \leqq u$ then v forces φ. (Exercise.)

6426. METATHEOREM. For each pure SNF φ the following is provable in **TSS**′:

Let $\boldsymbol{a}$ be a separatively ordered set and let Z be a complete ultrafilter on $\boldsymbol{a}$ which is a total support. If q, $\bullet$ are sets and r, $\bullet$ are $\boldsymbol{a}$-sets then $\varphi(q, \bullet, r\text{"}Z, \bullet)$ holds iff some $u \in Z$ forces φ at $q^{(\boldsymbol{a})}$, $\bullet$, r, $\bullet$.

Demonstration. We proceed by induction the assertion being clear for atomic formulas. We consider the case of negation. Suppose $\neg\varphi$ holds. Then no $u \in Z$ forces φ and by 4302 there exists $u \in Z$ such that no $v \leqq u$ forces φ; hence u forces $\neg\varphi$. Conversely, if there exists $u \in Z$ such that no $v \leqq u$ forces φ, then $\neg\varphi$ holds; for if φ holds then some $v \in Z$ forces φ and there exists $w \in Z$ such that $w \leqq v$ and $w \leqq u$ and w forces both φ and $\neg\varphi$, a contradiction.

We now consider quantification. Consider a fixed $u \in Z$. We have u forces $(\forall x)\,\varphi$ iff u forces φ at $z^{(\boldsymbol{a})}$ for any set z but the latter holds iff $(\forall x)\,\varphi$. The remaining cases are treated similarly.

6427. DEFINITION. Let $\boldsymbol{a}$ be a separatively ordered set and let $u \in a$. If r is an $\boldsymbol{a}$-set then we define

$$r \mid_{\boldsymbol{a}} u = r \restriction \mathbf{Seg}_{\boldsymbol{a}}(u)\,.$$

6428. METATHEOREM. For each pure SNF φ the following is provable in **TSS**′:

Let $\boldsymbol{a}$ be a separatively ordered set and let $v \leqq u \in a$. If r, $\bullet$ are $\boldsymbol{a}$-sets then v forces φ at r, $\bullet$ in $\boldsymbol{a}$ iff v forces φ at $r \mid_{\boldsymbol{a}} u$, $\bullet$ in $\langle \mathrm{Seg}_{\boldsymbol{a}}(u), \leqq\rangle$.

The demonstration is by induction and is left as an exercise.

6429. METATHEOREM. For each pure SNF φ the following is provable in **TSS**′, E2:

Let $\boldsymbol{b}$ be a complete Boolean algebra and set $\boldsymbol{a} = \mathring{\boldsymbol{b}}$. If r, $\bullet$ are $\boldsymbol{b}$-sets and s, $\bullet$ are $\boldsymbol{a}$-sets such that $\langle x, v\rangle \in s \equiv v \leqq r_x$ then the φ-value of r, $\bullet$ in $\boldsymbol{b}$ is the join of all $u \in \boldsymbol{a}$ which force φ at s, $\bullet$ in $\boldsymbol{a}$.

Demonstration. Working in the theory **TSS**′, E2, we let

$$u = [\![\varphi]\!]' \langle r, \bullet, \boldsymbol{b} \rangle$$

and

$$v = \bigvee\{w; \langle w, a, s, \bullet\rangle \in \mathbf{Forc}_{\varphi}\}\ .$$

The assumption $u < v$ leads to a contradiction. Indeed, by considering the algebra $\boldsymbol{b} \,\big|\, (v - u)$ we may assume $u = 0_{\boldsymbol{b}}$ and $v = 1_{\boldsymbol{b}}$; by Metatheorem 5418 we may assume that there is a total Boolean support on $\boldsymbol{b}$; hence by Metatheorem 6406 we have φ while on the other hand by Metatheorem 6426 we have $\neg\varphi$. Similarly the assumption $u > v$ leads to a contradiction.

Thus $u = v$, as was to be shown.

Bibliography

Abbreviations: BAPS = Bulletin de l'Académie Polonaise des Sciences, CMUC = = Commentationes Mathematicae Universitatis Carolinae (Prague), NHPC = North Holland Publishing Company, Proc. Symp. = Proceeding of Symposia in Pure Mathematics, Volume XIII - Axiomatic Set Theory (Part 1 — 1971; Part 2 — to appear).

J. L. Bell & A. B. Slomson, Models and ultraproducts. NHPC 1969.

P. Bernays & A. A. Fraenkel, Axiomatic set theory. NHPC 1958.

L. Bukovský, The continuum problem and powers of alephs. CMUC 6 (1965) 181—197.

P. J. Cohen, Set theory and the continuum hypothesis. W. A Benjamin 1966

W. B. Easton, Powers of regular cardinals. Annals of mathematical logic 1 (1970) 139—178.

K. Gödel, The consistency of the axiom of choice and of the generalized continuum hypothesis. Princeton University Press 1940.

P. Hájek, Logische Kategorien. Archiv für Math. Logik und Grundlagenforschung 13 (1970) 168—193.

A. Hajnal, Consistency theorem connected with the generalized continuum problem. Acta Math. Sci. Hung. XII (1961) 321—376.

T. Jech & A. Sochor, On Θ-model of the set theory & Applications of the Θ-model. BAPS 14 (1966) 297—303 & 351—355.

H. J. Keisler, Limit ultrapowers. Trans. Amer. Math. Soc. 107 (1963) 382—408.

A. Lévy & R. Vaught, Principles of partial reflection in the set theories of Zermelo and Ackermann. Pacif. Journ. Math. 11 (1961) 1045—1062.

A. R. D. Matthias, Surrealistic landscape with figures. In: Proc. Symp. Part. 2.

K. McAloon, Consistency results about ordinal definability, Annals of mathematical logic 2 (1970) 449—467.

A. Mostowski, An undecidable arithmetical statement. Fundamenta Math. 36 (1949) 143—164.

J. Myhill & D. Scott, Ordinal definability. In: Proc. Symp. Part. 1, 271—278.

J. B. Rosser, Simplified independence proofs — Boolean valued models of set theory. Academic Press 1969.

D. Scott & R. Solovay, Boolean valued models of set theory. In: Proc. Symp. Part. 2.

J. R. SHOENFIELD, Mathematical logic. Addison-Wesley 1967.
R. SIKORSKI, Boolean algebras. Springer Verlag 1964 (Second edition.)
E. SPECKER, Zur Axiomatik der Mengenlehre. Zeitschrift für Math. Logik und Grundlagen der Mathematik 3 (1957) 173—210.
A. TARSKI & A. MOSTOWSKI & R. M. ROBINSON, Undecidable theories. NHPC 1953.
P. VOPĚNKA, General theory of ∇-models. CMUC 8 (1967) 145—170.
P. VOPĚNKA & B. BALCAR, On complete models of the set theory. BAPS 15 (1967) 839—841.
P. VOPĚNKA & P. HÁJEK, Permutation submodels of the model ∇. BAPS 19 (1965) 611—614.

Index

Index of Symbols